D1535798

THE COLLECTOR'S GUIDE TO

Antique Radios

SECOND EDITION

— by Marty & Sue Bunis —

COLLECTOR BOOKS
A Division of Schroeder Publishing Co., Inc.

The current values in this book should be used only as a guide. They are not intended to set prices, which vary from one section of the country to another. Auction prices as well as dealer prices vary greatly and are affected by condition as well as demand. Neither the Authors nor the Publisher assume responsibility for any losses that might be incurred as a result of consulting this guide.

On the cover:

Magic Box by Bill Bell, copyright, 1992, used with permission.
A full color copy of this print is available from:

Frank Krantz
100 Osage Avenue
Somerdale, NJ 08083

Cartoon on page 6:

Copyright Ron Boucher, 1991, used by permission.

An 8 x 10 color version of this cartoon is available for $8.50 postpaid from:

Vintage Radio Services
P.O. Box 541
Goffstown, NH 03045

Additional copies of this book may be ordered from:

Collector Books
P.O. Box 3009
Paducah, KY 42002-3009

or

Marty and Sue Bunis
RR 1, Box 36
Bradford, NH 03221
(603) 938–5051

@ $17.95. Add $2.00 for postage and handling.

Copyright: Marty and Sue Bunis, 1992.

This book or any part thereof may not be reproduced without the written consent of the Authors and Publisher.

1 2 3 4 5 6 7 8 9

Printed by IMAGE GRAPHICS, INC., Paducah, Kentucky

Dedication

This book is dedicated to radio collectors everywhere.

Acknowledgments

Once again, we extend our most sincere thanks to the following people, almost all of them members of the New England Antique Radio Club, who provided the radios, photographs, and some of the facts and figures that made this book possible. We couldn't have done this without your help and generosity and we thank you.

Len Arzoomanian, Frank Atwood/Nostalgia Radio, Merrill Bancroft, John Bayusik, Ron Boucher, Peter Burton, Warren Chamberland, Dave Crocker, Bob Davidson, Al DeCristofano, John DeLoria, Karl Doerflinger, Ward Eaton, John Ellsworth, Richard Foster, Dave Froehlich, Ernie Gilford, John Hayes, Ted Hutson, George and Julia Kay, David Lent, Ray and Rose Lumb, Dick Mackiewicz and Alice Villard, Al Monroe, Michael Morin, Herb Parsons, Scott and Sue Phillips, Sr., Dave Pope, Harry Poster, Frank Rasada, Steve and Yvonne Slabe, David Smith, Glenn Smith, John Terrey, Ron Trench, Jr., Ralph Williams, Chet Wisner, Wally Worth.

Special thanks to the Leandre Poisson Collection for the following photos:

Airline 25BR-1542R, Bendix 111, Delco R-1228, Emerson 520, Firestone 4-A-26, Philco 38-14, Philco 41-290, Philco 42-KR3, Philco 48-200, Philco 48-460, Philco 53-563, Radiola 61-8, RCA 15X, RCA 96-X-1, Sentinel 284-NI, Silvertone 4500A, Silvertone 7204, Sonora WALL-243, Stewart Warner 07-51H, Sylvania 1102, Zenith 6-D-311.

Sincere thanks also to all those who loaned radio-related paper goods, catalogs, encouragement and constructive suggestions. We appreciate your help and support.

Introduction

Once again, welcome to the wonderful world of Antique Radios!

This second edition of *The Collector's Guide to Antique Radios* features all of the information from the first book along with hundreds of new model numbers, updated pricing and new photographs. Whether you're a newcomer to the fascinating hobby of vintage radio collecting or an "old timer" who has been building a radio collection for many years, we hope you find this book informative and helpful.

Feel free to write us at any time with anything you feel would be helpful for future books. To maintain accuracy, we only list a model number if we have actually seen the radio – either in person or from a photograph or old company advertising – so we always welcome new information from all sources.

You will find there are many more transistors listed in this second edition. As the older sets get harder to find and more expensive, more and more people begin to collect newer, more affordable radios, and transistors are fast becoming a popular part of radio collecting. With a few exceptions, they are still affordable and fairly easy to find at yard sales and flea markets, so keep your eyes open – there is still time to build a good collection before transistor prices hit the Catalin heights!

Be sure to check the back of the book for current information on antique radio clubs throughout the country. Radio collecting is growing at a rapid pace and there are new clubs forming all the time. We recommend joining a club near you – it's a great way to meet others with the same interests.

Explanation of Pricing Section

If you are familiar with the first edition, you will find prices are generally higher across the board, with dramatic increases occurring on some of the Deco, streamlined, novelty and hard-to-find sets. As more and more people become involved in this fast-growing hobby, prices will continue to rise according to supply and demand, with the "good" sets always commanding the premium prices.

The pricing information in this book is provided for the potential buyer and has been gathered from many different sources: regional and national radio meets, flea markets, auctions, private collectors, various antique radio classified magazines, etc. Our method is to gather as many prices as possible on each set and take an average to arrive at a fairly current figure. Keep in mind there is no "suggested retail price" for old radios and pricing is extremely variable and depends on many factors.

The primary factor when considering whether an old radio is "worth it" is CONDITION. We have tried to provide pricing that reflects sets in "good" condition, using the following guidelines:

1. Electronically complete. Although the set may not be working, all the electrical components, including tubes, are there and a minimal amount of repair would bring it to operating condition.

2. Cabinet in good condition. All knobs and other removable parts are in place. There are no cracks, chips or other highly visible damage to the case.

These prices in no way reflect sets in "MINT" condition – very few old radios fall into that category. There may be an occasional set that was very seldom or never used that may be close to or even mint, but most radios were a vital and functional part of the household and they will usually bear a few nicks and scratches and show some signs of wear.

A secondary factor in determining a fair price for an old set is where you happen to be at the time. Prices generally tend to be higher on the West Coast than the East Coast, auctions are usually more expensive than yard sales, etc.

Naturally if you have a radio for sale, you want to receive the best possible price. Keep in mind that radio prices are very volatile and are more dependent on condition than age. For example, we receive many phone calls and letters from people who have sets for sale, and we <u>always</u> ask very specifically about the condition. We have learned this the hard way – too many times we have been told that a radio has been "in storage" for years and is in "good" condition only to arrive and find a dirty set that has been sitting in a damp basement or hot attic for years, with a grill cloth full of cat scratches, peeling veneer, missing knobs, no tubes and the sellers expect to get top dollar because the set is OLD! Remember, "OLD" does not necessarily mean "GOOD!" To avoid disappointment, the best policy is to be realistic about your radio's condition when you contact a potential buyer.

A special note to Catalin collectors:
Catalin radios are the most avidly sought after and hardest-to-find-in-good-condition radios around and, despite the predictions (usually made by battery set collectors who are shocked that a 1940's radio can be worth thousands of dollars) that the bottom has to fall out of the Catalin market any day now, they still continue to command high prices. Because there are so many factors that affect the price of Catalin sets – color, the smallest stress cracks, heat discoloration, the part of the country you are in – it is imposible to establish firm prices and we have chosen to generalize the pricing in this book. You will find most Catalin prices followed by a "+," which means you should expect to pay at least this much for a set in good condition, depending on the above variables.

Basic Radio Terms and Descriptive Information

1. Addresses – We have included addresses for as many companies as possible, along with brief histories of a few of the well-known companies. This is not necessarily complete information; many radio companies, especially during the 1920's, went in and out of business so fast, or merged with one another so often, or moved so frequently, that it is sometimes impossible to pinpoint exact addresses and historical background. Note that a few of the addresses for transistors may be for suppliers rather than manufacturers, as some of these sets were imports.

2. Model Numbers – Model numbers are tricky critters! There seems to be no consistency to them sometimes; even within a given company's advertising, model numbers can be written several different ways. For example: A model #AR36-809 might be written AR-36-809 or AR36809 or even AR36-809! The best plan is to use a little creativity when searching for model numbers in this price guide. We have tried to list them in a logical numerical sequence, so try different combinations until you find the right one.

3. Style – There are many different types or styles of radios available. We have used various common descriptive terms in this price guide. The following is a basic description of each:

Breadboard – An early radio with the tubes and other component parts attached to a rectangular board that usually has "breadboard" ends. Breadboard radios have no cases or covers, all parts and tubes are exposed.

Cathedral – Sometimes referred to as "beehives," cathedrals are taller than wide with a top that is usually rounded or sometimes comes to a rounded peak, generally thought to resemble a cathedral window in shape. The cathedral is the "classic" radio shape.

Chairside – A console radio with a low shape made to sit beside an armchair, usually with the controls on the top within easy reach.

Console – The console is also called a floor model. It can consist of a cabinet on legs or just a rectangular (higher than wide) case that sits directly on the floor.

Portable – A radio made to be used in the home as well as outdoors or in a location with no electricity. Most portables run on battery power, although a great many are three-way sets, using electricity or batteries.

Table – A general term used to describe many shapes and sizes of radios that are table-top sets.

Tombstone – These are sets that are rectangular in shape, higher than they are wide and are flat on top; generally they resemble the shape of a "tombstone." They are sometimes called "upright table" models.

C – Clock

N - Novelty – This is a very broad term and covers all radios that are highly collectible due to their unusual case, whether they sport a Mickey Mouse, a Charlie McCarthy, a built-in camera, look like a baseball, a bar, a bottle or any other design where one would not expect to find a radio.

R/P – Radio/phono combination.
R/P/Rec – Radio/phono/recorder combination.
R/P/REC/PA – Radio/phono/recorder/public address system combination.

4. Year – We have tried to list the correct year for most of the sets listed; however, there is some flexibility here as many manufacturers overlapped models from one year to the next and many popular models were made for a number of years.

5. Description – We have included, wherever possible, the material of which the case is made, a general description of the set's shape, placement of the dial, grill and knobs, power source(s), bands, and any other information important to identifying each set. The following terms are frequently used in the descriptions:

AC – Alternating Current

Airplane Dial – A dial which moves in a 360° circle.

Bat – Battery

BC – Broadcast

DC – Direct Current

Highboy – A floor model radio whose legs can be as much as ½ the total height of the case.

Lowboy – A floor model radio whose legs are considerably lower than a highboy's; generally, less than ½ the height of the case.

LW – Long Wave.

Midget – A small table set, usually less than 7" in its longest measurement.

Movie Dial – A dial which features a projected image from a circular film rotating around a bulb.

Plastic – We have used the word "plastic" as a generic term to cover most, but not all, plastics. In the case of Catalin plastics, we have used the word "Catalin" specifically rather than the generic word "plastic," because of the much higher pricing structure of Catalin sets.

Repwood – A type of pressed wood popular in the 1930's and used to make various decorative items, including ornate radio cases that have a beautiful "carved" appearance.

Robot Dial – A feature of some Zenith radios, the robot dial is really three dials in one – standard broadcast, shortwave and police/amateur. Only one dial is visible at any time and can be changed by the turn of a lever.

Slide Rule Dial – A rectangular dial, usually horizontal, which feature a thin sliding indicator.

SW – Short Wave.

6. Pictures – We have tried, to the best of our ability, to photograph only sets that are as close to original as possible. There may be a few pictures included here that show radios with replacement knobs or grill cloths, but in most cases the replacements blend well with the sets and should not be considered a detriment.

Keep in mind that this is a GUIDE only and it was written to do just that – guide you with identification information and current pricing. Because there are so many variables to consider and radio prices are escalating rapidly, we make no guarantees that these prices are hard and fast but we do recommend that you use your judgment when considering the purchase of an old radio – if it is in good condition, the price seems fair according to the book and you like it – buy it!

We are always happy to hear from other radio collectors and we welcome your calls and letters. Please let us know how you like the book and your ideas to improve future editions – we welcome all comments, suggestions and constructive criticisms.

Marty and Sue Bunis
RR 1, Box 36
Bradford, NH 03221
(603) 938–5051

A-C DAYTON

The A-C Electrical Mfg. Co., Dayton, Ohio
The A-C Dayton Company, Dayton, Ohio

The A-C Electrical Manufacturing Company began business in Dayton, Ohio in 1901 as a manufacturer of electric motors. By 1922 they were manufacturing and selling radios and radio parts. The company was out of business by 1930.

AC-63, Table, 1928, Wood, low rectangular case, center front dial w/ large escutcheon, lift top, AC **$100.00**
AC-65, Table, 1928, Wood, low rectangular case, center front dial w/ large escutcheon, lift top, AC **$100.00**
AC-98 "Navigator", Table, 1929, Wood, low rectangular case, center front window dial w/escutcheon, lift top, bat . . **$100.00**
AC-9960 "Navigator", Console, 1929, Wood, lowboy, front window dial, metal escutcheon, round grill, 3 knobs, battery . **$115.00**
AC-9970 "Navigator", Console, 1929, Wood, lowboy, front window dial, metal escutcheon, sliding front doors, battery . . **$130.00**
AC-9980 "Navigator", Console, 1929, Wood, lowboy, front window dial, metal escutcheon, grill w/vertical bars, bat **$115.00**
AC-9990 "Navigator", Console, 1929, Wood, highboy, inner front window dial, metal escutcheon, 3 knobs, doors, bat . **$145.00**
R-12, Table, 1924, Mahogany finish, low rectangular case, 3 dial front panel, 4 tubes, battery **$115.00**
XL-5, Table-Glass , 1925, Plate glass, low see-though rectangular case, 3 dial front panel, 5 tubes, battery **$250.00**
XL-5, Table, 1925, Wood, low rectangular case, 3 dial front panel, lift top, 5 tubes, battery **$100.00**

XL-10, Table, 1924, Wood, low rectangular case, 3 dial front panel, lift top, 5 tubes, battery **$110.00**
XL-15, Console, 1923, Wood, highboy, 3 dial front panel, storage, 5 tubes, battery **$150.00**
XL-20, Table, 1926, Wood, high rectangular case, slanted 3 dial front panel, lift top, battery **$110.00**
XL-25, Table, 1926, Wood, low rectangular case, 2 dial front panel, lift top, 5 tubes, battery **$125.00**
XL-30, Table, 1925, Wood, low rectangular case, 2 dial panel w/ escutcheons, fluted columns, bat **$120.00**
XL-50, Table, 1927, Walnut, low rectangular case, 1 front dial w/ metal escutcheon, 2 knobs, AC **$70.00**
XL-60, Console, 1927, Wood, low rectangular case, center front dial & knobs, fold-down door, battery storage, battery . . . **$190.00**
XL-61, Table, 1928, Wood, low rectangular case, center front dial w/ large escutcheon, battery **$80.00**
XL-71 "Navigator", Table, 1929, Wood, low rectangular case, center front window dial w/metal escutcheon, bat **$85.00**

ABBOTWARES
Los Angeles, California

Z477, Table-N, Metal horse w/saddle, stands on radio base, 2 knobs, horizontal grill bars **$285.00**

ACE
The Precision Equipment Co., Inc., Peebles Corner, Cincinnati, Ohio

In 1922, Crosley bought Precision Equipment but continued to use the old Precision "Ace" trademark on this line until 1924.

3B, Table, 1923, Wood, low rectangular case, 1 dial black front panel, lift top, battery **$180.00**
V, Table, 1923, Wood, low rectangular case, 1 dial black front panel, lift top, 1 tube, battery **$175.00**

ACRATONE
Federated Purchasers, Inc.
25 Park Place, New York, New York

87, Console-R/P , 1934, Wood, upper front dial, lower grill w/cut-outs, lift top, inner phono, feet **$150.00**

ADDISON
Addison Industries, Ltd
Toronto, Ontario, Canada

5F, Table, 1940, Catalin, shouldered top, slide rule dial, 5 vertical grill bars, 3 knobs, AM/SW, AC **$1,800.00**
A2A, Table (Catalin), 1940, Catalin, Deco, right front dial, left vertical wrap-over grill bars, 2 knobs, AC **$1,150.00**
A2A, Table (Plastic), 1940, Plastic, Deco, right front dial, left vertical wrap-over grill bars, 2 knobs, AC **$300.00**
B2E, Table, Plastic, Deco, right front dial, left vertical grill bars, 2 "pin-wheel" knobs, AC **$300.00**

ADLER
Adler Manufacturing Co.
881 Broadway, New York City, New York

199 "Royal", Table, 1924, Wood, low rectangular case, 3 dial front panel, 5 tubes, battery **$150.00**
201-A "Royal", Table, 1924, Wood, low rectangular case, 3 dial front panel, 5 tubes, battery **$125.00**
324, Console, 1930, Wood, highboy, inner front dial, grill w/cut-outs, doors, stretcher base **$150.00**
325, Console, 1930, Wood, highboy, inner front dial, grill w/cut-outs, doors, stretcher base **$150.00**

ADMIRAL
Continental Radio & Television Corporation
3800 W. Cortland Street, Chicago, Illinois
Admiral Corporation, 3800 W. Cortland St., Chicago, Illinois

Admiral was founded as the Continental Radio & Television Corporation in 1934 by a group of four investors, some of whom literally sold their belongings to raise the initial capitol. Growing rapidly due to the production of good products at affordable prices, the company ranked #5 in sales volume by 1939. Because of quality control problems and competition from Japanese TV manufacturers, the company was forced to close its last plant in 1979.

4D11, Portable, 1948, Plastic, right side dial, vertical grill bars, stand-up handle, 2 knobs, BC, battery **$45.00**
4P21, Portable, 1957, Black transistor, right dial, perforated grill, thumbwheel knob, handle, AM, bat **$30.00**
4P22, Portable, 1957, Red transistor, right dial, perforated grill, thumbwheel knob, handle, AM, bat **$40.00**
4P24, Portable, 1957, Tan transistor, right dial, perforated grill, thumbwheel knob, handle, AM, bat **$30.00**
4P28, Portable, 1957, Turquoise transistor, right dial, perforated grill, thumbwheel knob, handle, bat **$40.00**
4W19, Portable, 1951, Plastic, raised front dial, lattice grill, flex handle, right button, BC, AC/DC/bat **$40.00**
5A32/16, Table-C, 1953, Plastic, metal front panel, right round dial, left alarm clock, 4 knobs, BC, AC **$30.00**
5A33, Table-C, 1952, Plastic, round clock & round dial faces, gold front, AC **$30.00**
5E22, Table, 1951, Plastic, center front round dial/inner perforated grill, handle, 2 knobs, BC, AC/DC **$25.00**
5F11, Portable, 1949, Plastic, inner right dial, lattice grill, flip-up front, handle, BC, AC/DC/battery **$35.00**
5G22, Table-C, 1951, Plastic, right square dial, left alarm clock, vertical front bars, 4 knobs, BC, AC **$30.00**

5G32N, Table, Plastic, raised top, upper front slide rule dial, lower horizontal bars, 2 knobs, BC **$30.00**

5J21, Table, 1951, Plastic, right trapezoid dial, left horizontal grill openings, 2 knobs, BC, AC/DC **$40.00**
5J23, Table, 1951, Painted plastic, right trapezoid dial, left horizontal grill openings, 2 knobs, BC **$40.00**
5L21, Table-C, 1952, Plastic, right dial w/inner perforations, left alarm clock, vertical front lines, AC **$30.00**
5M21, Table-R/P, 1952, Plastic, outer front round dial/louvers, 2 knobs, 3/4 lift top, inner phono, BC, AC **$45.00**
5R11, Table, 1949, Plastic, right square dial, left square checker-board grill, 2 knobs, BC, AC/DC **$30.00**
5S22AN, Table, 1953, Plastic, right front curved dial, off-center circular louvers, 2 knobs, BC, AC **$35.00**
5T12, Table-R/P, 1949, Plastic, slide rule dial, criss-cross grill, 3 knobs, lift top, inner phono, BC, AC **$45.00**
5W12, Table-R/P, 1949, Plastic, large round dial/grill bars, 2 knobs, 3/4 lift top, inner phono, BC, AC **$50.00**
5X12, Table, 1949, Plastic, right front round dial over horizontal grill bars, 2 knobs, BC, AC/DC **$30.00**
5X12N, Table, 1949, Plastic, right front round dial over horizontal grill bars, 2 knobs, BC, AC/DC **$30.00**
5X13, Table, 1949, Plastic, right front round dial over horizontal grill bars, 2 knobs, BC, AC/DC **$30.00**
5X22, Table-C, 1953, Plastic, slide rule dial, large center alarm clock, side grill, 4 knobs, BC, AC **$35.00**
5X23, Table-C, 1953, Plastic, slide rule dial, large center alarm clock, side grill, 4 knobs, BC, AC **$35.00**
5Y22, Table-R/P, 1952, Plastic, outer front round dial/louvers, 2 knobs, 3/4 lift top, inner phono, BC, AC **$45.00**
5Z, Table, Plastic, right front gold dial, left grill w/chrome bars, decorative lines, 2 knobs **$85.00**
6A22, Table, 1950, Plastic, right round dial w/center crest, horizontal grill bars, 2 knobs, BC, AC/DC **$30.00**
6C11, Portable, 1949, Plastic, inner dial, lattice grill, flip-up front, molded handle, BC, AC/DC/battery **$35.00**

6C22, Table, 1954, Plastic, large center gold metal dial over cloth grill, feet, center knob, AC/DC **$45.00**
6C22AN, Table, 1954, Plastic, large center gold metal dial over cloth grill, feet, center knob **$45.00**
6C71, Console-R/P, 1946, Wood, slide rule dial, 4 knobs, 6 pushbuttons, pull-out phono, doors, BC, 3SW, AC . . . **$90.00**
6J21, Table-R/P, 1949, Plastic, round dial w/circular louvers, 2 knobs, 3/4 lift top, inner phono, BC, AC **$55.00**
6J21N, Table-R/P, 1950, Plastic, round dial w/circular louvers, 2 knobs, 3/4 lift top, inner phono, BC, AC **$55.00**
6N26, Console-R/P, 1952, Wood, inner pull-out radio/phono, slide rule dial, 3 knobs, double doors, BC, AC **$50.00**
6P32, Portable, 1946, "Alligator", palm tree grill, 2 knobs, 2 handles, fold down front, BC, AC/DC/bat **$75.00**
6Q12, Table, 1949, Plastic, right front round dial over horizontal louvers, 2 knobs, AM, FM, AC/DC **$30.00**
6R11, Table-R/P, 1949, Plastic, outer slide rule dial, criss-cross grill, 3 knobs, lift top, BC, FM, AC **$45.00**

6RT42A, Table-R/P, 1947, Wood, outer slide rule dial, 3 horizontal grill bars, 2 knobs, lift top, BC, AC $35.00

6RT43-5BL, Table-R/P, 1946, Wood, lower slide rule dial, horizontal louvers, 3/4 lift up lid, 2 knobs, BC, AC $35.00

6RT44, Table-R/P, 1947, Wood, outer dial, horiz grill bars, 4 knobs, 3/4 lift top, inner phono, BC, SW, AC $35.00

6RT44A-7B1, Table-R/P, 1947, Wood, lower slide rule dial, upper horiz louvers, 4 knobs, lift top, inner phono, AC . . $35.00

6S12, Table-R/P, 1950, Plastic, outer front round dial/louvers, 2 knobs, 3/4 lift top, inner phono, BC, AC. $55.00

6T01, Table, 1946, Plastic, lower front slide rule dial, horizontal louvers, 2 knobs, BC, AC/DC $30.00

6T02, Table, 1946, Plastic, lower front slide rule dial, molded horiz louvers, 2 knobs, BC, AC/DC $35.00

6T06, Table, 1946, Wood, lower slide rule dial, grill w/circular cutouts, 2 knobs, BC, 2SW, battery $45.00

6V12, Table-R/P, 1949, Plastic, outer front slide rule dial, criss-cross grill, 3 knobs, lift top, BC, AC $45.00

6W12, Table-R/P, 1949, Plastic, outer front slide rule dial, criss-cross grill, 3 knobs, lift top, BC, FM, AC $45.00

6Y18, Portable, 1949, Leatherette, inner dial, lattice grill, fold-down front, handle, BC, AC/DC/battery $35.00

7C60M, Console-R/P, 1948, Wood, inner slide rule dial, 4 knobs, phono, lift top, lower record storage, BC, AC $60.00

7C62, Console-R/P, 1947, Wood, outer dial w/large escutcheon, 4 knobs, fold-down phono door, BC, AC $60.00

7C63-UL, Console-R/P, 1947, Wood, outer dial w/large escutcheon, 4 knobs, fold-down phono door, BC, SW, AC $60.00

7C65W, Console-R/P, 1948, Wood, inner right dial, 4 knobs, left pull-down phono door, 2 section grill, BC, AC $60.00

7C73, Console-R/P, 1948, Wood, right tilt-out dial & 4 knobs, pull-out phono, criss-cross grill, BC, FM, AC $75.00

7G14, Console-R/P, 1949, Wood, inner right dial, 3 knobs, left fold-down phono door, lower storage, BC, AC $65.00

7L12, Portable, 1958, Red transistor, right side dial knob, front grill w/stylized "V", handle, AM, bat $45.00

7L14, Portable, 1958, Tan transistor, right side dial knob, front grill w/ stylized "V", handle, AM, bat $35.00

7L16, Portable, 1958, Yellow transistor, right side dial knob, front grill w/stylized "V", handle, AM, bat $35.00

7L18, Portable, 1958, Turquoise transistor, right side dial, front grill w/stylized "V", handle, AM, bat $45.00

7M12, Portable, 1958, White/red transistor, right front dial over perforated grill, handle, AM, battery $40.00

7M14, Portable, 1958, White/tan transistor, right front dial over perforated grill, handle, AM, battery $30.00

7M16, Portable, 1958, White/yellow transistor, right front dial over perforated grill, handle, AM, bat. $30.00

7M18, Portable, 1958, White/turquoise transistor, right front dial over perforated grill, handle, AM, bat. $40.00

7P33, Portable, 1947, Briefcase-style, inner slide rule dial, fold-down front, handle, BC, AC/DC/bat. $30.00

7P33-4, Portable, 1947, Black leatherette, inner slide rule dial & plastic grill, fold-down front, handle $30.00

7RT41-N, Table-R/P, Bakelite, slide rule dial, cloth grill w/metal cutouts, 4 knobs, inner phono, AC $45.00

7RT42, Table-R/P, 1947, Wood, outer front slide rule dial, 4 knobs, lift top, inner phono, BC, AC $30.00

7RT42N, Table-R/P, 1947, Wood, outer front slide rule dial, 4 knobs, lift top, inner phono, BC, AC $30.00

7T01C-N, Table, 1948, Plastic, lower front slide rule dial, upper horizontal louvers, 2 knobs, BC $35.00

7T01M-UL, Table, 1948, Plastic, lower front slide rule dial, upper horizontal louvers, 2 knobs, BC $35.00

7T04-UL, Table, 1948, Two-tone, lower front slide rule dial, upper cloth grill, 2 knobs, BC, AC/DC $35.00

7T10, Table, 1947, Plastic, right front square dial, left horizontal louvers, 2 knobs, BC, AC/DC $35.00

7T10E-N, Table, 1947, Plastic, right front square dial, left horizontal louvers, 2 knobs, BC, AC/DC $35.00

7T12, Table, 1947, Plastic, slanted upper slide rule dial, large lower grill area, 2 knobs, BC, battery $30.00

8D15, Console-R/P, 1949, Wood, right tilt-out black dial, 3 knobs, left fold-down phono door, BC, FM, AC $80.00

9B14, Console-R/P, 1948, Wood, right tilt-out black dial, 4 knobs, left fold-down phono door, BC, FM, AC $80.00

9E15, Console-R/P, 1949, Wood, right tilt-out black dial, 4 knobs, left fold-down phono door, BC, FM, AC. $80.00

12-B5, Table, 1941, Ebony plastic, right front dial, left horizontal louvers, 2 knobs, 2 bands, AC/DC. $35.00

13-C5, Table, 1940, Mahogany plastic, right front dial, left horizontal louvers, handle, 2 knobs, AC/DC. $40.00

14-B5, Table, 1941, Ivory plastic, right dial, left horizontal louvers, handle, 2 knobs, 2 bands, AC/DC $40.00

14-C5, Table, 1940, Ivory plastic, right front dial, left horizontal louvers, handle, 2 knobs, AC/DC **$40.00**

15-B5, Table, 1940, Walnut plastic, streamline, right dial, left wrap-around louvers, 2 knobs, AC/DC **$80.00**

15-D5, Table, 1940, Mahogany plastic, streamline, right dial, wrap-around louvers, 2 knobs, AC/DC **$80.00**

16-B5, Table, 1940, Chartreuse/ivory plastic, streamline, left wrap-around louvers, 2 knobs, AC/DC **$80.00**

16-D5, Table, 1940, Ivory plastic, streamline, right dial, left wrap-around louvers, 2 knobs, AC/DC **$80.00**

17-B5, Table, 1940, Walnut wood, right front dial, left horizontal louvers, handle, 2 bands, AC/DC **$40.00**

18-B5, Table, 1940, Walnut wood, right front dial, left horizontal louvers, 2 knobs, 2 bands, AC/DC **$35.00**

20-A6, Table, 1940, Walnut plastic, streamline, right dial, left wrap-around louvers, 2 bands, AC/DC **$80.00**

21-A6, Table, 1940, Ivory plastic, streamline, right dial, left wrap-around louvers, 2 knobs, AC/DC **$80.00**

22-A6, Table, 1940, Walnut wood, right front dial, left vertical grill bars, 2 knobs, 2 bands, AC/DC **$40.00**

23-A6, Table, 1940, Walnut wood, right front dial, left vertical grill bars, 2 knobs, 2 bands, AC/DC **$40.00**

25-Q5, Table, 1940, Walnut wood, right front dial, left horizontal louvers, 4 pushbuttons, AC/DC **$45.00**

28-G5, Portable, 1942, Inner front dial, horizontal louvers, fold-open front door, handle, AC/DC/battery **$50.00**

33-F5, Portable, 1940, Upper front slide rule dial, lower square grill, 2 knobs, handle, AC/DC/battery **$20.00**

34-F5, Portable, 1940, Leatherette, slide rule dial, lower horizontal louvers, 2 knobs, handle, AC/DC/bat **$25.00**

35-G6, Portable, 1940, Brown leatherette, inner slide rule dial, detachable cover, handle, AC/DC/battery **$25.00**

37-G6, Portable, 1940, Upper front slide rule dial, horizontal louvers, 2 knobs, handle, AC/DC/battery **$25.00**

44-J5, Table, 1941, Ebony plastic, right dial, left horizontal louvers, 3 bullet knobs, BC, SW, AC/DC **$40.00**

45-J5, Table, 1941, Ivory plastic, right dial, left horizontal louvers, 3 bullet knobs, BC, SW, AC/DC **$40.00**

47-J55, Table, 1940, Walnut wood, right front dial, left horizontal louvers, 3 knobs, 2 bands, AC/DC **$35.00**

48-J6, Table, 1940, Mahogany plastic, right dial, left louvers, pushbuttons, 3 knobs, 2 bands, AC/DC **$50.00**

49-J6, Table, 1940, Ivory plastic, right dial, horizontal louvers, pushbuttons, 3 knobs, 2 bands, AC/DC **$50.00**

50-J6, Table, 1940, Walnut wood, right dial, left wrap-around grill bars, pushbuttons, 3 knobs, AC/DC **$50.00**

51-J55, Table, 1940, Mahogany plastic, right dial, left horizontal louvers, 3 knobs, 2 bands, AC/DC **$40.00**

51-K6, Table, 1940, Mahogany plastic, right dial, left wrap-around louvers, 3 knobs, 2 bands, AC **$40.00**

52-J55, Table, 1940, Ivory plastic, right front dial, left horizontal louvers, 3 knobs, 2 bands, AC/DC **$40.00**

52-K6, Table, 1940, Ivory plastic, right front dial, left wrap-around louvers, 3 knobs, 2 bands, AC **$40.00**

53-K6, Table, 1940, Walnut wood, right front dial, left wrap-around louvers, 3 knobs, 2 bands, AC **$40.00**

54-XJ55, Table-R/P, 1940, Wood, outer right dial, left horizontal louvers, lift top, inner phono, 2 band, AC **$35.00**

55-A7, Table, 1940, Walnut wood, right dial, left grill w/bars, pushbuttons, 3 knobs, 2 bands, AC **$45.00**

56-A77, Console, 1940, Walnut, slanted front dial, vertical grill bars, pushbuttons, 3 knobs, 2 bands, AC **$100.00**

57-B7, Console-R/P, 1940, Wood, inner dial & pushbuttons, phono, front vertical bars, BC, SW, AC **$100.00**

58-A11, Console-R/P, 1940, Wood, front grill w/3 vertical bars, tuning eye, pushbuttons, BC, SW, AC **$135.00**

59-A11, Console-R/P, 1940, Wood, inner dial, pushbuttons, tuning eye, phono, front grill, BC, SW, AC **$150.00**

63-A11, Console, 1940, Walnut, slanted front dial, tuning eye, lower vertical grill bars, 5 bands, AC **$125.00**

70-K5, Table-R/P, 1941, Wood, inner right dial, left phono, lift top, outer horizontal wrap-around bars, AC **$35.00**

71-M6, Console-R/P, 1941, Wood, upper front dial, center pull-out phono behind double doors, 2 bands, AC **$100.00**

76-P5, Portable, 1941, Brown plastic, right dial, left horizontal grill bars, 2 knobs, handle, AC/DC/bat **$30.00**

79-P6, Portable, 1941, Leatherette, inner right dial, left grill, fold-down front, handle, AC/DC/battery **$30.00**

113-5A, Table, 1938, Black plastic, right magnifying lens dial, left vertical grill bars, pushbuttons, AC **$100.00**

114-5A, Table, 1938, Walnut plastic, right magnifying lens dial, vertical grill bars, pushbuttons, AC **$100.00**

115-5A, Table, 1938, Ivory plastic, right magnifying lens dial, left vertical grill bars, pushbuttons, AC **$100.00**

123-5E, Table, 1938, Black plastic, midget, right round dial knob, vertical wrap-over grill bars, AC **$75.00**

124-5E, Table, 1938, Walnut plastic, midget, right round dial knob, vertical wrap-over grill bars, AC **$75.00**

125-5E, Table, 1938, Ivory plastic, midget, right round dial knob, left vertical wrap-over grill bars, AC **$75.00**

126-5E, Table, 1938, Red plastic, midget, right round dial knob, left vertical wrap-over grill bars, AC **$95.00**

139-11A, Console, 1939, Wood, upper slide rule dial, pushbuttons, cloth grill w/vertical bars, 3 bands, AC **$100.00**

141-4A, Table, 1939, Two-tone wood, right slide rule dial, left cloth grill w/2 horiz bars, 2 knobs, bat **$35.00**

142-8A, Console-R/P, 1938, Wood, inner slide rule dial, pushbuttons, Deco front grill w/discs, 3 bands, AC **$150.00**

153-5L "The Gypsy", Portable-R/P, 1939, Fold-down front, inner right dial, 2 knobs, lift top, inner phono, handle, AC . . **$30.00**

159-5L, Table-R/P, 1939, Wood, outer right front dial, left wrap-around louvers, lift top, inner phono, AC **$40.00**

162-5L, Table, 1939, Plastic, right front dial, left horizontal wrap-around louvers, AC/DC **$40.00**

202, Portable, 1958, Leatherette, right front half-moon dial, perforated grill, handle, BC, AC/DC/bat **$40.00**

218, **Portable, 1958, Leatherette, right front dial, metal perforated grill w/logo, handle, 2 knobs $40.00**

221, Portable, 1958, Black transistor, right dial, perforated grill w/ stylized "A" logo, handle, AM, bat **$40.00**

227, Portable, 1958, Tan transistor, right dial, perforated grill w/ stylized "A" logo, handle, AM, bat **$40.00**

228, Portable, 1958, Turquoise transistor, right dial, perforated grill w/stylized "A" logo, handle, AM **$45.00**

237, Portable, 1959, Transistor, right front dial, perforated grill w/ stylized "A" logo, handle, battery **$40.00**

242, Table, 1958, Plastic, right front round dial, horizontal front bars, "Admiral" logo, BC, AC/DC **$20.00**

284, Table-C, 1958, Plastic, lower right front dial, horizontal louvers, left alarm clock, feet, BC, AC **$20.00**

292, Table-C, 1958, Plastic, lower left front dial, upper alarm clock, right horizontal louvers, BC, AC $20.00

303, Table, 1959, Wood, lower front slide rule dial, large upper grill, 2 knobs, AM, FM, AC/DC $20.00

331-4F, Portable, 1939, Striped, upper front dial, lower square grill, 2 flat bakelite knobs, handle, bat $25.00

361-5Q, Table, 1939, Plastic, right front dial, left horizontal wrap-around louvers, 2 knobs, AC $45.00

384-5S, Table, 1939, Walnut wood, right front dial, left horizontal grill bars, handle, AC/DC $45.00

396-6M, Table, 1940, Walnut plastic, streamline, right dial, wrap-around louvers, pushbuttons, AC/DC $95.00

397-6M, Table, 1940, Ivory plastic, streamline, right dial, wrap-around louvers, pushbuttons, AC/DC $95.00

398-6M, Table, 1940, Onyx plastic, streamline, right dial, wrap-around louvers, pushbuttons, AC/DC $95.00

399-6M, Table, 1940, Walnut wood, right dial, left horizontal louvers, 4 pushbuttons, 2 bands, AC/DC $45.00

512-6D, Table, 1938, Two-tone wood, front slide rule dial, cloth grill w/Deco cut-outs, 4 knobs, DC $50.00

516-5C, Table, 1938, Plastic, right front dial, raised left vertical grill bars, 2 knobs, 2 bands, AC/DC $45.00

521, Portable, 1959, Transistor, right front dial, center grill cut-outs, handle, AM, battery $35.00

521-5C, Table-R/P, 1938, Two-tone wood, right dial, left grill w/3 horiz bars, lift top, inner phono, AC $35.00

521-5F, Table-R/P, 1938, Two-tone wood, right dial, left grill w/horiz bars, lift top, inner phono, AC $35.00

531, Portable, 1959, Transistor, upper right dial, center grill cut-outs, antenna in handle, AM, battery $35.00

549-6G, Console-R/P, 1939, Walnut, inner slide rule dial, 8 pushbuttons, phono, large front grill, 2 band, AC $85.00

561, Table, 1959, Plastic, transistor, right dial over horizontal front bars, 3 knobs, feet, AM, bat $25.00

566, Table, 1959, Plastic, transistor, right dial over horizontal front bars, 3 knobs, feet, AM, bat $25.00

581, Portable, 1959, Transistor, right front dial knob, perforated grill, right side knob, AM, battery $25.00

582, Portable, 1959, Transistor, right front dial knob, perforated grill, right side knob, AM, battery $25.00

692 "Deluxe-5", Portable, 1960, Transistor, right front round dial, center lattice grill, swing handle, BC, battery $30.00

703 "Super-7", Portable, 1960, Transistor, right front round dial, center lattice grill, swing handle, BC, battery $30.00

717 "Imperial 8", Portable, 1960, Transistor, right front round dial, center lattice grill, swing handle, BC, battery $30.00

739, Portable, 1960, Leather case, transistor, right round dial over lattice grill, handle, BC, battery $25.00

742, Portable, 1960, Leather case, transistor, right front round dial, lattice grill, handle, BC, battery $25.00

751, Portable, 1960, Black leather case, transistor, right front dial, lattice grill, handle, BC, battery $30.00

757, Portable, 1960, Tan leather case, transistor, right front dial, lattice grill, handle, BC, battery $30.00

801, Portable, 1959, Plastic, transistor, right front dial, perforated grill, wire stand, BC, battery $30.00

811B "Super 8", Table-C, 1960, Transistor, right front round dial/left clock over horizontal bars, feet, BC, bat $25.00

816B "Super 8", Table-C, 1960, Transistor, right front round dial/left clock over horizontal bars, feet, BC, bat $25.00

909 "All World", Portable, 1960, Transistor, inner dial, fold-down front, telescoping antenna, handle, 9 bands, bat $75.00

920-6Q, Tombstone, 1937, Wood, center front dial, upper horizontal grill bars, 4 knobs, BC, SW, battery $65.00

930-16R, Console, 1937, Wood, top front slide rule dial, lower cloth grill w/vertical "ladder", BC, SW, AC $150.00

935-11S, Console, 1937, Walnut, Deco, upper slide rule dial, large cloth grill w/3 vert bars, BC, SW, AC $110.00

940-11S, Console, 1937, Walnut, slide rule dial, rounded sides, cloth grill w/3 vertical bars, BC, SW, AC $125.00

945-8K, Console, 1937, Walnut, Deco, upper round front dial, cloth grill w/center vertical bar, BC, SW, AC $115.00

945-8T, Console, 1937, Walnut, Deco, upper round front dial, cloth grill w/center vertical bar, BC, SW, AC $115.00

950-6P, Console, 1937, Wood, Deco, upper front dial, 4 knobs, cloth grill w/center vert bar, BC, SW, bat $80.00

955-8K, Chairside, 1937, Walnut, Deco, top dial, streamline semi-circular front w/ashtray, BC, SW, AC $180.00

955-8T, Chairside, 1937, Walnut, Deco, top dial, streamline semi-circular front w/ashtray, BC, SW, AC/DC $180.00

960-8K, Table, 1937, Wood, right round dial, left horiz louvers, pushbuttons, 4 knobs, feet, BC, SW, AC $60.00

960-8T, Table, 1937, Wood, right round dial, left horiz louvers, pushbuttons, 4 knobs, feet, BC, SW, AC $60.00

965-6P, Table, 1937, Wood, right front dial, left grill w/3 horizontal bars, 4 knobs, BC, SW, battery $45.00

965-7M, Table, 1937, Wood, right front dial, left grill w/3 horizontal bars, 4 knobs, BC, SW, AC $55.00

975-6W, Table, 1937, Wood, right front dial, left cloth grill w/3 horizontal bars, 3 knobs, BC, SW, AC $55.00

980-5X, Table, 1937, Wood, right front dial, left cloth grill w/Deco cut-outs, 3 knobs, BC, SW, AC $55.00

985-5Z, Table, 1937, Ivory & gold plastic, can be used horizontally or vertically, 2 grill bars, BC, AC $70.00

985-6Y, Table, 1937, Ivory & gold bakelite, plays horizontally or vertically, 2 grill bars, BC, SW, AC $70.00

990-5Z, Table, 1937, Ebony bakelite & chrome, plays horizontally or vertically, 2 grill bars, BC, AC $85.00

990-6Y, Table, 1937, Ebony bakelite & chrome, plays horizontally or vertically, 2 grill bars, BC, SW, AC $85.00

4202-B6, Table, 1941, Plastic, upper front slide rule dial, horizontal grill bars, 2 knobs, BC, AC/DC $40.00

4204-B6, Table, 1941, Two-tone wood, front slide rule dial, cloth grill w/lyre cut-out, 2 knobs, BC, SW $60.00

4207-A10, Console-R/P, 1941, Walnut, front dial, pushbuttons, 4 knobs, inner pull-out phono, doors, BC, SW, AC $125.00

39411-B, Console, 1939, Wood, Deco, slide rule dial, pushbuttons, cloth grill w/center bar, BC, SW, AC $135.00

AM6, Console, 1936, Wood, rounded front w/oval 4 band dial, cloth grill w/vertical bars, 4 knobs, AC $135.00

AM786, Console, 1936, Wood, round dial w/oval escutcheon, 3 vert grill bars, 4 knobs, BC, LW, SW, AC $110.00

AM787, Console, 1936, Wood, round dial w/oval escutcheon, curved top, 4 grill bars, 4 knobs, BC, SW, AC $135.00

M169, Table, 1936, Rectangular case, left front round dial, right round grill w/cut-outs, AC $65.00

R58-B11, Console-R/P/Rec, 1940, Wood, front grill w/3 vertical bars, tuning eye, pushbuttons, recorder, BC, SW, AC $140.00

R59-B11, Console-R/P/Rec, 1940, Wood, inner dial & pushbuttons, phono, wire recorder, front grill, BC, SW, AC $150.00

Y-2023 "Super 7", Table, 1960, White transistor, front off-center dial, right/left lattice grills, BC, battery $25.00

Y-2027 "Super 7", Table, 1960, Beige transistor, front off-center dial, right/left lattice grills, BC, battery $25.00

Y-2028 "Super 7", Table, 1960, Green transistor, front off-center dial, right/left lattice grills, BC, battery $25.00

Y-2127 "Imperial 8", Portable, 1959, Leather case, transistor, right round dial, left lattice grill w/knob, handle, bat . . . $35.00

ADVANCE ELECTRIC
Advance Electric Company, 1260 West Second Street, Los Angeles, California

Advance Electric was founded in 1924 by Fritz Falck. He had previously manufactured battery chargers, transformers and did repair work and rewinding of electric motors. By 1924, the company was producing radios and continued to do so until 1933, when it dropped the radio line and continued in business with the production of electrical relays and electronic parts.

4, Table, 1924, Wood, high rectangular case, 2 dial slant front panel, 3 exposed tubes, lift top **$150.00**
4 Junior, Table, C1924, Wood, high rectangular case, 2 dial slant front panel, 2 exposed tubes, lift top **$150.00**
69, Cathedral, 1930, Two-tone wood, center window dial, scalloped grill w/cut-outs, 3 knobs, AC **$200.00**
88, Cathedral, 1930, Wood, center front window dial, 3 knobs, upper scalloped grill w/ cut-outs, AC, ... **$200.00**
89, Cathedral, 1930, Wood, center front window dial, 3 knobs, upper scalloped grill w/cut-outs, AC, ... **$200.00**
F, Cathedral, 1932, Wood, lower half-moon dial w/escutcheon, upper round grill w/cut-outs, 3 knobs, ... **$175.00**

AERMOTIVE
Aermotive Equipment Corp., 1632-8 Central Street, Kansas City, Missouri

181-AD, Table, 1947, Wood, right black dial w/airplane, left round grill w/bars, 2 knobs, BC, AC/DC **$45.00**

AERODYN
**Aerodyn Co.,
1780 Broadway, New York City, New York**

Special, Table, 1925, Wood, rectangular case, slanted three dial black front panel, 5 tubes **$125.00**

AIRADIO
Airadio, Inc., Stamford, Connecticut

3049, Table-R/P, Wood, slide rule dial, horizontal grill bars, 4 knobs, lift top, inner phono, BC, AC **$40.00**
3100, Table, 1948, Wood, high rectangular cabinet, slide rule dial, 4 knobs, FM only, AC/DC **$25.00**

AIR CASTLE
Spiegel, Inc., 1061- 1101 West 35th Street, Chicago, Illinois

7B, Console-R/P, 1948, Wood, inner right slide rule dial, 4 knobs, left pull-out phono drawer, BC, FM, AC **$80.00**
9, Table, 1948, Plastic, slanted upper slide rule dial, horizontal louvers, 3 knobs, BC, FM, AC/DC **$35.00**
102B, Portable, 1950, Plastic, lower slide rule dial, horizontal louvers, handle, 2 knobs, BC, AC/DC/bat **$30.00**
106B, Table, 1947, Plastic, streamline, right square dial, left horizontal louvers, 2 knobs, BC, AC/DC **$85.00**
121, Console, Wood, upper dial, pushbuttons, tuning eye, lower grill w/vertical bars, BC, SW, AC **$125.00**

153, Console-R/P, 1951, Wood, inner slide rule dial, 4 knobs, center pull-out phono, double doors, BC, AC **$50.00**
171, Table, 1950, Plastic, lower front slide rule dial, horizontal grill bars, 2 knobs, BC, AC/DC **$40.00**
179, Portable, 1948, Upper front slide rule dial, lattice grill, handle, 2 knobs, BC, battery **$35.00**
180, Portable, 1948, Upper front slide rule dial, lattice grill, handle, 2 knobs, BC, AC/DC/battery **$35.00**
198, Table, 1950, Wood, lower slide rule dial, recessed upper grill, feet, 2 knobs, AM, FM, AC **$30.00**
201, Table, 1950, Plastic, lower slide rule dial, recessed checkerboard grill, 2 knobs, BC, AC/DC **$45.00**
211, Table, 1949, Plastic, right round dial, center front lattice grill, 2 knobs, BC, AC/DC **$30.00**
212, Table, 1949, Plastic, lower slide rule dial, upper vertical grill bars, 2 knobs, BC, FM, AC **$35.00**
213, Portable, 1949, Plastic, lower slide rule dial, lattice grill, handle, 2 knobs, BC, AC/DC/battery **$35.00**
350, Console-R/P, 1951, Wood, inner slide rule dial, 4 knobs, pull-out phono, double doors, BC, FM, AC **$50.00**
472-053VM, Console-R/P, 1952, Wood, pull-out front drawer contains right dial, 3 knobs, left phono, BC, AC **$45.00**
472.254, Console-R/P, 1953, Wood, inner right slide rule dial, 4 knobs, pull-out phono, double doors, BC, AC **$50.00**
568, Table, 1947, Top right front slide rule dial, left round grill, 3 recessed knobs, BC, SW, AC/DC **$35.00**
572, Console-R/P, 1949, Wood, inner right dial, 4 knobs, left phono, lift top, front grill, BC, AC **$50.00**
603.880, Table-R/Rec, 1954, Leatherette, inner left dial, disc recorder, lift top, outer grill, handle, BC, AC **$35.00**
603-PR-8.1, Table-R/P/Rec, 1951, Leatherette, inner left dial, phono, recorder, mike, lift top, handle, BC, AC **$35.00**
606-400WB, Table, 1951, Wood, right rectangular dial, left cloth grill w/cut-outs, 2 knobs, BC, battery **$35.00**
607.299, Table, 1952, Plastic, right half-moon dial, left horizontal grill bars, 2 knobs, BC, AC/DC **$40.00**
607-314, Table, 1951, Plastic, large front dial w/horizontal decorative lines, 2 knobs, BC, AC/DC **$35.00**
607-316-1, Table, 1951, Plastic, right round dial, diagonal grill w/ lower vertical bars, 2 knobs, BC, AC/DC **$30.00**

611-1, Tombstone, Wood, lower front cylindrical dial, upper cloth grill, 6 pushbuttons, 4 knobs **$85.00**
629, Table, 1937, Wood, right slide rule dial, pushbuttons, tuning eye, left grill w/cut-outs, 4 knobs **$100.00**
651, Table, 1947, Plastic, upper slide rule dial, lower horizontal louvers, 2 knobs, BC, AC/DC **$35.00**

652.505, Table-R/P, 1952, Leatherette, inner right round dial, left phono, lift top, handle, BC, AC **$20.00**

652.5X5, Table-R/P, 1955, Wood, right side dial knob, large front grill, 3/4 lift top, inner phono, BC, AC. **$30.00**

659.511, Table-C, 1952, Plastic, center front round dial, right checkered grill, left alarm clock, BC, AC **$30.00**

659.520E, Table-C/N, 1952, Plastic, clock/radio/lamp, center round dial knob, left alarm clock, BC, AC **$75.00**

751, Table, Wood, right black slide rule dial, left grill w/diagonal bars, 3 knobs, black top **$50.00**

782.FM-99-AC, Table, 1955, Plastic, lower front slide rule dial, 2 knobs, large upper grill, AM, FM, AC. **$35.00**

935, Table-C, 1951, Plastic, right round dial, horizontal center bars, left clock, 4 knobs, BC, AC **$30.00**

1200, Table, Two-tone wood, right dial, 4 pushbuttons, left grill w/cut-outs, 5 knobs, battery **$50.00**

227I, Table, 1950, Plastic, right square dial, left horizontal wraparound louvers, 2 knobs, BC **$45.00**

5000, Table, 1947, Plastic, right dial, left grill w/3 horiz bars, stepdown top, 2 knobs, BC, AC/DC **$50.00**

5001, Table, 1947, Plastic, right front square dial, left cloth grill, 2 knobs, BC, AC/DC **$35.00**

5002, Table, 1947, Plastic, right square dial, left vertical wrap-over louvers, 2 knobs, BC, AC/DC **$50.00**

5003, Table, 1947, Plastic, slide rule dial, wrap-around louvers, 2 knobs on top of case, BC, AC/DC **$40.00**

5008, Table, 1948, Wood, slanted upper slide rule dial, burl side panels, base, 3 knobs, BC, AC/DC **$45.00**

5011, Table, 1947, Two-tone wood, slanted upper slide rule dial, cloth grill, 4 knobs, BC, SW, AC/DC **$50.00**

5015.1, Table, 1950, Plastic, raised top slanted slide rule dial, horizontal louvers, 2 knobs, BC, AC/DC **$45.00**

5020, Portable, 1947, Luggage-style, right square dial, left cloth grill, 3 knobs, handle, BC, AC/DC/bat **$25.00**

5022, Portable, 1951, "Snakeskin" w/plastic front panel, right dial, handle, 3 knobs, BC, AC/DC/bat **$45.00**

5024, Table, 1948, Wood, slanted slide rule dial, cloth grill, top burl veneer, 3 knobs, BC, AC/DC/bat **$45.00**

5025, Portable, 1947, "Snakeskin", top dial, horizontal louvers, 3 knobs, handle, BC, AC/DC/battery **$35.00**

5027, Portable, 1948, Leatherette, top slide rule dial, horiz louvers, handle, 3 knobs, BC, AC/DC/bat **$30.00**

5028, Portable, 1948, "Alligator", right dial, left horizontal louvers, handle, 3 knobs, BC, AC/DC/bat **$35.00**

5029, Portable, 1948, "Alligator", right square dial, left horizontal louvers, handle, 2 knobs, BC, battery **$30.00**

5035, Table-R/P, 1948, Leatherette, outer "horse-shoe" dial, horizontal louvers, 3 knobs, lift top, BC, AC **$25.00**

5036, Table-R/P, 1949, Wood, outer front slide rule dial, 3 knobs, lift top, inner phono, BC, AC. **$30.00**

5044, Table-R/P, 1951, Wood, outer front slide rule dial, 3 knobs, lift top, inner phono, BC, AC. **$30.00**

5050, Table, 1948, Right front dial, left large square cloth grill, 2 knobs, BC, AC/DC **$30.00**

5052, Table, 1948, Plastic, right dial, left horizontal grill bars, stepdown top, 2 knobs, BC, AC/DC **$50.00**

5056-A, Table, 1951, Small case, right front dial, left checkered grill, 2 knobs, BC, AC/DC **$35.00**

6042, Table-R/P, 1949, Wood, outer front slanted slide rule dial, 4 knobs, lift top, inner phono, BC, AC. **$35.00**

6050, Table-R/P, 1949, Wood, outer front slanted slide rule dial, 4 knobs, lift top, inner phono, BC, AC. **$35.00**

6053, Table-R/P, 1950, Wood, outer front slanted slide rule dial, 4 knobs, lift top, inner phono, BC, AC. **$35.00**

6514, Table, 1947, Wood, slanted slide rule dial, cloth grill w/ 5 horizontal bars, 2 knobs, BC, AC/DC **$40.00**

6541, Table-R/P, 1947, Wood, outer slide rule dial, horizontal louvers, 4 knobs, lift top, BC, AC **$35.00**

6547, Table-R/P, 1947, Wood, outer slide rule dial, horizontal louvers, 4 knobs, lift top, BC, AC **$35.00**

6634, Table-R/P, 1947, Leatherette, slide rule dial w/ small cover, 4 knobs, lift top, handle, BC, SW, AC **$45.00**

7553, Table, 1948, Plastic, slide rule dial, wrap-around horiz louvers w/ cross, 2 knobs, BC, AC/DC **$55.00**

9008W, Table, 1950, Plastic, right dial, left vertical wrap-over grill bars, 2 knobs, BC, AC/DC **$45.00**

9009W, Table, 1950, Plastic, streamline, right dial, horiz wraparound louvers, 2 knobs, BC, AC/DC **$65.00**

9012W, Table, 1950, Plastic, Deco, right dial, left wrap-around horizontal louvers, 2 knobs, BC, AC/DC **$65.00**

9151-W, Table-C, 1951, Plastic, right half-moon dial, left alarm clock, perforated center grill, BC, AC **$30.00**

9904, Tombstone, 1934, Wood, lower round dial, upper grill w/cutouts, 7 tubes, 4 knobs, BC, SW, AC **$140.00**

10002, Table, 1949, Plastic, upper slide rule dial, horizontal wraparound louvers, 3 knobs, BC, AC/DC **$40.00**

10003-I, Table, 1949, Plastic, streamline, square dial, left horizontal wrap-around louvers, BC, AC/DC **$65.00**

10005, Table, 1949, Plastic, slide rule dial, horizontal wrap-around louvers, 4 knobs, AM, FM, AC/DC **$40.00**

10023, Table-R/P, 1949, Outer right front round dial, 3 knobs, handle, lift top, inner phono, BC, AC **$25.00**

108014, Table, 1949, Plastic, slanted upper slide rule dial, horizontal grill bars, 4 knobs, BC, AC. **$35.00**

121104, Console-R/P, 1949, Wood, right tilt-out dial, 4 knobs, left pull-out phono drawer, BC, FM, AC **$70.00**

121124, Console-R/P, 1949, Wood, outer right slide rule dial, 4 knobs, left lift top, inner phono, BC, FM, AC **$60.00**

127084, Console-R/P, 1949, Wood, outer front dial, 4 knobs, lower grill, lift top, inner phono, BC, AC. **$50.00**

131504, Table, 1949, Plastic, slanted upper slide rule dial, horiz grill bars, 4 knobs, BC, FM, AC/DC **$35.00**

132564, Table, 1949, Wood, right square dial, left cloth grill w/ crossed bars, 2 knobs, BC, battery **$30.00**

138104, Console-R/P, 1949, Wood, outer right front slide rule dial, 5 knobs, left lift top, inner phono, BC, AC. **$60.00**

138124, Console-R/P, 1949, Wood, right tilt-out slide rule dial, 5 knobs, left pull-out phono drawer, BC, AC. **$75.00**

147114, Portable, 1949, Inner right dial, lattice grill, 2 knobs, flip-up front, handle, BC, AC/DC/battery. **$40.00**

149654, Table, 1949, Plastic, slanted upper slide rule dial, horizontal grill bars, 4 knobs, BC, FM, AC **$35.00**

150084, Console-R/P, 1949, Wood, slide rule dial, 4 knobs, lift top, inner phono, criss-cross grill, BC, FM, AC. **$50.00**

A-2000 , Table, Wood, center cylindrical dial, pushbuttons, tuning eye, upper cloth grill, 4 knobs **$95.00**

G-516, Table-R/P, 1948, Wood, outer slanted slide rule dial, lower grill, 4 knobs, lift top, inner phono, AC **$35.00**

G-521, Portable, 1949, Leatherette, slide rule dial, tambour top, telescope antenna, BC, SW, AC/DC/bat **$45.00**

G-722, Console-R/P, 1948, Wood, inner right dial, pushbuttons, door, left pull-out phono drawer, AC **$80.00**

G-724, Table, 1948, Wood, slanted top slide rule dial, large criss-cross grill, 4 knobs, BC, FM, AC/DC **$35.00**

G-725, Console-R/P, 1948, Wood, inner right dial, 4 knobs, door, left pull-out phono drawer, BC, FM, AC/DC **$75.00**

PX, Table, 1947, Wood, slanted upper slide rule dial, cloth grill w/side cut-outs, 2 knobs, BC, bat **$30.00**

REV248, Table, 1951, Plastic, curved upper slide rule dial, front horizontal louvers, 3 knobs, BC, AC/DC **$50.00**

WEU-262, Table, 1950, Plastic, right dial, left cloth grill, decorative case lines, 4 knobs, BC, FM, AC/DC **$45.00**

AIR KING
Air King Products Co., Inc., 1523 63rd Street, Brooklyn, New York

5H110 , Table, 1946, Plastic, right front round dial w/gold pointer, left lattice grill, 2 knobs, AC **$45.00**

52, Tombstone, 1933, Plastic, Deco, center window dial, upper insert w/Egyptian figures, 3 knobs, AC . . . . . . **$4,000.00+**

66, Tombstone, 1935, Plastic, Deco, center front round dial, upper insert w/globes, 3 knobs, 2 band, AC . . **$3,000.00+**

222, Table, 1938, Plastic, midget, right front dial, left grill, 2 knobs, BC, AC/DC **$60.00**

770, Tombstone, C1935, Plastic, Deco, center front square dial, upper cloth grill w/vertical bars, 4 knobs **$1,500.00**

800, Console-R/P, 1949, Wood, right tilt-out slide rule dial, 4 knobs, pull-out phono drawer, BC, FM, AC **$70.00**

911, Table, 1938, Wood, right front dial, left horizontal louvers, pushbuttons, 3 knobs, BC, AC/DC **$45.00**

4129, Table-R/P, 1941, Wood, lower front slide rule dial, upper grill, 2 knobs, lift top, inner phono, AC **$30.00**

4603, Table, 1946, Wood, upper slanted slide rule dial, horizontal louvers, 2 knobs, BC, AC/DC **$40.00**

4604, Table, 1946, Wood, upper slanted slide rule dial, criss-cross grill, 4 knobs, BC, SW, AC **$40.00**

4604-D, Table, 1946, Wood, upper slanted slide rule dial, lower grill, fluted columns, 4 knobs, BC, SW **$35.00**

4608, Table, 1946, Plastic, right vertical slide rule dial, left perforated grill, 2 knobs, BC, AC/DC **$45.00**

4609, Table, 1947, Wood, right vertical slide rule dial, left horizontal louvers, 2 knobs, BC, AC/DC **$35.00**

4610, Table, 1947, Plastic, right vertical slide rule dial, left horizontal louvers, 2 knobs, BC, AC/DC **$35.00**

4700, Console-R/P/Rec, 1948, Wood, inner right slide rule dial, left phono, lift top, criss-cross grill, BC, AC **$55.00**

4704, Table-R/P, 1947, Wood, outer top slide rule dial, 4 knobs, louvers, inner phono, lift top, BC, AC **$40.00**

4705, Table, 1946, Plastic, right round dial w/gold pointer, left lattice grill, 2 knobs, BC, AC/DC **$45.00**

4706, Table, 1946, Painted plastic, right round dial w/gold pointer, left lattice grill, 2 knobs, AC **$45.00**

A-400, Table, 1947, Plastic, right half-moon dial, left checkerboard grill, 2 knobs, BC, AC/DC **$40.00**

A-403 "Court Jester", Table-R/P, 1947, Wood, outer right dial, cloth grill w/ 3 bars, 3 knobs, open top phono, BC, AC **$30.00**

A-410, Portable/Camera, 1948, "Alligator", lower dial, perforated grill, 2 knobs, inner camera, strap, BC, bat **$125.00**

A-426, Portable, 1948, Inner metal grill, louvers, 2 thumbwheel knobs, flip-open door, handle, BC, bat **$45.00**

A-450, Table, Plastic, midget, raised top, right half-round dial, left checkered grill, 2 knobs, AC **$60.00**

A-502, Table, 1948, Plastic, right front dial, left lattice grill, 3 knobs, BC, 2SW, AC/DC **$45.00**

A-510, Portable, 1947, Leatherette, right front dial, horiz grill bars, 2 knobs, handle, BC, AC/DC/battery **$25.00**

A-511, Table, 1947, Plastic, right front dial, left lattice grill, 2 knobs, BC, AC/DC **$45.00**

A-520, Portable, 1948, Plastic, recessed lower right dial, vertical grill bars, handle, BC, AC/DC/bat **$35.00**

A-600 "The Dutchess", Table, 1947, Two-tone Catalin, lower slide rule dial, recessed lattice grill, 2 knobs, BC,AC/DC. **$900.00+**

A-604, Table, 1950, Wood, slanted upper slide rule dial, horizontal louvers, 4 knobs, BC, SW, AC **$30.00**

A-625, Table, 1948, Two-tone plastic, slanted slide rule dial, horizontal louvers, 2 knobs, BC, AC/DC **$50.00**

A-650, Table, 1948, Two-tone plastic, slide rule dial, horizontal louvers, 2 knobs, BC, FM, AC/DC **$50.00**

AIR KNIGHT
Butler Brothers, Randolph & Canal Streets, Chicago, Illinois

CA-500, Table, 1947, Wood, right square dial, left cloth wrap-around grill, 2 knobs, BC, AC/DC **$35.00**

N5-RD291, Table-R/P, 1947, Wood, outer square dial, 2 cloth grills, 2 knobs, lift top, inner phono, BC, AC **$35.00**

AIRITE

3000, Table-N, 1936, Desk set radio, center radio w/top grill bars, right pen/inkwell, left clock $385.00

───────────────

AIRLINE
Montgomery Ward & Co.,
619 Chicago Avenue, Chicago, Illinois

Airline was the brand name used for Montgomery Ward's radio line. Airlines were second only to Sear's Silvertones in mail order radio sales. In the 1930's, Airline sets were made for Montgomery Ward by several companies: Wells-Gardner & Co., Davidson-Hayes Mfg. Co., and US Radio & Television Corp.

04BR-397A, Table, Wood, large center front multi-band dial w/ escutcheon, side grill bars, 4 knobs $55.00
04BR-566A, Portable, Cloth covered, inner right slide rule dial, left grill, 2 knobs, fold-in front, handle $30.00
04BR-609A, Table, Wood w/inlay, right slide rule dial, pushbuttons, left grill w/horiz bars, 4 knobs $60.00

04WG-754C, **Table, Wood, center front cylindrical dial, 6 pushbuttons, upper grill w/cut-outs, bat** $50.00
05B-A, Table, 1934, Right front dial, center cloth grill w/ shield-shaped cut-out, flared base, handle $75.00

05GAA-992A, Table-R/P, 1951, Outer right front dial, 3 knobs, switch, 3/4 lift top, inner phono, handle, BC, AC $25.00
05GCB-1541 "Lone Ranger", Table-N, 1951, Plastic, Lone Ranger & Silver on rounded left front, right round dial knob . . $550.00
05GCB-1541A "Lone Ranger", Table-N, 1951, Plastic, Lone Ranger & Silver on rounded left front, right dial knob, BC, AC/DC $550.00
05GHM-1061A, Portable, 1951, Leather case, right front round dial knob over grill, handle, BC, AC/DC/battery $25.00
05WG-1813A, Table, 1951, Wood, right front rectangular dial over large cloth grill, 4 knobs, AM, FM, AC $30.00
05WG-2748F, Console-R/P, 1951, Wood, inner right slide rule dial, 4 knobs, pull-out phono, double doors, BC, FM, AC . . $65.00
05WG-2749D, Console-R/P, 1951, Wood, inner right slide rule dial, 4 knobs, left pull-out phono drawer, BC, FM, AC $80.00
05WG-2752, Console-R/P, 1950, Wood, inner right slide rule dial, 4 knobs, left pull-out phono drawer, BC, FM, AC $65.00
4BR-511A, Table, 1946, Plastic, lower slide rule dial, upper horizontal louvers, rounded top, 2 knobs $50.00
5D8-1, Table, Plastic, right front dial, left horizontal louvers, 2 knobs, handle, AC/DC $40.00

14BR-514B, **Table, 1946, Painted plastic, Deco, right slide rule dial, pushbuttons, left horiz louvers, 2 knobs** . . . $100.00
14BR-521A, Table, 1941, Plastic, small case, lower front dial, upper horizontal grill bars, 2 knobs $70.00
14BR-522A, Table, 1941, Plastic, lower front slide rule dial, upper horizontal louvers, rounded top, 2 knobs $70.00
14WG-806A, Table, 1941, Wood, right slide rule dial, curved left vertical grill bars, pushbuttons, 4 knobs $70.00

15BR-1535B, **Table, Plastic, lower front slide rule dial, upper quarter-moon louvers, 4 knobs** $50.00

15BR-1536B, Table, 1951, Plastic, right slide rule dial, left lattice grill, 6 pushbuttons, 2 knobs, BC, AC/DC **$45.00**

15BR-1544A, Table, 1951, Plastic, lower front slide rule dial, upper lattice grill, 2 knobs, BC, AC/DC **$35.00**

15BR-1547A, Table, 1951, Plastic, lower front curved slide rule dial, upper lattice grill, 4 knobs, BC, AC/DC **$40.00**

15GAA-995A, Table-R/P, 1952, Leatherette, outer front dial, 3 knobs, switch, 3/4 lift top, inner phono, BC, AC **$25.00**

15GHM-934A, Table-R/P, 1952, Suitcase-style, inner right dial, left phono, lift top, handle, BC, AC **$25.00**

15GHM-1070A, Portable, 1952, Suitcase-style, right front round dial over grill, handle, BC, AC/DC/bat **$35.00**

15GSE-2764A, Console-R/P, 1952, Wood, front dial, 4 knobs, pull-out phono drawer, embossed front panel, BC, AC . . . **$65.00**

15WG-1545A, Table, 1952, Plastic, top slide rule dial, horizontal wrap-around louvers, 4 knobs, BC, FM, AC **$55.00**

15WG-2745C, Console-R/P, 1951, Wood, inner right slide rule dial, pull-out phono, storage, double doors, BC, FM, AC . . **$60.00**

15WG-2758A, Console-R/P, 1951, Wood, inner right slide rule dial, 4 knobs, pull-out phono, double doors, BC, FM, AC . . **$50.00**

17A80, Console-R/P, 1941, Wood, upper front slide rule dial, 4 knobs, cloth grill w/vertical bars, inner phono **$90.00**

20, Cathedral, 1931, Wood, small case, left front half-moon dial, cloth grill w/cut-outs, scalloped top **$195.00**

25BR-1542A, Table, 1953, Plastic, lower front slide rule dial, upper lattice grill, 4 knobs, BC, AC/DC **$45.00**

25BR-1549B, Table-C, 1953, Plastic, perforated front panel, right dial, left clock, horizontal bars, BC, AC **$30.00**

25GAA-996A, Table-R/P, 1952, Outer center front round dial & knobs, 3/4 lift top, inner phono, handle, AC **$30.00**

25GSE-1555A, Table, 1952, Plastic, right front round dial over large woven grill area, "Airline" logo, 2 knobs **$30.00**

25GSG-2016A, Table-R/P, 1953, Wood, right front dial over large grill, 3 knobs, 3/4 lift top, inner phono, BC, AC **$30.00**

25GSL-1560A, Table-C, 1952, Plastic, right side dial, large front rectangular alarm clock, 4 knobs, BC, AC **$30.00**

25GSL-1814A, Table, 1953, Wood, front half-moon dial over criss-cross grill, 2 knobs, BC, AC/DC **$30.00**

25GSL-2000A, Table-R/P, 1953, Plastic, outer right front round dial, left vertical grill bars, open top, BC, AC **$30.00**

25WG-1573A, Table, 1953, Plastic, right front slide rule dial, left perforated grill, 4 knobs, BC, SW, AC **$40.00**

35GAA-3969A, Table-R/Rec, 1954, Leatherette, inner left dial, disc recorder, lift top, outer grill, handle, BC, AC **$20.00**

35GSE-1555D, Table, 1952, Plastic, right front round dial over large woven grill area, "Airline" logo, 2 knobs **$30.00**

35GSL-2770A, End Table-R/P, 1954, Wood, step-down top, slide rule dial, 4 knobs, front pull-out phono drawer, BC, AC **$125.00**

35WG-1573B "Global", Table, 1954, Plastic, right front slide rule dial, left perforated grill, 4 knobs, BC, SW, AC **$40.00**

54BR-1501A, Table, 1946, Plastic, lower slide rule dial, upper horizontal louvers, 2 knobs, BC, AC/DC **$45.00**

54BR-1503A, Table, 1946, Plastic, upper slide rule dial, lower horizontal louvers, 2 bullet knobs, BC, AC/DC **$45.00**

54BR-1503B, Table, 1946, Plastic, curved slide rule dial, horizontal louvers, bullet knobs **$35.00**

54BR-1505B, Table, 1946, Plastic, large center dial, side louvers, 2 knobs, 5 pushbuttons, BC, AC/DC **$45.00**

54BR-1506A, Table, 1946, Plastic, large center front dial, right & left side louvers, 5 pushbuttons, 2 knobs **$45.00**

54KP-1209B, Table, 1946, Two-tone wood, upper slide rule dial, lower cloth grill, 2 knobs, BC, battery **$30.00**

54WG-2500A, Console-R/P, 1946, Wood, slanted front slide rule dial, cloth grill w/vert bars, 4 knobs, BC, SW, AC **$65.00**

54WG-2700A, Console-R/P, 1946, Wood, slant front slide rule dial, tilt out phono unit, 4 knobs, BC, SW, AC **$75.00**

62-84, Console, Wood, lowboy, upper half-moon dial, lower cloth grill w/cut-outs, 6 legs, 3 knobs, ... **$150.00**

62-114, Tombstone, Wood, shouldered, lower front round dial, upper grill w/3 vertical bars, 3 knobs $85.00

62-123, Console, Two-tone wood, upper round dial, lower cloth grill w/cut-outs, 4 knobs, fluting $150.00

62-131, Tombstone, 1935, Wood, shouldered, lower airplane dial, upper grill w/3 vertical bars, 4 knobs $85.00

62-148, Tombstone, Wood, rounded shoulders, lower round dial, upper cloth grill w/cut-outs, 2 knobs $100.00

62-177, Tombstone, Wood, lower round dial, upper cloth grill w/black cut-outs, 4 knobs, horiz fluting $110.00

62-288 "Miracle", Table, Plastic, right dial, 6 pushbuttons, finished all sides, tuning eye, right side knob $125.00

62-306, Table, C1938, Wood, right front round "telephone" dial, tuning eye, left wrap-around louvers, AC $75.00

62-316, Table, Wood, off-center oval dial, tuning eye, left cloth grill w/Deco cut-outs, 3 knobs $95.00

62-318, Table, 1935, Wood, off-center "movie dial", tuning eye, left grill w/cut-outs, 4 knobs, BC, 2SW $90.00

62-336, Table, C1939, Wood, center front oval dial, left grill w/horiz bars, right horiz lines, 3 knobs $50.00

62-376, Table, 1939, Wood, right oval dial, tuning eye, wrap-around grill bars, 4 knobs, BC, SW, bat $50.00

62-425, Table, 1936, Wood, right front round dial, left cloth grill w/ free-form cut-outs, 2 knobs $60.00

62-437, Table, 1936, Wood, off-center "movie dial", left cloth grill w/ Deco cut-outs, 4 knobs $80.00

62-476, Table, 1941, Plastic, right front telephone dial, tuning eye, left vertical grill bars, 2 knobs $85.00

62-553, Table, Wood, right slide rule dial, pushbuttons, left round grill w/Deco cut-outs, 2 knobs $125.00

62-606, Table, 1938, Right front telephone dial, tuning eye, left vertical wrap-over grill bars, 2 knobs $85.00

64BR-1051A, Portable, 1946, Luggage style, small upper slide rule dial, handle, 2 knobs, BC, AC/DC/bat $25.00

64BR-1205A, Table, 1946, Plastic, center dial, horizontal wrap-around louvers, 2 knobs, BC, battery $30.00

64BR-1208A, Table, 1947, Wood, slide rule dial, wrap around grill, 4 knobs, 6 pushbuttons, BC, SW, battery $35.00

64BR-1501A, Table, 1946, Plastic, lower slide rule dial, upper horizontal grill bars, 2 knobs, rounded top $50.00

64BR-1514A, Table, Painted plastic, lower slide rule dial, 6 pushbuttons, upper metal grill, 4 knobs $45.00

64BR-1808, Table, 1947, Wood, left slide rule dial, right criss-cross grill, pushbuttons, 3 knobs, BC, 2SW $45.00

64BR-1808A, Table, 1947, Wood, left slide rule dial, criss-cross grill, 3 knobs, 8 pushbuttons, BC, 4SW, AC $45.00

64WG-1050A, Portable, 1946, Inner right half-moon dial, square grill, left volume, flip-up lid, BC, AC/DC/bat $40.00

64WG-1052A, Portable, 1946, Inner slide rule dial, 3 knobs, fold-down front, handle, BC, AC/DC/battery $35.00

64WG-1207B, Table, 1947, Wood, upper slide rule dial, large center cloth grill, 2 knobs, BC, battery $25.00

64WG-1511A, Table, 1946, Plastic, lower slide rule dial, recessed cloth grill, 2 knobs, BC, AC/DC $30.00

64WG-1801C, Table, 1946, Wood, right square dial, criss-cross grill, round sides, 3 knobs, BC, AC/DC $35.00

64WG-1804B, Table, 1946, Wood, lower slide rule dial, cloth grill, rounded corners, 2 knobs, BC, AC/DC $40.00

64WG-1807A, Table, 1946, Wood, lower slide rule dial, recessed cloth grill, 4 knobs, BC, SW, AC $35.00

64WG-1809A, Table, 1946, Wood, lower slanted slide rule dial, upper recessed cloth grill, 2 knobs, base $35.00

64WG-2007B, Table-R/P, 1946, Wood, outer right dial, 2 knobs, criss-cross grill, inner phono, lift top, BC, AC $30.00

64WG-2009A, Table-R/P, 1946, Wood, outer slide rule dial, 3 knobs, cloth grill, inner phono, lift top, BC, AC $30.00

74BR-1053A, Portable, 1948, Upper slide rule dial, lower horizontal louvers, handle, 2 knobs, BC, AC/DC/bat $30.00

74BR-1055A, Portable, 1948, Upper slide rule dial, lower lattice grill, handle, 2 knobs, BC, AC/DC/battery $30.00

74BR-1501B, Table, 1946, Walnut plastic, lower slide rule dial, upper horiz louvers, metal back, 2 knobs $80.00

74BR-1502B, Table, 1946, Ivory plastic, lower slide rule dial, upper horizontal louvers, metal back, 2 knobs $80.00

74BR-1514B, Table, 1947, Plastic, lower slide rule dial, large grill, 4 knobs, 6 pushbuttons, BC, SW, AC/DC $45.00

74BR-1812B, Table, 1947, Wood, slanted lower slide rule dial, recessed cloth grill, 4 knobs, BC, FM, AC $45.00

74BR-2001A, Table-R/P, 1947, Right front slanted slide rule dial, left rounded grill, 3 knobs, open top, BC, AC $40.00

74BR-2001B, Table-R/P, 1947, Right front slanted slide rule dial, left rounded grill, 3 knobs, open top, BC, AC $40.00

74BR-2701A, Console-R/P, 1947, Wood, inner slide rule dial, pushbuttons, pull-out phono drawer, BC, 4SW, AC. . . **$85.00**

74BR-2702B, Console-R/P, 1947, Wood, inner slide rule dial, 4 knobs, phono, lift top, vertical grill bars, BC, FM, AC . **$65.00**

74KR-1210A, Table, 1948, Wood, lower slide rule dial, cloth grill, rounded sides, 2 knobs, BC, AC/DC/bat **$35.00**

74KR-2706B, Console-R/P, 1948, Wood, slide rule dial in front of grill, pull-out phono drawer, 3 knobs, BC, AC **$50.00**

74KR-2713A, Console-R/P, 1948, Wood, inner right slide rule dial, 3 knobs, left pull-out phono drawer, BC, AC **$70.00**

74WG-1054A, Portable, 1947, Leatherette, half-moon dial, perforated grill, handle, 2 knobs, BC, AC/DC/bat **$30.00**

74WG-1056A, Portable, 1947, Cloth covered, inner dial, grill, 3 knobs, fold-down front, handle, BC, AC/DC/bat **$25.00**

74WG-1057A, Portable, 1948, Leatherettte & plastic, inner dial, perforated grill, flip-up front, BC, AC/DC/bat **$35.00**

74WG-1510A, Table, 1947, Plastic, lower slide rule dial, upper cloth grill, 2 knobs, BC, AC/DC **$35.00**

74WG-1802A, Table, 1947, Wood, right round dial, large woven grill, rounded corners, 2 knobs, BC, AC/DC **$35.00**

74WG-2002A, Table-R/P, 1947, Wood, outer slide rule dial, cloth grill, 3 knobs, lift top, inner phono, BC, AC **$30.00**

74WG-2004A, Table-R/P, 1947, Wood, outer round dial, 2 knobs, lift top, inner phono, BC, AC. **$30.00**

74WG-2010B, Table-R/P, 1947, Wood, inner slide rule dial, 4 knobs, lift top, criss-cross grill, BC, SW, AC **$35.00**

74WG-2504A, Console, 1947, Wood, upper slide rule dial, 4 knobs, cloth grill w/3 horizontal bars, BC, SW, AC **$65.00**

74WG-2505A, Console, 1947, Wood, slanted slide rule dial, 4 knobs, 6 pushbuttons, tuning eye, BC, SW, FM, AC. **$75.00**

74WG-2704A, Console-R/P, 1947, Wood, slide rule dial, 4 knobs, tilt-out phono, cloth grill w/vert bars, BC, SW, AC **$70.00**

74WG-2709A, Console-R/P, 1947, Wood, upper slide rule dial, 4 knobs, tilt-out phono, vert grill bars, BC, SW, AC **$70.00**

83BR-351A, Table, 1938, Plastic, Deco, rounded right, pushbuttons, left wrap-over grill bars, side knob, AC **$95.00**

84BR-1065A, Portable, 1949, Slide rule dial, horizontal louvers, plays when front is opened, handle, AC/DC/bat **$35.00**

84BR-1065B, Portable, 1949, Slide rule dial, horizontal louvers, plays when front is opened, handle, battery **$35.00**

84BR-1502B, Table, 1946, Plastic, lower front slide rule dial, upper horizontal louvers, rounded top, 2 knobs **$50.00**

84BR-1815B, Table, 1949, Wood, large plastic recessed half-moon dial/circular louvers, 2 knobs, BC, AC/DC **$45.00**

84GCB-1062A, Portable, 1948, Leatherette & plastic, inner round dial, flip-up front, side handle, BC, battery **$45.00**

84GSE-2731A, Console-R/P, 1949, Wood, top dial, 4 knobs, lift cover, inner phono, lower storage, BC, AC **$50.00**

84HA-1810C, Table, 1949, Wood, slanted slide rule dial, large perforated grill/front, 4 knobs, BC, FM, AC/DC **$30.00**

84KR-1520A, Table, 1949, Metal, right vertical slide rule dial, left horizontal louvers, 2 knobs, BC, AC/DC **$50.00**

84KR-2510A, End Table, 1949, Wood, "2 drawer" end table, tilt-out front w/inner slide rule dial, 2 knobs, AC **$85.00**

84KR-2511A, End Table, 1949, Wood, "2 drawer" end-table, tilt-out front w/slide rule dial, 2 knobs, BC, AC/DC **$85.00**

84WG-1056B, Portable, 1949, Cloth covered, slide rule dial, 3 knobs, fold-down front, handle, BC, AC/DC/bat **$30.00**

84WG-1060A, Portable, 1948, Inner right dial, perforated grill, 2 knobs, flip-up front, handle, BC, AC/DC/bat **$30.00**

84WG-1060C, Portable, 1948, Inner right dial, perforated grill, 2 knobs, flip-up front, handle, BC, AC/DC/bat **$30.00**

84WG-2015A, Table-R/P, 1948, Wood, top slide rule dial, criss-cross grill, 4 knobs, 3/4 lift top, BC, FM, AC **$30.00**

84WG-2506B, Console, 1949, Two-tone wood, upper slide rule dial, 4 knobs, cloth grill w/bars, BC, FM, AC **$75.00**

84WG-2712A, Console-R/P, 1948, Wood, inner slide rule dial, 6 knobs, 6 pushbuttons, tuning eye, BC, SW, FM, AC . . **$90.00**

84WG-2714A, Console-R/P, 1948, Wood, upper slide rule dial, 4 knobs, front tilt-out phono, BC, FM, AC **$80.00**

84WG-2714F, Console-R/P, 1949, Wood, upper slide rule dial, 4 knobs, front tilt-out phono door, BC, FM, AC **$80.00**

84WG-2720A, Console-R/P, 1948, Wood, inner slide rule dial, 4 knobs, 6 pushbuttons, tuning eye, BC, SW, FM, AC . . **$85.00**

84WG-2721A, Console-R/P, 1948, Wood, inner right slide rule dial, 4 knobs, left pull-out phono drawer, BC, FM, AC **$80.00**

93BR-460A, Table, 1940, Wood, lower front slide rule dial, upper horizontal grill bars, 2 knobs, battery $45.00

93BR-563A, Table, Wood, right cylindrical dial, left herringbone grill, 6 pushbuttons, 2 knobs, bat. **$45.00**

93WG-604A, Table, 1946, Plastic, right dial, 2 thumbwheel knobs, 6 pushbuttons, left wrap-around louvers **$115.00**

94BR-1525A, Table, 1950, Plastic, upper curved slide rule dial, lower wrap-around louvers, 2 bullet knobs **$40.00**

94BR-1533A, Table, 1950, Plastic, slide rule dial, upper quarter-moon louvers, 4 knobs, BC, FM, AC/DC **$50.00**

94BR-1535A, Table, Plastic, lower front slide rule dial, upper quarter-moon louvers, 4 knobs, crest **$50.00**

94BR-2740A, Console-R/P, 1950, Wood, inner right slide rule dial, 4 knobs, left pull-out phono drawer, BC, FM, AC **$80.00**

94GCB-1064A, Portable, 1950, Leatherette & plastic, inner dial, dotted grill, fold-open front, handle, BC, battery **$40.00**

84HA-1810, Table, 1949, Wood, slanted slide rule dial, large metal perforated grill/front, 4 knobs, BC, FM $30.00

84HA-1810A, Table, 1949, Wood, slanted slide rule dial, large perforated grill/front, 4 knobs, BC, FM, AC/DC **$30.00**

94GSE-2735A, Console-R/P, 1949, Wood, top right slide rule dial, 4 knobs, lift top, inner phono, storage, BC, FM, AC **$50.00**
94HA-1528C, Table, 1949, Plastic, right half-round dial, large woven grill/front area, 2 knobs, BC, AC/DC **$25.00**
94HA-1529A, Table, 1950, Plastic, lower slanted slide rule dial, large upper grill, 4 knobs, BC, FM, AC/DC **$40.00**

94HA-1562 , Table, 1945, Plastic, right front round metal dial over large plastic woven grill, 2 knobs, AC **$25.00**
94WG-1059A, Portable, 1949, Leatherette, inner slide rule dial, 3 knobs, lift-open lid, handle, BC, AC/DC/bat **$40.00**
94WG-1804D, Table, 1950, Wood, recessed lower slide rule dial, upper cloth grill, 2 knobs, BC, AC/DC **$40.00**
94WG-1811A, Table, 1950, Wood, right round dial over woven grill, mitered corners, 4 knobs, AM, FM, AC **$35.00**
94WG-2742A, Console-R/P, 1949, Wood, inner right slide rule dial, 4 knobs, left pull-out phono drawer, BC, FM, AC **$75.00**
94WG-2745A, Console-R/P, 1949, Wood, inner slide rule dial, 4 knobs, pull-out phono, criss-cross grill, AM, FM, AC . . **$65.00**
94WG-2748A, Console-R/P, 1950, Wood, inner right dial, 4 knobs, pull-out phono, criss-cross grill, BC, FM, AC **$65.00**
345, Table, C1941, Wood, right telephone dial, tuning eye, left cloth grill w/Deco cut-outs, battery **$65.00**
GAA-990A, Table-R/P, 1956, Leatherette, right side dial & knobs, front grill, 3/4 lift-top, inner phono, BC, AC **$25.00**
GEN-1090A, Portable, 1957, Leather case w/grill, right round dial, handle, side thumbwheel knob, BC, bat **$25.00**
GEN-1120C, Portable, 1959, Transistor, right front round dial knob, lower grill w/logo, handle, AM, battery **$30.00**
GSL-1079-A, Portable, 1955, Leatherette, front slide rule dial, map, telescope antenna, BC, SW, AC/DC/bat **$65.00**
GTM-1108A, Portable, 1958, Leather case, transistor, right front dial, left perforated grill, handle, AM, bat **$25.00**
GTM-1109A, Portable, 1958, Plastic, 7 transistor, right round dial, front vertical grill bars, AM, battery **$30.00**
GTM-1200A, Portable, 1960, Transistor, upper front slide rule dial, handle, telescoping antenna, BC, SW, bat **$35.00**
GTM-1201A, Portable, 1960, Transistor, lower round dial over horiz bars, swing handle, side knob, AM, battery **$25.00**
GTM-1639B, Table, 1958, Plastic, lower front slide rule dial, upper vertical grill bars, 2 knobs, BC, AC/DC **$20.00**
Rudolph, Table-N, 1951, Plastic, right round dial knob, Rudolph on rounded left, side louvers, AC **$550.00**
WG-1637A, Table, 1957, Plastic, front recessed slide rule dial, upper grill w/crest, 3 knobs, BC, AC/DC **$25.00**

AIR-WAY
Air-Way Electric Appliance Corp,
Toledo, Ohio

The Air-Way Company began business in 1920 as a manufacturer of vacuum cleaners and electrical parts. They made radios and radio parts briefly during the mid twenties but by 1926 they had ceased radio production.

41, Table, 1924, Wood, low rectangular case, 2 dial front panel, lift top, 4 tubes, battery **$110.00**
51, Table, 1924, Wood, low rectangular case, 3 dial front panel, lift top, 5 tubes, battery **$145.00**
61, Table, 1925, Walnut, low rectangular case, 2 front window dials, fluted columns, 6 tubes, bat **$150.00**
62, Table, 1925, Walnut, high rectangular case, 2 front window dials, built-in loud speaker, bat **$165.00**
B, Table, 1922, Wood, high rectangular case, detector & 1 stage amp, 2 dial front panel, battery **$250.00**
C, Table, 1922, Wood, high rectangular case, detector & 2 stage amp, 3 dial front panel, battery **$275.00**
F, Table, 1923, Wood, high rectangular case, 2 dial front panel, 4 tubes, battery **$225.00**
G, Table, 1923, Wood, high rectangular case, 3 dial front panel, 5 tubes, battery **$275.00**

ALADDIN

Big 4, Table, Wood, low rectangular case, 2 dial front panel, battery **$140.00**

ALDEN
Alden, Inc.

1818, Portable, 1949, Inner slide rule dial, 2 knobs, fold-down front, handle, BC, AC/DC/bat **$35.00**

ALGENE
Algene Radio Corp.,
305 Throop Avenue, Brooklyn, New York

AR5U, Portable, 1947, "Cosmetic case", inner dial, horiz bars, 2 square knobs, mirror in lid, BC, AC/DC **$85.00**
AR6M, Portable, 1948, "Cosmetic case", inner dial, horizontal grill bars, 2 square knobs, mirror in lid **$85.00**
AR-6U, Portable, 1947, "Cosmetic case", square dial, louvers, 3 square knobs, mirrored lid, BC, AC/DC/bat **$85.00**
AR-404 "Jr.", Portable, 1948, "Cosmetic case", inner dial and knobs, front grill, fold-open lid, carrying strap **$55.00**
AR-406 "Middie", Portable, 1948, "Cosmetic case", "alligator", inner dial, 2 square knobs, fold-open lid, battery **$60.00**

AMBER
Amber Sales, Inc.,
112 Chambers St., New York, New York

512-C "Marv-o-dyne", Table, 1924, Low rectangular case, 3 dial front panel, Weston meter, 5 tubes, battery **$225.00**

AMC
Associated Merchants Corp., 1440 Broadway,
New York, New York

125-P, Table, 1946, Plastic, upper slide rule dial, 2 knobs, lower horizontal louvers, BC, AC/DC **$45.00**

126, Table, 1947, Catalin, slide rule dial, grill w/circular cut-outs, handle, 2 knobs, BC, AC/DC . . . . $2,000.00+

AMERICAN BOSCH
American Bosch Magneto Corporation, Springfield, Massachusetts
United American Bosch Corporation, Springfield, Massachusetts

The American Bosch Magneto Corporation was formed in 1919 selling magnetos and automobile parts. Radio sales began in 1925. In addition to their own models, American Bosch produced radios for other companies such as Sonora and Eveready. By the mid 1930's radio production had ceased.

04, Table, 1935, Wood, center round dial w/surrounding 4-section circular grill, 2 knobs, AC/DC $60.00

05, Table, 1935, Wood, lower round dial, large upper grill with cut-outs, 2 knobs, BC, SW, AC/DC $75.00

5A, Table, 1931, Wood, right window dial, round grill w/cut-outs, gold pinstriping, 2 knobs, AC $75.00

5C, Console, 1931, Wood, upper front dial, lower criss-cross grill, bowed front legs $125.00

10, Tombstone, C1935, Two-tone wood, small case, lower round dial, cloth grill w/cut-outs, 3 knobs $70.00

16 "Amborola", Table, 1925, Wood, 2 dial bakelite panel w/escutcheon, lift top, feet, burl front, 6 tubes, bat $225.00

18, Console, 1929, Wood, decorative case, inner dial, 3 knobs, grill with urn cut-outs, sliding doors $225.00

20-L, Console, 1931, Wood, ornate case, upper window dial, lower grill w/cut-outs, stretcher base $200.00

27 "Amborada", Console, 1926, Wood, plain cabinet looks like dresser, inner 2 dials & knobs, feet, battery $210.00

28, Table, 1928, Wood, low rectangular case, center front window dial w/escutcheon, 3 knobs, AC $135.00

35 "Cruiser", Table, 1926, Wood, upper front window dial, lift top, 3 knobs, feet, battery $135.00

35 "Imperial Cruiser", Console, 1926, Walnut, low cabinet, inner front dial, 3 knobs, double front doors, feet, battery . $165.00

35 "Royal Cruiser", Table, 1926, Walnut, upper front window dial, lift top, 3 knobs, feet, 5 tubes, battery $135.00

46 "Little Six", Table, 1927, Walnut, high rectangular case, right thumbwheel window dial, 2 lower knobs, bat $125.00

57, Console, 1927, Wood, lowboy, inner dial, fold-down front, lower double doors, inner grill, battery $145.00

58A, Console, 1930, Wood, upper front dial, lower cloth grill w/cut-outs, AC $140.00

66 "Cruiser", Table, 1927, Wood, front window dial w/escutcheon, 3 knobs, lift top, horizontal bands, bat $140.00

66AC, Table, 1927, Wood, front window dial w/escutcheon, 3 knobs, separate A & B power unit, AC $135.00

76 "Cruiser", Console, 1927, Wood, lowboy, inner window dial w/escutcheon, fold-down front door, battery $125.00

87 "Cruiser", Table, 1927, Wood, low rectangular case, center front window dial w/escutcheon, battery $110.00

96, Console, 1927, Wood, lowboy, inner front window dial w/escutcheon, fold-down door, AC $115.00

107, Console, 1927, Wood, lowboy, inner dial, 3 knobs, fold-down front, double doors, fancy grill, AC $165.00

116, Table, 1927, Wood, front window dial w/escutcheon, 3 knobs, right & left panels, feet, AC $90.00

126, Table, 1927, Wood, right front thumbwheel window dial, 2 lower knobs, lift top, A/C $110.00

200-A "Treasure Chest", Table, 1932, Wood, chest-style, inner dial & grill, lift top, fancy "carved" front, 2 knobs, AC $250.00

350, Table, 1933, Mahogany w/inlay, right front window dial, center scrolled grill, 4 knobs, AC $65.00

355, Table, 1933, Mahogany w/inlay, right front window dial, center scrolled grill, 4 knobs, AC/DC $65.00

360T, Tombstone, 1933, Wood, center front dial, upper cloth grill w/cut-outs, 4 knobs, BC, SW, AC $100.00

370S, Console, 1933, Wood, lowboy, front dial, lower grill w/cut-outs, 4 knobs, stretcher base, AC $135.00

370T, Tombstone, 1933, Wood, center front dial, upper cloth grill w/cut-outs, 4 knobs, BC, SW, AC $100.00

402, Table, 1934, Wood, right front window dial, center round cloth grill w/star cut-out, AC/DC $60.00

440-T, Tombstone, 1934, Wood, center front dial, upper cloth grill w/cut-outs, rounded top, BC, SW, AC $125.00

460-R, Console, 1934, Wood, inner dial, fold-back top, "horseshoe" grill w/splayed bars, BC, SW, AC $125.00

470-G, Console, 1935, Wood, upper slanted front dial w/escutcheon, lower grill w/scroll cut-outs, AC $120.00

470-U, Tombstone, 1935, Wood, lower front dial, upper grill w/cut-outs, right & left fluted columns, AC $115.00

480-D, Console, 1934, Wood, inner slanted dial, fold-up top, lower cloth grill w/cut-outs, BC, SW, AC $175.00

500, Table, 1933, Wood w/inlay, right front window dial, center grill w/cut-outs, 2 knobs, AC/DC $65.00

501, Table, 1933, Wood w/inlay, right front window dial, center grill w/cut-outs, 2 knobs, AC/DC $70.00

505, Table, 1935, Wood, right front square black dial, left cloth grill, center star, 3 knobs, AC $50.00

510, Tombstone, 1935, Wood, small case, lower round dial, cloth grill w/cut-outs, 3 knobs, BC, SW, AC $75.00

510-E, Console, 1935, Wood, upper front round dial, lower grill w/scroll cut-outs, 3 knobs, BC, SW, AC $100.00

565-W, Tombstone, 1935, Wood, center front round dial, upper cloth grill w/intersecting cut-outs, AC $85.00

575-F, Tombstone, 1935, Wood, lower front round dial, upper grill w/vertical bars, 4 knobs, BC, SW, AC $85.00

575-Q, Console, 1935, Wood, upper round dial, lower cloth grill w/vertical bars, 4 knobs, BC, SW, AC $110.00

585-Y, Tombstone, 1935, Wood, lower front round dial, upper grill w/scroll cut-outs, 4 knobs, BC, SW, AC $110.00

585-Z, Console, 1935, Wood, upper front round dial, lower cloth grill w/cut-outs, 4 knobs, BC, SW, AC $110.00

604, Table, 1935, Wood, right front dial, left cloth grill w/horizontal bars, 3 knobs, BC, SW, AC $40.00

610A2, Table, Wood, right front square airplane dial, left grill w/horizontal bars, 3 knobs $40.00

625, Console, 1936, Wood, upper front round dial, lower cloth grill w/center vertical bar, BC, SW, AC $95.00

650, Console, 1936, Wood, upper front round dial, lower grill w/center vert bars, 4 knobs, BC, SW, AC $95.00

660T, Table, 1936, Wood, right round dial, left & right grills w/horizontal bars, 4 knobs, BC, SW, AC $55.00

670C, Console, 1936, Wood, recessed front black dial, lower grill w/vertical bars, 4 knobs, BC, SW, AC $115.00

680, Console, 1934, Two-tone wood, upper front black dial, lower cloth grill w/vertical bars, 5 knobs $150.00

805, Table, Wood, right dial, center grill, two shield-shaped escutcheons, gold pin-striping $70.00

AMERICAN RADIO
American Radio Corp.,
6116 Euclid Ave., Cleveland, Ohio

3-A "Arc-Lininger", Table, 1925, Wood, low rectangular case, 2 dial front panel, 3 tubes, battery $110.00

4 "Arc-Lininger", Table, 1925, Wood, low rectangular case, 2 dial front panel, 4 tubes, battery $120.00

Super 5 "Arc-Lininger", Table, 1925, Wood, low rectangular case, 2 dial front panel, 5 tubes, battery $130.00

AMERICAN SPECIALTY
American Specialty Co.
Bridgeport, Connecticut

Standard "Electrola", Table, 1925, Wood, high rectangular case, 3 dial front panel, built-in speaker, 5 tubes, battery . . . $175.00

AMPLEX

C "Lectrosonic", Table, Metal, low rectangular case, center window dial, lift-off top, 2 knobs, switch $115.00

De Exer, Table, C1925, Wood, low rectangular case, brown bakelite 3 dial front panel w/gold trim $150.00

AMRAD
The Amrad Corporation,
Medford Hillside, Massachusetts
American Radio & Research Corporation,
Medford Hillside, Massachusetts

The name Amrad is short for American Radio & Research Corporation. The company began business manufacturing transmitters and receivers for the government during WW I and produced its first crystal set in 1921. By 1925 Amrad was in serious financial trouble and was bought out by Crosley although their radios still retained the Amrad label. A victim of the Depression, Amrad closed in 1930.

70 "Concerto", Console, 1928, Walnut, highboy, inner dial & escutcheon, double doors w/brass hardware, AC $150.00

70 "Nocturne", Console, 1928, Walnut, highboy, inner dial & escutcheon, double doors, stretcher base, AC $125.00

70 "Sonata", Console, 1928, Walnut, highboy, inner dial w/escutcheon, double doors, stretcher base, AC $140.00

81 "The Aria", Console, 1929, Walnut, inner front window dial, 3 knobs, lower grill, double doors, AC $110.00

81 "The Serenata", Console, 1929, Wood, highboy, inner front dial, double doors, stretcher base, AC $125.00

81 "The Symphony", Console, 1929, Wood, Moderne, highboy, inner front dial, double doors, stretcher base, AC . . . $120.00

81 "Duet", Console-R/P, 1929, Veneers, lowboy, inner dial & escutcheon, large double doors, stretcher base, AC . $120.00

2575, Table, 1922, Wood, square case, crystal receiver, black bakelite front panel with 1 center dial $250.00

2575/2776, Table, 1922, Wood, 2 units - crystal receiver and 2 stage amp, black front panels **$525.00**

2596/2634, Table, 1921, "Double-decker", detector/two-stage amp & shortwave tuner, bakelite panels **$475.00**

3366, Table, 1923, Wood, low rectangular case, crystal set, 2 dial bakelite panel, top screen, 1 tube **$425.00**

3500-1 (3475/2634), Table, 1923, Two unit double-decker, receiver & tuner, front bakelite panels, top screens, bat **$500.00**

3500-3 "Inductrole", Table, 1925, Wood, high rectangular case, upper 2 dial panel, lower storage, doors, 4 tubes, bat **$300.00**

3500-4 "Cabinette", Table, 1925, Wood, high rectangular case, 2 dial black front panel, 4 tubes, battery **$325.00**

3500-6 "Jewel", Table, 1925, Wood, chest-type case, inner 2 dial panel, double front doors w/carvings, battery **$375.00**

AC-5, Table, 1926, Wood, low rectangular case, 3 front window dials, front columns, 4 knobs, AC **$175.00**

AC-5-C, Console, 1926, Mahogany, lowboy, 3 inner window dials, 4 knobs, fold-down front door, AC **$200.00**

AC-6 "The Warwick", Table, 1927, Walnut, low rectangular case, center front dial w/escutcheon, 3 knobs, AC **$130.00**

AC-6-C "The Berwick", Console, 1927, Walnut, lowboy, inner dial w/escutcheon, 3 knobs, fold-down front, AC **$160.00**

AC-7 "The Windsor", Table, 1927, Wood, low rectangular case, center dial w/escutcheon, 3 knobs, AC **$130.00**

AC-7-C "The Hastings", Console, 1927, Wood, lowboy, inner dial, large double doors, stretcher base, AC **$160.00**

AC-9, Table, 1926, Mahogany, low rectangular case, 2 window dials, 5 knobs, front columns, AC **$140.00**

AC-9-C, Console, 1926, Wood, lowboy, 2 inner front dials, AC **$170.00**

Neutrodyne, Table, 1923, Wood, low rectangular case, 2 dial front panel, lift top, 5 tube, battery **$140.00**

S-522, Table, 1926, Wood, low rectangular case, 3 window dials, 4 knobs, lift top, front columns, bat **$125.00**

S-733, Table, 1926, Two-tone mahogany, low rectangular case, 2 front window dials, lift top, battery **$125.00**

S-733-C, Console, 1927, Wood, lowboy, 2 inner front window dials, fold-down front door, battery **$150.00**

ANDREA
Andrea Radio Corporation,
27-01 Bridge Plaza North,
Long Island City, New York

The Andrea Radio Corporation was begun by Frank D'Andrea in 1934 after the sale of his previous radio company - F. A. D. Andrea, Inc.

2-D-5, Table, 1937, Walnut, right front lighted dial, left grill w/diagonal bars, 3 knobs, BC, SW, AC **$55.00**

2-D-8, Table, 1937, Wood, slanted front, right dial, left cloth grill w/ wrap-over bars **$50.00**

5-E-11, Console, 1939, Wood, front square dial, 4 knobs, pushbuttons, grill w/horizontal bars, BC, SW, AC **$90.00**

6G63, Portable, 1940, Cloth, inner right dial, left grill, fold-up front, handle, 3 bands, AC/DC/battery **$25.00**

6G63A, Portable, 1940, Leatherette, inner right dial, left grill, fold-up front, handle, 3 bands, AC/DC/bat **$30.00**

6H44, Table, 1942, Two-tone walnut w/black trim, right front dial, left grill, 4 knobs, 3 bands **$45.00**

14-E-6, Table, 1938, Wood, right front vertical dial, left horizontal louvers, pushbuttons, BC, SW, AC **$45.00**

826, Upright Table, 1940, Wood, lower front rectangular dial, upper grill w/horizontal bars, 5 knobs **$65.00**

CO-UI5, Table-R/P, 1947, Wood, upper slanted slide rule dial, 5 knobs, lift top, inner phono, BC, SW, AC **$30.00**

P-I63, Portable, 1947, Luggage-type, inner black dial, 3 knobs, fold-up door, handle, BC, 2SW, AC/DC/bat **$30.00**

T-16, Table, 1947, Two-tone wood, slide rule dial, cloth grill w/2 vert bars, 4 knobs, BC, 2SW, AC **$50.00**

T-U15, Table, 1947, Plastic, slanted top slide rule dial, lattice grill, 4 knobs, BC, SW, AC/DC **$40.00**

T-U16, Table, 1947, Two-tone wood, slide rule dial, grill w/2 vert bars, 4 knobs, BC, 2SW, AC/DC **$50.00**

W69P "Spacemaster Deluxe", Portable, 1957, Leatherette, flip-up front w/map, telescope antenna, handle, 9 bands, AC/DC/bat **$80.00**

ANDREWS
Andrews Radio Co.,
327 S. LaSalle St.
Chicago, Illinois

De Luxe "Deresnadyne", Table, 1925, Wood, low rectangular case, 3 dial front panel, 5 tubes, battery **$120.00**

ANGEL

Boy's Radio, Portable, Plastic, transistor, large metal perforated grill, 2 thumbwheel knobs, battery **$30.00**

ANSLEY
Ansley Radio Corp.,
41 St. Joes Avenue
Trenton, New York

53, Console-R/P, 1947, Wood, inner dial, 5 knobs, 8 pushbuttons, phono, lift top, modern, BC, SW, FM, AC **$95.00**

105 "Dynatone", Piano Console-R/P, 1948, Upright piano-style, unusual piano/radio/record player combination ... **$1,500.00**

D-4, Table-R/P, 1933, Wood, outer front grill w/vertical bars, lift top inner phono, AC **$30.00**

D-10, Table-R/P, 1935, Wood, front dial, center grill, 4 knobs, lift top, inner phono, BC, SW, AC **$30.00**

D-10-A, Table-R/P, 1941, Wood, right front slide rule dial, front/right & left grills, lift top, inner phono, AC **$30.00**

D-17, Console-R/P, 1936, Wood, front tilt-out radio unit, lower grill, lift top, inner phono, BC, SW, AC **$90.00**

D-23, Chairside-R/P, 1937, Wood, inner dial & phono, "glider top" slides sideways, front grill, BC, SW, AC **$85.00**

APEX

**Apex Electric Manufacturing Company,
1410 West 59th Street, Chicago, Illinois
Apex Radio & Television Corp.
United States Radio & Television Corporation,
Chicago, Ilinois**

The Apex Electric Manufacturing Company began business selling parts for automobiles. By 1925 they were producing radios. Financial difficulties forced a merger with Case to form the United States Radio & Television Corporation in 1928.

5, Table, 1926, Wood, low rectangular case, single front window dial, feet, 3 knobs **$100.00**

6, Table, 1926, Wood, low rectangular case, inner window dial, fold-down front, 3 knobs **$120.00**

8A, Cathedral, 1931, Wood, center front dial, upper scalloped grill w/cut-outs, fluted columns **$225.00**

10B, Console, 1931, Wood, lowboy, upper front dial, lower grill w/cut-outs, doors, stretcher base, AC **$150.00**

11, Console, 1930, Wood, lowboy, upper front dial, lower scalloped grill w/cut-outs **$150.00**

12B, Console, 1932, Wood, front curved dial, grill w/gothic cut-outs, 6 legs, stretcher base, AC **$150.00**

60, Console, 1928, Walnut, upper front dial, 2 knobs, grill w/petal cut-outs, stretcher base, AC **$125.00**

70, Console, 1928, Walnut, highboy, upper dial, 2 knobs, cloth grill w/cut-outs, stretcher base, AC **$160.00**

89, Table, 1929, Metal, low rectangular case, center front window dial w/escutcheon, 2 knobs, AC **$70.00**

106, Console, 1926, Wood, highboy, inner window dial, fold-down front, lower storage, 3 knobs **$145.00**

116, Console, 1926, Wood, highboy, inner window dial, upper speaker grill, double doors, storage **$160.00**

120B, Console, 1932, Wood, lowboy, ornate cabinet, double front doors, 6 legs, stretcher base, AC **$160.00**

160, Console, 1930, Wood, highboy, lower window dial, upper grill w/cut-outs, stretcher base **$140.00**

Baby Grand, Console, 1925, Wood, looks like spinet piano, inner 3 dial panel, fold-down front door, battery **$225.00**

Corsair, Table, 1927, Wood, low rectangular case, center front dial w/escutcheon, fluted columns **$90.00**

Deluxe, Table, 1925, Wood, high rectangular case, 3 dial lower front panel, built-in speaker, battery **$125.00**

Lyric, Table, 1927, Wood, low rectangular case, inner window dial, fold-down front **$110.00**

Milan Electric, Console, 1927, Wood, lowboy, inner front dial, 2 knobs, lower gothic grill, double doors, AC **$120.00**

Minstrel, Console, 1927, Wood, inner dial, 2 knobs, lower gothic grill, double doors, stretcher base, AC **$125.00**

Music Chest, Table, 1928, Low rectangular case, center front dial & escutcheon, 2 knobs **$85.00**

Neutrodyne, Table, 1930, Walnut finish metal rectangular case, front illuminated dial, 2 knobs, AC **$70.00**

Super Five, Table, 1925, Wood, low rectangular case, 3 dial front panel, lift top, feet, battery **$125.00**

Troubadour, Console, 1927, Wood, inner front dial, grill w/gothic cut-outs, double doors, stretcher base, AC **$225.00**

APEX INDUSTRIES

**Apex Industries,
192 Lexington Avenue, New York, New York**

4B5, Table, 1948, Plastic, right square dial, left horizontal louvers, 2 knobs, BC, AC/DC **$35.00**

APPLEBY

**Appleby Mfg. Co.,
250 N. Juniper St., Philadelphia, Pennsylvania**

60, Table, Wood, low rectangular case, 2 dial metal front panel, meter, lift top **$165.00**

ARBORPHONE

27, Table, 1927, Wood, low rectangular case w/rounded front corners, metal 2 dial panel, battery **$150.00**

ARC

**ARC Radio Corp.,
523 Myrtle Avenue, Brooklyn, New York**

601, Portable, 1947, Shoulder bag-style, "alligator", 2 top thumbwheel knobs, strap, BC, battery **$55.00**

ARCADIA

**Whitney & Co.,
933 5th Avenue, San Diego, California**

37D14-600, Table, 1946, Wood, slanted upper slide rule dial, criss-cross grill, 3 knobs, BC, SW, AC/DC **$45.00**

ARIA
International Detrola Co.,
Detroit, Michigan

554-1-61A, Table-R/P, 1946, Wood, outer right top dial, 3 knobs, inner phono, lift top, BC, AC **$40.00**
571, Table, Plastic, right front slide rule dial, left horizontal louvers, 2 knobs, AC **$45.00**

572-10, **Table, Wood, right slide rule dial, left & right metal perforated grills, feet, 3 knobs, AC** **$45.00**
572-21, Table, Wood, right slide rule dial, cloth grill with chrome molding & center bar, 3 knobs **$55.00**

593, **Table, Wood, right front dial, left curved grill w/vert bars, 3 knobs (1 on grill), AC/DC** **$95.00**

ARKAY

421, Table, 1934, Wood, right front dial, center grill w/cut-outs, 2 knobs **$75.00**
633, Console, 1934, Wood, upper front round dial, lower cloth grill w/ vertical bars, fluting **$125.00**

ARTHUR ANSLEY
Arthur Ansley Mfg. Co.,
Doylestown, Pennsylvania

R-1, Table, 1953, Wood, inner vertical slide rule dial, left grill, 4 knobs, fold-up front, AM, FM, AC **$50.00**

ARTONE
Affiliated Retailers, Inc.,
Empire State Building, New York, New York

524, Table-C, 1949, Slide rule dial, center front alarm clock, 2 knobs, grill bars on case top, BC, AC **$40.00**

ARVIN
Noblitt-Sparks Industries, Inc.,
13th Street & Big Four R. R., Columbus, Indiana
Arvin Industries, Inc., Columbus, Indiana

The Arvin line was produced by Noblitt-Sparks Industries which was founded in 1927 for the manufacture of automobile parts. The company began making automobile radios in 1933 and by 1934 was producing radios for the home. They also made radios for Sears which were sold under the Silvertone brand name.

40 "Mighty Mite", **Table, 1938, Metal, midget, right front dial, left horiz louvers, rounded top, 2 knobs, AC/DC** **$80.00**
58, Table, 1939, Black plastic, right front dial, left round grill w/ horizontal bars, 2 knobs, AC/DC **$50.00**
58A, Table, 1939, Ivory plastic, right front dial, left round grill w/ horizontal bars, 2 knobs, AC/DC **$50.00**

60R23, **Portable, 1960, Plastic, transistor, large right round dial over vert bars, swing handle, BC, bat** **$25.00**

60R49, Portable, 1960, Transistor, off-center dial under plastic, horiz. grill bars, metal handle, BC, bat **$30.00**

68, Table, 1938, Brown plastic, right pushbutton tuning, left round grill w/horiz bars, 3 knobs, AC **$65.00**

78, Table, 1938, Wood, right dial, left grill w/horizontal bars, 4 pushbuttons, 3 knobs, BC, SW, AC **$55.00**

88, Table-R/P, 1938, Wood, outer right dial, left grill w/vertical bars, 2 knobs, lift top, BC, SW, AC **$35.00**

89, Table, 1939, Wood, right dial, left grill w/vertical bars, pushbuttons, tuning eye, 4 knobs, AC **$60.00**

91, Console, 1939, Walnut, slanted front rectangular dial, pushbuttons, tuning eye, 4 knobs, BC, SW **$120.00**

140-P, Portable, 1947, Two-tone, slide rule dial, 2 knobs on top, lattice grill, handle, BC, AC/DC/bat **$35.00**

150TC, Table-R/P, 1948, Mahoghany veneer, slide rule dial, horiz grill bars, 4 knobs, 3/4 lift top, BC, AC **$35.00**

151TC, Table-R/P, 1948, Walnut veneer, slide rule dial, horizontal grill bars, 4 knobs, 3/4 lift top, BC, AC **$35.00**

152T, Table, 1948, Plastic, left half-moon dial, horizontal wrap-around bars, 2 knobs, BC, AC/DC **$35.00**

160T, Table, 1948, Plastic, slide rule dial, lower metal perforated grill, case top is ribbed, 3 knobs **$35.00**

182TFM, Table, 1948, Wood, slide rule dial w/clear escutcheon, perforated grill, 4 knobs, BC, FM, AC/DC **$40.00**

240-P, Portable, 1948, Plastic, center vert slide rule dial, checkered grill, 2 knobs, handle, BC, battery **$40.00**

241P, Portable, 1949, Plastic, center vert slide rule dial, checkered grill, handle, 2 knobs, AC/DC/bat. **$40.00**

242T, Table, 1948, Metal, small, right dial, vertical grill bars w/oblong cut-outs, 2 knobs, BC, AC/DC **$70.00**

250-P, Portable, 1948, Metal, upper slide rule dial, vert lattice louvers, handle, 2 knobs, BC, AC/DC/bat **$50.00**

253T, Table, 1949, Metal, right front round dial over vertical grill bars, left volume knob. **$40.00**

255T, Table, 1949, Plastic, right round dial knob, vertical grill openings & bars, BC, AC/DC **$40.00**

264T, Table, 1949, Wood, lucite covered slide rule dial on top of case, large grill, 4 knobs, BC, AC/DC **$40.00**

280TFM, Table, 1948, Wood, slide rule dial w/lucite escutcheon, large grill, 4 knobs, BC, FM, AC/DC **$40.00**

341T, Table, 1950, Metal, midget, right round dial knob, left horiz louvers, rounded top, BC, AC/DC **$75.00**

350P, Portable, 1949, Plastic, right dial part of circular grill bars, handle, 2 knobs, BC, AC/DC/battery **$40.00**

350PL, Portable, 1950, Plastic, right dial part of circular grill bars, handle, 2 knobs, BC, AC/DC/battery **$40.00**

356T, Table, 1949, Plastic, right round dial, vertical grill slats & front bars, 2 knobs, BC, AC/DC **$40.00**

358T, Table, 1948, Plastic, left half-moon dial, right & left wrap-around louvers, 2 bullet knobs. **$45.00**

360TFM, Table, 1949, Plastic, raised top with curved slide rule dial, large grill, 3 knobs, BC, FM, AC/DC **$40.00**

402, Table, 1939, Walnut metal, midget, right dial, left wrap-around louvers, 2 knobs, feet, AC **$80.00**

402-A, Table, 1939, Ivory metal, midget, right front dial, left wrap-around louvers, 2 knobs, feet, AC **$80.00**

417 "Rhythm Baby", Tombstone, 1936, Wood, lower round dial, oval grill w/3 splayed bars, 2 knobs, BC, SW, AC . . . **$175.00**

422, Table, 1941, Brown metal, midget, right dial, left wrap-around louvers, 2 knobs, BC, AC/DC **$85.00**

422-A, Table, 1941, Ivory metal, midget, right dial, left wrap-around louvers, 2 knobs, BC, AC/DC **$85.00**

440T, Table, 1950, Metal, midget, right front round dial knob, left round checkered grill, BC, AC/DC **$75.00**

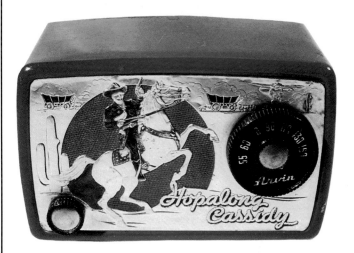

441-T "Hopalong Cassidy", Table-N, 1950, Metal, aluminum front Hopalong Cassidy, right dial, rear "lariatenna", BC, AC/DC . . . . Black $300.00; Red $425.00

442, Table, 1948, Metal, midget, right dial, left horizontal louvers, rounded top, 2 knobs, BC, AC/DC **$70.00**

444, Table, 1946, Metal, midget, right front dial, rounded corners, horiz louvers, 2 knobs, BC, AC/DC **$80.00**

444A, Table, 1946, Metal, midget, right front dial, horizontal louvers, raised top, 2 knobs, BC **$80.00**

444AM, Table, 1947, Metal, midget, right front dial, horizontal louvers, raised top, 2 knobs, BC, AC/DC **$80.00**

446P, Portable, 1950, Plastic, right side dial knob, center front vertical louvers, handle, BC, battery **$35.00**

451T, Table, 1950, Plastic, center round dial w/inner metal perforated grill, 2 knobs, BC, AC/DC **$50.00**

451-TL, Table, 1950, Plastic, center round dial w/inner metal perforated grill, 2 knobs, BC, AC/DC **$50.00**

460T, Table, 1950, Plastic, right see-through dial, left horizontal grill bars, 3 knobs, BC, AC/DC **$30.00**

462-CM, Console-R/P, 1950, Wood, outer front dial, 3 knobs, center front pull-out phono drawer, BC, AC $50.00

467 "Rhythm Belle", Table, 1936, Wood, left front round dial, right round grill, 3 knobs, BC, SW, AC. $70.00

480TFM, Table, 1950, Plastic, right see-through dial, left horizontal grill bars, 3 knobs, AM, FM, AC. $35.00

502, Table, 1939, Metal, midget, right front square dial, left wrap-around louvers, 2 knobs, AC $75.00

517 "Rhythm Junior", Tombstone, 1936, Wood, center round dial, oval grill w/3 splayed bars, 4 knobs, BC, SW, AC . . . $190.00

518 "Phantom Baby", Table, 1937, Wood, lower front round dial, upper cloth grill w/horizontal bars, BC, SW, AC $70.00

522, Table, 1941, Brown metal, midget, right front square dial, left horizontal louvers, 2 knobs, AC $75.00

522A, Table, 1941, Ivory metal, midget, right front square dial, left horizontal louvers, 2 knobs, AC. $75.00

527 "Rhythm Senior", Console, 1936, Wood, upper front round dial, oval grill w/3 splayed bars, 4 knobs, BC, SW, AC . . . $140.00

532, Table, 1941, Two tone Catalin, left dial, raised right w/vertical grill bars, 2 knobs, AC/DC . . . . $1,000.00+

532A, Table, 1941, Two tone Catalin, left dial, raised right w/vertical grill bars, 2 knobs, AC/DC . . . . $1,000.00+

540T, Table, 1951, Painted metal, large right round dial knob over lattice grill, BC, AC/DC. $80.00

541TL, Table, 1950, Plastic, large center front round dial w/stars & inner perforated grill, 2 knobs $50.00

542J, Table, 1947, Metal, right front round dial knob, horizontal grill bars, lower left knob. $65.00

544, Table, 1946, Plastic, left dial, right wrap-over vertical grill bars, 2 knobs, rounded corners $60.00

544A, Table, 1946, Plastic, left dial, right vert grill bars, rounded corners, 2 knobs, BC, AC/DC $60.00

547, Table, 1948, Plastic, left dial, right vertical wrap-over grill bars, 2 knobs, BC, AC/DC $60.00

555, Table, 1947, Walnut plastic, upper slide rule dial, lower vert grill bars, 2 knobs, BC, AC/DC $45.00

555A, Table, 1947, Ivory plastic, upper slide rule dial, vertical grill bars, 2 knobs, BC, AC/DC $45.00

558, Table-R/P, 1946, Wood, lower front slide rule dial, 4 knobs, upper cloth grill, lift top, BC, AC/DC. $30.00

568A "Phantom Blonde", Table, 1938, Lower round dial, upper wrap-over vertical grill bars, 4 knobs $65.00

580TFM, Table, 1951, Plastic, right front dial, left horizontal bars, 3 knobs, AM, FM, AC $35.00

581TFM, Table, 1953, Plastic, right front dial, left cloth grill, 3 knobs, feet, AM, FM $35.00

602, Table, 1939, Walnut plastic, right dial, left grill w/horizontal bars, handle, 3 knobs, AC/DC $60.00

602A, Table, 1939, Ivory plastic, right dial, left grill w/horizontal bars, handle, 3 knobs, AC/DC $60.00

617 "Rhythm Maid", Tombstone, 1936, Wood, center front round dial, oval grill w/3 splayed bars, BC, SW, AC $235.00

622, Table, 1941, Walnut plastic, midget, right dial, left horizontal grill bars, rounded top, 2 knobs $75.00

622-A, Table, 1941, Ivory plastic, midget, right dial, left horizontal grill bars, rounded top, 2 knobs $75.00

627 "Rhythm Master", Console, 1936, Wood, upper front round dial, lower oval grill w/3 splayed bars, BC, SW, AC. . $155.00

628CS "Phantom Bachelor", Chairside, 1937, Two-tone wood, Deco, top dial, rounded front w/vertical bars, BC, SW, AC $200.00

632, Table, 1941, Walnut, right front rectangular dial, left cloth grill w/ vertical bars, 2 knobs, AC $45.00

638CS, Chairside, 1938, Wood, top dial, step-down front w/horizontal louvers, lower storage, BC, SW, AC $150.00

650-P, Portable, 1952, Large center front round dial, right & left side knobs, flex handle, BC, AC/DC/bat $30.00

655SWT, Table, 1952, Plastic, right clear dial over horiz. bars, center strip, 3 knobs, BC, SW, AC/DC $35.00

657-T, Table-C, 1952, Plastic, lower right slide rule dial, left square alarm clock, 2 side knobs, BC, AC $25.00

664, Table, 1947, Plastic, right square dial, left vertical grill bars, handle, 3 knobs, BC, AC/DC $50.00

665, Console-R/P, 1947, Wood, inner right slide rule dial, 4 knobs, phono, lift top, legs, BC, AC. $50.00

669, Table-R/P, Walnut, lower front dial, pushbuttons, 4 knobs, horizontal grill bars, lift top, AC $35.00

702, Table, 1940, Walnut, right dial, left louvers, step-down top w/ pushbuttons, 3 knobs, AC/DC $75.00

722, Table, 1941, Walnut plastic, right square dial, left vertical grill bars, handle, 3 knobs, AC $45.00

722-A, Table, 1941, Ivory plastic, right front square dial, left vertical grill bars, handle, 3 knobs, AC. $45.00

732, Table, 1941, Walnut, right front dial, left horizontal louvers, 3 knobs, AC $40.00

741T, Table, 1953, Plastic, right front round dial, left criss-cross grill, BC, AC/DC $30.00

746P, Portable, 1953, Plastic, top dial and volume knobs, large front metal lattice grill w/logo, handle $75.00

753T, Table, 1953, Plastic, right front clear round dial over oblong cloth grill, 2 knobs, BC, AC/DC $35.00

758T, Table-C, 1953, Plastic, right vertical slide rule dial, left clock, woven grill, 5 knobs, BC, AC $25.00

760T, Table, 1953, Plastic, right front dial, left cloth grill, 3 knobs on lower strip, feet, BC, AC/DC $30.00

802, Portable, 1939, Cloth w/lower stripe, right front dial, left grill, 2 knobs, handle, AC/DC/battery $25.00

840T, Table, 1954, Metal, right round plastic dial knob over vertical bars, lower left plastic knob $65.00

842J, Table, Metal, right front round dial knob over horizontal grill bars, left volume knob $65.00

850T, Table, 1955, Plastic, large right round dial over oval metal perforated grill, BC, AC/DC $35.00

857T, Table-C, 1955, Plastic, modern, lower slide rule dial, center front alarm clock, 5 knobs, BC, AC $25.00

927 "Rhythm Queen", Console, 1936, Wood, large upper front round dial, oval grill w/3 splayed bars, BC, SW, AC . $195.00

950T2, Table, 1958, Plastic, upper right front round dial over checkered grill, BC, AC/DC $30.00

952P1, Portable, 1956, Plastic, right front round dial, left octagonal grill, top thumbwheel knob, handle $35.00

954P, Portable, 1955, Plastic, right front round dial, left octagonal grill, top thumbwheel knob, handle $35.00

958T, Table-C, 1956, Plastic, lower slide rule dial, center alarm clock w/day-date, 5 knobs, BC, AC $30.00

1127 "Rhythm King", Console, 1936, Wood, large upper front round dial, oval grill w/3 vertical bars, BC, SW, AC . . $200.00

1237 "Phantom Prince", Console, 1937, Wood, upper front round telephone dial, horseshoe grill w/vertical bars, AC . . $150.00

1247 "Phantom Queen", Bookcase, 1937, Wood, upper front round dial, right & left bookcases, BC, SW, AC $175.00

1247D, Bookcase, 1938, Wood, center front round dial, pushbuttons, right & left 3 shelf bookcases $175.00

1427 "Phantom King", Console, 1937, Walnut, upper front front dial, pushbuttons, tuning eye, BC, SW, AC . . . $225.00

1581, Table, 1958, Plastic, right & left side knobs-right is dial, front lattice grill, BC, AC/DC $25.00

2410P, Portable, 1948, Plastic, vert. slide rule dial, checkerboard grill, handle, 2 knobs, BC, AC/DC/bat $40.00

2581, Table, 1959, Plastic, right & left side knobs-right is dial, front lattice grill, BC, AC/DC $30.00

2598, Portable, 1960, Transistor, large center front round dial over horizontal grill bars, AM, battery $25.00

3561, Table, 1957, Plastic, center slide rule dial & tone control, lattice grills, 2 knobs, BC, AC/DC $35.00

3582, Table, 1959, Plastic, lower slide rule dial, raised upper grill w/hi/fi logo, 4 knobs, BC, AC/DC $40.00

3586, Table, 1959, Plastic, lower slide rule dial, raised upper grill, 4 knobs, pushbuttons, AM, FM, AC $45.00

3588, Table, 1959, Transistor, lower left slide rule dial, large upper grill, 3 knobs, feet, AM, bat $30.00

5561, Table-C, 1957, Pink plastic, lower left dial w/alarm clock, right front checkered grill, BC, AC $25.00

5581, Table-C, 1957, Green plastic, lower left dial w/alarm clock, right front checkered grill, BC, AC $25.00

6640, Table, 1948, Walnut, right square dial, left cloth grill w/Deco cut-outs, 3 knobs, BC, AC/DC $50.00

7595, Portable, 1960, Two-tone plastic, transistor, right front round dial, swing handle, AM, battery $30.00

8572, Portable, 1958, Leather, right side dial, front perforated grill w/lines, handle, BC, AC/DC/bat $35.00

8576, Portable, 1958, Transistor, upper right dial, lower perforated grill w/random lines, AM, battery $35.00

8584, Portable, 1959, Leatherette, transistor, right thumbwheel dial, horiz grill bars, handle, AM, bat $35.00

9562, Portable, 1957, Transistor, dial & 2 knobs on case top, front perforated grill, handle, AM, battery $40.00

9574-P, Portable, 1956, Leatherette, transistor, front lattice grill w/star, side knobs, handle, AM, bat $40.00

9577, Portable, 1957, Plastic, modern, transistor, right and left side thumbwheel knobs, AM, battery $35.00

9594, Portable, 1960, Plastic, transistor, front off-center thumbwheel dial, AM, battery $25.00

9595, Portable, 1960, Plastic, transistor, right round dial over large front grill, swing handle, AM, bat $30.00

9598, Portable, 1960, Transistor, multi-band dial, "woven" front grill, telescoping antenna, handle, bat $35.00

RE-200, Table, 1946, Metal, midget, raised top, right front dial, horizontal louvers, 2 knobs $75.00

ATLANTIC
Atlantic Radio Corp., Brooklyn, New York

31AC, Grandfather Clock, 1931, Wood, Colonial-style grandfather clock, front dial, knobs and grill, AC $350.00

ATLAS
Atomic Heater & Radio Co.,
102-104 Park Row, New York, New York

AB-45, Table, 1947, Wood, right square dial, left cloth grill w/ vertical bars, 3 knobs, BC, SW, AC/DC $40.00

ATWATER KENT
Atwater Kent Manufacturing Company,
4703 Wissahickon Avenue,
Philadelphia, Pennsylvania

Atwater Kent began business manufacturing electrical items and parts for automobiles. In 1922 the Atwater Kent Manufacturing Company began to market component parts for radios and the first of their large line of breadboards. By 1924 the company produced the first of their cabinet sets. Both the Pooley Company and the Red Lion Cabinet Company were among several furniture manufacturers who made radio cabinets for Atwater Kent. From 1925 to 1927, the company sponsored the Atwater Kent Radio Hour, a popular Sunday night show of radio music. A victim of the Depression, Atwater Kent was out of business by 1936.

1, Breadboard, 1922, Small rectangular wooden board, 1 left & one center dial, 2 right side tubes, bat **$825.00**

2, Breadboard, 1922, Small rectangular wooden board, 1 left & one center dial, 3 right side tubes, bat **$800.00**

5, Breadboard, 1923, Small rectangular wooden board, 1 left side dial, 5 right side tubes, battery **$6,000.00**

9, Breadboard, 1923, Rectangular wooden board, one left & one center dial, 4 tubes, battery **$850.00**

9A, Breadboard, 1923, Rectangular wooden board, one left and one center dial, 4 tubes, battery **$650.00**

9C, Breadboard, 1923, Rectangular wooden board, one left & one center dial, 4 tubes, battery **$650.00**

10, Breadboard, 1923, Rectangular wooden board, one left & 2 centered dials, 5 tubes, battery **$900.00**

10, Console, 1923, Model 10 breadboard unit built into large, ornate "Valley Tone" cabinet, battery **$1,000.00**

10A, Breadboard, 1923, Rectangular wooden board, one left & 2 centered dials, 5 tubes, battery **$775.00**

10B, Breadboard, 1924, Rectangular wooden board, one left & two centered dials, 5 tubes, battery **$700.00**

10C, Breadboard, 1924, Rectangular wooden board, one left and two centered dials, 5 tubes, battery **$750.00**

12, Breadboard, 1924, Rectangular wooden board, one left & two centered dials, 6 tubes, battery **$1,000.00**

19, Table, 1924, Wood, low rectangular case, black 2 dial front panel w/AK logo, lift top, bat **$400.00**

20 "Big Box", Table, 1924, Wood, low rectangular case, 3 dial black front panel w/ AK logo, lift top, battery **$100.00**

20C, Table, 1925, Wood, low rectangular case, 3 dial front panel w/center AK logo, lift top, battery **$145.00**

21, Table, 1925, Wood, low rectangular case, 3 dial black front panel w/AK logo, lift top, battery **$200.00**

24, Table, 1925, Wood, low rectangular case w/overhanging lid, 3 dial front panel, button feet, bat **$300.00**

30, Console, 1926, 1 dial front panel w/center AK logo, battery, built into various console cabinets **$175.00**

30, Table, 1926, Wood, low rectangular case, 1 dial front panel w/center AK logo, lift top, battery **$130.00**

32, Table, 1926, Wood, low rectangular case, 1 dial front panel w/center AK logo, lift top, battery **$110.00**

33, Console, 1927, 1 dial front panel w/center AK logo, battery, built into various console cabinets **$150.00**

33, Table, 1927, Wood, low rectangular case, 1 dial metal panel w/center AK logo, lift top, bat **$100.00**

35, Table, 1926, Metal, low rectangular case, right front dial, top gold AK logo, battery **$65.00**

36, Console, 1927, Model 36 built into various console cabinets, left front dial, center AK logo, AC **$150.00**

36, Table, 1927, Metal, low rectangular case, left front dial, center AK logo, AC **$110.00**

37, Table, 1927, Metal, low rectangular case, left front dial, top gold ship logo, lift-off top, AC **$80.00**

38, Table, 1927, Metal, low rectangular case, left front dial, lift-off top with AK ship logo, AC **$60.00**

40, Console, 1928, Model 40 built into various console cabinets, left front dial, AC **$150.00**

40, Table, 1928, Metal, low rectangular case, left front dial, lift-off top with gold AK logo, AC **$65.00**

41, Table, 1928, Metal, low rectangular case, left front dial, lift-off top, feet **$75.00**

42, Table, 1928, Metal, low rectangular case, left front dial, lift-off top w/gold AK logo, AC **$65.00**

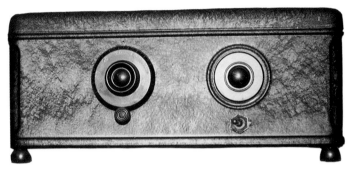

42F, Table, 1928, Metal, low rectangular case, left front dial, lift-off top, feet **$65.00**

43, Table, 1928, Metal, low rectangular case, left front dial, lift-off top, feet, AC **$85.00**

44, Console/Bar, 1928, Model 44 built into Pooley bar cabinet, lift top, inner bar supplies, 8 tubes, AC **$975.00**

44, Table, 1928, Metal, low rectangular case, left front dial, lift-off top w/AK logo, feet, AC **$75.00**

44F, Table, 1928, Metal, low rectangular case, left front dial, lift-off top w/gold AK logo, feet **$75.00**

45, Table, 1929, Metal, low rectangular case, left front dial, lift-off top w/AK logo, AC **$75.00**

46, Console, 1929, Model 46 built into various console cabinets, left front dial, AC **$165.00**

46, Table, 1929, Metal, low rectangular case, left front dial, lift-off top w/AK logo, AC **$85.00**

47, Console, 1929, Model 47 built into various console cabinets, left front dial, AC **$175.00**

47, Table, 1929, Metal, low rectangular case, left front dial, center gold AK logo, lift-off top, AC **$70.00**

48, Table, 1928, Wood, low rectangular case, front panel w/left dial & center AK logo, lift top, bat **$75.00**

49, Table, 1928, Wood, low rectangular case, metal front panel w/left dial, lift top, 6 tubes, AC **$85.00**

50, Table, 1928, Wood, low rectangular case, front panel w/left dial, shielded interior, battery **$900.00**

52, Console, 1928, Metal, upper left front dial, lower large center round "caned" grill, legs, AC **$120.00**

53, Console, 1929, Metal, upper left front dial, lower round "caned" grill, center AK logo, legs, AC **$120.00**

55, Console, 1929, Center front window dial w/escutcheon, 3 knobs, AC, various console cabinets **$150.00**

55, Kiel Table, 1929, Inner window dial w/escutcheon, 3 knobs, AC, lift-top wooden 6 legged table **$275.00**

55, Table, 1929, Metal, low rectangular case, center front window dial w/escutcheon, 3 knobs, AC. **$100.00**

55-C, Kiel Table, 1929, Inner front window dial w/escutcheon, 3 knobs, AC, 6 legged lift-top wood table **$275.00**

55-C, Console, 1929, Model 55-C built into various console cabinets, window dial w/escutcheon, AC **$150.00**

56, Console, 1929, Metal, upper left front dial, lower round grill w/7 circular cut-outs, legs, AC **$100.00**

57, Console, 1929, Metal, upper left front dial, lower round grill w/7 circular cut-outs, legs, AC **$100.00**

60, Table, 1929, Metal, low rectangular case, center front window dial w/escutcheon, 3 knobs, AC. **$70.00**

60, Console, 1929, Model 60 built into various console cabinets, window dial w/escutcheon, AC **$150.00**

60, Kiel Table, 1929, Inner front window dial w/escutcheon, 3 knobs, AC, 6 legged lift-top wood table **$300.00**

60C, Kiel Table, 1929, Inner front window dial w/escutcheon, 3 knobs, AC, 6 legged lift-top wood table **$300.00**

61, Table, 1929, Metal, low rectangular case, window dial w/escutcheon, 3 knobs, lift-off top, DC **$65.00**

66, Console, 1929, Wood, inner window dial, 3 knobs, upper cloth grill w/cut-outs, sliding doors, AC. **$175.00**

67, Table, 1930, Metal painted to look like wood, center window dial, lift-off top, 3 knobs, bat **$75.00**

70, Console, 1930, Wood, lowboy, quarter-round dial, cloth grill, 3 knobs, available as AC, DC or bat **$145.00**

72, Console, 1930, Wood, highboy, quarter-round dial, 3 knobs, lower cloth grill, stretcher base, AC **$160.00**

74, Console, 1930, Wood, lowboy, quarter-round dial, cloth grill, 3 knobs, available as AC or DC **$160.00**

75, Console-R/P, 1930, Wood, lowboy, quarter-round dial, cloth grill, 3 knobs, lift top, inner phono, AC. **$195.00**

76, Console, 1930, Wood, highboy, inner dial and knobs, double front doors, available as AC, DC or bat **$150.00**

80, Cathedral, 1931, Wood, half-round dial, upper cloth grill w/cut-outs, twisted columns, 3 knobs, AC **$450.00**

81, Console, 1932, Wood, lowboy, inner quarter-round dial, double doors, stretcher base, 6 legs, AC **$165.00**

82, Cathedral, 1931, Wood, half-round dial, grill w/gothic cut-outs, 3 knobs, twisted columns, AC. **$475.00**

82D, Cathedral, 1931, Wood, half-round dial, grill w/gothic cut-outs, 3 knobs, twisted columns, DC **$475.00**

82Q, Cathedral, 1931, Wood, half-round dial, grill w/gothic cut-outs, 3 knobs, twisted columns, bat **$475.00**

84, Cathedral, 1931, Wood, center half-round dial, upper cloth grill w/gothic cut-outs, 3 knobs, AC **$450.00**

84, Grandfather Clock, 1931, Wood, center front quarter-round dial, lower grill, upper clock face, 3 knobs **$650.00**

84B, Cathedral, 1931, Wood, center front half-round dial w/escutcheon, upper grill w/cut-outs, 3 knobs **$450.00**

84D, Cathedral, 1931, Wood, center half-round dial, upper cloth grill w/gothic cut-outs, 3 knobs, DC **$450.00**

85Q, Console, 1931, Wood, lowboy, upper quarter-round dial, lower cloth grill w/cut-outs, battery **$135.00**

87D, Console, 1931, Wood, available as highboy or lowboy, quarter-round dial, grill cut-outs, DC **$135.00**
89, Console, 1931, Wood, highboy, inner quarter round dial, 3 knobs, double sliding front doors, AC **$175.00**
90, Cathedral, 1931, Wood, half-round dial, upper cloth grill w/cut-outs, twisted columns, 3 knobs, AC **$450.00**

92, Cathedral, 1931, Wood, half-round dial, upper cloth grill w/ cut-outs, twisted columns, 3 knobs, AC $425.00
94, Console, 1931, Wood, available as highboy or lowboy, quarter-round dial, grill cut-outs, AC **$135.00**
96, Console, 1931, Wood, available as highboy or lowboy, quarter-round dial, grill cut-outs, AC **$165.00**
99, Console, 1932, Wood, available as highboy or lowboy, quarter-round dial, grill cut-outs, AC **$165.00**
112N, Console, 1934, Wood, front quarter-round dial, lower cloth grill w/scrolled cut-outs, BC, SW, AC **$200.00**
112S, Console, 1934, Wood, front quarter round dial, lower cloth grill w/scrolled cut-outs, BC, SW, AC **$175.00**

145, Tombstone, 1934, Wood, lower round dial, upper cloth grill w/cut-outs, fluting, 4 knobs, BC, SW, AC $145.00
155, Table, 1933, Wood, right front window dial, large center cloth grill w/cut-outs, 2 knobs, AC **$80.00**

165, Cathedral, 1933, Wood, right window dial, cloth grill w/scrolled cut-outs, 3 knobs, BC, SW, AC **$225.00**

165Q, Cathedral, 1933, Wood, right window dial, cloth grill w/ scrolled cut-outs, 3 knobs, BC, SW, bat $225.00
184, Tombstone, 1935, Wood, lower right front window dial, upper cloth grill w/cut-outs, 3 knobs, AC **$150.00**
185, Cathedral, 1934, Wood, right window dial, 3-section cloth grill w/scrolled cut-outs, 3 knobs, AC **$225.00**
188, Console, 1932, Wood, lowboy, upper quarter-round dial, lower cloth grill w/cut-outs, 8 tubes, AC **$145.00**
206, Cathedral, 1934, Wood, round dial, upper grill w/cut-outs, 4 knobs, fluted columns, BC, SW, AC **$250.00**
217, Table, 1933, Wood, window dial, center grill w/cut-outs, round top, 4 knobs, BC, SW, AC **$350.00**

225, Tombstone, 1932, Wood, lower front round airplane dial, upper cloth grill w/cut-outs, 3 knobs, AC $135.00
228, Cathedral, 1932, Wood, center half-round dial, upper cloth grill w/gothic cut-outs, 3 knobs, AC **$350.00**
246, Table, 1933, Wood, right front window dial, center grill w/cut-outs, rounded top, 3 knobs, AC **$275.00**

260, Console, 1932, Wood, lowboy, upper quarter-round dial, cloth grill w/cut-outs, 4 knobs, AC **$155.00**

266, Console, 1933, Wood, small upper right window dial, lower cloth grill w/cut-outs, 3 knobs, AC **$135.00**

275, Table, 1933, Wood, right window dial, center cloth grill w/ Deco cut-outs, 4 knobs, BC, SW, AC $195.00

305, Tombstone, 1935, Wood, lower front round airplane dial, upper cloth grill w/cut-outs, 4 knobs, DC **$135.00**

310, Console, 1933, Wood, front quarter-round dial, grill w/scrolled cut-outs, 5 knobs, 6 legs, BC, SW **$235.00**

317, Console, 1935, Wood, upper round dial, lower cloth grill w/ vertical bars, 4 knobs, BC, SW, AC **$160.00**

318, Console, 1934, Wood, front quarter-round dial, lower grill w/cut-outs, 5 knobs, BC, SW, AC **$200.00**

318K, Console, 1934, Wood, front quarter-round dial, lower grill w/ cut-outs, 6 legs, BC, SW, AC **$200.00**

325E, Console, 1934, Wood, lowboy, round dial, lower grill w/cut-outs, 6 legs, 4 knobs, BC, SW, AC **$175.00**

328, Console, 1935, Wood, upper quarter-round dial, lower cloth grill w/cut-outs, 5 knobs, BC, SW, AC **$175.00**

337, Tombstone, 1935, Wood, lower round dial, grill w/cut-outs, fluted columns, 4 knobs, BC, SW, AC **$250.00**

356, Tombstone, 1935, Wood, lower round airplane dial, upper grill w/cut-outs, 4 knobs, fluting, AC $225.00

387, Cathedral, 1934, Wood, center front half-round dial, upper cloth grill w/cut-outs, 3 knobs **$275.00**

427, Console, 1933, Wood, small right window dial, lower cloth grill w/cut-outs, 4 knobs, BC, SW, AC **$135.00**

435, Console, 1935, Wood, upper front round dial, lower cloth grill w/ cut-outs, 3 knobs, BC, SW, AC **$160.00**

447, Tombstone, 1934, Wood, rounded shoulders, quarter-round dial, grill cut-outs, 5 knobs, BC, SW, AC **$225.00**

448, Console, 1933, Wood, front quarter-round dial, lower cloth grill w/cut-outs, 4 knobs, 6 legs, AC **$235.00**

456, Tombstone, 1936, Wood, lower front round dial, upper cloth grill w/cut-outs, 4 knobs, AC **$250.00**

465Q, Tombstone, 1934, Wood, rounded top, lower round airplane dial, upper grill cut-outs, 4 knobs, bat $235.00

469, Console, 1932, Wood, lowboy, quarter-round dial, grill cut-outs, 4 knobs, available AC, DC or bat **$155.00**

509, Console, 1935, Wood, center quarter-round dial, upper "tune-o-matic" clock, 5 knobs, BC, SW, AC **$250.00**

510, Console, 1933, Wood, modern, quarter-round dial, lower grill w/ cut-outs, 5 knobs, BC, SW, AC **$200.00**

511W, Console, 1934, Wood, center quarter-round dial, upper "tune-o-matic" clock, 5 knobs, BC, SW, AC **$250.00**

521N, Console, 1930, Metal, upper left front dial, large lower center "caned" grill, legs **$125.00**

535, Console, 1936, Wood, upper front round dial, lower cloth grill w/ center vert bars, 3 knobs, AC **$150.00**

545, Tombstone, 1935, Wood, lower round airplane dial, upper grill w/cut-outs, 3 knobs, BC, SW, AC **$150.00**

555, Table, 1933, Inlaid walnut, chest-style, lift top, inner metal panel, front grill w/cut-outs, AC $375.00

558, Cathedral, 1932, Wood, half-round dial, cloth grill w/cut-outs, 4 knobs, available as AC, DC or bat $350.00

725, Tombstone, 1936, Wood, lower front round dial w/globes, upper grill w/cut-outs, 3 knobs, BC, SW $165.00

567, Cathedral, 1932, Wood, half-round dial, grill w/gothic cut-outs, front carved columns, 3 knobs, AC $350.00
612, Console, 1932, Wood, lowboy, upper quarter-round dial, cloth grill w/cut-outs, 4 knobs, AC $200.00

735, Cathedral, 1935, Wood, lower round dial, upper cloth grill w/cut-outs, fluted columns, 4 knobs, AC $250.00
808, Console, 1933, Wood, upper front dial, lower cloth grill w/cut-outs, 6 legs, AC $225.00
810, Console, 1935, Wood, front quarter-round dial, lower cloth grill w/cut-outs, 5 knobs, BC, SW, AC $200.00

627, Cathedral, 1932, Wood, center half-round dial, upper grill w/ cut-outs, front columns, 3 knobs, AC $400.00
637, Tombstone, 1932, Wood, lower front round multi-colored dial, upper cloth grill w/cut-outs, 4 knobs $175.00
649, Console, 1935, Wood, upper quarter-round dial, lower cloth grill w/cut-outs, 5 knobs, BC, SW, AC $160.00
667, Console, 1933, Wood, modern, front window dial, lower grill w/ cut-outs, 4 knobs, BC, SW, AC $225.00
676, Console, 1936, Wood, upper round "rainbow" dial, lower cloth grill w/vertical bars, 4 knobs, AC $150.00
708, Table, 1933, Wood, right front dial, cloth grill w/cut-outs, 4 knobs, rounded top, BC, SW, AC $275.00
711, Console, 1933, Wood, lowboy, inner quarter-round dial, double doors, 6 legs, 5 knobs, BC, SW, AC $260.00

854, Tombstone, 1935, Wood, small right window dial, upper cloth grill w/cut-outs, 3 knobs, AC $140.00
856, Tombstone, 1935, Wood, lower round airplane dial, upper grill w/cut-outs, 4 knobs, BC, SW, AC $130.00

944, Cathedral, 1934, Two-tone wood, right window dial, upper cloth grill w/cut-outs, 2 knobs, AC $300.00

AUDAR
Audar, Inc., Argos, Indiana

AV-7T, Table-R/P, 1952, Wood, inner front dial & 5 knobs, fold-down front door, lift top, inner phono $40.00

PR-6, Table-R/P, 1947, Inner dial, phono & knobs, lift top, outer grill, handle, BC, AC $20.00

PR-6A, Table-R/P, 1947, Inner dial, phono & knobs, lift top, outer grill, handle, BC, AC $20.00

RER-9 "Telvar", Console-R/P/Rec, 1949, Wood, vertical dial & 4 knobs on top of case, left inner phono, BC, AC $45.00

AUDIOLA
Audiola Radio Co.
Chicago, Illinois

5W, Table, 1933, Wood, front dial, center grill w/cut-outs, 2 knobs, AC/DC $75.00

517, Cathedral, 1932, Wood, ornate scrolled front, lower half-moon dial, upper grill w/cut-outs, BC, SW $250.00

610, Cathedral, 1931, Wood, center front quarter-moon dial, upper grill w/scrolled cut-outs, 3 knobs $185.00

612, Console, 1931, Wood, lowboy, upper front quarter-moon dial, lower grill w/scrolled cut-outs $125.00

811, Cathedral, 1932, Wood, center front quarter-moon dial, upper cloth grill, fluted columns, 3 knobs $165.00

814, Console, 1931, Wood, lowboy, upper quarter-moon dial, lower cloth grill w/scrolled cut-outs $140.00

1168, Console, 1932, Wood, lowboy, upper half-moon dial, lower cloth grill w/cut-outs, 6 legs, AC $140.00

10300D, Console, 1933, Wood, lowboy, upper half-moon dial, lower cloth grill w/cut-outs, 6 legs $150.00

AUTOCRAT
Autocrat Radio Co.,
3855 N. Hamilton Ave.
Chicago, Illinois

101, Table, 1938, Plastic, right dial panel over horiz grill bars, decorative case lines, BC, AC/DC $60.00

AUTOMATIC
Automatic Radio Mfg. Co., Inc.,
122 Brookline Avenue
Boston, Massachusetts

The Automatic Radio Manufacturing Company began in 1920. The company is well-known for its line of car radios and farm tractor sets as well as for their "Tom Thumb" line of home radios. The company is still in business as manufacturers of solid state products and test equipment.

8-15, Table, 1937, Plastic, Deco, right front square dial, left grill w/ Deco design, 2 knobs $75.00

141, Table-R/P, 1941, Wood, outer right front dial, left grill w/diagonal bars, 3 knobs, lift top, AC $35.00

145, Table-R/P, 1941, Walnut, outer right front dial, left grill, 3 knobs, lift top, inner phono, AC $35.00

152, Table-R/P, 1941, Wood, right front dial, left grill w/Deco bands, 2 knobs, open top, AC $35.00

434-A, Table-R/P, 1940, Walnut, center front square dial, right & left vertical grill bars, lift top, AC $35.00

458, Table, 1939, Wood, rectangular dial, horizontal louvers, tapered cylindrical sides, 3 knobs $100.00

601, Table, 1947, Plastic, right square dial, left vertical louvers, 2 knobs, curved sides, BC, AC/DC $60.00

602, Table, 1947, Plastic, right front square dial, left vertical louvers, curved sides, 2 knobs, AC $60.00

612X, Table, 1946, Wood, right front round convex dial, "S" curve on top of case, 2 knobs, BC, AC/DC $75.00

614X, Table, 1946, Plastic, right round starburst dial, left square grill, 2 knobs, BC, AC/DC $60.00

620, Table, 1947, Two-tone wood, slanted slide rule dial, criss-cross grill, 2 knobs, BC, AC/DC $50.00

640, Table-R/P, 1946, Wood, outer right square dial, 3 knobs, criss-cross grill, lift top, BC, AC $30.00

662, Table, 1947, Wood, upper slanted slide rule dial, horizontal grill bars, 3 knobs, BC, SW, AC/DC $45.00

677, Table-R/P, 1947, Wood, inner dial, 4 knobs, phono, lift top, wooden grill, BC, AC $30.00

720, Table, 1947, Wood, right black dial, left criss-cross grill, rounded sides, 2 knobs, BC, AC/DC $45.00

933 "Tom Thumb", Table, 1939, Catalin, small case, right front dial, left grill w/Deco cut-outs, 2 knobs, AC . . . . $2,000.00+

1975, Table, 1935, Wood, right front slide rule dial, left horizontal wrap-around louvers, 3 knobs, AC $50.00

ATTP "Tom Thumb", Portable, 1947, Leatherette & plastic, inner half-moon dial & grill, door, handle, BC, AC/DC/bat . . $85.00

B-44 "Tom Thumb", Portable-Bike, 1949, Crackle finish bike radio, telescope antenna, slide rule dial, handle, BC, bat . . $100.00

C-51, Portable, 1952, Leatherette, center front round metal dial, handle, 2 knobs, BC, AC/DC/battery $30.00

C-60, Portable, 1946, Luggage style, right dial, left horizontal louvers, 3 knobs, handle, BC, AC/DC/bat $35.00

C-60X, Portable, 1947, Two-tone leatherette, right dial, horizontal louvers, handle, BC, AC/DC/battery $40.00

C-65, Portable, 1942, Leatherette, right front dial, left horizontal louvers, handle, AC/DC/battery $35.00

CL-100, Table-C, 1959, Plastic, right square dial, left alarm clock, center vert wrap-over bars, BC, AC $35.00

CL-152B, Table-C, 1953, Plastic, right square dial, center horizontal grill bars, left alarm clock, BC, AC $30.00

CL-175, Table-C, Wood, right front dial, left alarm clock, center lattice grill panel, 2 knobs, AC $40.00

F-790, Console-R/P, 1947, Wood, modern, inner half-moon dial, 4 knobs, pull-out phono drawer, BC, AC $100.00

P-64, Portable, Leatherette, front plastic panel, round dial w/ inner perforations, handle, 2 knobs **$40.00**
P-72, Portable, 1939, Cloth covered, upper front dial, lower grill, handle, 2 knobs, AC/DC/battery **$30.00**

PTR-15B, Portable, 1958, Leather case, transistor, right front dial knob, lattice grill cut-outs, handle, bat **$25.00**
Tom Boy, Portable, 1947, Two-tone leatherette, right dial, criss-cross grill, handle, 2 knobs, BC, battery **$35.00**
Tom Thumb Buddy, Portable, 1949, Leatherette & plastic, slide rule dial, flip-open front, handle, BC, AC/DC/bat **$60.00**
Tom Thumb Camera, Portable, 1948, Leatherette, reflex camera, lower grill bars, top rear dial, strap, 2 knobs, BC, bat **$175.00**
Tom Thumb Jr, Table, 1933, Black bakelite w/chrome trim, Deco, left dial, center grill, chrome ball feet, AC **$160.00**
Tom Thumb, Jr, Portable, 1947, "Snakeskin", right square dial, left horizontal grill bars, 2 knobs, handle, BC, bat **$35.00**
Tom Thumb Portable, Portable, 1929, Leather case, inner engraved metal one dial panel, hinged front cover, handle, battery **$300.00**
TT528, Portable, 1957, Plastic, right front dial, left lattice grill, top right thumbwheel knob, handle **$45.00**
TT600, Portable, 1957, Plastic, tubes & transistors, right front dial, left checkered grill, handle, battery **$75.00**

AVIOLA
Aviola Radio Corp., Phoenix, Arizona

509, Table-R/P, 1946, Wood, front dial, 3 knobs, octagonal grill, feet, lift top, inner phono, BC, AC **$35.00**
601, Table, 1947, Plastic, upper slanted slide rule dial, checkerboard grill, 2 knobs, BC, AC/DC **$40.00**
608, Table-R/P, 1947, Wood, inner slide rule dial, 3 knobs, phono, lift top, criss-cross grill, BC, AC **$40.00**

612, Table, 1947, Plastic, upper slanted slide rule dial, horizontal louvers, 2 knobs, BC, AC/DC **$40.00**

B. F. GOODRICH
B. F. Goodrich Co., Akron, Ohio

92-523, Table, 1951, Plastic, right front round dial over rectangular grill bars, 2 knobs, BC, AC/DC **$35.00**

BALKEIT
Balkeit Radio Corporation,
Clinton & Randolph Streets, Chicago, Illinois

The Balkeit Radio Company was formed as subsidiary of the Pfanstiehl Radio Company in 1929.

44, Portable, 1933, Two-tone Deco case, center front grill w/vertical cut-outs, carrying case **$65.00**
A-3, Table, 1928, Metal, low rectangular case, center front slanted thumbwheel dial, AC **$130.00**
A-5, Table, 1928, Wood, low rectangular case, center slanted thumbwheel dial, fluted columns, AC **$140.00**
A-7, Console, 1928, Wood, highboy, inner slanted thumbwheel dial, lower grill, double front doors, AC **$175.00**
B-7, Console, 1928, Wood, inner slanted thumbwheel dial, upper speaker grill, double front doors, AC **$225.00**
B-9, Console-R/P, 1928, Wood, inner thumbwheel dial, speaker grill, double doors, lift top, inner phono, AC **$200.00**
C, Console, 1929, Wood, inner window dial, upper cloth grill w/ scalloped cut-outs, double doors, AC **$200.00**

BARBAROSSA

Beer Bottle, Table-N, 1934, Bakelite, looks like large Barbarossa beer bottle, base with switch, AC $350.00

BELLTONE
Jewel Radio Corp.,
583 Sixth Avenue, New York, New York

500, Table, 1946, Wood, upper slanted slide rule dial, horizontal grill bars, 2 knobs, BC, AC/DC **$40.00**

─────────────────────

BELMONT
Belmont Radio Corp.,
5921 West Dickens Avenue
Chicago, Illinois

4B17, Table, 1946, Wood, lower front slide rule dial, upper vert grill openings, 2 knobs, BC, battery **$30.00**

4B112, Table, 1946, Plastic, right front dial, left horizontal wrap-around louvers, 2 knobs, BC, bat **$40.00**

4B115, Table, 1948, Plastic, rounded top, half-moon dial curves over checkered grill, 2 knobs, BC, bat **$75.00**

5D110, Table-R/P, 1947, Wood, front slide rule dial, cloth grill, 2 knobs, 3/4 lift top, inner phono, BC, AC **$30.00**

5D118, Table, 1948, Plastic, center front airplane dial inside concentric circular grill, 2 knobs, AC **$75.00**

5D128, Table, 1946, Plastic, streamline, right dial, pushbuttons, left horiz bars, side knob, BC, AC/DC **$150.00**

5P19, Portable, 1946, Luggage-style, slide rule dial, 2 knobs on top of case, handle, BC, AC/DC/bat **$30.00**

5P113 "Boulevard", Portable, 1947, Very small, dial & 2 knobs on top of case, ear plug, BC, battery **$75.00**

6D111, Table, 1946, Plastic, streamlined, right dial, pushbuttons, left horizontal louvers, side knob **$250.00**

6D120, Table, 1947, Plastic, streamline, half-moon dial, louvers, 2 knobs, 6 pushbuttons, BC, AC/DC **$150.00**

8A59, Console-R/P, 1946, Wood, slide rule dial, 4 knobs, pushbuttons, door, phono in drawer, BC, 4SW, AC **$75.00**

401, Cathedral, Two-tone wood, lower front round airplane dial, upper grill w/cut-outs, 3 knobs **$195.00**

407, Portable, 1939, Striped cloth, luggage-style, top dial & knobs, front grill, handle, battery **$30.00**

509, Table, 1940, Wood, lower front slide rule dial, pushbuttons, upper grill w/vertical bars **$70.00**

510, Table, 1938, Plastic, streamline, right front dial, left horiz wrap-around louvers, 2 knobs, AC **$75.00**

519, Table, 1939, Plastic, streamline, right front dial, pushbuttons, left circular grill w/horiz bars **$135.00**

525, Table, 1933, Wood, Deco case lines, right front dial, center cloth grill w/cut-outs, 2 knobs **$75.00**

526, Table, 1938, Plastic, Deco, right front "Bel-Monitor" tuning system, left vertical grill bars **$75.00**

533-D, Table-R/P, 1941, Wood, right front square dial, left grill, 2 knobs, lift top, inner phono, AC **$30.00**

534, Table, 1940, Plastic, streamline, right dial, pushbuttons, left horizontal louvers, side knob, AC **$135.00**

571, Table-C, 1940, Walnut, lower front slide rule dial, upper electric clock, ribbed sides, 2 knobs, AC **$55.00**

575, Tombstone, 1934, Wood, shouldered, center front round dial, upper cloth grill w/cut-outs, AC **$110.00**

602 "Scotty", Table, 1937, Plastic, right front dial, rounded right top, left wrap-over vert grill bars, AC/DC **$85.00**

636, Table, 1939, Plastic, right front dial, 5 pushbuttons, left wrap-around louvers, side knob **$85.00**

675, Tombstone, 1934, Wood, center front round dial, upper cloth grill w/cut-outs, BC, SW, AC **$115.00**

675E, Console, 1934, Two-tone wood, upper front round dial, lower cloth grill w/cut-outs, BC, SW, AC **$115.00**

792, Console, 1939, Wood, upper front rectangular dial, pushbuttons, lower grill, 4 knobs, AC **$100.00**

797, Console-R/P/Rec, 1940, Wood, inner right rectangular dial, pushbuttons, left phono, lift top, BC, SW, AC **$100.00**

840, Console, 1937, Wood, oval dial, tuning eye, lower grill w/vertical bars, 4 knobs, BC, 2SW, AC **$135.00**

1170, Console, 1936, Wood, front oval dial, tuning eye, lower cloth grill w/vertical bars, 4 knobs **$135.00**

A-6D110, Table, 1947, Plastic, slide rule dial, vertical grill bars, 2 knobs, 6 pushbuttons, BC, AC/DC **$95.00**

─────────────────────

BENDIX
Bendix Radio/Bendix Aviation,
Baltimore, Maryland

The company began in 1937 as a division of Bendix Aviation. During World War II, Bendix was a major supplier of radio-related aircraft equipment for the British and American governments.

0516A, Table, 1946, Plastic, Deco case, upper slide rule dial, vertical louvers, 2 knobs, BC, AC/DC **$65.00**

0526A, Table, 1946, Plastic, Deco case, upper slide rule dial, vertical louvers, 2 knobs, BC, AC/DC **$65.00**

0526B, Table, 1946, Plastic, Deco case, upper slide rule dial, vertical louvers, 2 knobs, BC, AC/DC **$65.00**

0526E, Table, 1946, Wood, slanted slide rule dial, vert grill bars, rounded sides, 2 knobs, BC, AC/DC **$50.00**

55L2, Table, 1949, Ivory plastic, slide rule dial, vert grill bars, rear hand-hold, 2 knobs, BC, AC/DC **$40.00**

55L3, Table, 1949, Ivory plastic, slide rule dial, wood-grained grill, rear hand-hold, 2 knobs, AC/DC **$45.00**

55P2, Table, 1949, Walnut plastic, slide rule dial, vert grill bars, rear hand-hold, 2 knobs, BC, AC/DC **$40.00**

55P3, Table, 1949, Walnut plastic, slide rule dial, wood-grained grill, rear hand-hold, 2 knobs, AC/DC **$45.00**

55P3U, Table, 1949, Walnut plastic, slide rule dial, wood-grained grill, rear hand-hold, 2 knobs, AC/DC **$45.00**

55X4, Portable, 1949, Plastic, slide rule dial, horizontal louvers, 2 knobs, flip-up front, BC, AC/DC/bat $40.00

65P4, Table, 1949, Plastic, upper slide rule dial, metal grill, top hand hold, 3 knobs, BC, AC/DC $40.00

65P4U, Table, 1949, Plastic, upper slide rule dial, metal grill, top hand hold, 3 knobs, BC, AC/DC $40.00

69B8, Console-R/P, 1949, Blonde wood, inner right radio/phono, left storage, double front doors, BC, FM, AC $75.00

69M8, Console-R/P, 1949, Mahogany, inner right radio/phono, left storage, double front doors, BC, FM, AC $75.00

69M9, Console-R/P, 1949, Wood, inner right slide rule dial, 3 knobs, pull-out phono, door, BC, FM, AC $75.00

75B5 "Fairfax", Console-R/P, 1949, Blonde wood, top left dial & knobs, right front pull-out phono drawer, BC, FM, AC $100.00

75M8 "Heritage", Console-R/P, 1949, Mahogany, upper dial and knobs, inner pull-out phono drawer & storage, BC, FM, AC $100.00

75P6, Table, 1949, Walnut plastic, slide rule dial, wood-grained grill, rear hand-hold, AM, FM, AC/DC $45.00

75P6U, Table, 1949, Walnut plastic, slide rule dial, wood-grained grill, rear hand-hold, AM, FM, AC/DC $45.00

75W5 "York", Console-R/P, 1949, Walnut wood, top left dial & knobs, right front pull-out phono drawer, BC, FM, AC $100.00

79M7, Console-R/P, 1949, Wood, inner right slide rule dial, 3 knobs, lower pull-out phono, door, BC, FM, AC $80.00

95B3 "Boulevard", Console-R/P, 1949, Blonde wood, right tilt-out radio, 3 knobs, left pull-out phono drawer, BC, FM, AC $95.00

95M3 "Wiltondale", Console-R/P, 1949, Mahogany, right tilt-out radio, 3 knobs, left pull-out phono drawer, BC, FM, AC $95.00

95M9 "Wayne", Console-R/P, 1949, Mahogany, right tilt-out radio, 3 knobs, left pull-out phono drawer, BC, FM, AC $95.00

110 , Table, 1948, Walnut plastic, slide rule dial, vert grill bars, rear hand-hold, 2 knobs, BC, AC/DC $65.00

110W, Table, 1948, Ivory plastic, slide rule dial, vert grill bars, rear hand-hold, 2 knobs, BC, AC/DC $65.00

111, Table, 1949, Walnut plastic, slide rule dial, vert grill bars, rear hand-hold, 2 knobs, BC, AC/DC $40.00

111W, Table, 1949, Ivory plastic, slide rule dial, vert grill bars, rear hand-hold, 2 knobs, BC, AC/DC $40.00

112, Table, 1948, Walnut, upper slide rule dial, large lower front grill area, 2 knobs, BC, AC/DC $40.00

114, Table, 1948, Tan & brown plastic, slide rule dial, wrap-around grill bars, 2 knobs, BC, AC/DC $275.00

115, Table, 1948, Ivory & burgundy plastic, slide rule dial, wrap-around grill bars, 2 knobs, AC/DC $275.00

300, Table, 1948, Brown plastic, slanted slide rule dial, vertical grill bars, 3 knobs, BC, AC/DC $40.00

300W, Table, 1948, Ivory plastic, slanted slide rule dial, vertical grill bars, 3 knobs, BC, AC/DC $40.00

301, Table, 1948, Wood, upper slanted slide rule dial, horizontal grill bars, 3 knobs, BC, AC/DC $35.00

302, Table, 1948, Wood, upper slanted slide rule dial, horizontal grill bars, 3 knobs, BC, AC/DC $35.00

416A, Table, 1948, Wood, upper slide rule dial, horizontal louvers, small base, 2 knobs, BC, battery $35.00

526C, Table, 1946, Catalin, green & black, upper slide rule dial, lower horizontal louvers, BC, AC/DC $575.00

526MB, Table, 1947, Plastic, Deco case, upper slide rule dial, vertical grill bars, 2 knobs, BC, AC/DC $65.00

613, Table-R/P, 1948, Wood, slide rule dial, 4 knobs, lift top, inner phono, BC, AC $30.00

626-A, Table, 1947, Plastic, upper slanted slide rule dial, vertical louvers, 3 knobs, BC, SW, AC/DC $40.00

626-C, Table, 1947, Plastic, upper slanted slide rule dial, vertical louvers, 3 knobs, BC, SW, AC/DC $40.00

636A, Table, 1947, Plastic, upper slanted slide rule dial, vertical louvers, 3 knobs, BC, AC/DC $40.00

636B, Table, 1947, Wood, upper slanted slide rule dial, lower woven grill, 3 knobs, BC, AC/DC $75.00

636C, Table, 1947, Wood, upper slanted slide rule dial, lower woven grill, 3 knobs, BC, AC/DC $75.00

636D, Table, 1947, Wood, slanted front slide rule dial, large center woven metal grill, 3 knobs, AC $75.00

646A, End Table, 1946, Wood, drop leaf end table, dial lights up across lower panel, 4 knobs, BC, AC/DC $125.00

656A, Table-R/P, 1946, Wood, inner right dial, 3 knobs, phono, lift top, outer louvers, BC, AC $35.00

676D, Console-R/P, 1946, Wood, inner vert slide rule dial, 4 knobs, phono, lift top, front grill, BC, SW, AC $65.00

687A, Portable, 1949, Leatherette, inner slide rule dial, 4 knobs, drop-front, handle, BC, AC/DC/battery $35.00

697A, End Table-R/P, 1947, Wood, step-down end table, radio in top, phono in base with sliding door, BC, AC $125.00

736-B, Console-R/P, 1946, Wood, inner right vertical dial, 4 knobs & left phono, lift top, BC, 2SW, AC $100.00

753F, Table-C, 1953, Wood, lower slide rule dial, large front alarm clock, side louvers, handle, feet, AC $65.00

753M, Table-C, 1953, Wood, lower small slide rule dial, large front glass clock face, 5 knobs, AC $65.00

847-B, Console-R/P, 1947, Wood, inner right dial, 4 knobs, pushbuttons, left phono, lift top, BC, FM, AC $75.00

1217B, Console-R/P, 1947, Wood, inner dial, pushbuttons, lift top, pull-out phono drawer, BC, SW, FM, AC $70.00

1217D, Console-R/P, 1948, Wood, slide rule dial, lift top, left front pull-out phono, oval grill, BC, SW, FM, AC $75.00

1518, Console-R/P, 1948, Mahogany, inner dial, pushbuttons, lift-up lid, left pull-out phono, BC, FM, AC $70.00

1519, Console-R/P, 1948, Walnut, inner dial, pushbuttons, lift-up lid, left pull-out phono, BC, FM, AC $70.00

1521, Console-R/P, 1948, Wood, inner slide rule dial, 4 knobs, door, lift top, inner phono, BC, FM, AC $75.00

1524, Console-R/P, 1948, Mahogany, inner dial, pushbuttons, lift-up lid, left pull-out phono, BC, FM, AC $70.00

1525, Console-R/P, 1948, Walnut, inner dial, pushbuttons, lift-up lid, left pull-out phono, BC, FM, AC $70.00

PAR-80, Portable, 1948, Luggage style, slide rule dial, fold-down front, handle, BC, SW, LW, AC/DC/bat $35.00

PAR-80A, Portable, 1948, Luggage style, slide rule dial, fold-down front, handle, BC, SW, LW, AC/DC/bat $35.00

BENRUS
Benrus Watch Co., Inc.,
50 West 44th Street
New York, New York

10B01B15B, Table-C, 1955, Metal, right side dial knob, large front clock face, left volume knob, BC, AC $50.00

BEST

221, Table, 1936, Wood, rounded right, oval dial, left grill w/chrome wrap-around bars, 3 knobs $110.00

BETTS & BETTS
Betts & Betts Corp.,
643 W. 43rd St., New York, New York

T8 "Trans-Continental", Table, 1925, Wood, high rectangular case, 2 dial front panel, 8 tubes, battery $200.00

BOSWORTH
The Bosworth Mfg. Co., Cinncinati, Ohio

B-2, Table, 1926, Wood, rectangular case, 2 dial fancy slanted metal panel, lift top, 5 tubes, bat $150.00

BOWMAN
A. W. Bowman & Co.,
Cambridge, Massachusetts

Airophone, Table, 1923, Wood, low rectangular case, 2 dial black bakelite front panel, battery $210.00

BRANDEIS
J. F. Brandeis Corp.,
35½ Oxford St., Newark, New Jersey

Brandola, Table, 1925, Wood, low rectangular case, one center front dial, storage, battery $185.00

BREMER-TULLY
Bremer-Tully Manufacturing Company,
520 South Canal Street, Chicago, Illinios

Bremer-Tully was begun in 1922 by John Tully and Harry Bremer. The company started in business manufacturing radio parts and kits and by 1925 they were selling fully assembled radios. The company was sold to Brunswick in 1928.

6-22 "Counterphase", Table, 1927, Two-tone wood, rectangular case, center front dial w/escutcheon, battery $125.00

6-35 "Counterphase", Table, 1927, Wood, rectangular case w/ slant front, 2 center front dials w/escutcheon, battery $135.00

6-37 "Counterphase", Console, 1927, Wood, lowboy, 2 inner front dials w/escutcheon, 4 knobs, fold-down front, bat . . . $150.00

6-40, Table, 1928, Wood, rectangular case w/emblems, recessed center front dial w/escutcheon, AC $135.00

6-40R, Table, 1928, Wood, low rectangular case w/emblems, center front dial w/escutcheon, 3 knobs $130.00

6-40S, Table, 1928, Wood, low rectangular case w/emblems, center front dial w/escutcheon, 2 knobs $135.00

7-70, Table, 1928, Wood, low rectangular case, recessed center front dial w/escutcheon, AC $135.00

7-71, Console, 1928, Wood, highboy, center front dial w/escutcheon, upper grill w/cut-outs, AC $175.00

8 "Counterphase", Table, 1926, Wood, rectangular case w/slant front, center dial w/escutcheon, lift top, battery $135.00

8-12 "Counterphase", Table, 1927, Wood, rectangular case w/ slant front, center dial w/escutcheon, lift top, battery . $135.00

8-20, Table, 1928, Wood, low rectangular case, recessed center front dial w/escutcheon, lift top, AC $130.00

8-21, Console, 1928, Wood, upper front dial, lower cloth grill w/cut-outs, AC $135.00

fully-assembled radios. Their business slowly decreased until the partners went their separate ways in 1937.

4-R, Table, Wood, low rectangular case, 2 dial black front panel, lift top, 4 tubes $140.00

5-R, Table, 1926, Wood, low rectangular case, 2 dial black front panel, lift top, 5 tubes, battery $150.00

81, Console, 1929, Wood, highboy, center dial, upper grill w/oval cut-out, stretcher base,3 knobs, AC $150.00

82, Console, 1929, Walnut, highboy, center dial, upper grill w/cut-outs, double doors, 3 knobs, AC $165.00

BREWSTER
Meissner Mfg. Div.,
Maguire Industries, Inc., Mt Carmel, Illinois

9-1084, Table, 1946,Plastic, recessed right, slide rule dial, left horiz bars, step-down top, 3 knobs. $55.00

9-1086, Table, 1946, Plastic, recessed right, slide rule dial, step-down top, 3 knobs, BC, SW, AC/DC $55.00

BROWNING-DRAKE
Browning-Drake Corporation,
353 Washington Street
Brighton, Massachusetts

Frederick Drake and Glenn Browning created the Browning-Drake circuit in 1924. In 1925 their Browning-Drake Company was selling

6-A, Table, 1927, Wood, low rectangular case, inner front dial, 5 knobs, double doors, lift top, bat $110.00

7-A, Table, 1927, Walnut, low rectangular case, inner dial & knobs, double front doors, lift top, bat $115.00

30, Table, 1928, Wood, low rectangular case, center front dial, 3 knobs, AC $130.00

32, Console, 1928, Wood, lowboy, upper front dial, lower cloth grill w/gothic cut-outs, 3 knobs, AC $180.00

34, Table, 1928, Wood, low rectangular case, center front window dial w/escutcheon, 3 knobs, AC. $125.00

53, Table, 1929, Wood, low rectangular case, center window dial with escutcheon, 3 knobs, AC. $115.00

54, Console, 1929, Wood, lowboy, window dial w/escutcheon, round grill w/cut-outs, 3 knobs, AC $150.00

57, Console, 1929, Wood, window dial w/escutcheon, lower scalloped grill w/cut-outs, 3 knobs, AC $150.00

84, Console, 1929, Wood, lowboy, upper front window dial, lower round grill w/cut-outs, battery. $135.00

B-D Junior, Table, 1925, Wood, low rectangular case, 2 dial front panel, lift top, 5 tubes, battery $120.00

B-D Senior, Table, 1925, Wood, lower radio w/2 dial front panel, upper speaker w/scroll grill, 6 tubes, bat $200.00

B-D Standard, Table, 1925, Mahogany, low rectangular case, 2 dial front panel, lift top, 5 tubes, battery $140.00

BRUNSWICK

5KR, Table, 1928, Wood, low rectangular case, center front dial, lift top, 7 tubes, 2 knobs, AC $135.00

5-WO, Table, 1928, Wood, low rectangular case, center window dial w/escutcheon, lift top, 9 tubes $150.00

11, Upright Table, 1931, Wood, shouldered case, center front window dial, upper grill w/cut-outs, AC $165.00

14, Console, 1929, Wood, upper front dial, lower grill w/cut-outs, 3 knobs, stretcher base, AC $145.00

15, Console, 1930, Wood, upper front window dial, lower scalloped grill, 2 knobs, stretcher base, AC $145.00

16, Console, 1931, Wood, lowboy, upper front window dial, lower cloth grill w/cut-outs, AC $125.00

21, Console, 1929, Wood, highboy, inner front dial & knobs, double doors, stretcher base, AC $150.00

22, Console, 1930, Wood, inner front dial & knobs, French doors, stretcher base, AC $150.00

31, Console-R/P, 1929, Wood, inner front dial & knobs, double doors, phono, arched stretcher base, AC $150.00

33, Console-R/P, 1931, Wood, upper front dial, lower grill w/cut-outs, lift top, inner phono, AC $135.00

50, Console, Wood, upper window dial, lower tapestry grill w/cut-outs, medallions, 2 knobs $145.00

1559, Side Table, 1939, Wood, French Provincial styling, inner dial & grill, double front doors, AC/DC $135.00

1580, Table, Wood, rectangular case, inner right dial, 2 knobs, left grill w/cut-outs, doors $100.00

1669, Side Table, 1939, Wood, Hepplewhite styling, inner dial/grill/pushbuttons, double front doors, AC $150.00

2559, Side Table, 1939, Wood, Early American styling, inner dial & grill, double front doors, AC/DC $110.00

2689, Side Table, 1939, Wood, Duncan Phyfe styling, inner dial/tuning eye/pushbuttons, double doors, AC $175.00

3689, Side Table, 1938, Wood, French styling, half-round table, front dial, pushbuttons, BC, SW, LW, AC $150.00

4689, Console, 1939, Wood, Queen Anne styling, inner dial/tuning eye/pushbuttons, double doors, AC. $200.00

5000, Console-R/P, 1948, Wood, inner front slide rule dial, 4 knobs, fold-back door, BC, FM, AC. $100.00

8109, Console-R/P, 1939, Wood, Queen Anne styling, inner dial/grill/tuning eye/pushbuttons, lift top, AC $225.00

BJ-6836 "Tuscany", End Table-R/P, 1947, Wood, step-down end table, dial & 5 knobs in top, doors, phono in base, BC, AC $125.00

D-1000, Console-R/P, 1949, Wood, inner right slide rule dial, 4 knobs, left pull-out phono drawer, BC, FM, AC $100.00

BUCKINGHAM

80, Table, Metal, low rectangular case, large escutcheon w/window dial, lift top, 3 knobs $150.00

BUICK

981970 "Trans-Portable", Car/Portable, 1958, Metal case, right dial over lattice grill w/crest, plugs in for use as car radio $200.00

BULOVA
Electronics Guild, Inc.,
Sunrise Highway, Valley Stream,
Long Island, New York

100, Table-C, 1957, Plastic, left dial & clock, right metal grill, side knobs, step-down top, BC, AC $40.00

110, Table-C, 1957, Plastic, left dial & clock, right grill, step-down top, side knobs, feet, BC, AC $40.00

120, Table-C, Plastic, left dial & clock, right grill, step-down top, side knobs, feet, BC, AC $40.00

250, Portable, Plastic, transistor, upper right round dial, lower perforated grill, BC, battery $250.00

260 Series, Portable, 1957, Leather case, transistor, round dial, side thumbwheel knob, handle, AM, battery $45.00

270, Portable, 1957, Leatherette, transistor, right round dial knob, plastic grill w/crest, BC, battery $35.00

278, Portable, 1958, Leather case, transistor, right front dial, left checkerboard grill, handle, AM, bat $35.00

300, Table-C, Plastic, center front square dial in middle of concentric squares, feet, 2 knobs **$40.00**

M-701, Cathedral-C, 1932, Wood, center front "rainbow" dial, upper grill w/round clock & vertical bars **$375.00**

BUSH & LANE
Bush and Lane Piano Company, Holland Michigan

11K, Console, 1930, Wood, lowboy, Deco, upper dial, lower square grill w/cut-outs, varigated veneers **$195.00**

34, Console, 1929, Wood, upper front dial, lower scalloped grill w/ cut-outs, stretcher base, AC **$145.00**

40, Console, 1929, Wood, inner front dial & knobs, double doors, stretcher base, AC **$165.00**

CALRAD
Burstein-Applebee Co.
1012-14 McGee St., Kansas City, Missouri

60A183, Portable, 1960, Transistor, left window dial, lower perforated grill, thumbwheel knobs, AM, bat **$35.00**

CAPEHART
Farnsworth Television & Radio Corp., Fort Wayne, Indiana

1P55, Portable, 1955, Plastic, center front round dial over horizontal grill bars, handle, BC, AC/DC/bat **$35.00**

2P56, Portable, 1956, Center front round dial over grill, top thumbwheel knob, handle, BC, AC/DC/bat **$30.00**

3T55E, Table, 1954, Plastic, right front round dial over large recessed woven grill, feet, BC, AC/DC **$25.00**

10, Portable, 1952, Plastic, upper front half-round dial over lattice grill, thumbwheel knob, handle **$30.00**

17RPQ155F, Console-R/P/Rec, 1955, Wood, inner center slide rule dial, right phono, left recorder, lift top, BC, FM, AC . . . **$65.00**

29P4, Console-R/P, 1949, Wood, inner right black dial, 5 knobs, left pull-out phono, double doors, BC, FM, AC **$90.00**

33P9, Console-R/P, 1949, Wood, inner right dial, 4 knobs, phono, lift top, grill w/cut-outs, storage, BC, FM, AC **$80.00**

75C56, Table-C, 1956, Right side dial knob, center front square alarm clock over recessed grill, BC, AC **$25.00**

88P66BNL, Portable, 1956, Leatherette, flip-up front w/map, telescope antenna, handle, 8 band, AC/DC/bat **$75.00**

115P2, Console-R/P, 1949, Wood, inner right dial, left lift top, woven front grill, 25 tubes, BC, SW, FM, AC **$275.00**

413P, Console-R/P, 1949, Wood, inner right dial, door, left lift top, inner phono, 28 tubes, BC, SW, FM, AC **$300.00**

1002F, Console-R/P, 1951, Wood, inner right slide rule dial, 4 knobs, inner left pull-out phono, BC, FM, AC **$90.00**

1006-M, Console-R/P, 1951, Wood, inner right slide rule dial, 4 knobs, left pull-out phono drawer, BC, AC **$70.00**

1007AM, Console-R/P, 1951, Wood, inner right slide rule dial, 4 knobs, left pull-out phono drawer, BC, FM, AC **$75.00**

C14, Table-C, Plastic, lower slide rule dial, large upper clock face, step-down top, feet, 4 knobs **$35.00**

P-213, Portable, 1954, Plastic & metal, top dial & 2 knobs, V-shaped plaid grill, handle, BC, AC/DC/bat **$45.00**

RP-152, Console-R/P, 1953, Wood, inner front slide rule dial, 4 knobs, phono, large double doors, BC, AC **$50.00**

T-30, Table, 1951, Plastic, right half-moon dial, center raised patterned grill, 2 knobs, BC, AC/DC **$40.00**

T-522, Table, 1953, Plastic, large front dial w/inner perforations, 2 knobs, slanted feet, BC, AC/DC **$40.00**

TC-20, Table-C, 1951, Plastic, right side dial knob, left front alarm clock, step-down top, BC, AC **$30.00**

TC-62, Table-C, 1953, Plastic, right side dial knob, left front alarm clock, step-down top, BC, AC **$30.00**

CAPITOL
Capitol Radio Corp.

UN-61 "Music Master", Table, 1948, Upper slanted slide rule dial, lower horizontal louvers, 2 knobs, BC, AC/DC **$40.00**

UN-72, Table, 1949, Two-tone wood, upper slanted slide rule dial, lower horizontal louvers, 4 knobs **$45.00**

UN-72P "High Fidelity Symphonic", Table-R/P, 1948, Wood, outer slanted slide rule dial, horiz louvers, lift top, inner phono, BC, SW **$35.00**

CARDINAL
Cardinal Radio Manufacturing Co., 2812 South Main St., Los Angeles, California

60, Cathedral, 1931, Wood, center front quarter-round dial, upper 4-section grill, 3 knobs **$275.00**

60, Tombstone, 1931, Wood, arched top, center quarter-round dial, upper grill w/cut-outs, 3 knobs **$150.00**

CARLOYD
Carloyd Electric & Radio Co., 342 Madison Ave., New York, New York

Mark II "Malone-Lemmon", Table, 1925, Wood, high rectangular case, 3 dial slanted front panel, lift top, 5 tubes, bat **$250.00**

CASE
Indiana Mfg. & Elec. Co., 570 Case Avenue, Marion, Indiana

60A, Table, 1926, Mahogany, low rectangular case, 2 dial front panel, lift top, 6 tubes, feet, bat **$95.00**

61C, Console, 1926, Two-tone walnut, center front dial, upper
  speaker grill w/cut-outs, battery **$150.00**
62C, Console, 1927, Wood, center front dial w/escutcheon, upper
  speaker grill w/cut-outs, AC **$150.00**
66A, Table, 1928, Wood, low rectangular case, center illuminated
  window dial w/escutcheon, AC **$115.00**
73B, Console, 1928, Wood, highboy, upper window dial w/escutch-
  eon, lower grill w/cut-outs, 3 knobs **$135.00**
73C, Console, 1928, Wood, lowboy, inner dial & knobs, double front
  doors, stretcher base, AC **$165.00**
90A, Table, 1927, Walnut, low rectangular case, center front dial, left
  side loop antenna, battery **$175.00**
90C, Console, 1927, Wood, lowboy, large case, inner dial & knobs,
  front doors, battery **$175.00**
500, Table, 1925, Mahogany, low rectangular case, 3 dial front panel,
  lift top, 5 tubes, battery **$95.00**
503, Table, 1926, Wood, low rectangular case, front panel w/3
  pointer dials, 6 tubes, battery **$100.00**
510, Upright Table, 1935, Wood, center front "Tell-Time Jumbo Dial",
  upper cloth grill with cut-outs **$135.00**
701, Console, 1926, Walnut, inner left 3 dial panel, right grill, fold-
  down front, fold-up top, battery **$215.00**
710, Upright Table, 1935, Wood, center front "Tell-Time Jumbo Dial",
  upper cloth grill with cut-outs **$135.00**
1017, Console, 1935, Wood, upper front round "Tell-Time Jumbo
  Dial", lower cloth grill w/cut-outs **$150.00**

CAVALIER
Hinners-Galanek Radio Corp.,
2514 Broadway,
Long Island City, New York

4CL4, Table-C, 1955, Plastic, right side dial knob, center front clock
  over checkered grill, BC, AC **$30.00**
5C1, Table-C, 1954, Plastic, right side dial knob, center front clock
  over checkered grill, BC, AC **$30.00**
603, Table, 1957, Plastic, right side dial, large perforated grill w/
  crown logo & "V", BC, AC/DC **$40.00**

CBS-COLUMBIA
CBS-Columbia Inc.,
3400 47th Avenue,
Long Island City, New York

515A, Table, 1953, Plastic, large right front round dial, left lattice grill,
  lower left knob **$35.00**
525, Portable, 1953, Plastic, lower right front dial knob, left lattice grill,
  thumbwheel knob, handle **$35.00**
541, Table-C, 1953, Plastic, right round dial, left alarm clock, center
  checkered panel, BC, AC **$30.00**
5165, Table, Plastic, large half-round metal dial w/inner horiz lines,
  large dial pointer, feet **$40.00**
5220, Portable, 1954, Plastic, right side dial knob, horizontal upper
  front bars, handle, BC, AC/DC/bat. **$25.00**
5440, Table-C, 1956, Plastic, right round dial, left round alarm clock,
  center panel, BC, AC **$30.00**

CHANCELLOR
Radionic Equipment Co.,
170 Nassau Street, New York, New York

35P, Portable, 1947, Leatherette, right dial, plastic front panel, 2
  knobs, handle, BC, AC/DC/battery **$25.00**

CHANNEL MASTER
Channel Master Corp.,
Ellenville, New York

6500, Portable, 1960, Ivory plastic, transistor, right dial over horiz grill
  bars, feet, handle, BC, bat **$25.00**
6501, Portable, 1959, Transistor, right front thumbwheel dial, left
  perforated grill, BC, battery **$35.00**
6505, Portable, 1960, Two-tone plastic, transistor, upper right dial,
  large lattice grill, handle, BC, bat **$35.00**
6506, Portable, 1960, Plastic, transistor, right front dial, left metal
  perforated grill, BC, battery **$40.00**
6507, Portable, 1960, Plastic, transistor, right dial, lattice grill, telescoping antenna, BC, SW, battery . . . . . , ... **$35.00**
6508, Portable, 1960, Plastic, transistor, upper window dial, perforated grill, swing handle, BC, bat **$35.00**

**6509, Portable, 1960, Plastic, transistor, upper window dial,
  perforated grill, swing handle, BC, bat $35.00**
6510, Portable, 1960, Two-tone plastic, transistor, upper right dial,
  large lattice grill, handle, BC, bat **$35.00**

**6511, Table, 1960, Plastic, transistor, right front slide rule dial,
  left horizontal bars, feet, BC, bat $40.00**
6512, Portable, 1960, Plastic, transistor, right front dial, left grill,
  detachable antenna, BC, SW, bat **$55.00**
6514, Portable, 1960, Plastic, transistor, right front dial, left grill,
  detachable antenna, BC, SW, bat **$55.00**
6515, Portable, 1960, Plastic, transistor, right front square dial, left
  perforated grill, BC, battery **$45.00**
6516, Portable, 1960, Plastic, transistor, upper window dial, perforated grill, swing handle, BC, battery **$35.00**
6517, Portable, 1960, Plastic, transistor, slide rule dial, perforated
  grill, built-in compass, BC, LW, bat **$50.00**
6518, Portable, 1960, Plastic, transistor, slide rule dial, perforated
  grill, telescoping antenna, AM, FM **$50.00**
6519, Portable, 1960, Transistor, slide rule dial, perforated grill,
  telescoping antenna, 3 band, battery **$50.00**

6520, Portable, 1960, Two-tone plastic, transistor, slide rule dial, lattice grill, handle, BC, battery **$35.00**
6521, Table-C, 1960, Transistor, slide rule dial, perforated grill, raised top alarm clock, BC, SW, bat **$40.00**
6523, Portable, 1960, Transistor, slide rule dial, perforated grill, telescoping antenna, BC, 2SW, bat **$55.00**
6524, Portable, 1960, Transistor, slide rule dial, perforated grill, dual antennas, handle, AM, FM, bat **$35.00**
6526, Portable, 1960, Transistor, slide rule dial, perforated grill, dual antennas, handle, AM, FM, bat **$35.00**
6527, Portable, 1960, Plastic, transistor, upper right window dial, lower lattice grill, BC, battery **$35.00**
6528, Portable, 1960, Two-tone plastic, transistor, right front window dial, lower lattice grill, BC, bat **$35.00**
6531, Portable, 1960, Plastic, transistor, upper right window dial, lower lattice grill, BC, battery **$35.00**
6532, Table, 1960, Two-tone plastic, center front dial, right & left speakers, 2 knobs, tubes, AM **$20.00**
6533, Table-C, 1960, Two-tone plastic, center alarm clock & dial, twin speakers, snooze bar, tubes, AM **$25.00**
6535, Table, 1960, Walnut, two right front slide rule dials, left lattice grill, 4 knobs, tubes, AM, FM **$25.00**
6536, Table-C, 1960, Walnut, right clock & slide rule dial, left lattice grill, 4 knobs, tubes, AM, FM **$25.00**

CHELSEA
Chelsea Radio Co.,
150 Fifth Street, Chelsea, Massachusetts

102, Table, 1923, Wood, high rectangular case, 2 dial black front panel, 3 tubes, battery **$120.00**
107 "Regenodyne", Table, 1925, Wood, low rectangular case, 2 dial front panel, 4 tubes, battery **$110.00**
122, Table, 1925, Wood, high rectangular case, slant front black bakelite panel, lift top, battery **$110.00**

Super Five, Table, 1925, Wood, slant front, 3 oval window dials, 5 bakelite knobs, curved sides, battery $110.00
Super Five, Table, 1925, Wood, three dials, slant front, wooden panel, 5 bakelite knobs, lift top, battery **$110.00**
Super Five, Table, 1925, Wood, metal front panel, three window dials, w/brass etcutcheons, 5 knobs, battery **$120.00**
Super Six, Table, 1925, Wood, high rectangular case, slant front, three window dials, curved sides, battery **$115.00**

CISCO
Cities Service Oil Co., New York, New York

1A5, Table, 1948, Wood, right front square dial, left horizontal louvers, 2 knobs, BC, AC/DC **$40.00**

9A5, Table, 1947, Plastic, right square dial, left horizontal louvers, 2 knobs, handle, BC, AC/DC **$75.00**

CLAPP-EASTHAM
Clapp-Eastham Company,
136 Main Street, Cambridge, Massachusetts

The Clapp-Eastham Company was formed in 1908 to manufacture X-ray and wireless parts and eventually produced complete radio sets. The company declined during the early 1920's and was out of business by 1929.

Baby Emerson, Table, 1927, Small case, 1 dial front panel, Emerson Multivalve, battery **$475.00**

DD Radak, Table, 1925, Leatherette or wood, low rectangular case, 2 dial black front panel, 3 tubes, bat $225.00
Gold Star, Table, Wood, low rectangular case, black front panel w/ 3 pointer dials and gold trim **$150.00**

HR, Table, 1922, Wood, high rectangular case, 2 dial black bakelite front panel, 1 tube, battery $325.00
HR/HZ, Table, 1922, Wood, 2 units w/black bakelite front panels, battery **$675.00**
R-3 Radak, Table, 1923, Wood, high rectangular case, black bakelite front panel, battery **$250.00**
R-4 Radak, Table, 1924, Wood, high rectangular case, black bakelite front panel, 1 tube, battery **$350.00**
RZ Radak, Table, 1922, Wood, rectangular case, black bakelite front panel, 3 tubes, battery **$600.00**

CLARION
Warwick Mfg. Corp.,
4640 West Harrison Street
Chicago, Illinois

61, Cathedral, 1931, Wood, small case, off-center window dial, upper scalloped grill w/cut-outs, AC $165.00
80, Tombstone, 1931, Wood, shouldered top, front half-moon dial, upper grill w/cut-outs, 3 knobs $175.00
81, Console, 1931, Wood, lowboy, upper front half-moon dial, lower cloth grill w/cut-outs, 7 tubes $135.00
90, Tombstone, 1931, Wood, shouldered top, half-moon dial, upper grill w/cut-outs, 3 knobs, 8 tubes $160.00
91, Console, 1931, Wood, lowboy, upper front half-moon dial, lower cloth grill w/cut-outs, 8 tubes $150.00
320 Clarion Jr, Tombstone, Wood, lower window dial w/escutcheon, upper cloth grill w/cut-outs, 3 knobs $90.00
400, Table, 1937, Wood, step-down top, right front dial, center grill w/cut-outs, 2 knobs, AC/DC $90.00
422, Table, 1933, Wood, rounded top, right front dial, center grill w/cut-outs, 5 tubes, AC/DC $75.00
450, Table, 1933, Wood, ribbed sides, right front dial, center grill w/cut-outs, 6 tubes, AC/DC $90.00
470, Cathedral, 1933, Wood, small case, center front window dial, upper grill w/cut-outs, 4 knobs, AC $175.00
691, Table, 1937, Two-tone wood, right front round telephone dial, left horizontal louvers, 4 knobs $75.00
770, Table, 1937, Wood, large right front dial, left cloth grill w/2 horizontal bars, 3 knobs, BC, SW $45.00
11011, Portable, 1947, Leatherette, slide rule dial, horizontal louvers, 2 knobs, handle, BC, AC/DC/bat $25.00
11305, Table-R/P, 1947, Wood, outer dial, 2 knobs, criss-cross grill, lift top, inner phono, BC, AC $30.00
11411-N, Portable, 1947, Inner right dial, lattice grill, 2 knobs, 1 switch, flip-up cover, BC, AC/DC/bat $35.00
11801, Table, 1947, Plastic, slanted lower slide rule dial, horizontal louvers, 2 knobs, BC, AC/DC $50.00
12110M, Console-R/P, 1949, Wood, right tilt-out dial, 4 knobs, left pull-out phono drawer, BC, FM, AC $70.00
12310W, Console-R/P, 1948, Wood, upper slanted slide rule dial, 5 knobs, pull-out phono drawer, BC, SW, AC $90.00
12708, Console-R/P, 1948, Wood, slide rule dial, criss-cross grill, 4 knobs, lift top, inner phono, BC, AC $50.00
12801, Table, 1949, Plastic, right square dial, left lattice grill, ridged base, 2 knobs, BC, AC/DC $50.00
13101, Table, 1948, Plastic, upper slanted slide rule dial, horiz grill bars, 4 knobs, AM, FM, AC/DC $40.00
13201, Table, 1949, Plastic, right square dial, left horizontal wrap-around louvers, 2 knobs, BC, bat $30.00
14601, Table, 1949, Plastic, right square dial, checkerboard grill, ridged base, 2 knobs, BC, AC/DC $50.00
14965, Table, 1949, Plastic, upper slanted slide rule dial, horizontal grill bars, 4 knobs, BC, FM, AC $40.00
AC-51, Console, 1929, Wood, lowboy, front window dial, lower scalloped grill, stretcher base, 3 knobs $125.00
AC-53, Console, 1929, Wood, lowboy, front window dial, lower scalloped grill, stretcher base, 3 knobs $125.00
C100, Table, 1946, Plastic, right square dial, left louvered wrap-around grill, 2 knobs, BC, AC/DC $45.00
C101, Table-R/P, 1946, Wood, front square dial, 3 knobs, horizontal louvers, inner phono, lift top, BC, AC $35.00
C102, Table, 1946, Plastic, slide rule dial, wrap-around horizontal grill bars, 2 knobs, BC, AC/DC $40.00
C103, Table, 1946, Plastic, upper slide rule dial, cloth grill w/horizontal bars, 4 knobs, BC, AC $40.00
C104, Table, 1946, Wood, upper curved dial, lower grill w/horizontal louvers, 4 knobs, BC, AC/DC $45.00
C105-A, Console-R/P, 1946, Wood, front slide rule dial, horizontal louvers, 4 knobs, doors, lift top, BC, AC $65.00
C108, Table, 1946, Wood, upper slanted slide rule dial, cloth grill w/harp cut-out, 2 knobs, BC, bat $50.00

AC-60 "Junior", Cathedral, 1932, Walnut, off-center window dial, upper scalloped cloth grill w/cut-outs, 3 knobs $175.00
TC-2, Console, 1934, Wood, Deco, front quarter-moon dial, lower grill w/geometric cut-outs, 3 knobs $130.00

CLARK'S
Clark's,
30 Boylston St., Boston, Massachusetts

Acme Reflex, Table, Wood, low rectangular case, 2 dial black front panel, lift top, battery $130.00

CLEARSONIC
U. S. Television Mfg. Co.,
3 West 61st Street, New York, New York

5C66, Table, 1947, Wood, right square dial, left cloth grill w/horizontal bars, 2 knobs, BC, AC/DC $35.00
5D66, Table-R/P, 1947, Wood, front dial, 2 knobs, criss-cross grill, lift top, inner phono, BC, AC/DC $35.00

CLEARTONE
Cleartone Radio Company,
2427 Gilbert Avenue, Cincinnati, Ohio

60 "Goldcrest", Table, 1925, Wood, low rectangular case, front panel w/3 pointer dials, 4 tubes, battery $100.00
70 "Clear-O-Dyne", Table, 1925, Wood, low rectangular case, front panel w/2 pointer dials, 4 tubes, battery $100.00
72 "Clear-O-Dyne", Console, 1925, Wood, inner panel w/2 pointer dials, drop front, built-in speaker, 4 tubes, battery . . . $200.00
80 "Super Clear-O-Dyne", Table, 1925, Wood, low rectangular case, front panel w/3 pointer dials, 5 tubes, battery . $115.00
82, Console, 1925, Wood, inner panel w/3 pointer dials, drop front, built-in speaker, 5 tubes, bat $200.00

100, Table, 1926, Wood, high rectangular case, slant front panel, lift top, 5 tubes, battery $125.00
Mayflower, Console, 1926, Wood, upper front dial w/escutcheon, lower built-in speaker w/scrolled grill, AC $175.00

CLEVELAND
Cleveland Products Co.,
714 Huron Rd., Cleveland, Ohio

A-5, Table, 1925, Wood, low rectangular case, 3 dial front panel, 5 tubes, battery $125.00

CLIMAX
Climax Radio & Tel. Co., Inc.,
513 South Sangamon Street, Chicago, Illinois

Emerald, Table, 1937, Walnut veneer, streamlined, right oval con-vex dial, left wrap-around grill bars $115.00
Ruby, Table, 1937, Walnut veneer, ultra streamlined, right oval dial, left horiz grill bars, tuning eye $225.00

CLINTON
Clinton Mfg. Co.,
1217 West Washington Boulevard,
Chicago, Illinois

216, Table, Wood, rounded sides, right dial, grill w/Deco cut-outs, 3 knobs, BC, SW, AC/DC $100.00
254, Portable, 1937, Leatherette, right front dial, left round grill, fold-open front door, handle $65.00
1102, Console, 1937, Wood, upper front large round dial, tuning eye, lower cloth grill w/vertical bars $135.00

COCA COLA
Point of Purchase Displays, Inc.

Coke Cooler, Table-N, 1949, Red plastic, looks like Coca Cola cooler, upper front slide rule dial, 2 knobs $800.00

COLBLISS
The Colbliss Radio Company,
827 South Hoover St., Los Angeles, California

500 "Petite", Cathedral, 1931, Wood, lower front window dial, upper scalloped grill w/cut-outs, 3 knobs $225.00

COLONIAL
Colonial Radio Corporation,
Buffalo, New York

16, Table, 1925, Wood, low rectangular case, 3 dial front panel, 5 tubes, battery $125.00
17, Table, 1925, Wood, low rectangular case, 2 dial front panel, 4 tubes, battery $125.00

39, Cathedral, 1931, Wood, lower front window dial, upper scalloped grill w/cut-outs, AC $175.00
41C, Grandfather Clock, 1931, Wood, front window dial, lower grill w/cut-outs, rectangular clock face $350.00
300, Table, 1933, Plastic, Deco, right dial, center circular chrome cut-outs, ribbed sides, feet $225.00
656, Tombstone, 1934, Wood, lower round dial, upper cloth grill w/cut-outs, 3 knobs, BC, SW, AC $150.00
New World, Table-N, 1933, Plastic, world globe on stand, front window dial, side knobs, top vents, AC/DC $925.00

T345, Cathedral, Wood, lower front quarter-round dial, upper cloth grill w/cut-outs, 3 knobs $175.00

COLUMBIA
CBS Electronics,
100 Endicott St., Danvers, Massachusetts

400B, Portable, 1960, Black plastic, transistor, right window dial, circular grill w/vert bars, AM, bat $20.00
400G, Portable, 1960, Grey plastic, transistor, right window dial, circular grill w/vert bars, AM, bat $20.00
400R, Portable, 1960, Red plastic, transistor, right window dial, circular grill w/vert bars, AM, bat $25.00
600BX, Portable, 1960, Plastic, transistor, right thumbwheel dial, circular perforated grill, AM, battery $25.00
600G, Portable, 1960, Plastic, transistor, right thumbwheel dial, circular perforated grill, AM, battery $25.00
610G "Transistor Convertible", Table/Portable, 1960, Grey plas-tic, transistor, radio unit detaches for use as portable, AM, battery $45.00
610R "Transistor Convertible", Table/Portable, 1960, Red plastic, transistor, radio unit detaches for use as portable, AM, battery $50.00

COLUMBIA PHONOGRAPH
Columbia Phonograph Company, Inc.,
55 Fifth Avenue, New York, New York

C-1, Table, 1928, Wood, center window dial w/escutcheon, lift top, 7 tubes, 3 knobs, switch, AC $130.00

C-81, Upright Table, 1932, Two-tone walnut, window dial, upper grill w/cut-outs & notes, 3 knobs, AC $185.00

C-83, Console, 1932, Walnut, lowboy, upper window dial, lower grill w/cut-outs & notes, 6 legs, AC $140.00

C-84, Console, 1932, Walnut, highboy, upper window dial, lower twin grills w/cut-outs, 6 legs, AC $150.00

C-85, Console-R/P, 1932, Walnut, center dial, lower grill w/cut-outs & notes, lift top, inner phono, AC $150.00

C-93, Console, 1932, Walnut, lowboy, upper front dial, lower grill w/ vertical bars & notes, 6 legs, AC $175.00

C-95, Console, 1932, Walnut, upper window dial, lower twin grills w/ cut-outs, 6 legs, 3 knobs, AC $150.00

C-103, Console, 1932, Walnut, upper front dial, lower grill w/scalloped cut-outs & notes, 3 knobs, AC $130.00

C-150, Table, Wood, right front dial, center cloth grill w/cut-outs, left volume, 2 knobs $65.00

COLUMBIA RECORDS
Columbia Records,
799 Seventh Avenue
New York, New York

530, Console-R/P, 1957, Wood, inner right dial & knobs, left phono, lift top, front grill, legs, BC, FM, AC $40.00

TR-1000, Portable, 1958, Leather case w/front grill, transistor, inner right dial, lift-up lid, strap, AM, bat $50.00

COMMONWEALTH

170, Cathedral, 1933, Wood, center front quarter-round dial, upper cloth grill w/cut-outs, 7 tubes $225.00

CONCERT
Concert Radiophone Co.,
626 Huron Rd., Cleveland, Ohio

Concert Grand, Table, 1925, Wood, low rectangular case, 3 dial front panel, 4 tubes, battery $110.00

Concert Sr., Portable, 1925, Leatherette, inner 2 dial panel, built-in loop antenna, handle, 2 tubes, battery $200.00

CONCORD
Concord Radio Company,
901 West Jackson Boulevard, Chicago, Illinois

1-403, Table, 1948, Plastic, streamline, right dial, left wrap-around louvers, 2 knobs, BC, AC/DC $65.00

1-411, Table, 1948, Plastic, streamline, half-moon dial, wrap-around louvers, 2 knobs, BC, AC/DC $50.00

1-504, Table, 1949, Plastic, upper slanted slide rule dial, horizontal grill bars, 3 knobs, BC, 2SW, AC $35.00

1-513, Portable, 1949, Plastic, right rectangular dial, patterned grill, handle, 3 knobs, BC, AC/DC/bat $35.00

1-514, Table, 1949, Plastic, right front airplane dial, left horizontal louvers, 2 knobs, AC/DC $35.00

1-516, Table-R/P, 1948, Leatherette, slanted front slide rule dial, lower grill, 2 knobs, open top, BC, AC $30.00

1-518, Portable, 1948, Two-tone case, right front oval recessed dial area, handle, AC/DC/battery $30.00

1-606, Table, 1948, Plastic, upper slanted slide rule dial, horiz grill bars, 3 knobs, BC, FM, AC/DC $35.00

1-611, Portable, 1948, Leatherette, slide rule dial, horizontal grill bars, handle, 2 knobs, BC, AC/DC/bat $30.00

6C51B, Table, 1947, Plastic, streamline, right dial, left wrap-around louvers, 2 knobs, BC, AC/DC $65.00

6D61P, Table-R/P, 1947, Wood, outer slide rule dial, 3 knobs, lift top, inner phono, BC, AC $35.00

6E51B, Table, 1947, Plastic, upper slanted slide rule dial, horiz louvers, 2 knobs, handle, BC, AC/DC $40.00

6F26W, Table, 1947, Wood, slanted slide rule dial, cloth grill, 4 knobs, burl veneer, BC, SW, AC/DC $40.00

6T61W, Table, 1947, Wood, right front square dial, left horizontal louvers, 3 knobs, BC, AC/DC $35.00

7E71PR, Portable, 1949, "Alligator", right front airplane dial, left grill, 2 knobs, handle, AC/DC/battery $30.00

CONTINENTAL
Spiegel, Inc.,
1061 West 35th Street, Chicago, Illinois

150, Portable, 1959, Transistor, right front dial w/thumbwheel tuning, left lattice grill, AM, battery $35.00

160, Portable, 1959, Transistor, upper left round dial, lower perforated grill, swing handle, AM, bat $40.00

1000, Table, Plastic, streamline, right front dial, left wrap-around horizontal bars, 2 knobs $95.00

C-45, Table-C, 1957, Plastic, center round dial knob, right circular louvers, left alarm clock, BC, AC $30.00

K6, Table, Plastic, right front dial, left wrap-around horizontal louvers, 3 knobs, BC, SW $40.00

Piano, Table-N, 1940, Radio shaped like grand piano, lift-up lid covers dial & G clef grill $350.00

R-20 "Star Raider", Console, 1929, Walnut finish, highboy, inner dial & knobs, double front doors, stretcher base, AC .$135.00

R-30 "Star Raider", Console, 1929, Wood, lowboy, inner front dial & knobs, double doors, stretcher base, AC $135.00

SW-7, Portable, 1960, Transistor, upper slide rule dial, perforated grill, swing handle, BC, 2SW, battery $35.00

TR-100, Portable, 1960, Transistor, window dial, lower round perforated grill, swing handle, AM, battery $25.00

TR-200, Portable, 1960, Transistor, upper right wedge-shaped dial, lower perforated grill, AM, battery $25.00

TR-208, Portable, 1959, Transistor, diagonally divided front, window dial, checkered grill, AM, battery $25.00

TR-215, Portable, 1960, Transistor, right front dial, left perforated grill, 2 thumbwheel knobs, AM, bat $30.00

TR-300, Portable, 1960, Transistor, right slide rule dial, left perforated grill, thumbwheel knobs, AM, bat $25.00

CONTINENTAL ELECTRONICS
Continental Electronics, Ltd.,
81 Pine Street, New York, New York

82 "Sky Weight", Table-R/P, 1947, Luggage-style, "alligator", lift top, inner phono, 3 knobs, handle, BC, AC **$25.00**

CO-OP
National Cooperatives Inc.,
343 South Dearborn, Chicago, Illinois

6A47WT, Table, 1949, Wood, lower slide rule dial, large recessed grill, 4 knobs, BC, SW, AC **$35.00**

CORONADO
Coronado,
Minneapolis, Minnesota/
Los Angeles, California

05RA1-43-7755A, Console-R/P, 1950, Wood, upper front slide rule dial, 4 knobs, pull-out phono drawer, BC, FM, AC **$50.00**

05RA1-43-7901A, Console-R/P, 1950, Wood, inner right slide rule dial, 4 knobs, left pull-out phono drawer, BC, FM, AC .. **$60.00**

05RA2-43-8230A, Table, 1952, Plastic, upper front slide rule dial, lower metallic grill, 2 knobs, BC, AC/DC **$35.00**

05RA2-43-8515A, Table, 1950, Plastic, lower center raised round dial, large recessed grill, 2 knobs, AM, FM, AC **$35.00**

05RA4-43-9876A, Portable, 1950, Leatherette, front slide rule dial, metal grill, handle, 2 knobs, BC, AC/DC/battery **$30.00**

05RA33-43-8120A, Table, 1950, Plastic, oblong case, right round dial, round grill w/horizontal bars, BC, AC/DC **$55.00**

05RA37-43-8360A, Table, 1950, Two-tone plastic, right raised round dial, lattice louvers, 2 knobs, BC, AC/DC **$40.00**

15RA1-43-7654A, Console-R/P, 1951, Wood, upper front slide rule dial, 4 knobs, lower pull-out phono, BC, FM, AC **$45.00**

15RA1-43-7902A, Console-R/P, 1951, Wood, inner right slide rule dial, 4 knobs, left pull-out phono drawer, BC, FM, AC .. **$70.00**

15RA2-43-8230A, Table, 1952, Plastic, upper front slide rule dial, lower metallic grill, 2 knobs, BC, AC/DC **$35.00**

15RA33-43-8246A, Table, 1952, Plastic, center front round dial over checkered grill, 2 knobs, BC **$30.00**

15RA33-43-8365, Table, 1952, Plastic, lower front round raised dial, large upper grill, 2 knobs, BC, AC **$35.00**

15RA37-43-9230A, Table-R/P, 1952, Wood, right front square dial, left grill, 2 knobs, left 2/3 lift top, inner phono, AC **$30.00**

35RA-43-9856A, Portable, 1953, Center front round dial over criss-cross grill, 2 knobs, handle, BC, AC/DC/bat **$30.00**

35RA4-43-9856A, Portable, 1953, Center front round dial over criss-cross grill, 2 knobs, handle, BC, AC/DC/bat **$30.00**

35RA33-43-8125, Table, 1953, Plastic, right front square dial, left horizontal grill bars, 2 knobs, BC, AC/DC **$30.00**

35RA33-43-8145 "Ranger", Table, 1953, Plastic, right front round dial, left lattice grill, 2 knobs, BC, AC/DC **$30.00**

35RA33-43-8225, Table, 1953, Plastic, right dial, left vertical grill bars, center horizontal strip, BC, AC/DC **$35.00**

35RA37-43-8355, Table, 1953, Plastic, recessed slide rule dial, upper recessed lattice grill, 3 knobs, BC, AC/DC **$35.00**

35RA40-43-8247A, Table-C, 1954, Plastic, square case, right side dial knob, large front alarm clock, BC, AC **$30.00**

43-6301, Table, 1946, Wood, upper slanted slide rule dial, cloth grill, 2 knobs, curved sides, BC, battery **$40.00**

43-6321, Table, 1948, Wood, lower front slide rule dial, large upper cloth grill, 2 knobs, BC, battery **$35.00**

43-6451, Table, 1946, Two-tone wood, upper slanted slide rule dial, large cloth grill, 2 knobs, BC, bat **$35.00**

43-6485, Table, 1948, Wood, upper slanted slide rule dial, large cloth grill, 2 knobs, BC, battery **$35.00**

43-6927, Console, 1948, Wood, upper slanted slide rule dial, large lower cloth grill area, 4 knobs, AC **$75.00**

43-6951, Console, 1948, Wood, upper slanted slide rule dial, large lower cloth grill, 4 knobs, AM/FM, AC **$75.00**

43-7601, Console-R/P, 1946, Wood, upper slide rule dial, 4 knobs, lower pull-out phono drawer, BC, SW, AC **$80.00**

43-7651, Console-R/P, 1946, Wood, inner slide rule dial, 4 knobs, pushbuttons, pull-out phono, BC, 4 SW, AC **$85.00**

43-7652, Console-R/P, 1948, Wood, inner right slide rule dial & pushbuttons, left pull-out phono, BC, SW, AC **$85.00**

43-7851, Console-R/P, 1948, Wood, inner right slide rule dial, 4 knobs, left pull-out phono drawer, BC, SW, AC **$70.00**

43-8160, Table, 1947, Plastic, center round dial, circular louvers around dial, 2 knobs, base, BC, AC/DC **$90.00**

43-8178, Table, 1947, Plastic, right half-moon dial, wrap-around horizontal bands, 2 knobs, BC, AC/DC **$65.00**

43-8180, Table, 1946, Plastic, right dial, left horizontal wrap-around louvers, 2 knobs, BC, AC/DC **$45.00**

43-8190, Table, 1947, Plastic, contrasting plastic oblong encircles 2 knobs & grill, handle, BC, AC/DC **$200.00**

43-8213, Table, 1946, Wood, upper slanted slide rule dial, lower criss-cross grill, 2 knobs, BC, AC/DC **$40.00**

43-8240, Table, 1947, Plastic, upper slanted slide rule dial, lower horiz louvers, 2 knobs, BC, AC/DC **$45.00**

43-8305, Table, 1946, Plastic, lower slide rule dial, upper cloth grill, 2 knobs, curved base, BC, AC/DC **$55.00**

43-8312, Table, 1946, Plastic, lower slide rule dial, large upper cloth grill, curved base, BC, AC/DC **$55.00**

43-8312A, Table, 1946, Plastic, lower slide rule dial, large upper cloth grill, curved base, BC, AC/DC **$55.00**

43-8330, Table, 1947, Wood, lower slide rule dial, recessed horizontal louvers, 2 knobs, BC, AC/DC **$45.00**

43-8351, Table, 1947, Plastic, slide rule dial, horizontal louvers, 2 knobs, 6 pushbuttons, BC, AC/DC **$50.00**

43-8354, Table, 1947, Plastic, slide rule dial, cloth grill w/2 bars, 2 knobs, 5 pushbuttons, BC, AC/DC **$45.00**

43-8420, Table, 1947, Wood, lower slide rule dial, cloth grill w/2 horizontal bars, 2 knobs, BC, AC/DC **$35.00**

43-8470, Table, 1946, Wood, lower slide rule dial, cloth grill w/2 horizontal bars, 2 knobs, BC, AC/DC **$35.00**

43-8471, Table, 1946, Wood, lower slide rule dial, cloth grill w/2 horizontal bars, 2 knobs, BC, AC/DC **$35.00**

43-8576, Table, 1946, Wood, lower slide rule dial, large upper cloth grill, 4 knobs, BC, SW, AC **$35.00**

43-8685, Table, 1947, Wood, lower front slide rule dial over cloth grill, 3 knobs, BC, AC **$40.00**

43-9196, Table-R/P, 1947, Wood, front slide rule dial, horizontal louvers, 4 knobs, 3/4 lift top, BC, AC **$35.00**

43-9201, Table-R/P, 1947, Wood, front slide rule dial, large grill, 2 knobs, 3/4 lift top, inner phono, BC, AC **$35.00**

43-9865, Portable, 1948, Inner lower right dial, lattice grill, flip-up front, handle, BC, AC/DC/battery **$40.00**

94RA1-43-6945A, Console, 1949, Wood, upper front slide rule dial, large lower cloth grill, 4 knobs, BC, FM, AC **$70.00**

94RA1-43-7605A, Console-R/P, 1949, Wood, upper front slide rule dial, 4 knobs, tilt-out phono, BC, SW, AC **$70.00**

94RA1-43-7656A, Console-R/P, 1949, Wood, inner right slide rule dial, 4 knobs, left pull-out phono drawer, BC, FM, AC .. **$80.00**

94RA1-43-7751A, Console-R/P, 1950, Wood, upper dial, 4 knobs, front pull-out phono drawer, lower grill, BC, FM, AC ... **$65.00**

94RA1-43-8510A, Table, 1949, Plastic, upper slanted slide rule dial, off-center lattice grill, 4 knobs, BC, FM, AC **$45.00**

94RA1-43-8510B, Table, 1949, Plastic, upper slanted slide rule dial, off center lattice grill, 4 knobs, BC, FM, AC **$45.00**

94RA1-43-8511B, Table, 1949, Plastic, upper slanted slide rule dial, off center lattice grill, 4 knobs, BC, FM, AC **$45.00**

94RA4-43-8130A, Table, 1949, Plastic, lower slide rule dial, upper horizontal louvers, 2 knobs, BC, AC/DC **$30.00**

94RA31-43-8115A "Cub", Table, 1950, Plastic, right round dial, lattice wrap-around grill, 2 knobs, BC, AC/DC **$45.00**

94RA31-43-9841A, Portable, 1949, Inner slide rule dial, 2 knobs, fold-down front panel, handle, BC, AC/DC/bat **$35.00**

94RA33-43-8130C, Table, 1949, Plastic, lower front dial, large upper grill w/horizontal bars, 2 knobs, BC **$30.00**

675, Tombstone, 1934, Wood, front airplane dial w/center hemispheres, 6 tubes, BC, SW **$100.00**

686, Table, 1935, Wood, right front oval dial, left cloth grill w/ horizontal bars, 4 knobs, BC, 2SW **$75.00**

RA12-8121-A, Table, 1957, Plastic, right front round dial, lower knob, left horizontal bars, BC, AC/DC **$20.00**

RA37-43-9240A, Table-R/P, 1955, Wood, right side dial, large front grill, lift top, inner phono, BC, AC **$25.00**

RA37-43-9855, Portable, 1954, Plastic, right top dial knob, front vertical grill bars, handle, BC, AC/DC/battery **$30.00**

RA42-9850A, Portable, 1955, Plastic, large center front round dial over horizontal grill bars, handle, BC, bat **$35.00**

RA48-8157A, Table, 1958, Plastic, right front half-moon dial, horizontal grill bars, feet, BC, AC/DC **$20.00**

RA48-8158A, Table, 1958, Plastic, lower front slide rule dial, large upper grill, 2 knobs, feet, BC, AC/DC **$30.00**

RA48-8159A, Table, 1958, Plastic, lower front slide rule dial, large upper grill, 2 knobs, feet, BC, AC/DC **$30.00**

RA48-8342A, Table, 1957, Plastic, recessed front, right dial over lattice grill pattern, feet, BC, AC/DC **$30.00**

RA48-8351A, Table, 1957, Plastic, right side dial knob, front perforated grill w/crest & "V", BC, AC/DC **$45.00**

RA48-8352A, Table, 1957, Plastic, lower slide rule dial, upper grill w/ center divider, 2 knobs, BC, AC/DC **$35.00**

RA48-9898A, Portable, 1959, Leather case, transistor, right dial, lattice grill cut-outs, handle, AM, battery **$30.00**

RA48-9903A "66", Portable, 1960, Plastic, transistor, right front dial, horizontal bars, handle, AM, battery **$25.00**

RA48-9905A, Portable, 1960, Plastic, transistor, right front dial, horizontal bars, handle, AM, battery **$25.00**

RA50-9900A, Portable, 1960, Plastic, transistor, upper right dial, perforated grill, thumbwheel knobs, AM, bat **$20.00**

RA50-9902A, Portable, 1960, Plastic, transistor, wedge-shaped right dial over perforated grill, AM, battery **$30.00**

CORONET
Crystal Products, Co.,
1519 McGee Trafficway,
Kansas City, Missouri

Boy's Radio, Portable, Transistor, plastic case w/metal grill, "crown" logo/dial, screw-in antenna, bat **$25.00**

C-2, Table, 1946, Two-tone, slide rule dial & 2 knobs on top of case, cloth grill w/ bars, BC, AC/DC **$50.00**

COTO-COIL
Coto-Coil Co.,
87 Willard Ave., Providence, Rhode Island

Coto Symphonic, Table, 1925, Wood, low rectangular case, 3 dial front panel, 4 tubes, battery **$150.00**

CROSLEY
The Precision Equipment Company, Inc.,
Peebles Corner, Cincinnati, Ohio
Crosley Manufacturing Company, Cincinnati, Ohio
The Crosley Radio Corporation, Cincinnati, Ohio
Crosley Corporation,
1329 Arlington Street, Cincinnati, Ohio

The Crosley company was begun in 1921 by Powel Crosley, who believed his radio company should offer the consumer a good quality product at the lowest possible price and he called his sets the "Model T's" of radio. In 1923, Crosley bought out The Precision Equipment Company. The Crosley Radio Corporation enjoyed success until the late 1920's. One of the most sought-after of Crosley sets today is the "Pup", a small 1 tube set made to sell for only $9.75 in 1925 and known as the "Sky Terrier".

1-N "Litlfella", Cathedral, 1932, Wood, small case, center window dial, 3-section grill, scalloped base, 3 knobs, AC **$190.00**

4-29, Table, 1926, Two-tone wood, high rectangular case, slanted 2 dial panel, wood lid, 4 tubes, bat $150.00

4-29, Portable, 1926, Leatherette, inner 2 dial front panel, fold-back top, fold-down front, handle, bat **$185.00**

5-38, Table, 1926, Wood, 2 styles, high rectangular case, slanted front panel w/3 dials, 5 tubes, bat **$100.00**

5-50, Table, 1926, Wood, high rectangular case, right front thumbwheel dial, 5 tubes, battery **$85.00**

5-75, Console, 1926, Mahogany, highboy, upper right front dial, built-in speaker, 5 tubes, battery **$160.00**

5M3, Tombstone, 1934, Wood, center front window dial w/metal escutcheon, upper grill w/cut-outs, AC **$135.00**

6-60, Table, 1927, Wood, high rectangular case, slanted front panel, right thumbwheel dial, battery **$85.00**

6H2, Tombstone, 1933, Wood, center front round dial, upper cloth grill w/cut-outs, 3 knobs, BC, SW, AC **$110.00**

8H1, Console, 1934, Wood, upper front round dial, lower cloth grill w/ cut-outs, BC, SW, AC **$125.00**

9-101, Table, 1949, Plastic, upper curved slide rule dial, "boomer-ang" front louvers, 3 knobs, BC, bat **$55.00**

9-102, Table, 1948, Plastic, upper curved slide rule dial, "boomer-ang" front louvers, 3 knobs, BC, AC **$55.00**

9-103, Table, 1949, Plastic, lower slide rule dial, large perforated grill area, 3 knobs, BC, AC/DC **$40.00**

9-104W, Table, 1949, Plastic, lower slide rule dial, large upper grill area w/crest, 3 knobs, BC, AC/DC **$40.00**

9-105, Table, 1949, Plastic, lower slide rule dial, large upper grill area, 3 knobs, BC, SW, AC/DC **$40.00**

9-106W, Table, 1949, Plastic, lower slide rule dial, large upper grill w/crest, 3 knobs, BC, SW, AC/DC **$40.00**

9-113, Table, 1949, Plastic, lower slanted slide rule dial, metal perforated grill, 2 knobs, BC, AC/DC **$40.00**

9-114W, Table, 1949, Plastic, lower slanted slide rule dial, metal perforated grill, 2 knobs, BC, AC/DC **$40.00**

9-117, Table, 1948, Plastic, large right square dial, left vertical wrap-over grill bars, 3 knobs, bat **$35.00**

9-118W, Table, 1948, Plastic, upper curved slide rule dial, "boomer-ang" front louvers, 3 knobs, BC, AC **$55.00**

9-119, Table, 1948, Plastic, right round dial, left cloth grill w/horizon-tal bars, 2 knobs, BC, AC/DC **$45.00**

9-120W, Table, 1948, Plastic, right round dial, left cloth grill w/ horizontal bars, 2 knobs, BC, AC/DC **$45.00**

9-121, Table, 1949, Plastic, raised top, slide rule dial, horiz wrap-around louvers, 2 knobs, BC, AC/DC **$45.00**

9-122W, Table, 1949, Plastic, raised top, slide rule dial, horiz wrap-around louvers, 2 knobs, BC, AC/DC **$45.00**

9-201, Console-R/P, 1948, Wood, 2 upper slide rule dials, center pull-out phono, lower grill, BC, FM, AC **$90.00**

9-202M, Console-R/P, 1948, Wood, inner right slide rule dial, 4 knobs, left pull-out phono drawer, AM, FM, AC **$75.00**

9-203B, Console-R/P, 1948, Wood, inner right dial, left pull-out phono, lower storage & grill, BC, FM, AC **$70.00**

9-204, Console-R/P, 1949, Wood, inner right dial, left pull-out phono, lower storage & grill, BC, FM, AC **$70.00**

9-205M, Console-R/P, 1949, Wood, right tilt-out slide rule dial, 4 knobs, pull-out phono drawer, BC, FM, AC **$70.00**

9-207M, Console-R/P, 1949, Wood, tilt-out slide rule dial, 5 knobs, left pull-out phono drawer, BC, SW, FM, AC **$75.00**

9-209, Console-R/P, 1949, Walnut, inner slide rule dial, pull-out phono drawer, criss-cross grill, BC, AC **$75.00**

9-209L, Console-R/P, 1949, Walnut, inner slide rule dial, pull-out phono drawer, criss-cross grill, BC, AC **$75.00**

9-212M, Console-R/P, 1949, Mahogany, inner slide rule dial, pull-out phono drawer, criss-cross grill, BC, AC **$75.00**

9-212ML, Console-R/P, 1949, Mahogany, inner slide rule dial, pull-out phono drawer, criss-cross grill, BC, AC **$75.00**

9-213B, Console-R/P, 1949, Blonde, inner slide rule dial, pull-out phono drawer, criss-cross grill, BC, AC **$75.00**

9-214M, Console-R/P, 1949, Wood, right tilt-out slide rule dial, 4 knobs, pull-out phono drawer, BC, FM, AC **$70.00**

9-214ML, Console-R/P, 1949, Wood, right tilt-out slide rule dial, 4 knobs, pull-out phono drawer, BC, FM, AC **$70.00**

9-302, Portable, 1948, "Alligator", metal front panel w/ slide rule dial, handle, 2 knobs, BC, AC/DC/bat **$35.00**

10-135, Table, 1950, Plastic, center circular dial with inner perforated grill, chrome trim, 2 knobs **$100.00**

10-136E, Table, 1950, Plastic, center front circular dial with inner perforated grill, 2 knobs **$100.00**

10-137, Table, 1950, Plastic, center front circular dial with inner perforated grill, 2 knobs **$100.00**

10-138, Table, 1950, Plastic, center front circular dial w/inner perfo-rated grill, 2 knobs, BC, AC/DC **$100.00**

10-145M, Table-R/P, 1949, Wood, center round dial, right & left grills, 3 knobs, lift top, inner phono, AC **$40.00**

10-307M, Portable, 1949, Plastic, upper slide rule dial, horiz grill bars, handle, 2 knobs, BC, AC/DC/bat **$35.00**

11, Table, Plastic, right front dial, left horizontal wrap-around louvers, 2 knobs, BC **$45.00**

11-100U, Table, 1951, Plastic, large center round dial w/inner circular louvers, side "fins", 2 knobs **$115.00**

11-101U, Table, 1951, Plastic, large round dial w/inner circular louvers, side "fins", 2 knobs, AC/DC **$115.00**

11-102U, Table, 1951, Plastic, large center round dial w/inner circular louvers, side "fins", 2 knobs **$115.00**

11-104U, Table, 1951, Plastic, large center round dial w/round louvers, side "fins", 2 knobs, BC, AC/DC **$115.00**

11-107U, Table, 1952, Plastic, center front dial w/inner checkered grill & crest, 2 knobs, BC, AC/DC **$60.00**

11-109U, Table, 1952, Plastic, center front dial w/inner checkered grill & crest, 2 knobs, BC, AC/DC **$60.00**

11-114U, Table, 1951, Plastic, round dial knob, left circular louvers w/horiz center ridge, BC, AC/DC **$100.00**

11-116U, Table, 1951, Plastic, round dial knob, left circular louvers w/horiz center ridge, BC, AC/DC **$100.00**

11-126U, Table, 1951, Plastic, right front dial, left round grill w/crest, 3 knobs, AM, FM, AC/DC **$45.00**

11-207MU, Console-R/P, 1951, Wood, inner right slide rule dial, 3 knobs, left pull-out phono drawer, BC, AC **$50.00**

11-301U, Portable, 1951, Plastic, flip up semi-circular front w/crest, inner dial, handle, BC, AC/DC/bat **$45.00**

11-550MU, Console-R/P, 1951, Wood, inner front round metal dial, lower pull-out phono, double doors, BC, AC **$75.00**

11AB, Table, Plastic, right front dial, left horizontal wrap-around louvers, base, 2 knobs **$50.00**

14AG, Table, Wood, right front square dial, left grill w/horiz bars, handle, 3 knobs, AC/DC **$55.00**

15-16, Console, 1937, Wood, large upper round dial, lower cloth grill w/3 vert bars, 15 tubes, 5 knobs **$175.00**

16AL, Table, 1940, Walnut, right front dial, pushbuttons, left vertical grill bars, 3 knobs, AC/DC $65.00

20AP "Fiver", Table, Wood, right front square dial, left cloth grill w/ diagonal bars, 3 knobs, BC, SW $45.00

21, Table, 1929, Metal, rectangular case, center window dial, Deco corner details, 3 knobs, bat $100.00

22, Console, 1929, Walnut veneer, highboy, window dial, lower round grill w/cut-outs, 3 knobs, bat $130.00

22CB, Console, 1941, Wood, upper front slanted "Giant Circle" dial, pushbuttons, BC, SW, FM, AC $200.00

50/50A, Table, 1925, Wood, low rectangular case, receiver/ amplifier in one unit, lift top, 3 tubes, bat $250.00

50 Portable, Portable, 1924, Wood, inner 1 dial black front panel, fold-back top, fold-down front, storage, bat $300.00

51, Table, 1924, Wood, low rectangular case, 1 dial black front panel, lift top, 2 tubes, battery $125.00

51A, Table-Amp, 1924, Wood, box, 2-stage amplifier $150.00

25AY, Console, 1940, Walnut, upper slanted dial, pushbuttons, lower vertical grill bars, BC, SW, AC $135.00

28AZ "Recordola", Console-R/P/Rec/PA, 1940, Wood, inner dial, phono, recorder, PA system, front grill w/3 vertical bars, AC $110.00

31, Table, 1929, Metal, rectangular case, center window dial, Deco corner details, 3 knobs, AC $100.00

33-BG, Table-R/P/Rec, 1940, Wood, outer right front square dial, left horizontal louvers, 3 knobs, lift top, AC $35.00

33-S, Console, 1929, Wood, lowboy, upper front window dial, lower grill w/cut-outs, 3 knobs, AC $110.00

34-S, Console, 1929, Wood, lowboy, inner dial & controls, double front doors, AC $125.00

35AK, Table-R/P, 1941, Wood, right front square dial, left grill, 3 knobs, lift top, inner phono, AC $30.00

41, Table, 1929, Metal, rectangular case, center window dial, Deco corner details, 3 knobs, AC $100.00

43BT, Table, 1940, Wood, rectangular case, right front square dial, left horizontal louvers, 3 knobs $45.00

46FB, Table, 1947, Wood, square center dial, side cloth wrap-around grills, 3 knobs, BC, SW, battery $40.00

48 "Widgit", Cathedral, 1931, Repwood case, ornate "carved" front, upper grill w/cut-outs, 2 knobs, 5 tubes $375.00

50, Table, 1924, Wood, 1 dial black bakelite front panel, lift top, 1 tube, battery $120.00

50A, Table-Amp, 1925, Wood, amplifier, black bakelite front panel, lift top $135.00

51 Portable, Portable, 1924, Leatherette, inner 1 dial front panel, fold-back top, fold-down front, handle, bat $165.00

51-S "Special", Table, 1924, Wood, slanted 1 dial black bakelite front panel, lift top, battery $115.00

51SD "Special Deluxe", Table, 1924, Wood, high rectangular case, left front half-moon pointer dial, lift top, battery $140.00

52, Table, 1924, Wood, low rectangular case, 1 dial black front panel, lift top, 3 tubes, battery **$125.00**

52P, Portable, 1924, Leatherette, inner 1 dial front panel, fold-down front, handle, battery **$175.00**

52-S "Special", Table, 1924, Wood, high rectangular case, slant front panel with left dial, lift top, battery **$130.00**

52SD "Special Deluxe", Table, 1924, Wood, high rectangular case, slanted panel, left half-moon pointer dial, battery **$130.00**

52TA, Table, 1941, Walnut, right front square dial, left cloth grill w/2 horizontal bars, AC/DC **$40.00**

54G "New Buddy", Table, 1930, Repwood "carved" front panel, lower right dial, upper cloth grill, 2 knobs, AC **$250.00**

56FA, Table, 1948, Plastic, large center dial, horizontal wrap-around louvers, 4 knobs, BC, SW, bat **$45.00**

56PA, Portable, 1946, Plastic, slide rule dial, horizontal louvers, 2 top knobs, handle, BC, AC/DC/bat **$45.00**

56PB, Portable, 1946, Upper front slide rule dial, lower horizontal louvers, handle, AC/DC/battery **$45.00**

56TC, Table, Wood, right front square dial, left grill w/horizontal bars, 3 knobs, BC, SW **$40.00**

56TD "Duette", Table, 1947, Plastic, modern, top slide rule dial, lower vertical grill bars, 3 knobs, BC, AC/DC **$125.00**

56TD-R "Duette", Table, 1947, Plastic, modern, top slide rule dial, lower vertical grill bars, 3 knobs, BC, AC/DC...**$125.00**

56TD-W "Duette", Table, 1947, Plastic, modern, top slide rule dial, lower front vertical grill bars, 3 knobs **$125.00**

56TG, Table, 1946, Plastic, upper front slide rule dial, horizontal louvers, 2 knobs, BC, AC/DC **$45.00**

56TJ, Table, 1946, Wood, upper slanted slide rule dial, center cloth grill, 2 knobs, BC, AC/DC **$35.00**

56TN, Table, 1948, Wood, right dial, left wrap around grill w/horiz bars, 3 knobs, BC, SW, AC/DC **$40.00**

56TN-L, Table, 1946, Wood, right dial, left wrap around grill w/horiz bars, 3 knobs, BC, SW, AC/DC **$40.00**

56TP, Table-R/P, 1946, Wood, right front black dial, cloth wrap-around grill, 3 knobs, lift top, BC, SW, AC **$35.00**

56TP-L, Table-R/P, 1948, Wood, right front black dial, cloth wrap-around grill, 3 knobs, lift top, BC, SW, AC **$35.00**

56TR, Table-R/P, 1948, Wood, center front slide rule dial, right/left wrap-around grills, lift top, BC, AC **$35.00**

56TS, Table-R/P, 1947, Wood, front slide rule dial, wrap-around cloth grills, 3 knobs, lift top, BC, SW, AC **$35.00**

56TU, Table, 1946, Plastic, upper slide rule dial, wrap-around louvers, 2 knobs, handle, BC, AC/DC **$55.00**

56TU-O, Table, 1949, Plastic, upper slide rule dial, wrap-around louvers, 2 knobs, handle, BC, AC/DC **$55.00**

56TV-O, Table, 1949, Wood, upper slide rule dial, large center front grill, BC, AC/DC **$40.00**

56TX, Table, 1946, Plastic, right front square dial, horizontal wrap-around grill bars, base, 3 knobs **$50.00**

56TY, Table, 1948, Wood, upper slide rule dial, lower horizontal grill bars, 2 knobs, BC, AC/DC **$50.00**

56TZ, Table-R/P, 1948, Two-tone wood, inner slide rule dial, 3 knobs, phono, 3/4 lift top, BC, AC **$30.00**

57TK, Table, 1948, Plastic, upper slide rule dial, lower vertical grill bars, 2 knobs, BC, AC/DC **$45.00**

57TL, Table, 1948, Plastic, upper slide rule dial, lower vertical grill bars, 2 knobs, BC, AC/DC **$45.00**

58 "Buddy Boy", Cathedral, 1931, Ornate Repwood case, thumbwheel tuning, cloth grill w/cut-outs, 2 knobs, AC $350.00
58TK, Table, 1948, Plastic, right round dial, cloth grill w/horizontal bars, 2 knobs, BC, AC/DC $35.00
58TL, Table, 1948, Plastic, upper slanted slide rule dial, vertical grill bars, 2 knobs, BC, AC/DC $45.00

58TW, Table, 1948, Plastic, slide rule dial, wrap around louvers, lucite handle, 2 knobs, BC, AC/DC............................$55.00
59 "Oracle", Grandfather Clock, 1931, Wood, window dial, ornate grill, raised top, Deco clock face, 3 knobs $450.00
59 "Show Boy", Cathedral, 1931, Ornate Repwood case, center window dial, upper grill area, 3 knobs, 5 tubes, AC $250.00
61, Tombstone, 1934, Wood, shouldered, center round dial, upper grill w/cut-outs, 3 knobs, BC, SW, AC $135.00

62-148, Tombstone, Wood, lower front round airplane dial, upper grill w/cut-outs, inlay, 2 knobs, AC $130.00
63TA "Victory", Table, 1946, Wood, large red, white & blue dial w/ stars & stripes, wrap-around grills, 4 knobs $150.00
64 MD, Console, 1934, Wood, upper front round dial, lower cloth grill w/cut-outs, BC, SW, AC...$110.00
66CS, Console-R/P, 1947, Wood, large inner dial, 4 knobs, phono, lift top, 2 horizontal grill bars, BC, SW, AC $75.00
66-T, Table, 1946, Wood, large center square dial, right & left wrap-around grills, 4 knobs, BC, SW $55.00
66TW, Table, 1946, Plastic, large center square black dial, 4 knobs, wrap-around louvers, BC, SW, AC $50.00

68CP, Console-R/P, 1948, Wood, inner slide rule dial, 4 knobs, phono, lift top, criss-cross grill, BC, SW, AC $80.00
68CR, Console-R/P, 1948, Wood, slide rule dial, center pull-out phono, lower grill, 4 knobs, BC, SW, AC $90.00
68TA, Table, 1948, Plastic, upper curved slide rule dial, front "boomerang" louvers, 3 knobs, BC, AC $55.00
68TW, Table, 1948, Plastic, upper curved slide rule dial, front "boomerang" louvers, 3 knobs, BC, AC $55.00
72AF, Tombstone, 1934, Wood, shouldered, lower round dial, upper cloth grill w/cut-outs, 3 knobs, AC $120.00
75, Table, Wood, low rectangular case, 3 upper front window dials, lift top $115.00
82-S, Console, 1929, Wood, highboy, inner dial & controls, double front doors, stretcher base, AC $130.00
86CR, Console-R/P, 1947, Walnut, top slanted dial, center pull-out phono, lower grill, storage, BC, SW, AC $125.00
86CS, Console-R/P, 1947, Mahogany, top slanted dial, center pull-out phono, lower grill, storage, BC, SW, AC $125.00
87CQ, Console-R/P, 1948, Wood, slanted slide rule dial, 4 knobs, pull-out phono drawer, BC, SW, FM, AC $110.00
88CR, Console-R/P, 1948, Wood, inner right dial & knobs, left pull-out phono, criss-cross grill, BC, SW, AC $85.00
88TA, Table, 1948, Plastic, upper curved slide rule dial, lower "boomerang" louvers, 3 knobs, BC, FM $55.00
88TC, Table, 1948, Wood, upper curved slide rule dial, lower cloth grill, 3 knobs, BC, FM, AC/DC $45.00

122 "Super Buddy Boy", Cathedral, 1931, Repwood case, lower window dial, upper sectioned grill, 3 knobs, 7 tubes, AC $350.00
124, Cathedral, 1931, Wood, low case, center window dial, 3 section grill, 3 knobs, fluted columns $250.00
124 "Playtime", Grandfather Clock, 1931, Wood, center front window dial, lower grill w/cut-outs, 3 knobs, BC, SW, AC $375.00
125, Cathedral, 1932, Wood, center front window dial, upper 3 section grill, side columns, 3 knobs $190.00
125, Console, Wood, upper window dial w/escutcheon, lower cloth grill w/cut-outs, 3 knobs $130.00
127, Tombstone, 1931, Wood, window dial, upper grill w/cut-outs, scalloped base, curved top, 3 knobs $260.00
129, Cathedral, Wood, center front window dial w/escutcheon, upper grill w/cut-outs, 3 knobs $250.00
140, Table-N, 1931, Looks like set of books, inner left window dial, right grill, 2 knobs, double doors $275.00
141 "Library Universal", Table-N, 1931, Looks like set of books, inner right window dial, center round grill, double doors $275.00
146, Cathedral, Wood, center window dial, upper 3 section cloth grill, fluted columns, 4 knobs $240.00

146CS, Console-R/P, 1947, Wood, inner half-round dial, pushbuttons, pull-out phono drawer, BC, 2SW, FM, AC **$95.00**
148CP, Console-R/P, 1948, Wood, inner half-round dial, pushbuttons, pull-out phono drawer, BC, 2SW, FM, AC **$95.00**
154, Cathedral, 1933, Wood, center window dial w/escutcheon, upper cloth grill with cut-outs, 2 knobs **$200.00**
157, Cathedral, 1933, Two-tone wood, center window dial, upper cloth grill w/cut-outs, 3 knobs, AC **$275.00**
160, Cathedral, Wood, center front window dial, upper cloth grill w/cut-outs, 4 knobs **$265.00**

167, Cathedral, 1936, Wood, center window dial, upper grill w/cut-outs, 3 knobs, 5 tubes, BC, SW, AC $210.00
167, Console, 1936, Wood, upper front round dial, lower cloth grill w/vertical bars, BC, SW, AC **$130.00**
169, Cathedral, 1934, Wood, center front window dial, upper sectioned cloth grill, 2 knobs, AC **$200.00**

176 "Travette", Table, Metal, Deco case design, right dial, center 4-section grill, 2 octagonal knobs $95.00
179, Tombstone, Wood, Deco, shouldered, center window dial, upper sectioned grill, silver flutings **$250.00**
182 "Travette Moderne", Table, Wood, Deco, right dial, center grill w/chrome cut-outs, 2 knobs, 5 tubes, AC/DC **$140.00**
401 "Bandbox Jr", Table, 1928, Metal, low rectangular case, center window dial w/escutcheon, battery **$80.00**

495, Table, 1936, Wood, Deco, off-center round dial, left oblong grill w/Deco cut-outs, BC, SW, AC **$90.00**
515, Tombstone, 1934, Two-tone wood, lower airplane dial, grill w/cut-outs, 4 octagonal knobs, BC, SW **$125.00**
516, Tombstone, 1936, Wood, small case, lower red, white & blue dial, upper grill, 4 knobs, BC, SW, AC **$85.00**
527A, Table, C1938, Two-tone wood, right front square dial, left grill w/horizontal bars, 2 knobs **$45.00**
555, Tombstone, 1935, Wood, lower front round dial, upper sectioned grill, side fluting, BC, SW **$125.00**
587, Tombstone, 1939, Wood, small case, lower front round dial, upper cloth grill w/2 vert bars, 2 knobs **$80.00**
601 "Bandbox", Table, 1927, Metal, low rectangular case, front dial w/escutcheon, 3 knobs, battery **$80.00**
608 "Gembox", Table, 1928, Metal, low rectangular case, center window dial w/escutcheon, 3 knobs, AC **$125.00**

609 "Gemchest", Console, 1928, Metal, upper front window dial, fancy Chinese grill & corner "straps", AC $350.00
610 "Gembox", Table, 1928, Metal, low rectangular case, front window dial w/escutcheon, 3 knobs, AC **$100.00**
612, Tombstone, Wood, shouldered, center front round dial, upper cloth grill w/cut-outs, 4 knobs **$110.00**
614EH, Tombstone, 1934, Wood, shouldered, center front round dial, upper sectioned grill, BC, SW, AC **$135.00**
614PG, Console, 1934, Wood, upper front round dial, lower cloth grill w/cut-outs, BC, SW, AC **$120.00**
637 "Super 6", Table, 1937, Wood, large round right dial, left cloth grill w/2 horizontal bars, 3 knobs **$60.00**
649, Console, 1937, Wood, upper front round dial, lower cloth grill w/2 vertical bars, 4 knobs, BC, SW **$125.00**
649A, Table, 1940, Plastic, center vertical dial, wrap-around side louvers, top pushbuttons, AC/DC **$85.00**
655, Tombstone, 1935, Wood, center round dial, upper cloth grill w/cut-outs, fluting, 4 knobs, BC, SW **$125.00**
704 "Jewelbox", Table, 1928, Metal, low rectangular case, center front dial w/escutcheon, 3 knobs, AC **$80.00**

704 "Perfecto", Side Table, 1928, Model 704 built into a 6 legged lift top wood table w/drop front panel, AC $275.00

705 "Showbox", Table, 1928, Metal, low rectangular case, front window dial w/escutcheon, 3 knobs, DC $100.00

706 "Showbox", Table, 1928, Metal, low rectangular case, front window dial w/escutcheon, 3 knobs, AC $100.00

714GA, Tombstone, 1934, Wood, shouldered, lower round dial, upper grill w/cut-outs, 3 knobs, 3 band, AC $125.00

714NA, Console, 1934, Wood, upper front round dial, lower cloth grill w/cut-outs, BC, SW, AC $120.00

725, Tombstone, Wood, center front round dial, upper cloth grill w/ cut-outs, 4 knobs, BC, SW $125.00

769, Console, 1937, Wood, upper front large round dial, lower cloth grill w/vertical bar, 7 tubes $135.00

804 "Jewelbox", Table, 1929, Metal, low rectangular case, center window dial w/escutcheon, 3 knobs, AC $95.00

814FA, Tombstone, 1934, Wood, center front round dial, upper sectioned grill, 3 knobs, BC, SW, AC $110.00

814QB, Console, 1934, Wood, upper front round dial, lower cloth grill w/cut-outs, BC, SW, AC $120.00

819M, Console, 1939, Wood, upper slanted dial, pushbuttons, 3 vertical grill bars, 4 knobs, BC, SW, AC $150.00

989, Console, 1937, Wood, upper large round dial, lower cloth grill with vertical bar, 9 tubes, 5 knobs $130.00

1014, Tombstone, C1935, Wood, shouldered, lower front round dial, upper cloth grill w/cut-outs, 4 knobs $95.00

1199, Console, 1937, Wood, upper large round dial, lower cloth grill with vert bar, 11 tubes, 5 knobs $150.00

1211, Console, 1937, Wood, upper large round dial, lower cloth grill with vert bar, 12 tubes, 5 knobs $160.00

1313, Console, 1937, Wood, upper large round dial, lower cloth grill with vert bars, 13 tubes, 5 knobs $170.00

3716 "WLW", Console/PA, 1936, Wood, massive console, round dial, 4 chassis, 6 speakers, 37 tubes, 4' 10" high ..**$12,000.00**

5519, Table, 1939, Plastic, center front vertical slide rule dial, vertical grill bars, handle, 2 knobs $50.00

5628-A, Table, Plastic, slide rule dial, pushbuttons, upper horiz louvers, slanted sides, 2 knobs $85.00

5628-B, Table, Plastic, slide rule dial, pushbuttons, horiz louvers, slanted sides, 2 knobs, BC, SW $85.00

AC-7, Table, 1927, Wood, high rectangular case, slanted front panel w/right thumbwheel dial, AC $100.00

AC-7C, Console, 1927, Wood, upper right front thumbwheel dial, thumbwheel knobs, lower speaker, AC $140.00

Ace 3B, Table, 1923, Wood, low rectangular case, 1 dial front panel, 3 tubes, battery $165.00

Ace 3C, Console, 1923, Wood, upper Ace 3C table model w/feet removed, lower storage area, battery $200.00

Ace 3C, Table, 1923, Wood, high rectangular case, inner 1 dial panel & speaker, double doors, battery $175.00

Arbiter, Console-R/P, 1930, Walnut veneer w/ornate repwood front panel, center dial, lift top, inner phono, AC $165.00

B-429A, Portable, 1939, Cloth covered, high case, right front square dial, left grill, handle, 2 knobs, bat $30.00

B-439A, Portable, 1939, Striped case, right front dial, left grill, handle, 2 knobs, battery $30.00

B-549A, Portable, 1939, Brown striped cloth, right front dial, left grill, handle, 2 knobs, AC/DC/battery $30.00

B-667A, Tombstone, 1937, Walnut, center front round dial, upper grill w/cut-outs, 4 knobs, AC/battery $100.00

B-5549-A, Portable-R/P, 1939, Cloth covered, right front dial, left grill, 2 knobs, crank phono, handle, AC/DC/bat $35.00

Centurion, Console, 1935, Wood, upper front round airplane dial, lower cloth grill w/cut-outs, BC, SW $135.00

Chum, Side Table, 1930, Walnut veneer w/inlay, 28 1/2" high, front window dial w/escutcheon, 3 knobs $160.00

Comrade, Side Table, 1930, Walnut veneer, 29 1/2" high, inner front window dial, double front doors $140.00

D10TN, Table, 1951, Plastic, large center round dial w/inner circular louvers, side "fins", 2 knobs $115.00

D-25BE, Table-C, 1953, Plastic, right round metal dial, left alarm clock, center crest, 5 knobs, BC, AC $110.00

D-25CE, Table-C, 1953, Plastic, right front round dial, left alarm clock, center crest, 5 knobs, BC, AC $110.00

D-25MN, Table-C, 1953, Plastic, right front round dial, left alarm clock, center crest, 5 knobs, BC, AC $110.00

D-25WE, Table-C, 1953, Plastic, right round metal dial, left alarm clock, center crest, 2 knobs, BC, AC $110.00

Director, Console, 1930, Walnut veneer w/ornate repwood front panel, lowboy, center window dial $165.00

E-10BE, Table, 1953, Plastic, large front dial with center pointer & inner criss-cross grill, 2 knobs $85.00

E15TN, Table, 1953, Plastic, upper front dial, perforated grill w/ horizontal bar, 2 knobs, BC, AC/DC $75.00

E-20-MN, Table, 1953, Plastic, large center front dial w/inner cloth grill & gold pointer, 2 knobs, AC/DC $75.00

E30BE, Table, 1953, Plastic, right dial, left round grill over checkered panel, 3 knobs, AM, FM, AC $35.00

E-90BK, Table-C, 1953, Plastic, right side dial knob, large front grill w/left square alarm clock, AC $30.00

Elf, Cathedral, 1931, Pressed wood case, ornate "carved" front, upper cloth grill w/cut-outs, 2 knobs $375.00
F5TWE "Musical Chef", Table-Timer, 1959, Plastic, kitchen radio/ timer, right side dial knob, front lattice grill, BC, AC $65.00
F-110BE, Portable, 1953, Plastic, right front round dial over lattice grill w/crest, handle, BC, battery $35.00

F-110BK "Skyrocket", Portable, 1953, Plastic, right front round dial over lattice grill w/crest, handle, BC, battery $35.00
G1465, Table, 1938, Catalin, right front dial, left horiz wrap-around louvers, flared base, 2 knobs .... $1,000.00+
Harko Senior, Table, 1922, Mahogany finish, low rectangular case, 1 dial front panel,1 tube, battery $300.00
JM-8BG "Musical Memories", Portable-N, 1956, Transistor, looks like small book, inner thumbwheel dial, metal grill, AM, battery $165.00
Mate, Console, 1930, Walnut veneer w/ornate repwood front panel, upper dial, stretcher base $165.00
Minstrel, Console, 1931, Wood, inner front window dial, lower cloth grill w/cut-outs, double front doors $135.00
Pal, Consolette, 1931, Wood w/ornate Repwood front & sides, lowboy, window dial, 5 tubes, 3 knobs, AC $185.00
Playmate, Console, 1930, Walnut veneer, low case, 29 1/2 " high, inner front window dial, double doors $140.00

Prestotune 11, Console, 1937, Walnut, upper press button tuning dial, lower horiz louvers, 5 knobs, BC, SW, AC $150.00
Prestotune 12, Console-Upright, 1937, Walnut, upper press button tuning dial, lower horiz louvers, 5 knobs, BC, SW, AC $150.00
Prestotune 12, Console-Low Profile, 1937, Mahogany w/inlay, top press button tuning dial, front horiz louvers, BC, SW, AC $160.00
Pup "Sky Terrier", Table, 1924, Metal, square case, one top exposed tube, side knobs, battery $325.00
RFL-60, Table, 1926, Mahogany, low rectangular case, 3 dial panel engraved w/woodland scene, bat $140.00
RFL-75, Table, 1926, Mahogany, low rectangular case, slanted ornately engraved 3 dial front panel, bat $185.00
RFL-90, Console, 1926, Mahogany, upper front double dial w/ escutcheon, lower built-in speaker, battery $175.00

Sheraton, Cathedral, 1933, Wood, Sheraton-style, window dial, pedimented top w/finial, 3 knobs, 5 tubes, AC $275.00
Travo Deluxe, Table, 1932, Deco case, right & left front knobs, center cloth grill w/cut-outs, top fluting $85.00
Trirdyne, Table, 1925, Wood, low rectangular case, 2 dial front panel, lift top, battery $125.00
Trirdyne 3R3, Table, 1924, Wood, low rectangular case, 2 dial front panel, lift top, 3 tubes, battery $130.00
Trirdyn 3R3 Special, Table, 1924, Walnut, low rectangular case, 2 dial front panel, lift top, feet, 3 tubes, battery $130.00
Trirdyn Newport, Table, 1925, Wood, slanted 2 dial black front panel, 3 tubes, feet, battery $125.00
Trirdyn Regular, Table, 1925, Wood, low rectangular case, 2 dial front panel, lift top, battery $95.00
Trirdyne Super Regular, Table, 1925, Wood, low rectangular case, front panel with 2 pointer dials, lift top, battery $130.00
Trirdyne Special, Table, 1925, Wood, low rectangular case, 2 dial front panel, lift top, battery $125.00
Trirdyne Super Special, Table, 1925, Wood, low rectangular case, slanted black panel w/2 pointer dials, lift top, bat $150.00
Twelve, Console, 1933, Wood, lowboy, window dial, lower grill w/ cut-outs, 6 legs, 4 knobs, BC, SW, AC $130.00
IV, Table-Amp, 1922, Wood, low rectangular case, amplifier, black bakelite front panel, lift top, bat $220.00
V, Table, 1922, Wood, low rectangular case, 1 dial black bakelite front panel, lift top, battery $200.00
VI, Table, 1923, Wood, low rectangular case, 2 dial black bakelite front panel, lift top, battery $200.00
VI Special, Table, 1923, Wood, low rectangular case, 2 dial front panel, lift top, rear storage, battery $200.00

VIII, Table, 1923, Wood, low rectangular case, 2 dial black bakelite front panel, lift top, battery $210.00

VIII Portable, Portable, 1923, Wood, inner 2 dial panel, storage, fold-down front, handle, 3 tubes, loop antenna, bat $425.00

X, Table, 1922, Wood, low rectangular case, 2 dial black bakelite front panel, lift top, battery $200.00

XJ, Table, 1923, Wood, low rectangular case, 2 dial black bakelite front panel, lift top, battery $210.00

XL, Table, 1923, Wood, high rectangular case, inner left 2 dial panel, double doors, battery $225.00

XV, Table, 1922, Wood, high rectangular case, upper 2 dial panel, lower built-in speaker, battery $225.00

CROWN
Shriro, Inc., 276 Fourth Ave., New York, New York

TR-333, Portable, 1959, Transistor, right front thumbwheel dial, lower perforated grill, AM, battery $20.00

TR-400, Portable, 1960, Transistor, right front thumbwheel dial, left lattice grill, swing handle, AM, bat $25.00

TR-610, Portable, 1959, Transistor, right front round dial, left perfo-rated grill, swing handle, AM, battery $25.00

TR-666, Portable, 1959, Transistor, right front thumbwheel dial, left diagonal perforated grill, AM, bat $20.00

TR-670, Portable, 1960, Transistor, front diamond-shaped window dial, lower perforated grill, AM, bat $25.00

TR-800, Portable, 1960, Transistor, right slide rule dial, left round perforated grill, swing handle, AM, bat **$30.00**

TR-820, Portable, 1959, Transistor, right front round dial, left lattice grill, swing handle, AM, battery $25.00

TR-830, Portable, 1959, Transistor, right front round dial over hori-zontal grill bars, AM, battery $20.00

TR-875, Portable, 1960, Transistor, slide rule dial, perforated grill, telescoping antenna, BC, SW, battery $25.00

CROYDON

530, Tombstone, Two-tone wood, lower front airplane dial, upper cloth grill w/cut-outs, 3 knobs $145.00

CUTTING & WASHINGTON
Cutting & Washington Radio Corporation, Minneapolis, Minnesota

Cutting & Washington was formed in 1917 by Fulton Cutting and Bowden Washington in the business of making radio transmitters. In 1922 the company marketed its first radio but was out of business by 1924 due to legal difficulites and marketing problems.

11, Table, 1922, Wood, low rectangular case, 2 dial front panel, lift top, battery $500.00

11A, Table, 1922, Wood, low rectangular case, 3 dial black front panel, lift top, battery $325.00

12A, Table, 1923, Wood, high rectangular case, 2 dial front panel, removable front, battery $325.00

CYARTS

B "Deluxe", Table, 1946, Lucite, modern "bullet" shape, slide rule dial, right round grill, 3 knobs, BC ....$1,000.00+

DAHLBERG
The Dahlberg Company, Minneapolis, Minnesota

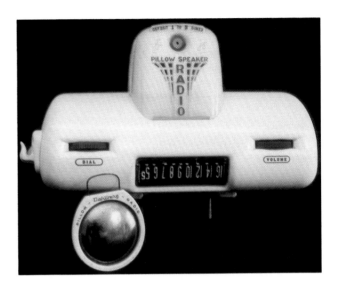

4130-D "Pillow Speaker", Coin Op-N, 1955, Plastic, mounts on motel bed headboard, moveable pillow speaker, AC $250.00

DALBAR
Dalbar Mfg. Co., 1314 Forest Avenue, Dallas, Texas

100-1000 Series, Table, 1946, Wood, right front dial, left horizontal grill bars, 4 knobs, BC, SW, AC/DC $35.00

400, Table, 1946, Wood, right square dial, left diamond-shaped grill, 2 knobs, BC, AC $35.00

Barcombo, Jr., Table-R/P, 1946, Wood, right front square dial, 4 knobs, diamond shaped grill, lift top, BC, AC $30.00

DAY-FAN
The Dayton Fan & Motor Company, Dayton, Ohio
Day-Fan Electric Company, Dayton, Ohio

Day-Fan began business as the Dayton Fan & Motor Company in 1889 producing fans. The company marketed a line of radio compo-nent parts in 1922 and by 1924 was producing complete radio sets. Day-Fan was bought out by General Motors in 1929.

5 (5049), Table, 1925, Wood, low rectangular case, 1 dial slant front panel, lift top, 5 tubes, battery $90.00

67 (5057), Table, 1927, Wood, low rectangular case, 1 dial front panel, left built-in speaker $125.00

Daycraft, Table, 1925, Mahogany, low rectangular case, 1 dial slant front panel, right speaker, battery $125.00

Daygrand, Console, 1926, Wood, inner 1 dial slanted panel, left speaker, fold-down front, storage, battery $150.00

Dayola (5112), Table, 1925, Mahogany, inner 3 dial slanted panel, fold-down front, feet, 4 tubes, battery $110.00

Dayradia (5107), Table, 1924, Wood, high rectangular case, slant front panel, upper built-in speaker, battery $185.00

Dayroyal, Console, 1926, Mahogany, desk-style, inner 1 dial panel, upper speaker & storage, doors, battery $150.00

Daytonia, Console, 1926, Wood, inner 1 dial slanted panel, fold-down front, right speaker, storage, battery **$150.00**

OEM-5, Table, Wood, low rectangular case, 3 dial black front panel, 5 tubes **$125.00**

OEM-7 (5106), Table, 1925, Wood, low rectangular case, 3 dial front panel, lift top, 4 tubes, battery **$110.00**

OEM-11 (5106) "Duo-Plex", Table, 1924, Wood, low rectangular case, 3 dial front panel, lift top, 3 tubes, battery **$100.00**

OEM-12, Table, 1925, Mahogany finish, low rectangular case, 3 dial front panel, lift top, 4 tubes, bat **$90.00**

DEARBORN
Dearborn Industries, Chicago, Illinois

100, Table-R/P, 1947, Wood, front square dial, cloth grill w/vertical bars, 2 knobs, open top, BC, AC **$25.00**

DEFOREST
The Radio Telephone Company
49 Exchange Place, New York, New York
Deforest Radio Telephone & Telegraph Company,
1415 Sedgwick Avenue, New York City, New York
Deforest Radio Company
Jersey City, New Jersey

Between 1900 and the end of WW I, Lee Deforest and his various companies produced much electrical equipment for the government. By 1922, the company was fully involved in the production of radio sets for the public. After many legal entanglements, the company was sold to RCA in 1933.

D-4 "Radiophone", Table, 1923, Wood, square case with fold-back top, inner 2 dial panel, 1 exposed tube, battery **$425.00**

D-6, Table, 1922, Wood, low rectangular case, 2 dial front panel, battery **$1,200.00**

D-7 "Radiophone", Table, 1922, Wood, square case, 2 dial front panel, top loop antenna, 3 tubes, battery **$700.00**

D-7A, Table, 1923, Wood, square case, 2 dial black front panel, top loop antenna, battery **$700.00**

D-10, Table, 1923, Leatherette or mahogany, inner 2 dial panel, doors, top loop antenna, 4 tubes, bat **$600.00**

D-12 "Radiophone", Table, 1924, Leatherette or mahogany cabinet, 2 dial front panel, loop antenna, 4 tubes, bat **$600.00**

D-14 "Radiophone", Table, 1925, Wood, tall case, inner slanted 2 dial panel w/5 exposed tubes, loop antenna, bat **$900.00**

D-17A, Table, 1925, Wood, upper 2 dial front panel, lower speaker, top loop antenna, battery **$600.00**

D-17L, Table, 1925, Leatherette, upper 2 dial front panel, lower speaker, top loop antenna, battery **$550.00**

D-17M, Table, 1925, Mahogany, upper 2 dial front panel, lower speaker, top loop antenna, battery **$600.00**

D-556A, Table, Plastic, upper slanted slide rule dial, large cloth grill w/4 vert bars, 2 knobs **$45.00**

DT-600 "Everyman", Table, 1922, Wood, box, crystal set, inner panel and storage, lift lid **$425.00**

DT-800, Table-Amp, 1922, Wood, box, two-step amplifier, inner panel w/2 exposed tubes **$400.00**

F-5 "Radiophone", Table, 1925, Wood, low rectangular case, 3 dial front panel, lift top, feet, 5 tubes, battery **$325.00**

F-5-AL, Table, 1925, Embossed leatherette, low rectangular case, 3 dial front panel, 5 tubes, battery **$275.00**

F-5-AW, Table, 1925, Walnut, low rectangular case, 3 dial front panel, 5 tubes, battery **$275.00**

F-5L, Table, 1925, Leatherette, upper 3 dial front panel, lower grill for enclosed speaker, battery **$225.00**

F-5M, Table, 1925, Mahogany, upper 3 dial front panel, lower grill for enclosed speaker, battery **$275.00**

MS-1, Table, 1921, 4 units - tuner, audion control, 1-step amp, transmitter, battery, "Interpanel" **$1,500.00**

DELCO
Delco Radio, Division of G. M. Corp.,
Kokomo, Indiana

608, Table, 1949, Wood, large multi-band slide rule dial over front grill, 4 knobs, BC, SW, AC **$45.00**

1102, Tombstone, 1935, Wood w/inlay, center dial w/escutcheon, upper grill w/vert bars, 6 tubes, 4 knobs $110.00
1106, Tombstone, 1935, Wood, lower airplane dial, upper cloth grill w/3 vert bars, 4 knobs, BC, SW, AC $125.00

1107, Tombstone, 1935, Wood, round airplane dial, upper cloth grill w/3 vert bars, 4 knobs, BC, SW, AC $125.00
R-1116, Table, Wood, right front round black dial, left cloth grill w/3 horizontal bars, 4 knobs $55.00
R-1227, Table, 1947, Plastic, slanted slide rule dial, horiz grill bars, decorative case lines, 3 knobs $55.00

R-1228, Table, 1947, Plastic, slanted slide rule dial, horiz grill bars, decorative case lines, 3 knobs $55.00
R-1229, Table, 1947, Two-tone wood, slanted slide rule dial, horizontal louvers, 3 knobs, BC, AC/DC $40.00
R-1230, Table, 1948, Ivory plastic, right dial, vert bars w/"ribbon candy" cut-out, 2 knobs, BC, AC/DC $90.00
R-1230A, Table, 1947, Ivory plastic, right dial, vert bars w/"ribbon candy" cut-out, 2 knobs, BC, AC/DC $90.00
R-1231, Table, 1948, Brown plastic, right dial, vert bars w/"ribbon candy" cut-out, 2 knobs, BC, AC/DC $90.00
R-1231A, Table, 1948, Brown plastic, right dial, vert bars w/"ribbon candy" cut-out, 2 knobs, BC, AC/DC $90.00
R-1232, Table, 1948, Walnut, right front square dial, left vertical grill bars, 2 knobs, BC, AC/DC $45.00

R-1233, Table, 1948, Plastic, right dial, vertical bars w/"ribbon candy" cut-out, 2 knobs, BC, AC/DC $90.00
R-1235, Table, 1946, Plastic, upper slanted slide rule dial, horizontal louvers, 3 knobs, BC, AC/DC $40.00
R-1236, Table, 1947, Plastic, upper slanted slide rule dial, horizontal louvers, 3 knobs, BC, AC/DC $40.00
R-1238, Table, 1948, Wood & plastic, right dial, left "circles" grill, lucite handle, 2 knobs, BC, AC/DC $60.00
R-1241, Table-R/P, 1949, Wood, front slide rule dial, large grill, 4 knobs, 3/4 lift top, inner phono, BC, AC $25.00
R-1242, Table-R/P, 1948, Wood, front square dial, 2 knobs, lift top, inner phono, BC, AC/DC $35.00
R-1243, Table, 1948, Wood, upper slanted slide rule dial, large lower grill area, 3 knobs, BC, AC $35.00
R-1244, Table-R/P, 1948, Walnut, lower front slide rule dial, 4 knobs, 3/4 lift top, inner phono, BC, AC $30.00
R-1245, Console-R/P, 1948, Walnut, upper slide rule dial & grill, center phono, lower storage, 4 knobs, BC, AC $65.00
R-1246, Console-R/P, 1948, Mahogany, upper slide rule dial & grill, center phono, storage, 4 knobs, BC, AC $65.00
R-1249, Console-R/P, 1949, Wood, inner right dial, 4 knobs, door, left pull-out phono drawer, BC, FM, AC $80.00
R-1251, Console-R/P, 1947, Wood, inner right dial, left pull-out phono, lower grill w/cut-outs, BC, SW, AC $80.00
R-1254, Console-R/P, 1948, Wood, slide rule dial, 6 knobs, pushbuttons, pull-out phono drawer, BC, SW, FM, AC $80.00
R-1408, Portable, 1947, Square dial, grill w/circle cut-outs, 3 knobs, recessed handle, BC, AC/DC/bat $35.00
R-1409, Portable, 1947, Square dial, grill w/circle cut-outs, 3 knobs, recessed handle, BC, AC/DC/bat $35.00
R-1410, Portable, 1948, Luggage-type, front slide rule dial, 2 knobs, handle, BC, AC/DC/battery $35.00

———————————————————————

DELMONICO
Delmonico International Corp.

TRS-6, Portable, 1959, Transistor, right window dial over perforated grill, thumbwheel knobs, AM, bat $25.00

———————————————————————

DETROLA
International Detrola Corp.,
1501 Beard Street, Detroit, Michigan

4D, Cathedral, 1934, Wood, small case, center front round dial, cloth grill w/cut-outs, 3 knobs $190.00
7A3, Tombstone, 1934, Wood, center front round dial, upper cloth grill w/vertical bars, BC, SW, AC $120.00
172A "Glen", Table, 1937, Wood, right front round dial, left grill w/horizontal bars, 3 knobs, BC, SW, AC $70.00
173EC "Lark", Console, 1937, Wood, upper front dial, lower grill w/vertical bars, side "louvers", BC, SW, AC $135.00
174EC "Martin", Console, 1937, Wood, upper front dial, cloth grill w/vert bars, side "louvers", BC, SW, AC/DC $135.00
218 "Pee Wee", Table, 1939, Swirled plastic, right front dial knob, left wrap-around louvers, BC, SW, AC $225.00
219 "Super Pee Wee", Table, 1938, Plastic, Deco, right front dial knob, left wrap-around louvers, BC, SW $225.00
258EPC, Console-R/P, 1938, Wood, front dial, lower grill w/horizontal bars & drawer, lift top, inner phono $150.00
302, Cathedral-C, 1938, Wood, lower right front dial, left cloth grill, upper round clock face, 2 knobs, AC $325.00
310, Table, Wood w/inlay, right front dial, left grill, 6 top pushbuttons, handle, BC, SW, AC $75.00
378, Portable, 1941, Leatherette & plastic, inner dial & knobs, fold-open door, handle, AC/DC/battery $50.00
383, Portable, 1941, Cloth covered, right front square dial, left grill, handle, 2 knobs, AC/DC/bat $30.00

558-1-49A, Table-R/P, 1946, Wood, front dial, horizontal louvers, 3 knobs, lift top, inner phono, BC, AC $30.00

568-13-221D, Table, 1946, Upper right slide rule dial, left round perforated grill, 3 knobs, BC, SW, AC/DC $30.00

571, Table, 1946, Right front slide rule dial, left horizontal louvers, 2 knobs $45.00

571L, Table, 1946, Wood, right slide rule dial, left cloth grill w/ horizontal bars, 2 knobs, BC, AC/DC $40.00

571X-21-94D, Table, 1946, Right front slide rule dial, left horizontal louvers, 2 knobs, BC, AC/DC $45.00

572-220-226A, Table, 1946, Wood, right slide rule dial, cloth grill w/ horiz bars, 3 knobs, feet, BC, SW, AC $40.00

576-1-6A, Table, 1946, Wood, upper slanted slide rule dial, cloth grill w/vert bars, 3 knobs, BC, AC/DC $60.00

579-2-58A, Table, 1946, Wood, slanted slide rule dial, cloth grill w/ 2 horizontal bars, 2 knobs, BC, AC/DC $45.00

582, Chairside, 1947, Wood, top dial, 3 knobs, lift top, lower record storage, feet, BC, SW, AC $85.00

610-A, Table, 1949, Wood, right round dial, 2 horizontal grill bars, round corners, 2 knobs, BC, battery $35.00

611-A, Table, 1948, Wood, right round dial, 2 horizontal grill bars, round corners, 4 knobs, BC, SW, bat $35.00

3281, Table-C, Wood, mantle clock style, lower right dial, left grill, upper round clock, 2 knobs $325.00

3861, Table-R/P, 1941, Walnut veneer, right front dial, left grill, 3 knobs, lift top, inner phono, AC $35.00

7156, Table-R/P, 1948, Wood, slide rule dial, horiz louvers, 4 knobs, 3/4 lift top, inner phono, BC, AC $30.00

7270, Table-R/P, 1947, Wood, front slide rule dial, 3 knobs, grill w/ cut-outs, lift top, BC, AC $35.00

DEWALD
Dewald Radio Mfg. Corp.,
440 Lafayette Street
New York, New York
Pierce-Aero, Inc., New York

54 "Dynette", Portable, Leatherette, inner wood panel w/right dial, center grill, removable front, handle $85.00

54A, Table, 1933, Right & left front knobs, center grill w/cut-outs, raised side panels, AC/DC/bat $90.00

60-3, Tombstone, 1933, Wood, center front window dial w/escutcheon, upper gothic grill, 3 knobs, AC/DC $125.00

60-42, Console, 1933, Wood, front window dial w/escutcheon, lower grill w/cut-outs, 6 legs, 3 knobs $145.00

406, Table, 1938, Plastic, small case, right round dial, left vertical wrap-over grill bars, "feet", AC $95.00

410, Portable, Leatherette, left side dial & knob, front rectangular metal grill, handle $30.00

511, Table-R/P, 1949, "Alligator", slide rule dial, 4 knobs, lift top, inner phono, handle, BC, SW, AC $30.00

522, Table, 1936, Wood, Deco, right dial, left grill w/wrap-over vertical bars, tuning eye, battery $65.00

530, Table, 1939, Plastic, right front dial, left grill w/checker board center, 2 knobs, BC, AC/DC $65.00

531, Table-R/P, 1939, Two-tone wood, right front dial, left grill, 2 knobs, lift top, inner phono, AC $35.00

532, Table-R/P, 1939, Two-tone wood, right front dial, left grill, 2 knobs, lift top, inner phono, AC/DC $35.00

533, Table, 1939, Wood, right recessed dial, left wrap-around horizontal louvers, 2 knobs, AC/DC $50.00

534, Table, 1939, Wood, right front recessed dial, curved left grill w/ vert bars, 2 knobs, AC/DC $50.00

550 "Dynette", Table, 1933, Walnut w/inlay, right dial, center grill w/ cut-outs, scalloped top, 2 knobs, AC/DC $90.00

551, Table, 1933, Walnut, right dial, center grill w/cut-outs, vertical front lines, 2 knobs, AC/DC $85.00

562 "Jewel", Table, 1941, Catalin, upper front slide rule dial, lower horizontal grill bars, handle, 2 knobs .... $750.00+

564 "Companionette", Portable, 1941, Leatherette, inner dial, vertical grill bars, fold-open front, handle, AC/DC/bat $50.00

565, Portable, 1941, Inner slide rule dial, horizontal louvers, 2 knobs, front cover, handle, AC/DC/bat $35.00

580, Table, 1933, Wood, lower front window dial, upper cloth grill w/ vertical bars, 3 knobs, AC $75.00

615, Table, 1935, Wood, lower round dial, upper cloth grill, rounded sides, 2 knobs, BC, SW, AC/DC $75.00

643, Table, 1939, Wood, right front dial, left grill w/horiz bars, tuning eye, 4 knobs, BC, SW, AC $55.00

645, Table, 1939, Wood, right slide rule dial, left horizontal wrap-around grill bars, 3 knobs, AC/DC $50.00

648, Table, 1939, Wood, right dial, curved left w/vertical bars, 6 pushbuttons, 4 knobs, AC/DC $65.00

649, Table, 1939, Wood, right dial, left wrap-around horiz grill bars, 6 pushbuttons, 4 knobs, AC $65.00

669 "Super Six", Table-R/P, 1941, Wood, right front dial, left horizontal louvers, 2 knobs, lift top, inner phono, AC $35.00

670, Table, 1941, Wood, upper slanted slide rule dial, lower grill, 4 knobs, BC, SW, AC/DC $45.00

701, Table, 1939, Wood, right dial, left grill w/3 horizontal bars, 6 pushbuttons, 3 knobs, AC/DC $50.00

708, Table, 1941, Two-tone wood, upper slanted slide rule dial, lower grill, 4 knobs, BC, SW, AC $40.00

802, Tombstone, 1934, Wood, Deco, center round dial, upper cloth grill w/triangular cut-out, BC, 3SW, AC $125.00

901, Tombstone, Wood, large case, lower round dial, upper cloth grill w/vert bars, 4 knobs $115.00

907, Console-R/P, 1940, Wood, inner right dial & knobs, left phono, lift top, front grill w/vertical bars $80.00

1200, Table, 1937, Wood, right front dial, left 3-sectioned cloth grill, BC, SW $50.00

A-500, Table, 1946, Plastic, upper slide rule dial, lower horizontal louvers, 2 knobs, BC, AC/DC $40.00

A-501 "Harp", Table, 1946, Catalin, harp-shaped, slide rule dial, cloth grill w/ 5 vertical bars, BC, AC/DC .... $700.00+

A-502, Table, 1946, Catalin, upper slide rule dial, lower horizontal louvers, 2 knobs, BC, AC/DC .... $700.00+

A-503, Table, 1946, Upper front slide rule dial, lower horizontal louvers, 2 knobs, BC, AC/DC $45.00

A-505, Table, 1947, Upper front slide rule dial, lower horizontal louvers, 4 knobs, BC, SW, AC/DC $40.00

A-507, Portable, 1947, Leatherette, slide rule dial, 2 knobs, fold-down front, handle, BC, AC/DC/bat $35.00

A-509, Table, 1948, Plastic, slide rule dial, horizontal louvers, 2 knobs front/1 side, BC, SW, AC/DC $40.00

A-514, Table, 1947, Plastic, right front dial, left horizontal wrap-around louvers, 2 knobs, BC, AC/DC $50.00

A-602, Table-R/P, 1947, Wood, inner slide rule dial, 4 knobs, fold-down front, lift top, handle, BC, AC $30.00

A-605, Table-R/P, 1947, Wood, slide rule dial over horiz louvers, 4 knobs, lift top, inner phono, BC, AC $30.00

B-400, Portable, 1948, Two-tone leatherette, right square dial, left cloth grill, handle, 2 knobs, BC, bat $30.00

B-401, Table, 1948, Plastic, right front round dial, left vertical grill bars, 2 knobs, BC, AC/DC $60.00

B-402, Portable, 1948, Plastic, lower slide rule dial, upper horizontal grill bars, handle, 2 knobs, BC, bat $35.00

B-403, Table-C, 1948, Catalin, harp-shaped, top slide rule dial, center clock, cloth grill, 2 knobs, BC, AC ....$700.00+

B-504, Portable, 1948, Plastic, lower slide rule dial, horizontal louvers, handle, 2 knobs, BC, AC/DC/bat $35.00

B-506, Table, 1948, Plastic, lower front slide rule dial, upper checkered louvers, 2 knobs, BC, AC/DC $35.00

B-510, Table, 1948, Plastic, upper slide rule dial, lower horizontal louvers, 3 knobs, BC, 2SW, AC/DC $35.00

B-512, Table-C, 1948, Catalin, slide rule dial, center clock, grill bars on top of case, 5 knobs, BC, AC ....$625.00+

B-515, Portable, 1949, Plastic, slide rule dial, horizontal grill bars, handle, 2 knobs, BC, SW, AC/DC/bat $35.00

B-614, Table-R/P, 1949, "Alligator", slide rule dial, 4 knobs, lift top, inner phono, handle, BC, AC $35.00

C-516, Table, 1949, Plastic, slide rule dial, checkered grill, 2 front knobs, 1 side knob, BC, SW, AC/DC $40.00

C-800, Table, 1949, Plastic, upper slanted slide rule dial, horizontal louvers, 4 knobs, BC, FM, AC/DC $35.00

D-508, Portable, 1950, "Alligator", front slide rule dial, handle, telescope antenna, BC, 2SW, AC/DC/bat $35.00

D-517, Portable, 1951, Plastic, lower slide rule dial, 2 knobs, perforated grill, handle, BC, AC/DC/bat $30.00

D-518, Table, 1950, Plastic, lower front slide rule dial, upper perforated grill, 2 knobs, BC, AC/DC $30.00

D-616, Table, 1950, Plastic, upper slide rule dial, perforated grill, top vents, 2 knobs, BC, AC/DC $35.00

D-E517A, Portable, 1952, Plastic, lower slide rule dial, checkered grill, handle, 2 knobs, BC, AC/DC/battery $35.00

E-520, Table, 1955, Plastic, upper front slide rule dial, horizontal louvers, 4 knobs, BC, 2SW, AC/DC $40.00

E-522, Table-R/P, 1951, Wood, upper front slide rule dial, horiz grill slats, 4 knobs, 3/4 lift top, BC, AC $30.00

F-404, Table, 1952, Plastic, right round dial knob, left horizontal wrap-around louvers, BC, AC/DC $45.00

F-405, Table, 1953, Plastic, slanted front slide rule dial, horizontal louvers, 4 knobs, BC, 2SW, bat $35.00

G-408, Portable, 1953, Plastic, lower round dial, vertical grill bars, thumbwheel knob, handle, BC, bat $35.00

H-528, Table-C, 1954, Plastic, right round dial, left clock, center checkered panel, 3 knobs, BC/ AC $30.00

H-533, Table-C, 1954, Plastic, right side dial knob, front clock, vertical wrap-over grill bars, BC, AC $45.00

H-537, Table, 1955, Plastic, right dial, left volume, center checkered grill, 3 knobs, BC, SW, AC/DC $40.00

J-802, Table, Plastic w/gold trim, slide rule dial, lower horizontal louvers, 2 knobs, BC, FM $45.00

K-412, Table, 1957, Plastic, right front dial, recessed left side w/ lattice grill, 2 knobs, BC, AC/DC $25.00

K-544, Portable, 1957, Leather case, transistor, right front horseshoe dial, left grill, handle, AM, battery $40.00

K-545, Table-C, 1957, Plastic, right front clock, recessed left side w/ lattice grill, side knobs, BC, AC $30.00

K-701-A, Portable, 1956, Plastic, transistor, lower front round dial, vertical grill bars, handle, AM, bat $40.00

K-701-B, Portable, 1956, Plastic, transistor, lower front round dial, vertical grill bars, side knob, handle $40.00

K-702B, Portable, 1957, Leather case w/front grill, transistor, round dial, side knob, handle, AM, battery $30.00

L-414, Portable, 1959, Leather case, transistor, right side dial, front vertical cut-outs, handle, AM, bat $35.00

L-546, Portable, 1959, Leather case, transistor, right front dial, left lattice cut-outs, handle, AM, bat $30.00

L-703, Portable, 1959, Leather case, transistor, right side dial, front "brick" cut-outs, handle, AM, bat $35.00

M-550, Table, 1959, Plastic, right front dial, recessed left side w/ lattice grill, 2 knobs, BC, AC/DC $25.00

———————————————————————————

DISTANTONE
Distantone Radio, Inc., Lynbrook, New York

C "Single Control", Table, 1926, Wood, slant front polished bakelite panel, one center dial, battery $115.00

———————————————————————————

DOOLITTLE
Doolittle Radio Corporation, New Haven, Connecticut

No model #, Table, C1925, Wood, low rectangular case, black 2 dial front panel, lift top, battery $160.00

———————————————————————————

DORON
Doron Bros. Electric Co., Hamilton, Ohio

R-5 "Super-Equidyne", Table, 1925, Wood, low rectangular case, 3 dial front panel, 5 tubes, battery $115.00

———————————————————————————

DUMONT
Allen B. Dumont Labs, Inc., 2 Main Street Passaic, New Jersey

1210, Portable, 1957, Leather case, transistor, right round dial, criss-cross grill, strap handle, BC, bat $40.00

RA-346, Table-C, 1956, Scrolled frame surrounds front criss-cross grill, center clock face, BC, AC/DC $100.00
RA-354 "Beachcomber", Portable, 1957, Leather case w/front criss-cross grill, right round dial, strap, BC, AC/DC/bat $35.00
RA-902, Portable, 1958, Leather case, transistor, right front dial, left criss-cross grill, strap, AM, bat $40.00

DYNAVOX
Dynavox Corp., 40-35 21st Street,
Long Island City, New York

3-P-801, Portable, 1948, Two-tone leatherette, inner dial, 3 knobs, flip-open door, handle, BC, AC/DC/bat $45.00

EAGLE
Eagle Radio Company,
16 Boyden Place, Newark, New Jersey

The Eagle Radio Company began business in 1922. After an initial surge of business, the company began to decline and in 1927 was sold to Wurlitzer.

A, Table, 1923, Wood, low rectangular case, 3 dial front panel, lift top, battery $110.00
B, "Neutrodyne", Table, 1924, Wood, low rectangular case, 3 dial front panel, 5 tubes, battery $120.00
C-1, Console, 1925, Wood, contains table model F receiver, fold-down front door, lower storage, bat $150.00
C-2, Console, 1925, Wood, contains table model F receiver, fold-down front door, lower storage, bat $150.00
C-3, Console, 1925, Wood, contains table model F receiver, fold-down front door, lower storage, bat $150.00
D, Table, 1925, Wood, low rectangular case, 3 dial black bakelite front panel, 5 tubes, battery $100.00
F "Neutrodyne", Table, 1925, Wood, low rectangular case, black bakelite front panel, 3 window dials, battery $110.00
H, Table, 1926, Wood, low rectangular case, 3 dial front panel, 5 tubes, battery $100.00
K, Console, 1926, Wood, inner 3 dial panel w/Eagle emblem, fold-down front door, storage, battery $175.00
K-2, Table, 1926, Wood, low rectangular case, inner 3 dial panel, fold-down front door, battery $125.00

ECA
Electrical Corp. of America,
45 West 18th Street, New York, New York

101, Table, 1946, Wood, upper curved slide rule dial, lower horizontal louvers, 2 knobs, BC, AC/DC $35.00
102, Table, 1947, Plastic, upper slide rule dial, lower vertical grill openings, 2 knobs, BC, AC/DC $45.00
104, Table-R/P, 1947, Wood, front slide rule dial, lower horizontal louvers, 4 knobs, lift top, BC, AC $30.00
105, Table-R/P, 1947, Leatherette, inner slide rule dial, 4 knobs, phono, lift top, BC, AC $25.00
106, Table-R/P, 1946, Wood, slide rule dial, contrasting horizontal louvers, 4 knobs, lift top, BC, AC $30.00
108, Table, 1946, Plastic, upper slide rule dial, shouldered top, lattice grill, 4 knobs, BC, AC/DC $50.00
121, Console-R/P, 1947, Wood, top left dial & 4 knobs, right front pull-out phono drawer, BC, AC $95.00
131, Table-R/P, 1947, "Alligator", slide rule dial, 4 knobs, fold-down side door, lift top, handle, BC, AC $45.00
132, Table, 1948, Wood, upper slanted slide rule dial, lower criss-cross grill, 4 knobs, BC, AC/DC $35.00
201, Table, 1947, Two-tone wood, slide rule dial, horizontal grill openings, 2 knobs, BC, AC/DC $50.00
204, Portable, 1948, "Alligator", inner dial, 2 knobs, fold-down front, handle, BC, AC/DC/battery $30.00

ECHOPHONE
Echophone Radio, Inc.,
1120 North Ashland Avenue, Chicago, Illinois
The Hallicrafter Co.,
2611 South Indiana Avenue, Chicago, Illinois
Radio Shop, Sunnyvale, California
Echophone Radio Mfg. Co.,
968 Formosa, Los Angeles, California

The trade name "Echophone" was used on many sets made by different maufacturers.

3, Table, 1925, Wood, high case, slanted two dial black panel w/ 3 exposed tubes, lift top, bat $225.00
4, Cathedral, 1932, Wood, front window dial w/escutcheon, upper cloth grill w/cut-outs, 2 knobs $235.00
4, Table, 1925, Wood, high rectangular case, 2 pointer dial slanted panel, 4 exposed tubes, bat......................................$225.00
6, Cathedral, Wood, right front thumbwheel dial, upper round grill w/ lyre cutout, 2 knobs $195.00

46, Console, 1928, Wood, rectangular cabinet on 4 legged stand, center front dial w/escutcheon $225.00

60, Cathedral, 1931, Walnut, quarter-round dial, ornate presssed wood grill, 6 tubes, 3 knobs, AC $295.00

80, Cathedral, 1931, Walnut, quarter-moon dial w/escutcheon, ornate pressed wood grill, 3 knobs, AC $275.00
81, Cathedral, 1931, Wood, quarter-moon dial w/escutcheon, ornate pressed wood grill, 3 knobs, AC $275.00
90, Console, 1931, Wood, lowboy, upper quarter-moon dial, lower ornate gothic grill, 3 knobs, AC $195.00
A, Table, 1923, Wood, high rectangular case, 2 dial front panel, battery $185.00
EC-113, Table, 1946, Wood, center square dial, right & left grills, 4 knobs, BC, 2 SW, AC/DC $35.00
EC-306, Table-R/P, 1947, Wood, center dial & 4 knobs, right & left grills, lift top, inner phono, BC, SW, AC $30.00
EC-403, Console-R/P, 1947, Wood, inner top right dial & knobs, center pull-out phono, 15 tubes, BC, SW, AC $225.00
EC-600, Table, 1946, Wood, upper slanted slide rule dial, cloth grill, round corners, 4 knobs, BC, bat $40.00
EX-102, Table, 1949, Plastic, center square dial, large vertical grill bars, 4 knobs, BC, 2SW, AC/DC $45.00
S-3, Cathedral, 1930, Wood, right front window dial, upper round grill w/lyre cut-out & vertical bars $195.00
S-5, Cathedral, 1930, Walnut, lower window dial, upper ornate pressed wood grill, 3 knobs, AC $295.00

V-3, Table, 1925, Wood, high rectangular case, slanted two dial front panel w/3 exposed tubes, bat $210.00

ECODYNE
Ecodyne Radio Co., Irwin, Pennsylvania

R-5, Table, 1925, Wood, low rectangular case, 4 dial front panel, 5 tubes, battery $120.00
RT-13, Table, 1925, Wood, low rectangular case, 4 dial front panel, 5 tubes, battery $115.00

EDISON

R-5, Console, 1929, Wood, inner dial, lower grill w/cut-outs, 3 knobs, sliding doors, 7 tubes, AC $195.00

EISEMANN
Eisemann Magneto Corporation,
165 Broadway, New York, New York

The Eisemann Magneto Corporation was formed in 1903 for the manufacture of automobile magnetos. The company began to produce component parts for radios in 1923 and in 1924 briefly manufactured only a few complete radios.

6-D, Table, 1924, Wood, low rectangular case, 3 dial front panel, feet, 5 tubes, battery $95.00
RF-2, Table, 1924, Wood, low rectangular case, 2 dial front panel, battery $95.00

ELCAR
Union Electronics Corp.,
38-01 Queens Blvd.,
Long Island City, New York

602, Table, 1946, Wood, upper slide rule dial, cloth grill w/2 wood slats, 2 knobs, BC, SW, AC/DC $35.00

ELECTONE
Northeastern Engineering Inc.,
Manchester, New Hampshire

T5TS3, Table, 1947, Right front half-round dial, left wrap-around grill bars, 2 knobs, BC, AC/DC $50.00

ELECTRO
Electro Appliances Mfg. Co.,
102-104 Park Row
New York, New York

B20, Table, 1947, Wood, right square black dial, cloth grill w/ vertical bars, 2 knobs, BC, AC/DC $40.00

ELECTROMATIC
**Electromatic Mfg. Corp.,
88 University Place
New York, New York**

26, Table, 1941, Ivory plastic, right dial, left grill, horiz & vert decorative lines, 3 knobs, 2 band $45.00
607A, Table-R/P, 1946, Wood, streamline, front slide rule dial, 2 knobs, louvers, lift top, BC, AC $85.00

ELECTRONIC LABS
**Electronic Laboratories Inc.,
122 West New York Street
Indianapolis, Indiana**

710-PC, Console-R/P, 1947, Wood, inner right vertical dial, 4 knobs, phono, lift top, lower doors, BC, AC/DC $60.00
710-W, Table, 1947, Wood, lower slide rule dial, wrap-around grills, center strip, 4 knobs, BC, AC/DC $50.00
2701, Table, 1946, Wood, lower front slide rule dial, large upper grill, 2 knobs, BC, AC/DC $30.00
3000 "Orthosonic", Table, 1948, Right front dial, 3 center graduated horizontal louvers, BC, AC/DC $30.00

ELECTRO-TONE
**Electro-Tone Corp.,
221 Hudson Street
Hoboken, New Jersey**

555, Table-R/P, 1947, Wood, inner left vertical dial, 4 knobs, lift top, outer horizontal louvers, BC, AC $30.00

ELKAY
**The Langbein-Kaufman Radio Co.,
511 Chapel Street
New Haven, Connecticut**

Senior, Table, 1926, Wood, low rectangular case, slanted front panel w/center thumbwheel dial, AC $125.00

EMERSON
**Emerson Radio & Phonograph Company,
111 8th Avenue, New York, New York**

The Emerson Radio & Phonograph Company was founded in 1923. To keep costs down during the Depression, the company began to produce "midget" radios and consistently geared their pricing to lower income customers. With this marketing strategy, Emerson became one of the leaders in radio sales.

17, Table, 1935, Bakelite, Deco, right dial, center cloth grill w/vert chrome bars, 2 knobs, AC/DC $110.00
19, Table, 1935, Bakelite, Deco, right dial, center cloth grill w/vert chrome bars, 2 knobs, AC/DC $110.00
20A, Table, 1933, Bakelite, small case, right front dial, center ornate molded grill, 2 knobs, AC $85.00
25A, Table, 1933, Wood, curved top, right front dial, center cloth grill w/cut-outs, 2 knobs, AC/DC $95.00

28, Tombstone, 1934, Walnut, lower window dial, upper grill w/ circular cut-outs, 3 knobs, BC, SW, AC $120.00
30, Portable, 1933, Burl walnut, inner dial, grill w/cut-outs, fold-down front, handle, 2 knobs, AC/DC $95.00

32, Table, 1934, Walnut, right dial, center cloth grill w/cut-outs, rounded top, 2 knobs, AC/DC $100.00
34-C, Tombstone, 1935, Walnut, center airplane dial, upper grill w/ vertical bars, 4 knobs, BC, SW, AC $110.00
34-F7, Tombstone, 1935, Walnut, center airplane dial, upper grill w/ vertical bars, 4 knobs, BC, SW, bat $90.00
35, Table, 1933, Walnut, Sheraton design, window dial, upper grill w/ cut-outs, pediment top, AC $195.00
36, Tombstone, 1934, Two-tone wood, center front dial, upper cloth grill w/3 vertical bars, BC, SW, AC $110.00
38, Table, 1934, Walnut, lower front round dial, upper cloth grill w/cut-outs, arched top, AC/DC $110.00
45, Tombstone, 1934, Walnut, center front airplane dial, upper grill w/cut-outs, 3 knobs, BC, SW, AC $100.00
49, Upright Table, 1934, Walnut, lower front round dial, upper cloth grill w/cut-outs, curved top, AC/DC $165.00
69, Console, 1934, Walnut, round dial, lower cloth grill with 3 vertical bars, step-down top, AC $115.00
71, Tombstone, 1934, Walnut, center front airplane dial, upper grill w/geometric cut-outs, 4 knobs, AC $140.00
77, Console, 1934, Walnut, upper window dial, grill w/Deco cut-outs, chrome trim, 3 knobs, AC $210.00
100, Console, 1934, Walnut, upper round dial, lower cloth grill w/ vertical bars, step-down top, AC $130.00
101, Console, 1935, Walnut, upper airplane dial, lower cloth grill w/ vert bars, 4 knobs, BC, SW, AC $120.00
101-F7, Console, 1935, Walnut, upper airplane dial, lower cloth grill w/vert bars, 4 knobs, BC, SW, bat $95.00
101-U, Console, 1935, Walnut, upper airplane dial, upper grill w/ vertical bars, 4 knobs, BC, SW, AC/DC $120.00
102, Console, 1935, Walnut, upper airplane dial, lower cloth grill w/ vert bars, 5 knobs, BC, SW, AC $125.00
102-LW, Console, 1935, Walnut, upper airplane dial, lower grill w/ vertical bars, 5 knobs, BC, SW, LW, AC $125.00
103, Tombstone, 1935, Walnut, upper airplane dial, lower cloth grill w/vertical bars, BC, SW, battery $90.00
104, Tombstone, 1935, Walnut, arched top, center airplane dial, grill cut-outs, 5 knobs, BC, SW, AC $130.00
104-LW, Tombstone, 1935, Walnut, arched top, center airplane dial, grill cut-outs, 5 knobs, BC, SW, LW, AC $130.00
105, Console, 1935, Walnut, upper airplane dial, lower grill w/cut-outs, step-down top, BC, SW, AC $160.00
105-LW, Console, 1935, Walnut, upper airplane dial, grill w/cut-outs, step-down top, BC, SW, LW, AC $160.00

106, Table, 1935, Wood, center front dial over horiz bars, finished front & back, BC, SW, AC/DC $125.00

107, Table, 1935, Walnut, lower dial, arched top, finished front & back, 4 knobs, BC, SW, AC/DC $135.00

107-LW, Table, 1935, Walnut, lower dial, arched top, finished front & back, 4 knobs, BC, SW, LW, AC/DC $135.00

108, Tombstone, 1935, Walnut plastic, airplane dial, cloth grill w/ vertical bars, 2 knobs, BC, SW, AC/DC $250.00

108-LW, Tombstone, 1935, Walnut plastic, airplane dial, cloth grill w/ vert bars, 2 knobs, BC, SW, LW, AC/DC $250.00

109, Table, 1934, Bakelite, center airplane dial over cloth grill w/ vertical bars, 2 knobs, AC/DC $80.00

110, Tombstone, 1935, Wood, lower airplane dial, upper cloth grill w/ vert bars, 2 knobs, BC, SW, AC/DC $95.00

110-LW, Tombstone, 1935, Wood, lower airplane dial, upper grill w/ vert bars, 2 knobs, BC, SW, LW, AC/DC $95.00

111, Table, 1935, Walnut, lower dial, upper grill w/cut-outs, arched top, 4 knobs, BC, SW, AC/DC $110.00

111-LW, Table, 1935, Walnut, lower dial, grill w/cut-outs, arched top, 4 knobs, BC, SW, LW, AC/DC $110.00

149, Table, Plastic, right dial, left cloth grill w/3 horiz bars, decorative case lines, 2 knobs $75.00

157, Table, Plastic, small case, upper front dial w/center pointer, decorative lines, 2 knobs $75.00

199, Table, 1938, Bakelite, round dial, cloth grill w/vertical bars, decorative lines, 2 knobs, AC $80.00

238, Table, 1939, Wood, chest-type, inner right dial, left grill w/horiz bars, lift top, BC, AC/DC $250.00

239, Consolette, 1939, Wood, square dial, right & left horiz louvers, lower shelves, 2 knobs, BC, AC/DC $125.00

240, Consolette, 1939, Wood, square dial, right & left horiz louvers, lower shelves, 2 knobs, BC, AC/DC $140.00

241, Table-R/P, 1939, Wood, front square dial, vertical grill bars, lift top, inner phono, 2 knobs, BC, AC $50.00

250, Table, 1933, Burl walnut, right front dial, center cloth grill w/cut-outs, arched top, AC $100.00

257, Kitchen Radio, 1939, White finish, right square dial, left horiz louvers, top shelf, 2 knobs, BC, AC/DC $80.00

368, Console, 1940, Walnut, upper dial, pushbuttons, lower grill w/ vertical bars, 4 knobs, BC, SW, AC $120.00

370, Console-R/P, 1940, Walnut, center dial, pushbuttons, horiz grill bars, tuning eye, lift top, BC, SW, AC $120.00

380, Portable, 1940, Pocket-size, top slide rule dial & knobs, front vertical grill bars, handle, battery $45.00

382, Portable-R/P, 1940, Leatherette, inner dial, knobs & phono, outer horiz louvers, lift top, handle, AC $30.00

383, Table-R/P, 1940, Walnut, right front square dial, left grill w/vert bars, lift top, inner phono, AC $35.00

384, Table-R/P/Rec, 1940, Right front square dial, left horiz louvers, lift top, inner phono, handle, AC $40.00

385, Portable, 1940, Cloth, inner right dial, left horiz louvers, fold-down front, handle, AC/DC/bat $30.00

400 "**Aristocrat**", Table, 1940, Catalin, right front round dial, left horizontal grill bars, handle, 2 knobs, AC ....$600.00+

400 "**Patriot**", Table, 1940, Red, white & blue plastic, round dial, wrap-around grill bars, handle, 2 knobs ....$800.00+

403, Table-R/P, 1942, Wood, right front dial, left horizontal louvers, 4 knobs, lift top, inner phono, AC $45.00

409 "**Mickey Mouse**", Table-N, 1933, Ivory & light green w/metal trim, center grill w/Mickey & cello, 2 knobs, AC $1,250.00

336, Table, 1940, Plastic, right front dial, left grill w/geometric cut-outs, feet, 2 knobs, AC/DC **$45.00**

343, Table, 1940, Plastic, right dial, left horizontal grill bars & center vertical bars, BC, SW, AC $50.00

357, Portable, 1940, Tan & maroon leatherette, right dial, left horizontal louvers, handle, 2 knobs $30.00

363, Portable, 1940, Blue leatherette, right dial, left horizontal louvers, handle, 2 knobs, AC/DC/bat $30.00

364, Table-R/P, 1940, Wood, center front dial, right & left horizontal louvers, lift top, inner phono, AC $35.00

365, Table, 1940, Walnut, center front slide rule dial, rounded sides, tuning eye, hifi, 4 knobs, AC $75.00

410 "**Mickey Mouse**", Table-N, 1933, Wood, black & silver w/metal trim, octagonal metal grill w/Mickey, 2 knobs, AC $1,250.00

411 "Mickey Mouse", Table-N, 1933, Pressed wood w/ side/top/front Mickey musical "carvings", 2 knobs, BC, AC/DC **$1,400.00**

414, Table, 1933, Pressed wood, ornate case w/fleur-de-lis designs, lower right dial, 2 knobs, AC **$135.00**

416, Table, 1934, Walnut, center octagonal grill w/cut-outs, horizontal fluting, 2 knobs, AC/DC **$65.00**

421, Table, 1941, Walnut plastic, slanted slide rule dial, lower horizontal louvers, 4 knobs, AC **$40.00**

423, Table-R/P, 1941, Walnut finish, front slide rule dial over horizontal louvers, 3 knobs, lift top, AC **$35.00**

424, Portable, 1941, Leatherette, right round dial, left horizontal louvers, handle, 2 knobs, AC/DC/bat **$40.00**

427, Portable, 1941, Leatherette, right round dial, left horiz louvers, handle, 2 knobs, BC, AC/DC/bat **$40.00**

447, Table-R/P, 1941, Walnut, front slide rule dial, horizontal louvers, 3 knobs, lift top, inner phono, AC **$35.00**

459, Table, 1941, Walnut, upper slanted slide rule dial, lower horiz louvers, 4 knobs, BC, SW, AC **$40.00**

462, Console-R/P, 1942, Wood, front dial, 4 knobs, upper slideaway tambour door, inner phono, 3 bands, AC **$150.00**

503, Table, 1946, Wood, right round dial over perforated front panel, 2 knobs, BC, AC/DC **$35.00**

504, Table, 1946, Wood, right round dial, lucite grill panel w/ circular cut-outs, 2 knobs, BC, AC/DC **$65.00**

505, Portable, 1946, Right front dial over perforated grill, handle, 2 knobs, BC, AC/DC/battery **$35.00**

506, Table-R/P, 1946, Wood, inner left vertical dial, 3 knobs, right automatic phono, lift top, BC, AC **$25.00**

507, Table, 1946, Plastic, right front dial, left grill w/Deco cut-outs & Emerson logo, feet, 2 knobs **$40.00**

508, Portable, 1946, Small case, inner thumbwheel knobs, flip open lid, handle, BC, battery **$65.00**

509, Table, 1946, Plastic, right front dial, left grill w/circular cut-outs, handle, 2 knobs **$45.00**

510, Table, 1946, Wood, right round dial, plastic grill panel w/ circular cut-outs, 2 knobs, BC, AC/DC **$60.00**

511 "Moderne", Table, 1946, Plastic, raised plastic dial over large grill, top hand-hold, 2 knobs, BC, AC/DC **$45.00**

512, Table, 1946, Wood, upper slanted slide rule dial, lower perforated grill, 4 knobs, BC, AC/DC **$30.00**

514, Table, 1947, Plastic, upper slanted slide rule dial, horizontal louvers, 4 knobs, BC, SW, AC/DC **$45.00**

515, Table, 1947, Plastic, upper slanted slide rule dial, horizontal louvers, 4 knobs, BC, AC/DC **$45.00**

516, Table, 1947, Plastic, upper slanted slide rule dial, horizontal louvers, 4 knobs, BC, AC/DC **$45.00**

517 "Moderne", Table, 1947, Plastic, raised plastic dial over large grill, top hand-hold, 2 knobs, BC, AC/DC $45.00

519, Table, 1947, Wood, right front round dial over perforated grill, 2 knobs, BC, AC/DC $35.00

520, Table, 1946, Two-tone Catalin, round dial, white panel w/ checkered grill, 2 knobs, BC, AC/DC $175.00

522, Table, 1946, Plastic, right front black dial, left Deco grill, 2 knobs, handle, feet, BC, AC/DC $50.00

523, Portable, 1946, Luggage style, cloth covered, right round dial, 2 knobs, handle, BC, AC/DC/battery $25.00

524, Table, 1947, Wood, upper slanted slide rule dial, lower perforated grill, 4 knobs, BC, 3SW, AC $50.00

525, Table-R/P, 1947, Wood, center front round dial, 4 knobs, lift top, inner phono, BC, AC $35.00

528, Table, 1947, Wood, upper slanted slide rule dial, lower perforated grill, 4 knobs, BC, FM, AC $50.00

530, Table, 1948, Wood, upper slanted slide rule dial, lower horiz grill bars, 4 knobs, BC, AC/DC $40.00

531, Table, 1947, Two-tone wood, slanted slide rule dial, perforated grill, 4 knobs, BC, battery $30.00

532, Table, 1947, Plastic, upper slanted slide rule dial, lower horizontal louvers, 4 knobs, BC, bat $30.00

535, Table, 1947, Wood, right front square dial over large perforated grill, 2 knobs, BC, AC/DC $35.00

536, Portable, 1947, Leatherette, slide rule dial, perforated grill, 2 top knobs, handle, BC, AC/DC/bat $30.00

536A, Portable, 1947, Leatherette, slide rule dial, large grill, 2 top knobs, handle, BC, AC/DC/battery $30.00

537, Console-R/P, 1947, Wood, inner slide rule dial, 4 knobs, phono, 2 lift tops, scalloped base, BC, FM, AC $110.00

539, Table, 1946, Wood, right front round dial over perforated grill, 2 knobs, BC, AC/DC $40.00

540 "Emersonette", Table, 1947, Plastic, midget, right vertical slide rule dial, checkered grill, 2 knobs, BC, AC/DC $80.00

540A, Table, 1947, Plastic, midget, right vertical slide rule dial, checkered grill, 2 knobs, BC, AC/DC $80.00

541, Table, 1947, Wood, right raised plastic dial over large perforated grill, 2 knobs, BC, AC/DC $45.00

543, Table, 1947, Plastic, right round dial over metal wrap-around grill, 2 knobs, handle, BC, AC/DC $50.00

544, Table, 1947, Two-tone wood, right round dial, left horizontal grill bars, 2 knobs, BC, AC/DC $50.00

546, Table-R/P, 1947, Wood, front slanted slide rule dial, horizontal grill bars, 4 knobs, lift top, BC, AC $35.00

547, Table, 1947, Plastic, lower front slide rule dial, upper louvered grill, 2 knobs, BC, AC/DC $40.00

547A, Table, 1947, Plastic, lower front slide rule dial, upper louvered grill, 2 knobs, BC, AC/DC $40.00

550, Table, 1947, Wood, upper slanted slide rule dial, horizontal louvers, 4 knobs, BC, AC/DC $40.00

551, Portable, 1947, Leatherette, inner dial, horiz louvers, 2 knobs, drop-front, handle, BC, AC/DC/bat $30.00

551A, Portable, 1947, Leatherette, inner dial, horiz louvers, 2 knobs, drop-front, handle, BC, AC/DC/bat $30.00

552, Table-R/P, 1947, Wood, center front round dial, 4 knobs, 3/4 lift top, inner phono, BC, AC $35.00

553, Portable, 1947, Leatherette, slide rule dial, horizontal grill bars, 2 knobs, handle, BC, AC/DC/bat $35.00

553A, Portable, 1947, Leatherette, slide rule dial, horizontal grill bars, 2 knobs, handle, BC, AC/DC/bat $35.00

555 "The All-American", Portable, 1959, Plastic, transistor, right dial, left lattice grill, thumbwheel knobs, AM, battery $35.00
556, Table, 1949, Plastic, lower slide rule dial, recessed woven grill, 4 knobs, BC, FM, AC/DC $40.00
557, Table, 1948, Plastic, lower front slide rule dial, recessed lattice grill, 4 knobs, BC, FM, AC/DC $45.00
557B, Table, 1948, Plastic, lower front slide rule dial, recessed lattice grill, 4 knobs, BC, FM, AC/DC $45.00
558, Portable, 1948, Plastic, inner slide rule dial, horiz louvers, 2 knobs, lift-up lid, handle, BC, bat $60.00
559, Portable, 1948, "Alligator", slide rule dial, vertical grill bars, 2 knobs, handle, BC, AC/DC/bat $40.00
559A, Portable, 1948, "Alligator", slide rule dial, vertical grill bars, 2 knobs, handle, BC, AC/DC/bat $40.00
560, Portable, 1947, Plastic, lower slide rule dial, vertical grill bars, handle, 2 knobs, BC, battery $40.00
561, Table, 1949, Plastic, lower curved slide rule dial, vertical grill bars, 2 knobs, BC, AC/DC $45.00
564, Table, 1940, Catalin, midget, right vertical slide rule dial, lattice grill, 2 knobs, BC, AC/DC .... $500.00+
565, Table, 1949, Wood, lower front slide rule dial, upper cloth grill, 4 knobs, AM, FM, AC $45.00
568, Portable, 1949, Plastic, curved slide rule dial, checkered half-moon grill, handle, BC, AC/DC/bat $35.00
568A, Portable, 1949, Plastic, curved slide rule dial, checkered half-moon grill, handle, BC, AC/DC/bat $35.00
569, Portable, 1948, Plastic, inner dial, horizontal grill bars, 2 knobs, flip-up front, BC, AC/DC/bat $60.00
569A, Portable, 1948, Plastic, inner dial, horizontal grill bars, 2 knobs, flip-up front, BC, AC/DC/bat $60.00

572, Table, 1949, Plastic, center square dial w/inner grill & large pointer, 2 knobs, BC, AC/DC $60.00
573B, Console-R/P, 1948, Inner right front dial, 4 knobs, left pull-out phono drawer, scalloped base, BC, AC $80.00
574 "Memento", Table, 1949, Wood, jewelry box-style, inner slide rule dial, 2 knobs, photo frame in lid, BC, bat $150.00
575, Portable, 1950, Plastic, curved slide rule dial, checkered half-moon grill, handle, BC, AC/DC/bat $35.00
576, Console-R/P, 1948, Wood, inner left slide rule dial, 4 knobs, pull-out phono, criss-cross grill, BC, AC $75.00
576A, Console-R/P, 1948, Wood, inner left slide rule dial, 4 knobs, pull-out phono, criss-cross grill, BC, AC $75.00
577, Table, 1948, Wood, center front slide rule dial, upper criss-cross grill, 4 knobs, BC, AC $50.00
577B, Table, 1948, Wood, center front slide rule dial, upper criss-cross grill, 4 knobs, BC, AC $50.00
578A, Table, 1946, Walnut, large gold plastic front grill w/slide rule dial, horiz louvers, 2 knobs, AC $50.00
579, Table-R/P, 1949, Plastic, lower front slide rule dial, vertical grill bars, 4 knobs, open top, BC, AC $30.00
579A, Table-R/P, 1949, Plastic, lower front slide rule dial, vertical grill bars, 4 knobs, open top, BC, AC $30.00
580 "Memento", Table, 1949, Jewelry box-style, inner slide rule dial, 2 knobs, photo frame in lid, BC, battery $135.00
581, Table, 1949, Plastic, recessed round dial, horizontal grill bars, handle, 2 knobs, BC, AC/DC $45.00
586, Console-R/P, 1949, Wood, inner right slide rule dial, 4 knobs, left pull-out phono drawer, BC, FM, AC $85.00
591, Table, 1949, Plastic, upper slanted slide rule dial, horizontal grill bars, 2 knobs, BC, AC/DC $40.00
596, Table-R/P, 1949, Wood, front slide rule dial, horiz louvers, 2 knobs, lift top, inner phono, BC, AC $35.00
597, Table, 1950, Plastic, lower front slide rule dial, upper lattice grill, 4 knobs, BC, 2SW, AC $40.00
599, Table, 1949, Plastic, lower slanted slide rule dial, recessed lattice grill, 4 knobs, BC, AC/DC $45.00
602, Table, 1949, Plastic, front cylindrical slide rule dial, lattice grill, 2 side knobs, FM, AC/DC $60.00
603, Console-R/P, 1949, Wood, inner right slide rule dial, 4 knobs, left pull-out phono drawer, BC, FM, AC $75.00
605, Console-R/P, 1949, Wood, center front dial, 4 knobs, fold-up front panel, pull-out phono, BC, FM, AC $55.00
610, Table, 1949, Plastic, front cylindrical slide rule dial, "woven" grill, 2 side knobs, BC, AC/DC $65.00
613A, Portable, 1949, Large round front dial/grill louvers, flex handle, 2 side knobs, BC, AC/DC/bat $35.00

570 "Memento", Table, 1949, Jewelry box-style, inner slide rule dial, 2 knobs, photo frame in lid, BC, battery $135.00

616A, Table, Wood, lower rounded slide rule dial, upper plastic hi-lo grill area, side knobs, BC $60.00
634B, Table-R/P, 1950, Wood, inner slide rule dial, 2 knobs, outer horizontal louvers, lift top, BC, AC $30.00

635, Table-R/P, 1950, Plastic, top slide rule dial, 2 knobs, vert wrap-over grill bars, lift top, BC, AC $60.00

636A, Table, 1950, Plastic, lower cylindrical slide rule dial, lattice grill, 2 side knobs, BC, AC/DC $65.00

640, Portable, 1950, Plastic, inner slide rule dial, horiz louvers, 2 knobs, flip-up front, handle, BC, bat $45.00

641B, Table, 1951, Plastic, right front slide rule dial, left geometric grill bars, 3 knobs, BC, AC/DC $35.00

642, Table, 1950, Plastic, right round dial over horizontal grill bars, handle, 2 knobs, BC, AC/DC $45.00

643A, Portable, 1950, Leatherette, slide rule dial, handle, telescoping antenna, BC, 3SW, AC/DC/battery $30.00

645, Portable, 1950, Plastic, round front dial/grill bars, flex handle, "fan-tenna" pulls up, BC, battery $45.00

646, Portable, 1950, Plastic, slide rule dial overlaps "woven" grill, handle, 2 knobs, AC/DC/battery $35.00

646A, Portable, 1950, Plastic, slide rule dial overlaps "woven" grill, handle, 2 knobs, BC, AC/DC/bat $35.00

652, Table, 1950, Plastic, right front slide rule dial, left round grill w/ checkered cut-outs, 2 knobs $35.00

656, Portable, 1950, Plastic, slide rule dial, half-moon grill, 2 thumbwheel knobs, handle, AC/DC/bat $35.00

657, Portable, 1950, Leatherette, slide rule dial, half-moon grill, thumbwheel knobs, handle, AC/DC/bat $35.00

659, Table, 1949, Plastic, right slide rule dial, left grill of concentric squares, 3 knobs, AM, FM $40.00

671B, Table-C, 1950, Plastic, left slide rule dial, horizontal grill bars, right clock, 5 knobs, BC, AC $35.00

672B, Table-R/P, 1951, Wood, inner slide rule dial, 2 knobs, phono, lift top, criss-cross grill, BC, AC $30.00

679B, Console-R/P, 1951, Wood, inner right slide rule dial, 3 knobs, left phono, double doors, BC, FM, AC $55.00

691B, Table, 1952, Plastic, lower slide rule dial, large upper lattice grill, 2 knobs, BC, SW, AC $35.00

695B, Table-C, 1952, Plastic, center vertical slide rule dial, right clock, left checkered grill, BC, AC $30.00

702B, Table, 1952, Wood, lower slide rule dial, upper right & left horizontal grill bars, 2 knobs, BC $30.00

703B, Table-R/P, 1952, Wood, inner dial, knobs & phono, lift top, front horizontal grill bars, BC, AC $35.00

704, Portable, 1952, Plastic, center front round dial, checkered grill, thumbwheel knob, handle, BC, bat $45.00

705, Portable, 1953, Plastic, upper front round dial knob, center checkered plastic grill panel, handle $45.00

706, Table, 1952, Plastic, small case, lower half-round dial over checkered grill, Emerson logo $50.00

706-B, Table, 1952, Plastic, small case, lower half-round dial over checkered grill, Emerson logo $50.00

707-B, Table, 1952, Plastic, "sun-burst" front, lower dial, upper "rayed" grill, 2 knobs, BC, AC/DC $70.00

713, Table, 1952, Wood, "sun-burst" front, lower dial, upper "rayed" grill, 2 knobs, BC, AC/DC $70.00

718B, Table-C, 1953, Plastic, left front square dial, right square alarm clock, 5 knobs, feet, BC, AC $25.00

724B, Table-C, 1953, Plastic, large half-moon dial w/inner round clock, side grill bars, 5 knobs, BC, AC $40.00

744B, Table, 1954, Plastic, large front raised dome-shaped dial, 2 top thumbwheel knobs, BC, AC/DC $200.00

745B, Portable, 1954, Leatherette, left front dial, right plaid grill, side knob, handle, BC, AC/DC/bat $35.00

746B, Portable, 1954, Left front dial, right plaid grill, handle, telescope antenna, BC, 2SW, AC/DC/bat $35.00

747, Portable, 1954, Plastic, right front dial, horizontal grill bars, handle, thumbwheel knob, BC, bat $35.00

756, Series B, Table, 1953, Plastic, right slide rule dial, left grill w/concentric squares, 3 knobs, BC, AC $35.00

778-B, Table, 1954, Plastic, center front round dial amid circular louvers, 2 knobs, feet, BC, AC/DC $45.00

788B, Table-C, 1954, Plastic, modern, center dial, right perforated grill, left clock, 3 knobs, BC, AC $60.00

790, Portable, 1954, Plastic, right front round dial knob, left wrap-around horizontal grill bars, handle $30.00

810B, Table, 1955, Plastic, lower front dial, upper pointer over large grill, side knobs, BC, AC/DC $40.00

811, Table, 1955, Plastic, small case, front half-round dial over horiz grill bars, feet, side knob $40.00

811B, Table, 1955, Plastic, small case, front half-round dial over horiz grill bars, feet, side knob **$40.00**

813, Table, 1955, Plastic, large round center dial over gold grill, upper left Emerson logo, feet **$35.00**
814B, Table-R/P, 1955, Wood, outer front round dial over large grill, 3 side knobs, 3/4 lift top, BC, AC **$25.00**

823, Table, Plastic, center slide rule dial, upper & lower horiz bars & vert dots, 3 knobs **$25.00**
832B, Table, 1956, Plastic, right side dial knob, front horizontal bars, "Emerson" band, BC, AC/DC **$20.00**

838, Portable, 1956, Plastic, tube & transistor, right round dial, left checkered grill, handle, BC, bat **$50.00**
842, Portable, 1956, Leather case w/grill, transistor, center round dial, side knob, handle, AM, bat **$40.00**
848, Portable, 1957, Plastic, top dial, front checkered grill, left thumbwheel knob, handle, AC/DC/bat **$30.00**

850, Portable, Plastic, front round dial knob, center lattice grill panel, thumbwheel knob, handle **$35.00**

852, Series B, Table, 1957, Plastic, lower right window dial over large recessed checkered grill, BC, AC/DC **$30.00**
855, Portable, 1957, Leather case w/front grill, transistor, center front round dial, handle, AM, bat **$45.00**
867B, Console-R/P, 1958, Wood, inner right dial, phono, lift top, lower front grill, feet, BC, AC **$40.00**
868, Portable, 1960, Plastic, transistor, front dial knob, trapezoid grill, handle, 2 knobs, AM, battery **$35.00**
883B, Table-C, 1958, Plastic, center window dial, right alarm clock, left horizontal louvers, BC, AC **$20.00**
888 "Atlas" , Portable, 1960, Plastic, transistor, upper dial, random-patterned grill, swing handle, AM, battery **$55.00**
888 "Explorer", Portable, 1960, Plastic, transistor, thumbwheel dial, diagonal metal grill, swing handle, BC, bat **$55.00**
888 "Pioneer", Portable, 1958, Plastic, transistor, upper dial, criss-cross metal grill, swing handle, AM, battery **$55.00**
888 "Transtimer", Portable-C, 1960, Leather case, transistor, inner round dial/alarm clock/speaker, handle, AM, bat **$100.00**
888 "Transtimer II", Portable-C, 1960, Leather case, transistor, inner round dial/alarm clock/speaker, handle, AM, bat **$100.00**

888 "Vanguard", **Portable, 1958, Plastic, transistor, upper dial, random patterned grill, swing handle, AM, battery** .**$55.00**
911 "Eldorado", Portable, 1960, Plastic, transistor, right front dial over checkered grill, swing handle, AM, bat **$40.00**

999 "Champion", Portable, 1959, Transistor, left front dial, checkered grill, right thumbwheel knob, AM, battery **$30.00**

1002, Table, 1947, Plastic, upper slanted slide rule dial, lower horiz louvers, 4 knobs, BC, AC/DC **$45.00**

A-130, Table, 1936, Wood, Ingraham case, right dial, left cloth grill w/horiz bars, 2 knobs, BC, SW, AC $45.00

A-132, Table, 1936, Wood, Repwood front panel, right dial, left cloth grill w/cut-outs, 2 knobs, AC $45.00

AA-204, Table, 1938, Wood, right front dial, left wrap-around horiz louvers, 3 knobs, BC, SW, AC/DC **$55.00**

AA-207, Table, 1938, Plastic, right dial, horiz louvers, decorative case lines, 3 knobs, BC, SW, AC/DC **$55.00**

AL-132, Table, 1936, Wood, Repwood front panel, right dial, left cloth grill w/cut-outs, 2 knobs, AC $45.00

AM-169, Table, 1937, Walnut, right front dial, rounded left side w/ vertical bars, 3 knobs, BC, SW, AC **$65.00**

AR-165, Chairside-R/P, 1937, Wood, Deco, step-down top, rounded front w/horizontal louvers, inner phono, AC **$175.00**

AU-190, Tombstone, 1938, Catalin, lower scalloped gold dial, 3 upper vertical sections, cloth grill, 3 knobs ...... $1,000.00+

AU-213, Table, 1938, Walnut, "tombstone" shape, half-moon dial, vert grill bars, 3 knobs, BC, AC/DC **$145.00**

AX-211 "Little Miracle", Table, 1938, Plastic, midget, right dial, horizontal grill bars & front lines, 2 knobs, AC/DC . $65.00

AX-212, Table, 1938, Wood, Deco, right dial, left round grill w/ concentric circle louvers, 2 knobs, AC/DC **$135.00**

AX-217, Table, 1938, Wood, right front square dial, left horizontal louvers, 2 bullet knobs, BC, AC/DC **$40.00**

AX-219, Table-R/P, 1938, Wood, right front square dial, left horiz louvers, 2 knobs, open top phono, BC, AC **$25.00**

AX-221, Table-R/P, 1938, Wood, right dial, curved left w/vertical bars, 2 knobs, lift top, inner phono, AC **$35.00**

AX-221-AC-DC, Table-R/P, 1938, Wood, right dial, curved left w/vert bars, 2 knobs, lift top, inner phono, AC/DC **$35.00**

AX-222, Portable-R/P, 1938, Cloth-covered, inner dial & grill, fold-down front, lift top, handle, BC, AC/DC **$30.00**

AX-232, Portable-R/P, 1938, "Alligator", inner dial/grill/2 knobs, fold-down front, lift top, handle, BC, AC **$30.00**

AX-232 AC-DC, Portable-R/P, 1938, "Alligator", inner dial/grill/2 knobs, fold-down front, lift top, handle, BC, AC/DC **$30.00**

AX-235 **"Little Miracle"**, Table, 1938, Catalin, right square dial, left horizontal louvers, flared base, 2 knobs, BC, AC/DC ... **$1,000.00+**

AX-238, Table, 1939, Wood, chest-type, inner right dial, left grill w/horiz bars, lift top, BC, AC/DC **$250.00**

B-131, Table, 1936, Wood, right front gold gemloid dial, left cloth grill w/horizontal bars, BC, SW, AC **$45.00**

BA-199, Table, 1938, Plastic, center front round dial, cloth grill w/ vertical bars, 2 knobs, AC/DC **$85.00**

BB-208, Table, 1938, Plastic, right front dial, left horizontal louvers, pushbuttons, 2 knobs, BC, AC/DC **$55.00**

BB-209, Table, 1938, Wood, right front dial, left horizontal louvers, pushbuttons, 2 knobs, BC, AC/DC **$50.00**

BD-197 "Mae West", Table, 1938, Wood, curved case, right front conical dial, left conical grill, 4 knobs, BC, SW, AC **$350.00**

BF-169, Table, 1938, Wood, right front dial, curved left w/vertical bars, 3 knobs, BC, SW, AC/DC **$65.00**

BF-204, Table, 1938, Wood, right front dial, left wrap-around horiz louvers, 3 knobs, BC, SW, AC/DC **$55.00**

BF-207, Table, 1938, Plastic, right dial, horiz louvers, decorative case lines, 3 knobs, BC, SW, AC/DC **$55.00**

BF-316, Table, 1938, Wood, right front gold dial, rounded left side w/ wrap-around grill bars, 3 knobs **$75.00**

BJ-200, Table, 1938, Plastic, right dial, left horiz louvers, decorative case lines, 2 knobs, BC, AC/DC **$55.00**

BJ-210, Table, 1938, Wood, right front dial, curved left w/vertical bars, 2 knobs, BC, AC/DC **$65.00**

BJ-214, Table, 1938, Wood, right dial, curved left w/horiz wrap-around louvers, 2 knobs, BC, AC/DC **$75.00**

BJ-218, Table-R/P, 1938, Wood, right front dial, curved left w/ horizontal bars, 2 knobs, lift top, BC, AC/DC **$40.00**

BJ-220, Table-R/P, 1938, Wood, right dial, left horiz grill bars, 2 knobs, lift top, inner phono, BC, AC/DC **$35.00**

BL-200, Table, 1938, Plastic, right gold dial, left horiz louvers, decorative lines, 2 knobs, BC, SW, AC **$55.00**

BL-210, Table, 1938, Wood, right front dial, curved left w/vertical bars, 2 knobs, BC, AC **$65.00**

BL-214, Table, 1938, Wood, right dial, curved left w/horizontal wrap-around louvers, 2 knobs, BC, AC **$75.00**

BL-218, Table-R/P, 1938, Wood, right front dial, curved left w/ horizontal bars, 2 knobs, lift top, BC, AC **$40.00**

BL-220, Table-R/P, 1938, Wood, right front dial, left horizontal bars, 2 knobs, lift top, inner phono, BC, AC **$35.00**

BM-206, Table, 1938, Plastic, right front dial, left horizontal louvers, 2 knobs, BC, AC/DC **$40.00**

BM-215, Table, 1938, Two-tone wood, right front gold dial, left horizontal louvers, 2 knobs, AC/DC **$45.00**

BM-216, Table-R/P, 1938, Wood, right front dial, left horizontal louvers, 2 knobs, open top phono, BC, AC **$25.00**

BM-242, Table-R/P, 1939, Wood, right front dial, left horizontal louvers, 2 knobs, lift top, inner phono, AC **$35.00**

BM-247 "Snow White", Table-N, 1938, Snow White and Dwarfs in pressed wood, right dial, left grill, 2 knobs, BC, AC/DC **$1,200.00**

BM-258, Table, 1937, Catalin, small case, right front dial, left inset horizontal louvers, 2 knobs, AC/DC .... **$1,000.00+**

BN-216, Table-R/P, 1938, Wood, right front dial, left horizontal louvers, 2 knobs, open top phono, BC, AC **$25.00**

BQ-223, Console-R/P, 1938, Wood, center conical dial, pushbuttons, 4 knobs, left lift top, BC, SW, AC **$175.00**

BQ-225, Console, 1938, Wood, center conical dial, pushbuttons, lower horiz louvers, 4 knobs, BC, SW, AC **$150.00**

BQ-228, Table, 1938, Wood, right conical dial, left horizontal louvers, pushbuttons, 4 knobs, BC, SW, AC **$100.00**

BR-224, Console-R/P, 1938, Wood, right conical dial, pushbuttons, horizontal louvers, left lift top, BC, SW, AC **$200.00**

BR-224-A, Console-R/P, 1938, Wood, right conical dial, pushbuttons, horizontal louvers, left lift top, BC, SW, AC **$200.00**

BR-226 "Symphony Grand", Console, 1938, Wood, center conical dial, pushbuttons, lower horiz louvers, 4 knobs, BC, SW, AC **$190.00**

BS-227 "Queen Anne", Console, 1938, Wood, lowboy, Queen Anne style, conical dial, pushbuttons, tuning eye, BC, SW, AC **$300.00**

BT-245, Tombstone, 1939, Catalin, scalloped dial, upper 3-section grill with horiz louvers, 3 knobs, AC/DC ....$1,000.00+

BU-229, Table, 1938, Wood, conical dial, horiz grill bars, pushbuttons, tuning eye, 4 knobs, BC, SW, AC **$115.00**

BU-230 "Chippendale", Console, 1938, Wood, conical dial, oval grill w/horiz louvers, pushbuttons, tuning eye, BC, SW, AC **$150.00**

BW-231, Table, 1938, Wood, right front conical dial, left horizontal louvers, 4 knobs, BC, SW, AC **$85.00**

BX-208, Table, 1939, Plastic, right front dial, left horizontal louvers, pushbuttons, 2 knobs, BC, AC/DC **$55.00**

CE-259, Portable, 1939, Luggage-style, center front square dial, side grill, handle, 2 knobs, BC, battery **$25.00**

CE-260, Table, 1939, Wood, center front square dial, right & left horizontal louvers, 2 knobs, BC, bat **$35.00**

CF-255 "Emersonette", Table, 1939, Plastic, midget, right vertical slide rule dial, horiz grill bars, 2 knobs, BC, AC/DC **$80.00**

CG-268, Table, 1939, Plastic, right dial, left horizontal wrap-around louvers, 3 knobs, BC, SW, AC/DC **$40.00**

CG-276, Table, 1939, Wood, right front square dial, left grill w/vertical bars, 3 knobs, BC, SW, AC/DC **$55.00**

CG-293, Console-R/P, 1939, Wood, inner right dial & knobs, center phono, lift top, front horiz bars, BC, SW, AC **$80.00**

CG-294, Console-R/P, 1939, Wood, inner right dial & knobs, left phono, lift top, 4 vert front bars, BC, SW, AC **$80.00**

CG-318, Table, 1939, Walnut w/inlay, right square dial, left vertical grill bars, 3 knobs, BC, SW, AC/DC **$60.00**

CH-246, Table, 1939, Plastic, right dial, raised left w/horiz wrap-around louvers, 2 knobs, BC, AC/DC **$60.00**

CH-253, Table, 1939, Embossed leatherette, right half-moon dial, left horiz louvers, 2 knobs, AC/DC **$65.00**

CH-256 "Strad", Table, 1939, Wood, violin shape, off-center dial, left horiz louvers, top cut-out, BC, AC/DC **$425.00**

CJ-238, Table, 1939, Wood, chest-type, inner dial, horiz louvers, lift top starts and stops radio, AC/DC **$250.00**

CJ-257, Kitchen Radio, 1939, White finish, right square dial, left horiz louvers, top shelf, 2 knobs, BC, AC/DC **$80.00**

CL-256 "Strad", Table, 1939, Wood, violin shape, off-center dial, left horiz louvers, top cut-out, BC, AC/DC $425.00

CQ-269, Table, 1939, Plastic, right dial, left horiz wrap-around louvers, 4 pushbuttons, 3 knobs, AC/DC **$50.00**

CQ-271, Table, 1939, Wood, right front dial, left vertical bars, 4 pushbuttons, 3 knobs, BC, SW, AC/DC **$65.00**

CQ-273, Table, 1939, Wood, right square dial, curved left w/vert bars, 4 pushbuttons, 3 knobs, AC/DC **$75.00**

CR-1-303, Portable-R/P, 1939, Leatherette, inner dial & louvers, fold-down front, lift top, inner phono, AC/DC **$35.00**

CR-261, Table, 1939, Wood, right front square dial, curved left w/vert grill bars, 2 knobs, BC, AC/DC **$65.00**

CR-262, Table, 1939, Wood, curved front w/recessed right dial/left vert grill bars, 2 knobs, BC, AC/DC **$75.00**

CR-274, Table, 1939, Plastic, right dial, raised left w/horiz wrap-around louvers, 2 knobs, BC, AC/DC **$45.00**

CR-297, Console-R/P, 1939, Wood, inner right dial & knobs, left phono, lift top, vertical front bars, BC, AC **$50.00**

CR-303, Portable-R/P, 1939, Leatherette, inner dial & louvers, fold down front, lift top, inner phono, AC **$35.00**

CS-268, Table, 1939, Plastic, right front dial, left wrap-around horiz louvers, 3 knobs, BC, SW, AC/DC **$45.00**

CS-270, Table, 1939, Wood, right front square dial, left vertical grill bars, 3 knobs, BC, SW, AC/DC **$65.00**

CS-272, Table, 1939, Wood, right square dial, curved left w/vertical bars, 3 knobs, BC, SW, AC/DC **$65.00**

CS-317, Table, 1939, Walnut, right front square dial, left vert grill bars, "waterfall" top & base, AC/DC **$60.00**

CS-320, Table, 1939, Two-tone walnut, right square dial, left vert grill bars, 3 knobs, BC, SW, AC/DC **$55.00**

CT-275, Portable, 1939, Cloth covered w/stripes, right front dial, left grill, handle, 2 knobs, battery **$25.00**

CU-265, Table, 1939, Plastic, right square dial over horizontal front bars, 5 tubes, 2 knobs, BC, AC/DC **$45.00**

CV-1-290, Portable-R/P, 1939, Cloth covered, inner dial & louvers, fold-down front, lift top, inner phono, AC/DC **$30.00**

CV-1-291, Table-R/P, 1939, Wood, right square dial, curved left w/ horiz bars, lift top, inner phono, AC/DC **$35.00**

CV-264, Table, 1939, Embossed design, right front square dial, left horiz louvers, 2 knobs, BC, AC/DC **$45.00**

CV-280, Portable, 1939, Cloth covered, inner right dial, left horiz louvers, fold-down front, handle, AC/DC **$30.00**

CV-289, Table-R/P, 1939, Wood, right front square dial, left vertical grill bars, lift top, inner phono, BC, AC **$35.00**

CV-290, Portable-R/P, 1939, Cloth covered, inner dial & louvers, fold-down front, lift top, inner phono, AC **$30.00**

CV-291, Table-R/P, 1939, Wood, right square dial, curved left w/ horizontal bars, lift top, inner phono, AC **$35.00**

CV-295, Table, 1939, Wood, right front square dial, curved left w/vert grill bars, 2 knobs, BC, AC/DC **$65.00**

CV-296 "Strad", Table, 1939, Wood, violin shape, right square dial, left horiz louvers, top cut-outs, BC, AC/DC ...**$425.00**

CV-298 "Strad", Table, 1939, Wood, violin shape, right square dial, left horiz louvers, top cut-outs, BC, AC/DC **$425.00**

CV-313, Table, 1939, Walnut, right square dial, left vertical grill bars, rounded sides, 2 knobs, AC/DC **$80.00**

CV-316, Table, 1939, Walnut, right square dial, left horiz louvers, rounded sides, 2 knobs, BC, AC/DC **$85.00**

CW-279, Table, 1939, Plastic, right dial, right & left wrap-around bars, 4 pushbuttons, 2 knobs, AC/DC **$50.00**

CX-263, Portable, 1939, Cloth covered, right front dial, left horizontal louvers, handle, 2 knobs, BC, bat **$25.00**

CX-283, Portable, 1939, Cloth covered, right front dial, left horizontal louvers, handle, 2 knobs, BC, bat **$25.00**

CX-284, Portable, 1939, Cloth covered, inner right dial, left louvers, slide-in door, handle, 2 knobs, bat **$30.00**

CX-285, Table, 1940, Wood, center front dial, right & left horizontal louvers, 2 knobs, BC, battery **$35.00**

CX-292, Portable-R/P, 1939, Cloth covered, right front dial, left louvers, lift top, inner phono, handle, BC, bat **$25.00**

CX-305, Portable, 1939, Walnut, right recessed dial, left horizontal grill bars, handle, 2 knobs, BC, bat **$45.00**

CY-269, Table, 1939, Plastic, right dial, left wrap-around horiz louvers, 4 pushbuttons, 3 knobs, AC/DC **$55.00**

CY-286, Table, 1939, Wood, right dial, left horiz grill bars, 4 pushbuttons, 3 knobs, BC, SW, AC/DC **$45.00**

CY-288, Table, 1939, Wood, right square dial, curved left w/horiz bars, 4 pushbuttons, 3 knobs, AC/DC **$75.00**

CZ-282, Table, 1939, Wood, curved right w/recessed pushbuttons, left vert bars, 2 knobs, BC, AC/DC **$85.00**

DA-287, Table, 1939, Two-tone wood, right front dial, left horizontal bars, 3 knobs, BC, SW, AC **$50.00**

DB-247 "Snow White", Table-N, 1939, Snow White & Dwarfs in pressed wood, right dial, left cloth grill, 2 knobs, AC/DC **$1,200.00**

DB-296, Table, 1939, Wood, right front dial, left horizontal louvers, handle, 2 knobs, BC, AC/DC **$45.00**

DB-301, Table, 1939, Plastic, right front dial, left horizontal louvers, handle, 2 knobs, BC, AC/DC **$45.00**

DB-315, Table, 1939, Walnut, right front dial, left conical grill, handle, 2 knobs, 5 tubes, BC, AC/DC **$55.00**

DC-308, Portable, 1939, Leatherette, right front dial, large left grill, handle, 2 knobs, battery **$25.00**

DF-302, Portable, 1939, Walnut w/inlay, right front dial, left horizontal louvers, handle, BC, AC/DC/bat **$50.00**

DF-306, Portable, 1939, Cloth covered, inner dial & louvers, fold-down front, handle, 2 knobs, AC/DC/bat **$30.00**

DH-264, Table, 1939, Embossed design, right front square dial, left horiz louvers, 2 knobs, BC, battery **$35.00**

DJ-200, Table, 1939, Plastic, right front dial, left horizontal louvers, decorative case lines, 2 knobs **$85.00**

DJ-310, Portable, 1939, Leatherette, right front dial, left horizontal louvers, handle, 2 knobs, AC/DC/bat **$30.00**

DM-331, Table, 1939, Two-tone wood, right front dial, left two-section grill w/vertical bars, 3 knobs **$40.00**

DQ-334, Table, Two-tone wood, right front square dial, left horizontal grill bars, 2 knobs **$45.00**

EC-301, Table, 1940, Plastic, right front gold dial, left horizontal louvers, handle, 2 knobs **$50.00**

EP-375 "5+1", Table, 1941, Catalin, right square dial, 5 left grill bars and 1 right bar, handle, 2 knobs, AC .... **$800.00+**

F-133, Table, 1936, Two-tone wood, right front gold dial, horizontal decorative lines, BC, SW, AC/DC **$85.00**

FU-427, Portable, 1941, Cloth covered, right front round dial, left horizontal louvers, handle, 2 knobs **$30.00**

L-143, Table-R/P, 1936, Wood, high case, lower right dial, left grill, lift top, inner phono, BC, SW, AC **$45.00**

L-150, Chairside, 1935, Walnut, oval, step-down top w/upper level drawer & lower dial & knobs, 2 band **$140.00**

L-458, Table, 1932, Two-tone walnut, right front window dial, center round grill w/cut-outs, 2 knobs **$65.00**

L-556, Cathedral, 1932, Two-tone wood, front window dial, upper cloth grill w/cut-outs, AC **$250.00**

L-559, Table, 1932, Wood, chest-type, inner top dial & knobs, lift-up lid, fancy front grill, AC **$225.00**

LA, Table, Plastic, decorative case design, right dial, center cloth grill w/cut-outs, 2 knobs **$95.00**

M-755, Cathedral, 1932, Two-tone burl walnut, center front window dial, upper cloth grill w/cut-outs **$265.00**

Q-236 "Snow White", Table-N, 1939, Snow White & Dwarfs in pressed wood, center front dial & cloth grill, 2 knobs, AC **$1,200.00**

U-5A, Tombstone, 1935, Plastic, Deco, round dial, cloth grill w/ vertical bars, 2 knobs, plastic back, AC **$250.00**

U-6D, Table, 1934, Wood, lower front round dial, upper cloth grill w/ cut-outs, 2 knobs **$135.00**

U-6F, Table, 1936, Wood, lower front dial, upper cloth grill w/cut-outs, rounded top, rounded sides **$135.00**

X-175, Console-R/P, 1937, Wood, Deco, right dial, raised and rounded left side w/horiz bars, BC, SW, AC **$285.00**

Z-159, Upright Table, 1937, Wood, slanted front panel, center dial, upper grill w/vertical bars, AC **$85.00**

EMOR
Emor Radio, Ltd.,
400 East 118th St., New York City, New York

100, Floor-N, 1947, Chrome, Deco, globe on adjustable floor stand, top grill bars, 5 tubes, BC, SW **$350.00**

EMPIRE
Empire Electrical Products, Co.,
102 Wooster Street, New York City, New York

30, Table, 1933, Wood, front dial, center cloth grill w/cut-outs, rounded top, 2 knobs, AC/DC **$85.00**

EMPRESS
Empire Designing Corp.,
1560 Broadway, New York, New York

55, Table-N, 1946, Wood, cottage-shaped, dial in door, 2 knobs in windows, lattice grill, BC, AC/DC **$200.00**

ERLA
Electrical Research Laboratories,
2500 Cottage Grove Avenue, Chicago, Illinois

The name Erla is short for Electrical Research Laboratories. The company began in 1921 selling component parts and radio kits and by 1923 they were marketing complete sets. After much financial difficulty, Erla was reorganized as Sentinel Radio Corporation in 1934.

22P, Cathedral-C, 1930, Wood, lower front horizontal dial, upper cloth grill w/cut-outs and clock face, AC **$350.00**

30, Console, 1929, Wood, highboy,lower front window dial w/escutcheon, upper grill w/cut-outs, AC **$165.00**

271, Cathedral, Wood, squared top, lower window dial, upper round grill w/cut-outs, 2 knobs, AC **$165.00**

De Luxe Super-Five, Table, 1925, Wood, low rectangular case, 3 dial front panel, scalloped molding, feet, battery **$145.00**

S-11, Table, 1925, Wood, low rectangular case, slanted panel w/ 3 pointer dials, button feet, battery $160.00

ESPEY
Espey Mfg. Co., Inc.,
33 West 46th Street, New York, New York

7-861, End Table, 1938, Mahogany or walnut, Duncan Phyfe style, drop-leaf, inner 6 tube radio, BC, SW, AC **$145.00**

18B, Table, 1950, Wood, upper front slide rule dial, large center cloth grill, 2 knobs, BC, AC/DC **$40.00**

31 "Roundabout", Table, 1950, Two-tone plastic cylindrical fluted case, right front dial, 2 knobs, BC, AC/DC **$150.00**

651, Table, 1946, Plastic, upper slanted slide rule dial, lower horiz louvers, 2 knobs, BC, AC/DC **$45.00**

861B, Portable, 1938, Inner right front dial, left grill, removable cover, handle, 3 knobs, BC, SW, AC **$25.00**

6545, Table-R/P, 1946, Wood, top slide rule dial, front criss-cross grill, 3 knobs, open top, BC, AC **$20.00**

6547, Table-R/P, 1946, Wood, front slide rule dial, lower horizontal louvers, 4 knobs, lift top, BC, AC **$30.00**

6613, Table, 1947, Wood, upper slanted slide rule dial, lower horiz louvers, 4 knobs, BC, SW, AC/DC **$40.00**

RR13, Table, 1947, Wood, slanted slide rule dial, cloth grill w/ "X" cut-out, 2 knobs, BC, SW, AC/DC **$50.00**

ESQUIRE
Esquire Radio Corp.,
51 Warren Street, New York, New York

60-10, Table, 1947, Plastic, right half-moon dial, large black grill/dial area, 2 knobs, BC, AC/DC **$65.00**

65-4, Table, 1947, Plastic, right half-moon dial, left horizontal louvers, 2 knobs, BC, AC/DC **$55.00**

511, Table-C, 1952, Plastic, center round dial knob, right checkered grill, left alarm clock, BC, AC **$30.00**

520, Table-N-C, 1952, Plastic clock/radio/lamp, center front round dial, left alarm clock, BC, AC **$65.00**

550, Table-C, 1952, Plastic, center round dial knob, right checkered grill, left alarm clock, BC, AC **$30.00**

550U, Table-C, 1952, Plastic, center round dial knob, right checkered grill, left round alarm clock, feet **$30.00**

ETHERPHONE
Radio Apparatus Co.,
40 West Montcalm Street, Detroit, Michigan

RX-1, Table, 1923, Wood, square case, lift top, inner black panel w/ 1 exposed tube **$400.00**

EVEREADY

1, Console, 1928, Wood w/maple finish, table model 1 on four legs, center window dial, 3 knobs, AC **$225.00**

1, Table, 1927, Gumwood w/maple finish, low rectangular case, center window dial, 3 knobs, AC **$175.00**

2, Console, 1928, Metal, table model 2 on aluminum legs, center front window dial, 3 knobs, AC **$125.00**

2, Table, 1928, Metal, low rectangular case w/decorative lines, center front dial, 3 knobs, AC **$95.00**

3, Table, 1928, Wood, low rectangular case, center front window dial w/escutcheon, 3 knobs **$95.00**

20, Table, 1928, Wood, high rectangular case, upper front window dial, lift top, 3 knobs, battery **$120.00**

31, Table, 1929, Wood, low rectangular case, center front dial in carved panel, 3 knobs, AC **$110.00**

32, Console, 1929, Wood, inner ornate panel w/upper window dial, sliding doors, 7 tubes, 3 knobs, AC **$175.00**

33, Console, 1929, Wood, inner ornate panel w/upper window dial, sliding doors, 7 tubes, 3 knobs, AC $185.00

54, Console, 1929, Wood, inner ornate front panel w/upper window dial, sliding doors, 3 knobs, AC $210.00

EXCEL
Excel Corp. of America,
9 Rockefeller Plaza, New York, New York

6T-2 "Aristocrat", Portable, 1959, Plastic, transistor, right front thumbwheel dial, left lattice grill, AM, battery $30.00

EXCELLO
Excello Products Corporation,
4822 West 16th St., Cicero, Illinois

154, Console-R/P, 1931, Wood, lower front window dial, upper grill w/cut-outs, lift top, inner phono $175.00

172, Console, 1931, Wood, highboy, carved, inner window dial, scrolled grill, doors, Queen Anne legs $250.00

D, Cathedral, 1931, Wood, center window dial, upper cloth grill w/cut-outs, scalloped top, 3 knobs $310.00

D-6, Console-R/P, 1931, Wood, front window dial, circular grill w/cut-outs, 3 knobs, lift top, inner phono $125.00

FADA
Fada Radio & Electric Company,
30-20 Thomson Avenue,
Long Island City, New York
F. A. D. Andrea, Inc.,
1581 Jerome Avenue, New York, New York

Fada was named for its founder - Frank Angelo D'Andrea. The company began business in 1920 with the production of crystal sets. Business expanded until the mid-twenties, but overproduction and internal problems took their toll and by 1932 the company had been sold and the name changed to Fada Radio & Electric Coporation. Fada continued to produce radios through the mid-1940's and is probably best known for the Fada Bullet, their ultra-streamlined Catalin radio made in the 1940's.

5F60, Table, 1939, Catalin, small, right square dial, left horiz louvers, 2 fluted knobs, BC, SW, AC ....$750.00+

6A39, Table, 1948, Plastic, right front dial, left horizontal wrap-around louvers, 2 knobs, BC, SW $75.00

7, Console, 1926, Wood, highboy, inner dial & 3 knobs, fold down front, optional side loop antenna $250.00

7 "Seven", Table, 1926, Wood, rectangular case, center front dials, loop antenna folds out of left side $190.00

10, Table, 1928, Low rectangular case, center front dial w/escutcheon, 3 knobs, switch, AC $110.00

11, Table, 1928, Wood, low rectangular case, center front window dial w/escutcheon, AC $135.00

12, Table, 1928, Low rectangular case, center front dial w/escutcheon, 2 knobs, switch, DC $100.00

16, Table, 1928, Low rectangular case, center front window dial, 3 knobs, AC $110.00

17, Table, 1928, Wood, low rectangular case, center front window dial w/escutcheon, AC $135.00

20, Table, 1929, Two-tone metal, low rectangular case, center front window dial, 3 knobs, AC $110.00

20-T, Table, 1938, Walnut, right front illuminated square dial, left grill w/Deco cut-outs, AC/DC $50.00

20-Z, Table, 1929, Two-tone metal, low rectangular case, center front window dial, 3 knobs, AC $110.00

22, Table, 1929, Wood, large front bronze escutcheon w/window dial, 3 knobs, 6 tubes, battery $115.00

25, Console, 1929, Walnut, recessed window dial w/escutcheon, lower grill w/cut-outs, 3 knobs, AC $140.00

25-Z, Console, 1929, Walnut, recessed window dial w/escutcheon, lower grill w/cut-outs, 3 knobs, AC $130.00

30, Console, 1928, Wood, highboy, inner window dial, 3 knobs, fold-down front, lower oval grill, AC $140.00

31, Console, 1928, Wood, inner window dial, 3 knobs, fold-down front, upper speaker w/doors, AC $190.00

32, Console, 1929, Walnut, inner window dial, "scallop shell" grill, 3 knobs, double front doors, AC $150.00

33 Series, Portable, 1940, Cloth covered, inner metal panel w/ vert grill bars, fold-open front, handle, bat $50.00

35, Console, 1929, Walnut, inner window dial, "scallop shell" grill, 3 knobs, double front doors, AC $150.00

35-B, Console, 1929, Walnut, inner window dial, "scallop shell" grill, 3 knobs, double front doors, AC $150.00

35-C, Console, 1929, Walnut, inner window dial, "scallop shell" grill, 3 knobs, double front doors, AC $150.00

35-Z, Console, 1929, Walnut, inner window dial, "scallop shell" grill, 3 knobs, double front doors, AC $150.00

41, Console, 1930, Wood, highboy, inner dial & knobs, double front doors, stretcher base $165.00

42, Console, 1930, Wood, lowboy, ornate center front panel w/upper dial, lower grill w/cut-outs $180.00

43, Cathedral, Wood, ornate pressed wood front, center window dial, upper grill, scalloped top $265.00

44, Console, 1930, Wood, lowboy, inner dial & knobs, double front sliding doors, stretcher base **$180.00**

45, Console, 1931, Wood, lowboy, upper front window dial, lower cloth grill w/cut-outs, 3 knobs **$150.00**

46, Console, 1930, Wood, highboy, ornate, inner dial & knobs, double doors, 6 legs, stretcher base **$185.00**

47, Console-R/P, 1930, Wood, lowboy, ornate, inner dial & knobs, double front doors **$235.00**

49, Console, 1931, Wood, highboy, inner window dial, 3 knobs, grill w/cut-outs, double doors, 6 legs **$180.00**

51, Cathedral, 1931, Wood, center front window dial, upper cloth grill w/cut-outs, front pillars **$225.00**

52, Table, 1938, Catalin, right square dial, left horizontal wrap-around louvers, 2 ribbed knobs ....**$850.00+**

60W, Table, Plastic, right dial, left grill w/Deco cut-outs, decorative case lines, 3 knobs $85.00

70, Console, 1928, Wood, lowboy, inner dial, fold-down front, lower speaker, doors, loop antenna, AC **$250.00**

72, Console-R/P, 1928, Wood, lowboy, ornate, inner window dial, lower grill, double front doors, AC **$250.00**

75, Console, 1929, Wood, lowboy, inner window dial, fold-down front, lower speaker w/doors, AC **$160.00**

77, Console-R/P, 1929, Wood, lowboy, inner window dial, grill cut-outs, double front doors, lift top, AC **$250.00**

79, Console, 1932, Wood, highboy, inner dial & knobs, double front doors, 6 legs, stretcher base, AC **$185.00**

100, Portable, "Alligator", inner slide rule dial & 3 knobs, fold-down front, handle, AC/DC/bat **$30.00**

115 "Bullet", Table, 1941, Catalin, bullet shape, right round dial, handle, 2 knobs, BC, AC/DC ....**$600.00+**

119, Table, Plastic, right dial, left cloth grill w/"T" design, decorative case lines, 3 knobs **$55.00**

130, Table, 1935, Wood, right front window dial, center cloth grill w/cut-outs, rounded top, AC **$80.00**

136, Table, 1941, Catalin, right dial, left wrap-around louvers, flared base, handle, 2 knobs, AC ....$1,000.00+
144MT, Table, 1940, Wood, right front square dial, left cloth grill w/ vertical cut-outs, 3 knobs **$60.00**

150-CA, Console, 1935, Wood, upper front round dial, lower cloth grill w/decorative lines, BC, SW, AC **$130.00**

155, Table, 1935, Two-tone wood, right dial, center cloth grill w/cut-outs, 4 tubes, 2 knobs, AC/DC **$75.00**

160 "One-Sixty", Table, 1923, Mahogany, low rectangular case, 3 dial black bakelite front panel, lift top, bat **$120.00**

160-A "One-Sixty", Table, 1925, Wood, low rectangular case, 3 dial black front panel, lift top, 4 tubes, battery **$120.00**

167, Table, 1936, Wood, lower front dial, upper cloth grill w/cut-outs, 6 tubes, BC, SW, AC/DC **$75.00**

168, Table, 1936, Wood, lower front dial, upper cloth grill w/cut-outs, 6 tubes, BC, SW, AC/DC **$75.00**

170-T, Tombstone, 1935, Wood, center front round dial, upper cloth grill w/cut-outs, 4 knobs, AC **$100.00**

175-A "Neutroceiver", Table, 1924, Mahogany, low rectangular case, 3 dial slanted front panel, 5 tubes, battery **$210.00**

175/90-A "Neutroceiver Grand", Console, 1925, Wood, model 175-A on a cabinet, 3 dial front panel, double doors, 5 tubes, bat **$250.00**

177, Table, Wood, right front 3 band slide rule dial, left vertical grill bars, base, 4 knobs **$60.00**

182G, Table-N, 1940, Looks like baby grand piano, 24 Kt gold-plated metal, 5 tubes, 2 knobs, BC, AC/DC **$600.00**

185-A "Neutrola", Table, 1925, Wood, inner 3 dial panel, fold-down front, upper speaker, 5 tubes, battery **$175.00**

185/90-A "Neutrola Grand", Console, 1925, Wood, model 185-A on a cabinet, inner 3 dial panel, speaker, 5 tubes, battery **$215.00**

192-A "Neutrolette", Table, 1925, Wood, rectangular case, 3 dial black slanted front panel, battery **$150.00**

195-A "Neutro-Junior", Table, 1924, Wood, low rectangular case, 2 dial black front panel, lift top, battery **$150.00**

252 "Temple", Table, 1941, Catalin, upper dial, horizontal grill bars, rounded corners, 2 knobs, BC, AC/DC ....**$550.00+**

260B, Table, 1936, Black bakelite, right front dial, left cloth grill w/ Deco bars, 2 knobs, BC, AC/DC **$85.00**

260D, Table, 1936, Black bakelite w/chrome, right dial, left grill w/ Deco bars, 2 knobs, BC, AC/DC **$145.00**

260G, Table, 1936, Ivory plastic w/gold, right dial, left cloth grill w/ Deco bars, 2 knobs, AC, BC/SW **$125.00**

260V, Table, 1936, Ivory plastic, right airplane dial, left cloth grill w/ Deco bars, 2 knobs, BC, AC/DC **$85.00**

263W, Table, 1936, Plastic, small right front dial, left Deco grill design, plastic back, 3 knobs $85.00
265-A "Special", Table, 1927, Mahogany, low rectangular case, 2 front window dials, 6 tubes, 3 knobs, battery **$135.00**

351, Table, Wood, right front gold dial, left cloth grill w/Deco cut-outs, 4 knobs **$55.00**

454V, Table, 1938, Ivory plastic, right square dial, left cloth grill w/ Deco cut-outs, 2 knobs, AC **$75.00**

480-B, Table, 1927, Walnut, large front escutcheon w/2 window dials, fold-out loop antenna, 8 tubes **$210.00**

602, Table-R/P, 1947, Wood, upper slide rule dial, lower grill, 4 knobs, lift top, inner phono, BC, AC $30.00

605, Table, 1946, Plastic, upper front slanted slide rule dial, lower horizontal louvers, 2 knobs $55.00

605W, Table, 1946, Plastic, upper front slide rule dial, lower horizontal louvers, 2 knobs, BC, AC/DC $55.00

609W, Table, 1946, Plastic, right side square dial, left vertical bars, 2 knobs, BC, AC/DC $50.00

637, Portable-R/P, 1947, Leatherette, inner slide rule dial, 4 knobs, phono, 2 fold-back covers, BC, AC $30.00

652 "Temple", Table, 1946, Catalin, upper dial, horizontal grill bars, rounded corners, 2 knobs, BC, AC/DC ....$650.00+

700, Table, 1948, Catalin, round dial, horizontal grill bars, handle, curved sides, 2 knobs, BC, AC/DC ....$600.00+

711, Table, 1947, Catalin, right front dial, left cloth grill, handle, rounded corners, 2 knobs, AC ....$650.00+

740, Table, 1947, Plastic, right dial, left horizontal wrap-around louvers, 2 knobs, BC, AC/DC $55.00

790, Table, 1949, Plastic, right square dial, left criss-cross grill w/"T", 4 knobs, BC, FM, AC/DC $85.00

795, Table, 1948, Plastic, upper slide rule dial, lower horizontal louvers, 2 knobs, FM tuner, AC $45.00

830, Table, 1950, Plastic, upper curved slide rule dial, lower horiz louvers, 2 knobs, BC, AC/DC $45.00

845, Table, 1950, Plastic, Deco, right round dial, horizontal grill bars, handle, 2 knobs, BC, AC/DC $185.00

855, Table, 1950, Plastic, lower slide rule dial, recessed horizontal grill bars, 2 knobs, BC, AC/DC $45.00

1000 "Bullet", Table, 1946, Catalin, streamline bullet shape, right round dial, handle, 2 knobs, BC, AC/DC ....$600.00+

1001, Table, 1947, Wood, slanted slide rule dial, horizontal louvers, 2 knobs, burl veneer, BC, AC/DC $60.00

1005, Table, Plastic, Deco, right front dial, left cloth grill, stepped top, handle, 2 knobs $75.00

1452A, Tombstone, 1934, Walnut, front round airplane dial, upper cloth grill w/Deco cut-outs, 4 knobs, AC $120.00

1452F, Console, 1934, Walnut, upper front round airplane dial, lower cloth grill w/cut-outs, 4 knobs, AC $135.00

1462D, Table, 1935, Wood, lower front round dial, upper cloth grill w/cut-outs, 4 knobs, BC, SW, AC $80.00

1470C, Tombstone, 1934, Wood, center front round dial, upper cloth grill w/cut-outs, 4 knobs, AC $120.00

1470E, Console, 1934, Wood, upper front round dial, lower cloth grill w/cut-outs, 4 knobs, 4 band, AC $130.00

A66SC, Console, 1939, Wood, upper front dial, pushbuttons, lower cloth grill w/vertical bars, BC, SW, AC $150.00

C-34, Portable, 1941, Leatherette, inner dial, vertical grill bars, flip-open front, handle, BC, AC/DC/bat $55.00

F55T, Table, 1939, Walnut, Deco, right front square dial, left horizontal grill bars, 2 knobs, AC $60.00

KG, Console, 1931, Wood, lowboy, ornate front, upper window dial, lower cloth grill w/cut-outs $225.00

L56 "Bell", Table, 1939, Catalin, right dial, left wrap-around louvers, flared base, handle, 2 knobs, AC ....$1,000.00+

L-96W, Table, 1939, Bakelite, right square dial, left cloth grill w/Deco cut-outs, handle, 3 knobs, AC $85.00

P-40, Portable, 1939, Cloth covered, right recessed dial, left 2-section grill, handle, 2 knobs, battery $25.00

P41, Portable, Leatherette, inner slide rule dial, square grill, 2 knobs, slide-out front, handle $25.00

P80, Portable, 1947, Plastic, inner right dial, Deco grill cut-outs, 3 knobs, flip-up lid, BC, AC/DC/bat $80.00

P82, Portable, 1947, Leatherette, inner slide rule dial, 3 knobs, slide-in door, handle, BC, AC/DC/bat $30.00

P-100, Portable, 1947, "Snakeskin", inner slide rule dial, 3 knobs, fold-down front, handle, BC, AC/DC/bat $35.00

P111, Portable, 1952, Plastic, inner right round dial, lattice grill, flip-up front, handle, BC, AC/DC/bat $45.00

P-130, Portable, 1951, Leatherette, inner dial, lift front, telescope antenna, handle, BC, 2SW, AC/DC/bat $65.00

PL24, Portable, 1940, Leatherette, inner dial, horiz louvers, slide-up door, handle, BC, SW, AC/DC/bat $30.00

PL50, Table-R/P, 1939, Wood, right front square dial, left horizontal louvers, lift top, inner phono, AC $30.00

PT-208, Table-R/P, 1941, Walnut, center front dial, right & left grills, lift top, inner phono, AC $30.00

FAIRBANKS-MORSE
Fairbanks-Morse Home Appliances,
430 South Green Street
Chicago, Illinois

6AC-7, Chairside, 1937, Two-tone wood, top "Great Circle" dial, tuning eye, front grill w/vertical bars, AC $120.00

9AC-4, Console, 1937, Wood, upper round dial w/automatic tuning, lower cloth grill w/vertical bars, AC $165.00

9AC-5, Console, 1937, Wood, upper round dial w/automatic tuning, lower cloth grill w/vertical bars, AC $165.00

12-C-6, Console, 1936, Wood, upper "semaphore dial", lower grill w/bowed vert bars, 12 tubes, BC, SW $185.00

57TO, Table, 1936, Wood, right front round dial, left cloth grill w/cut-outs, cabinet finished on back $85.00

58-T-1, Table, 1936, Wood, right front round dial, left grill w/Deco cut-outs, rounded left side, BC, SW $85.00

91-C-4, Console, 1936, Wood, upper "semaphore dial", lower grill w/center vertical bars, 9 tubes, BC, SW $150.00

814, Tombstone, 1934, Wood, Deco, lower half-moon dial, upper wrap-over vertical grill bars, 8 tubes, BC $125.00

5106, Cathedral, 1934, Wood, center front window dial, upper scalloped-top grill w/cut-outs, 3 knobs $225.00

5312, Tombstone, 1934, Walnut, Deco, lower front airplane dial, upper wrap-over vertical grill bars, BC $125.00

5341, Console, 1934, Wood, upper front round airplane dial, lower cloth grill w/cut-outs, 6 legs, AC $120.00

7014, Tombstone, 1934, Wood, Deco, lower front round dial, upper wrap-over vertical grill bars, 4 knobs $150.00

7040, Console, 1934, Wood, Deco, upper round dial, lower cloth grill w/vert bars, 6 legs, 4 knobs, AC $150.00

7117, Tombstone, 1935, Wood, lower front round dial, upper cloth grill w/cut-outs, 4 knobs $125.00

7146, Console, 1935, Wood, upper front round dial, lower cloth grill w/cut-outs, 3 band $120.00

FAIRVIEW
Fairview Electric Shop,
35 Fairview Avenue, Binghamton, New York

J-400 "Lasher Capacidyne", Table, 1925, Wood, high rectangular case, 3 dial front panel, storage, 5 tubes, battery $140.00

FARNSWORTH
Farnsworth Television & Radio Corp.,
Fort Wayne, Indiana

AC-55, Console, 1939, Wood, upper slide rule dial, pushbuttons, lower grill w/2 vert bars, 7 tubes, AC $120.00

AC-70, Console, 1940, Wood, 3 section dial, pushbuttons, lower grill w/vert bars, 4 knobs, 8 tubes, AC $125.00

AC-90, Console, 1939, Wood, upper slanted dial, pushbuttons, lower grill w/vert bars, 10 tubes, 4 knobs, AC $145.00

AC-91, Console, 1939, Wood, wide case, upper dial, pushbuttons, lower grill w/2 vert bars, 10 tubes, AC $150.00

AK-17, Table-R/P, 1939, Walnut, right front half-moon dial, left wrap-around grill, lift top, inner phono $35.00

AK-59, Console-R/P, 1939, Wood, inner dial & phono, pushbuttons, lower grill w/center vert bar, BC, SW, AC $135.00

AK-76, Console-R/P, 1939, Wood, inner dial & phono, pushbuttons, lower front grill w/vertical bars, 8 tubes $145.00

AK-86, Console-R/P, 1940, Wood, inner dial & phono, pushbuttons, lower front grill w/vertical bars, 8 tubes $145.00

AK-95, Console-R/P, 1939, Wood, inner dial & phono, pushbuttons, lower front grill w/vertical bars, 10 tubes $155.00

AK-96, Console-R/P, 1940, Wood, Chippendale design, inner dial & phono, pushbuttons, grill cut-outs, 10 tubes $135.00

AT-11, Table, 1939, Plastic, right half-moon dial, left wrap-around louvers, handle, 2 knobs, AC/DC $110.00

AT-12, Table, 1939, Plastic, right half-moon dial, left wrap-around louvers, handle, 2 knobs, AC/DC $110.00

AT-15, Table, 1939, Plastic, right half-moon dial, top pushbuttons, left horiz louvers, handle, 2 knobs $110.00

AT-16, Table, 1939, Wood, Deco, right half-moon dial, top pushbuttons, left grill w/horiz bars, AC/DC $85.00

AT-31, Portable, 1939, Cloth covered, inner slide rule dial, 3 knobs, fold-down front, handle, AC/DC/bat $25.00

AT-50, Table, 1939, Wood, lower slide rule dial, pushbuttons, upper grill w/horiz bars, 4 knobs, BC, SW $70.00

BC-81, Console, 1940, Wood, upper slide rule dial, pushbuttons, lower grill w/vert bars, 8 tubes, BC, SW, AC $125.00

BC-102, Console, 1940, Wood, upper slide rule dial, pushbuttons, lower grill w/vert bars, 10 tubes, BC, SW, AC $135.00

BK-69, Table-R/P, 1941, Wood, right front slide rule dial, pushbuttons, left grill, lift top, inner phono, AC $35.00

BK-73, Chairside-R/P, 1940, Walnut, top dial & knobs, front vertical louvers, lift top, inner phono, storage, AC $125.00

BKR-84, Table-R/P/Rec, 1940, Wood, right front slide rule dial, pushbuttons, left grill, lift top ,inner phono, AC $35.00

BT-20, Table, 1940, Plastic, right dial, raised left horiz louvers, 3 knobs, rounded top, BC, SW, AC/DC $50.00

BT-22, Table, 1940, Wood, right slide rule dial, pushbuttons, left grill w/vert bars, BC, SW, AC/DC $70.00

BT-40, Table, 1940, Two-tone wood, right front dial, left horizontal louvers, battery $40.00

BT-57, Table, 1940, Wood, right front slide rule dial, left horizontal grill bars, 3 knobs $45.00

BT-68, Portable, 1946, Leatherette, inner dial, 2 knobs & grill, fold-down front, handle, AC/DC/battery $45.00

BT-1010, Table, Wood, right slide rule dial w/plastic escutcheon, left horiz louvers, 4 knobs, AC $50.00

CK-58, Table-R/P, 1941, Wood, upper front slide rule dial, lower grill, 2 knobs, lift top, inner phono, AC $35.00

CK-73, Chairside-R/P, 1941, Wood, top dial & knobs, front horiz grill bars, lift top, inner phono, storage, AC $135.00

CT-41, Table, 1941, Mahogany plastic, right dial, left horiz wrap-around louvers, 2 knobs, BC, battery $50.00

CT-42, Table, 1941, Walnut, right slide rule dial, left grill w/Deco cut-outs, 2 knobs, BC, battery $45.00

CT-43, Table, 1941, Mahogany plastic, right dial, left horiz wrap-around louvers, 2 knobs, BC, battery $50.00

CT-50, Table, 1941, Mahogany plastic, upper front slide rule dial, lower horiz louvers, 2 knobs, AC $50.00

CT-51, Table, 1941, Ivory plastic, upper front slide rule dial, lower horizontal louvers, 2 knobs, AC $50.00

CT-52, Table, 1941, Mahogany plastic w/gold, slide rule dial, horizontal louvers, handle, 2 knobs, AC $50.00

CT-53, Table, 1941, Ivory plastic w/brown, upper slide rule dial, horiz louvers, handle, 2 knobs, AC $50.00

CT-54, Table, 1941, Walnut, upper slide rule dial, 3-section wrap-around grill, 2 knobs, BC, SW, AC $55.00

CT-59, Portable, 1941, Leatherette & plastic, inner dial & grill, flip-open front, handle, AC/DC/battery $65.00

CT-60, Portable, 1941, Luggage-type, inner right front dial, left grill, 2 knobs, fold-down front, handle $30.00

CT-61, Table, 1941, Mahogany plastic, upper front slide rule dial, lower grill, handle, 2 knobs, AC $50.00

CT-62, Table, 1941, Ivory plastic, upper front slide rule dial, lower grill, handle, 2 knobs, AC $50.00

CT-63, Table, 1941, Mahogany plastic, upper front slide rule dial, lower grill, 2 knobs, BC, SW, AC $45.00

CT-64, Table, 1941, Wood, upper slide rule dial, lower cloth grill w/ horizontal bars, 2 knobs, BC, SW $50.00

EC-260, Console, 1946, Wood, dial & 3 knobs on top of case, front cloth grill, BC, AC $100.00

EK-264, Console-R/P, 1946, Wood, dial & 3 knobs on top of case, criss-cross front grill, lift top, BC, AC $125.00

EK-264WL, Console-R/P, 1946, Wood, dial & knobs on case top, front criss-cross grill, lift top, inner phono, AC $125.00

ET-060, Table, 1946, Plastic, upper front slide rule dial, lower checkered grill, 2 knobs, BC, SW, AC/DC $50.00

ET-061, Table, 1946, Plastic, upper front slide rule dial, lower checkered grill, 2 knobs, BC, SW, AC/DC $60.00

ET-063, Table, 1946, Wood, upper slide rule dial, cloth grill w/ cutouts, curved sides, BC, SW, AC/DC $60.00

ET-064, Table, 1946, Plastic, upper pointer dial, cloth grill with chrome insert, 2 knobs, BC, AC/DC $90.00

ET-065, Table, 1946, Plastic, upper pointer dial, grill w/chrome insert, handle, 2 knobs, BC, AC/DC $90.00

ET-066, Table, 1946, Wood, upper pointer dial, cloth grill w/ "X's", scalloped base, 2 knobs, BC, AC/DC $60.00

ET-067, Table, 1946, Wood, upper slide rule dial, lower grill, vertical fluting on case top, 2 knobs, AC $45.00

GK-102, Console-R/P, 1947, Wood, inner slide rule dial, 4 knobs, pushbuttons, phono, lift top, BC, FM, AC $100.00

GK-111, Console-R/P, 1949, Wood, inner right dial, pushbuttons, 4 knobs, pull-out phono drawer, BC, FM, AC $80.00

GK-141, Console-R/P, 1947, Wood, inner slide rule dial, 5 knobs, 8 pushbuttons, phono, lift top, BC, SW, FM, AC $80.00

GP-350, Portable, Metal & leatherette, inner round dial, metal grill, 2 knobs, lift top, AC/DC $45.00

GT-050, Table, 1948, Plastic, Deco design, right round dial, wrap-around louvers, 2 knobs, BC, AC/DC $100.00

GT-051, Table, 1948, Plastic, Deco design, right round dial, wrap-around louvers, 2 knobs, BC, AC/DC $100.00

GT-064, Table, 1948, Plastic, upper slanted dial, lower cloth grill, vertical fluting, 2 knobs, BC, AC/DC $45.00

GT-065, Table, 1948, Plastic, upper slanted dial, cloth grill, vert fluting, handle, 2 knobs, BC, AC/DC $45.00

FEDERAL

Federal Telephone & Radio Corporation,
591 Broad Street, Newark, New Jersey
Federal Telephone & Telegraph Company,
Buffalo, New York
Federal Radio Corporation (Division of the Federal
Telephone Manufacturing Corp.), Buffalo, New York

Federal Telephone & Telegraph began in 1908 with the manufacture of telephones and soon began to produce radio parts as well. The company produced its first complete radio in 1921 and began to manufacture a high quality, relatively expensive line of sets. As time went on and less complicated radios began to appear on the market, Federal's business declined and by 1929 the company had been sold.

57, Table, 1922, Metal, 1 dial polished black bakelite front panel, top right door, 4 tubes, battery $500.00

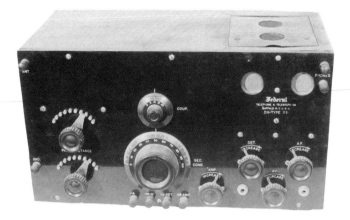

58 DX, Table, 1922, Metal or wood, polished black bakelite front panel, top right door, 4 tubes, bat $550.00

59, Table, 1923, Metal or wood, polished black bakelite front panel, lift top, 4 tubes, battery $900.00

61, Table, 1924, Metal or wood, polished black bakelite front panel, lift top, 6 tubes, battery $1,000.00
102, Portable, 1924, Wood, rectangular case, inner bakelite panel, removable front, handle, 4 tubes, bat $450.00

110, Table, 1924, Metal or wood, polished black bakelite front panel, lift top, 3 tubes, battery $550.00
141, Table, 1925, Mahogany, inner 2 dial bakelite front panel, double front doors, 5 tubes, battery $350.00
142, Table, 1925, Wood, rectangular case, right two dial panel, left built-in speaker, 5 tubes, bat $385.00
143, Console, 1925, Wood, lowboy, inner right two dial panel, left built-in speaker, 5 tubes, battery $400.00
144, Console, 1925, Wood, lowboy, inner right two dial panel, left built-in speaker, 5 tubes, battery $400.00
1024TB, Table, 1948, Black Catalin, upper slide rule dial, large lower wire mesh grill, 3 knobs, AC/DC $625.00

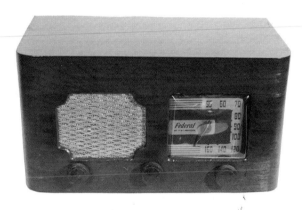

1028T, Table, Wood, right front dial with red pointer, left criss-cross grill, 3 knobs $45.00
1030T, Table, 1946, Wood, upper slanted slide rule dial, lower grill, 3 knobs, BC, SW, AC/DC $40.00
1040T, Table, 1947, Plastic, right front dial, horizontal wrap-around grill bars, 2 knobs, BC, AC/DC $45.00
1040-TB, Table, 1947, Plastic, right front dial, horizontal wrap-around grill bars, 2 knobs, BC, AC/DC $45.00
6001-PO, Portable, "Alligator", inner right slide rule dial, fold-down front, handle, BC, AC/DC/bat $40.00
A-10 "Orthosonic", Table, 1925, Wood, low rectangular case, slanted 3 dial front panel, 5 tubes, battery $125.00
B-20 "Orthosonic", Table, 1925, Mahogany, low rectangular case, slanted panel w/3 window dials, 5 tubes, bat $100.00
B-30 "Orthosonic", Table, 1925, Mahogany, slanted panel w/3 window dials, rounded top speaker, 5 tubes, battery ..$240.00
B-35 "Orthosonic", Console, 1925, Mahogany, inner 3 dial panel, built-in speaker, fold-down front, storage, battery $175.00
B-36 "Orthosonic", Console, 1925, Mahogany, inner 3 dial panel, built-in speaker, fold-down front, storage, battery $185.00
C-20 "Orthosonic", Table, 1925, Mahogany, low rectangular case, slanted front panel w/2 dials, 3 knobs, battery $100.00
C-30 "Orthosonic", Table, 1925, Mahogany, slanted front panel w/2 dials, rounded top speaker, 7 tubes, battery $250.00
E-10 "Orthosonic", Table, 1926, Mahogany, low rectangular case, center window dial w/escutcheon, 6 tubes, bat $110.00

H-10-60 "Orthosonic", Table, 1928, Wood, low rectangular case, center front window dial w/escutcheon, 3 knobs, AC $130.00

FERGUSON

J. B. Ferguson, 80 Beaver St.,
New York, New York

3, Table, 1925, Wood, low rectangular case, 3 dial front panel, 4 tubes, battery $125.00

4, Table, 1925, Wood, low rectangular case, slanted 2 dial front panel, 4 tubes, battery $125.00

6, Table, 1925, Wood, low rectangular case, 2 dial front panel, battery storage, battery $115.00

8, Table, 1925, Wood, rectangular case, slanted front panel w/ window dial, 2 knobs, 6 tubes, bat $135.00

12, Table, 1926, Wood, low rectangular case, right & left front window dials, 3 knobs, battery $115.00

FERRAR

Ferrar Radio & Television Corp.,
55 West 26th Street, New York, New York

C-81-B, Console, 1947, Two-tone wood, upper front slide rule dial, lower grill, 4 knobs, BC, SW, AC/DC $100.00

T-61-B, Table, 1948, Wood, right slide rule dial, left horizontal grill bars, 3 knobs, BC, SW, AC/DC $25.00

TA61B, Table, 1948, Wood, upper slanted slide rule dial, lower diamond shaped grill, BC, SW, AC $45.00

FIRESTONE

The Firestone Tire & Rubber Company,
1200 Firestone Parkway, Akron, Ohio

4-A-1 "Mercury", Table, 1948, Plastic, lower slanted slide rule dial, upper horizontal louvers, 2 knobs, BC, AC $45.00

4-A-2, Table, 1947, Plastic, upper slide rule dial, slant front, horizontal louvers, 2 knobs, BC, AC/DC $45.00

4-A-3 "Diplomat", Table, 1948, Plastic, right dial, left vertical grill bars w/lower loops, 2 knobs, 6 tubes, BC, AC/DC .. $65.00

4-A-10 "Reporter", Table, 1947, Plastic, upper slide rule dial, horizontal wrap-around louvers, 2 knobs, BC, AC/DC .. $60.00

4-A-11, Table, 1948, Plastic, slide rule dial, horizontal wrap-around louvers, 2 knobs, BC, AC/DC $50.00

4-A-12 "Narrator", Table, 1948, Plastic, elongated base, horizontal louvers, 2 thumbwheel knobs, AM, FM, AC/DC $65.00

4-A-15, Console-R/P, 1948, Wood, inner slide rule dial, pushbuttons, lift top, pull-out phono, BC, SW, FM, AC $95.00

4-A-17, Table-R/P, 1948, Wood, outer top dial, 3 knobs, grill w/cutouts, lift top, inner phono, BC, AC $40.00

4-A-20, Table, 1947, Wood, upper slanted slide rule dial, cloth grill w/ "X", 3 knobs, BC, SW, AC/DC $60.00

4-A-21, Table, 1947, Wood, slide rule dial, cloth grill w/cut-outs, 4 knobs, 6 pushbuttons, BC, SW, AC $80.00

4-A-23, Table, 1946, Wood, slanted slide rule dial, cloth grill, 4 knobs, 6 pushbuttons, BC, 2SW, AC $65.00

4-A-24, Table, 1947, Wood, upper slanted slide rule dial, lower cloth grill w/"X", 2 knobs, BC, battery $45.00

4-A-25 , Table, 1947, Plastic, upper slide rule dial, lower horizontal louvers, 2 knobs, BC, AC/DC $40.00

4-A-26 "Newscaster", Table, 1948, Plastic, upper slanted slide rule dial, lower horiz louvers, 2 knobs, BC, AC/DC . $45.00

4-A-27, Table, 1947, Plastic, lower slide rule dial, upper horizontal louvers, 2 knobs, BC, AC/DC $50.00

4-A-30, Console, 1949, Wood, upper slide rule dial, lower cloth grill w/vert bars, pushbuttons, 4 knobs $120.00

4-A-31, Console-R/P, 1947, Wood, upper slanted slide rule dial, 3 knobs, tilt-out phono drawer, BC, AC $95.00

4-A-37, Console-R/P, 1947, Wood, upper slide rule dial, 4 knobs, 6 pushbuttons, pull-out phono, BC, SW, AC $100.00

4-A-41, Table, 1948, Plastic, upper slanted slide rule dial, lower horiz louvers, 2 knobs, BC, AC/DC $50.00

4-A-42 "Georgian", Console-R/P, 1947, Wood, upper slide rule dial, 4 knobs, 6 pushbuttons, tilt-out phono, BC, FM, AC ... $100.00

4-A-60, Console-R/P, 1948, Wood, inner right slide rule dial, 4 knobs, left pull-out phono drawer, BC, FM, AC $85.00

4-A-61 "Cameo", Table, 1948, Plastic, lower slanted slide rule dial, horizontal louvers, 2 knobs, BC, AC/DC $50.00

4-A-62, Console-R/P, 1949, Wood, inner right black dial, 3 knobs, left lift top, inner phono, BC, FM, AC $60.00

4-A-64 "Contemporary", Console-R/P, 1949, Wood, upper right slanted dial, 3 knobs, left lift top, inner phono, BC, FM, AC $65.00

4-A-66, Console-R/P, 1949, Wood, inner top right slide rule dial, 4 knobs, front pull-down phono , BC, FM, AC $65.00

4-A-67, Table, 1949, Plastic, upper slanted slide rule dial, lower horizontal louvers, 2 knobs, BC, bat $40.00

4-A-68, Table, 1949, Plastic, right round dial knob, left checkerboard grill, ridged base, BC, AC/DC $45.00

4-A-69 "Sunrise", Table-C, 1949, Plastic, slide rule dial, center electric clock, grill on case top, 4 knobs, BC, AC $45.00

4-A-70, Table, 1951, Plastic, right half-moon dial w/large knob, left horizontal grill bars, BC, AC/DC $40.00

4-A-71, Table-R/P, 1949, Wood, center front dial, horizontal louvers, 4 knobs, lift top, inner phono, BC, AC $30.00

4-A-78, Table, 1950, Plastic, oblong panel, right semi-circular dial, left grill, 2 knobs, BC, AC/DC $45.00

4-A-86, Console-R/P, 1951, Wood, inner right slide rule dial, 4 knobs, pull-out phono drawer, BC, FM, AC $80.00

4-A-87, Console-R/P, 1951, Wood, upper front slide rule dial, 4 knobs, center pull-out phono drawer, BC, AC $50.00

4-A-89, Table, 1950, Plastic, large half-moon dial, lower panel w/ vertical bars & 2 knobs, BC, AC/DC $35.00

4-A-92, Table-C, 1951, Plastic, right front dial, left clock, center vertical grill bars, side knob, BC, AC **$35.00**

4-A-108, Table, 1953, Plastic, center front half-moon dial, perforated grill, 2 knobs, feet, BC, AC/DC **$35.00**

4-A-110, Table-C, 1953, Wood, right square dial, left alarm clock, center lattice grill, 5 knobs, BC, AC **$35.00**

4-A-115, Table, 1953, Wood, center front round dial over large woven grill, 2 knobs, BC, AC/DC **$40.00**

4-A-132, Table-R/P, 1956, Outer right side dial knob, front grill, lift top, inner phono, handle, BC, AC **$30.00**

4-A-134, Table-C, 1956, Plastic, right side round dial knob, left front clock, left lattice grill, BC, AC **$25.00**

4-A-143, Table, 1956, Plastic, lower front dial, large upper grill w/ center strip, 2 knobs, BC, AC/DC **$45.00**

4-A-149, Table, 1957, Plastic, right front round dial, lower knob, left horizontal bars, BC, AC/DC **$25.00**

4-A-152, Table-R/P, 1957, Right side dial knob, large front grill, 3/4 lift top, inner phono, handle, BC, AC **$20.00**

4-A-153, Table, 1957, Plastic, center slide rule dial & tone control, lattice grills, 2 knobs, BC, AC/DC **$20.00**

4-A-154, Table-C, 1957, Plastic, off-center vertical dial, right square clock, 3 knobs, BC, AC **$25.00**

4-A-159, Table, 1957, Plastic, right vertial slide rule dial, large left lattice grill, 2 knobs, BC, AC/DC **$20.00**

4-A-160, Table-C, 1957, Plastic, right side round dial knob, front alarm clock, left step-down top, BC, AC **$20.00**

4-A-162, Table-C, 1957, Plastic, lower dial, center front square clock w/day-date, 5 knobs, BC, AC **$35.00**

4-A-163, Table, 1957, Plastic, right dial knob over random patterned panel, left checkered grill, AC/DC $20.00

4-A-164, Table-C, 1957, Plastic, right side dial knob, off-center clock over checkered panel, BC, AC **$35.00**

4-A-167, Table-C, 1958, Plastic, right front round dial, left square clock, side half-moon cut-outs, BC, AC **$30.00**

4-A-175, Table-R/P, 1958, Right side dial knob, large front grill, 3/4 lift top, inner phono, handle, BC, AC **$25.00**

4-A-176, Table, 1958, Plastic, right front dial knob, left lattice grill, right side knob, BC, AC **$20.00**

4-A-179, Table-C, 1958, Plastic, right side dial knob, left front clock, right horizontal louvers, BC, AC **$20.00**

4-C-3, Portable, 1947, Leatherette, inner slide rule dial, 2 knobs, fold-back top, BC, AC/DC/battery **$35.00**

4-C-5, Portable, 1948, Top right dial knob, rounded center top, rounded sides, strap, BC, AC/DC/battery **$45.00**

4-C-13, Portable, 1949, Lower right front dial knob, upper horizontal grill bars, handle, BC, AC/DC/bat **$35.00**

4-C-16, Portable, 1951, Plastic, slide rule dial, vertical grill bars, flex handle, 2 knobs, BC, AC/DC/bat **$35.00**

4-C-18, Portable, 1950, Leatherette, center front slide rule dial, handle, 2 knobs, BC, AC/DC/battery **$25.00**

4-C-19, Portable, 1952, Plastic, "alligator" front panel w/circular louvers, flex handle, BC, AC/DC/bat **$40.00**

4-C-21, Portable, 1952, Two-tone leatherette, center front round dial, handle, 2 knobs, BC, AC/DC/bat **$35.00**

4-C-22, Portable, 1954, Plastic, lower front round dial, upper vertical grill bars, handle, BC, battery **$30.00**

4-C-29, Portable, 1956, Plastic, tubes and transistors, right dial, left checkered grill, handle, BC, bat **$50.00**

4-C-30, Portable, 1957, Leather case, slide rule dial, map, telescope antenna, handle, BC, SW, AC/DC/bat **$50.00**

4-C-31, Portable, 1957, Leather case w/front grill, upper right round dial, handle, helmet logo, BC, bat **$30.00**

4-C-32, Portable, 1957, Leather case w/grill, right front round dial, "helmet" logo, handle, BC, AC/DC/bat **$30.00**

4-C-33, Portable, 1958, Leather case w/front grill, transistor, right front round dial, handle, AM, battery **$35.00**

4-C-34, Portable, 1957, Case lies flat, transistor, front dial, 2 knobs, top lattice grill, handle, AM, bat **$30.00**

4-C-35, Portable, 1958, Leatherette, slide rule dial, telescope antenna, handle, BC, SW, AC/DC/battery **$50.00**

4-C-36, Portable, 1959, Transistor, upper right thumbwheel dial, horizontal front bars, handle, AM, bat **$25.00**

S-7403-2, Table, Plastic, right dial, left vertical grill bars w/lower loops, 2 knobs, 6 tubes **$65.00**

S-7426-6, Table, Plastic, right front dial w/statue & globe, horizontal grill bars, 2 knobs, AC **$45.00**

FLUSH WALL
**Flush Wall Radio Co.,
58 East Park Street
Newark, New Jersey**

5P, Wall Radio, 1947, Catalin front panel, right dial, left horizontal grill openings, 2 knobs, BC, AC/DC **$200.00**

FREED-EISEMANN
**Freed Radio Corporation,
200 Hudson Street, New York, New York
Freed-Eisemann Radio Corporation,
255 Fourth Avenue, New York, New York**

The Freed-Eisemann Radio Corporation began in 1922. Business was good for the company until 1924 when, with the onslaught of less expensive radios from the competition, the company began to decline. Most of the Freed-Eisemann stock was sold to the Freshman Company in 1928 and the final blow was the stock market crash in 1929. Two later companies also formed by the Freed Brothers were The Freed Television and Radio Corporation begun in 1931 and The Freed Radio Corporation begun in 1940.

10, Table, 1926, Wood, low rectangular case, 3 dial black front panel, 5 tubes, battery **$100.00**

11 "Electric 11", Console, 1927, Wood, fold-down front, inner window dial, inner grill, double doors, 6 tubes, AC **$135.00**

11 "Electric 11", Table, 1927, Wood, low rectangular case, inner window dial, fold-down front, 6 tubes, AC **$200.00**

26, Portable, 1937, Leatherette or cloth, inner right dial, left grill, fold-down front, handle, AC/DC **$30.00**

28, Table, 1937, Two-tone wood, right front square dial, left grill w/ horizontal bar, BC, SW, AC **$45.00**

29-D, Table, 1937, Walnut, right dial, tuning eye, left grill w/horiz bars, 4 knobs, BC, SW, AC/DC **$65.00**

30, Table, 1926, Mahogany, slanted front panel, 2 window dials w/ escutcheons, battery **$125.00**

30-D, Table, 1937, Wood, right front dial, tuning eye, left grill w/horiz bars, 3 knobs, BC, SW, AC/DC **$65.00**

40, **Table, 1926, Wood, high rectangular case, slanted front, center window dial w/escutcheon, bat** $115.00
46, Console-R/P, 1947, Wood, inner slide rule dial, 6 knobs, phono, lift top, doors, BC, 2SW, FM, AC $65.00
48, Table, 1926, Wood, inner center dial w/escutcheon, fold-down front, side columns, battery $125.00
50, Table, 1926, Mahogany, inner center dial w/escutcheon, fold-down front, side columns, battery $120.00
350, Table, 1933, Walnut w/black & silver trim, window dial, center grill w/cut-outs, 3 knobs, AC $85.00
800, Table, 1926, Wood, low rectangular case, slanted front w/center dial, loop antenna, 8 tubes $185.00
800-C-8, "Roller" Console, 1927, Wood, console on rollers, center dial, lower speaker, top loop antenna, 8 tubes $300.00
850, Console, 1926, Wood, tall case, inner dial, fold-down front, upper speaker w/doors, 8 tubes $275.00

FE-15, **Table, 1924, Wood, low rectangular case, 3 dial black front panel, lift top, 5 tubes, battery** $125.00
FE-24, Table, 1937, Wood, right front illuminated gold & blue scale dial, left grill, AC/DC $50.00
FE-28, Table, 1937, Wood, right front square dial, left cloth grill w/ horizontal bars, 3 knobs, BC, SW $50.00
FE-30, Table, 1925, Wood, low rectangular case, inner three dial panel, fold-down front, battery $135.00
FE-60, Table, 1936, Two-tone wood, right front dial, left grill w/bars, AC/DC $50.00

NR-5, **Table, 1923, Mahogany, low rectangular case, 3 dial black front panel, lift top, 5 tubes, bat** $110.00
NR-6, Table, 1924, Wood, low rectangular case, 3 dial front panel, lift top, 5 tubes, battery $120.00
NR-7, Table, 1925, Wood, low rectangular case, 3 dial black front panel, lift top, 6 tubes, battery $100.00

NR-8, Console, 1927, Wood, inner panel w/2 window dials, upper speaker, double doors, 6 tubes, bat $175.00
NR-8, Table, 1927, Mahogany, slanted front panel with 2 window dials, 6 tubes, battery $115.00
NR-9, Console, 1927, Wood, inner panel w/window dial, fold-down front, upper speaker, 6 tubes, bat $165.00
NR-9, Table, 1927, Wood, slant front panel w/window dial & brass escutcheon, 6 tubes, 3 knobs, bat $100.00
NR-12, Table, 1924, Wood, low rectangular case, 2 dial black front panel, 4 tubes, lift top, battery $200.00
NR-20, Table, 1924, Wood, low rectangular case, inner 3 dial panel, fold-down front, columns, battery $200.00

NR-45, **Table, 1925, Wood, low rectangular case (2 versions), 3 dial black front panel, battery** $195.00
NR-55, Console, 1929, Walnut veneer, upper front window dial, lower grill w/cut-outs, 3 knobs, AC $160.00
NR-60, Console, 1927, Wood, inner dial w/large escutcheon, 3 knobs, drop front, lower double doors, AC $140.00
NR-60, Table, 1927, Wood, low rectangular case, front window dial, 3 knobs, 7 tubes, AC $100.00
NR-66, Table, 1927, Wood, low rectangular case, front window dial, 6 tubes, 3 knobs, battery $100.00
NR-77, Table, 1927, Wood, low rectangular case, window dial, fold-out loop antenna, 7 tubes, battery $110.00
NR-78, Console, 1929, Walnut, highboy, upper window dial, lower grill w/cut-outs, 8 tubes, 3 knobs, AC $130.00
NR-79, Console, 1929, Walnut, highboy, center window dial, upper grill w/cut-outs, 8 tubes, 3 knobs, AC $135.00
NR-85, Table, 1928, Metal, rectangular case w/decorative decals, center front thumbwheel dial, AC $165.00
NR-90, Console, 1929, Walnut, upper front dial, lower cloth grill w/ cut-outs, stretcher base, 8 tubes, AC $150.00
NR-95, Console-AC, 1929, Walnut, highboy, inner window dial, upper grill w/cut-outs, doors, 9 tubes, AC $150.00
NR-95, Console-DC, 1929, Walnut veneer, highboy, inner window dial, upper grill w/cut-outs, doors, DC $135.00

FRESHMAN

Chas. Freshman Co., Inc.,
Freshman Building
240-248 West 40th Street,
New York, New York

In 1922 the Charles Freshman Company was formed for the manufacture of radio parts. The popular, low-cost Freshman Masterpiece was introduced in 1924 in both kit as well as a completed radio form. The company was plagued by quality control problems and by 1928, in spite of a merger with Freed-Eisemann, the stock market crash dealt the final blow.

5-F-2, **Table, 1925, Wood, low rectangular case, slanted 3 dial front panel, battery** $100.00
5-F-4, **Table, 1925, Mahogany, low rectangular case, slanted 3 dial front panel, battery** $100.00
5-F-5, **Table, 1925, Mahogany, low rectangular case, 3 dial front panel, left built-in speaker, battery** $145.00

6-F-1 "Masterpiece", Console, 1926, Mahogany, inner 3 dial panel, fold-down front, upper speaker, lower storage, bat **$185.00**

6-F-3 "Masterpiece", Console, 1926, Mahogany, right 3 dial front panel, left built-in speaker, lower storage, battery **$160.00**

6-F-6, Table, 1926, Wood, low rectangular case, 3 dial slanted front panel, 5 tubes, battery **$100.00**

6-F-11, Console, 1926, Mahogany, inner 3 dial panel, fold-down front, double doors, inner speaker, bat **$200.00**

7-F-2, Table, 1927, Wood, low rectangular case, center front window dial w/escutcheon, 6 tubes, bat **$85.00**

7-F-3, Console, 1927, Mahogany, upper front window dial, lower built-in speaker w/round grill, battery **$150.00**

7-F-5, Console, 1927, Mahogany, inner window dial, fold-down front, double doors, inner speaker, bat **$185.00**

21 "Earl", Table, 1929, Metal, low rectangular case, front window dial w/escutcheon, 8 tubes, AC **$85.00**

22 "Earl", Console, 1929, Walnut finish, highboy, upper window dial, lower grill w/circular cut-outs, AC **$185.00**

31 "Earl", Console, 1929, Walnut finish, upper window dial, lower grill w/cut-outs, stretcher base, AC **$185.00**

32 "Earl", Console, 1929, Walnut finish, inner front window dial, double doors, stretcher base, 8 tubes, AC **$200.00**

41 "Earl", Console, 1929, Wood, highboy, inner window dial and speaker grill w/cut-outs, double doors, AC **$185.00**

Concert, Table, 1925, Mahogany, tall case, inner 3 dial panel, fold-down front, upper built-in speaker **$165.00**

F-1 "Equaphase", Table, 1927, Wood, low rectangular case, center front dial, 6 tubes, battery **$95.00**

G-4 "Equaphase", Console, 1927, Wood, inner front dial & round speaker grill, double doors, lower storage, AC **$185.00**

G-7 "Equaphase", Console, 1927, Mahogany, inner dial & speaker, fold-down front, lower double doors, AC **$165.00**

M, Table, Metal, center front window dial w/escutcheon, lift-off top, 2 knobs, switch, AC **$85.00**

Masterpiece, Table, 1925, Wood, 2 styles, low rectangular case, 3 dial front panel, lift top, 5 tubes, battery **$125.00**

Masterpiece w/speaker, Table, 1925, Wood, low rectangular case w/ built-in speaker, 3 dial panel, lift top, battery **$145.00**

N-12, Console, 1928, Walnut, lowboy, upper window dial, lower grill w/cut-outs, stretcher base, AC **$165.00**

Q-15 "Little Giant of the Air", Table, 1928, Metal, center front dial knob, lift-off top, AC **$80.00**

GAROD
Garod Radio Corporation,
70 Washington Street, Brooklyn, New York
Garod Corporation,
8 West Park Street, Newark, New Jersey

The name Garod is short for the original Gardner-Rodman Corporation which manufactured crystal sets in 1921. The name was changed to the Garod Corporation in 1923 when the company began to produce radios. Due to legal and quality control problems, the original company was out of business by 1927 but the Garod name had appeared once again in radio manufacturing by 1933.

1B55L, Table, 1940, Catalin, right square dial over horizontal grill bars, recessed handle, 2 knobs ....**$900.00+**

4B-1, Portable, 1948, Plastic, inner left dial, geometric grill bars, flip-up front, handle, BC, battery **$35.00**

5A1 "Ensign", Table, 1947, Plastic, slide rule dial, Deco grill design, rounded corners, 2 knobs, BC, AC/DC **$55.00**

5A2, Table, 1946, Plastic, upper slide rule dial, inset vertical grill design, 2 knobs, BC, AC/DC **$50.00**

5A2-Y, Table, 1946, Plastic, upper slide rule dial, inset vertical grill design, 2 knobs, BC, AC/DC **$50.00**

5A3, Table, 1948, Plastic, right dial, horizontal louvers, rounded corners, 2 knobs, BC, AC/DC **$35.00**

5A4 "Thriftee", Table, 1948, Plastic, upper slanted slide rule dial, lower horiz grill bars, 2 knobs, BC, AC/DC **$40.00**

5AP1-Y "Companion", Table-R/P, 1947, Plastic, upper slide rule dial, lower vertical grill bars, 2 knobs, open top, BC, AC **$55.00**

5D-2, Portable, 1947, "Alligator", inner right dial, 2 knobs, louvers, flip-up front, BC, AC/DC/battery **$35.00**

5D-3A, Portable, 1947, Metal, inner right triangular slide rule dial, 2 knobs, flip-up lid, BC, AC/DC/bat **$35.00**

5D-5, Portable, 1948, Metal, inner right triangular slide rule dial, 2 knobs, flip-up lid, BC, AC/DC/bat **$35.00**

5RC-1 "Radalarm", Table-C, 1948, Plastic, right front dial, left alarm clock, horizontal louvers, 5 knobs, BC, AC **$35.00**

6A-2, Table, 1947, Wood, upper slide rule dial, lower horizontal louvers, 2 knobs, BC, AC/DC **$40.00**

6AU-1 "Commander", Table, 1946, Catalin, upper slide rule dial, lower horiz louvers, handle, 2 knobs, BC, AC/DC ....**$700.00+**

6BU-1A, Table, 1947, Plastic, lower slide rule dial, horizontal louvers, 2 knobs, wide base, BC, AC/DC **$50.00**

6DPS-A, Console-R/P, 1947, Wood, inner right slide rule dial, 4 knobs, pull-out phono drawer, legs, BC, SW, AC **$80.00**

11FMP, Console-R/P, 1948, Wood, slide rule dial, 4 knobs, pull-out phono drawer, high legs, BC, SW, FM, AC **$80.00**

62B, Table, 1947, Plastic, slide rule dial, horizontal louvers, wide base, 4 knobs, BC, SW, AC/DC **$50.00**

126, Table, 1940, Catalin, upper dial, cloth grill w/overlapping circle cut-outs, handle, 2 knobs ....**$1,800.00+**

306, Table-R/P, 1948, Wood, center front round dial, 4 knobs, 3/4 lift top, inner phono, BC, AC **$30.00**

711-P, Table-R/P, 1941, Walnut veneer, right front dial, left vertical grill bars, lift top, inner phono, AC **$45.00**

769, Table, 1935, Wood, Deco, right dial, pushbuttons, step-down left grill w/vert bars, BC, SW **$115.00**

1450, Table, 1940, Catalin, right front square dial, left cloth grill w/ horiz bars, handle, 2 knobs ....**$850.00+**

4159, Table, 1938, Wood, right front slide rule dial, pushbuttons, left rounded grill w/horiz bars, AC **$65.00**

EB, Console-R/P, 1926, Wood, inner 3 dial panel, fold-down front, lower storage & phono, doors, battery **$185.00**

M, Table, 1925, Wood, low rectangular case, 3 dial slanted front panel **$95.00**

RAF, Table, 1923, Mahogany, low rectangular case, 3 dial front panel, lift top, 5 tubes, battery **$195.00**

V, Table, 1923, Mahogany, low rectangular case, 3 dial slanted front panel, feet, 5 tubes, battery **$100.00**

GENERAL ELECTRIC
General Electric Co., Electronics Department, Bridgeport, Connecticut

GE began manufacturing radios in 1919 and marketed them through RCA until 1930, when they began to use their own General Electric trademark on their products.

41 "**Musaphonic**", Console-R/P, 1948, Wood, slide rule dial, 5 knobs, 9 pushbuttons, pull-out phono, BC, 3SW, 2FM, AC **$95.00**

42 "**Musaphonic**", Console-R/P, 1948, Wood, slide rule dial, 5 knobs, 9 pushbuttons, pull-out phono, BC, 3SW, 2FM, AC **$95.00**

43 "**Musaphonic**", Console-R/P, 1948, Wood, slide rule dial, 5 knobs, 9 pushbuttons, pull-out phono, BC, 3SW, 2FM, AC **$95.00**

44 "**Musaphonic**", Console-R/P, 1948, Wood, slide rule dial, 5 knobs, 9 pushbuttons, pull-out phono, BC, 3SW, 2FM, AC **$95.00**

45 "**Musaphonic**", Console-R/P, 1948, Wood, slide rule dial, 5 knobs, 9 pushbuttons, pull-out phono, BC, 3SW, 2FM, AC **$95.00**

50, Table-C, 1946, Plastic, lower right front dial, upper checkered grill, left square clock, 4 knobs **$45.00**
50W, Table-C, 1948, Ivory plastic, lower right front dial, upper checkered grill, left clock, 4 knobs **$45.00**

54, Table, 1940, Plastic, Deco, right vertical slide rule dial, left round grill w/horiz bars, 4 knobs **$85.00**
60, Table-C, 1948, Mahogany plastic, upper thumbwheel dial, left clock, horizontal grill bars, BC, AC **$45.00**
62, Table-C, 1948, Ivory plastic, upper thumbwheel dial, left clock, horizontal grill bars, BC, AC **$45.00**

65, Table-C, 1949, Plastic, thumbwheel dial knob, left round alarm clock, right vertical grill bars, AC **$40.00**
66, Table-C, 1949, Plastic, upper thumbwheel dial, left alarm clock, right horizontal louvers, BC, AC **$40.00**
100, Table, 1946, Plastic, lower half-moon dial, upper horizontal grill bars, 2 knobs, BC, AC/DC **$45.00**
102, Table, 1948, Brown plastic, lower slide rule dial, upper horiz grill bars, 2 knobs, BC, AC/DC **$45.00**
102W, Table, 1948, Ivory plastic, lower slide rule dial, upper horiz grill bars, 2 knobs, BC, AC/DC **$45.00**
103, Table, 1946, Wood, lower center black dial, upper grill, 2 knobs, BC, AC/DC **$50.00**
106, Table-R/P, 1946, Wood, right front round black dial, 3 knobs, lift top, inner phono, BC, AC **$30.00**
107, Table, 1948, Brown plastic, lower slide rule dial, upper horizontal louvers, 2 knobs, BC, AC/DC **$45.00**
107W, Table, 1948, Ivory plastic, lower slide rule dial, upper horizontal louvers, 2 knobs, BC, AC/DC **$45.00**
113, Table, 1948, Plastic, right front half-moon dial, left horizontal grill bars, 2 knobs, BC, AC/DC **$35.00**
114, Table, 1946, Brown plastic, lower slide rule dial, upper horizontal louvers, 2 knobs, BC, AC/DC **$55.00**
114W, Table, 1946, Ivory plastic, lower slide rule dial, upper horizontal louvers, 2 knobs, BC, AC/DC **$55.00**
115, Table, 1948, Brown plastic, slide rule dial, large plastic perforated grill, 2 knobs, BC, AC/DC **$50.00**
115W, Table, 1948, Ivory plastic, slide rule dial, large plastic perforated grill, 2 knobs, BC, AC/DC **$50.00**
119W, Console-R/P, 1948, Wood, inner right dial, 3 knobs, phono, lift top, criss-cross grill, BC, AC **$55.00**
135, Table, 1950, Plastic, lower slide rule dial, upper recessed vert grill bars, 2 knobs, BC, AC/DC **$40.00**
136, Table, 1950, Plastic, lower slide rule dial, upper recessed vertical grill bars, 2 knobs, BC **$40.00**

140, Portable, 1947, Metal & plastic, inner dial & grill, flip-open door, pull-up handle, BC, AC/DC/bat **$40.00**
143, Portable, 1949, Plastic, slide rule dial, horizontal louvers, 2 thumbwheel knobs, BC, AC/DC/bat **$45.00**
145, Portable, 1949, Inner slide rule dial, lattice grill, 2 knobs, lift-up top, handle, BC, AC/DC/bat **$35.00**
150, Portable, 1949, Leatherette, slide rule dial, horizontal grill bars, handle, 2 knobs, BC, AC/DC/bat **$35.00**
160, Portable, 1949, Plastic, slide rule dial, lattice grill, handle, 2 thumbwheel knobs, BC, AC/battery **$30.00**
165, Portable, 1950, Maroon plastic, slide rule dial, horiz grill bars, 2 knobs, handle, BC, AC/DC/bat **$35.00**
180, Table, 1947, Wood, right dial, cloth grill w/horizontal bars, 2 knobs, rounded corners, BC, bat **$30.00**

200, Table, 1946, Plastic, lower front black dial, upper horizontal grill bars, 2 knobs, BC, AC/DC $45.00

201, Table, 1946, Plastic, lower front black rectangular dial, upper metal criss-cross grill, 2 knobs $45.00

202, Table, 1947, Plastic, lower center rectangular dial, upper metal criss-cross grill, 2 knobs, BC $45.00

203, Table, 1946, Wood, lower black rectangular dial, upper metal woven grill, 2 knobs, BC, AC/DC $45.00

210, Table, 1948, Mahogany plastic, right round dial, recessed cloth grill, 2 knobs, BC, FM, AC/DC $30.00

211, Table, 1948, Ivory plastic, right round dial, recessed cloth grill, 2 knobs, BC, FM, AC/DC $30.00

212, Table, 1948, Wood, right dial over plastic wrap-around panel, cloth grill, 2 knobs, AM, FM, AC/DC $45.00

218, Table, 1951, Plastic, right dial, lattice grill, lower diagonal panel w/3 knobs, BC, FM, AC/DC $35.00

220, Table, 1946, Plastic, slide rule dial, checkered grill, top vert fluting, 2 knobs, BC, SW, AC/DC $40.00

221, Table, 1946, Wood, large lower slide rule dial, criss-cross grill, 2 knobs, BC, SW, AC/DC $50.00

226, Table, 1950, Plastic, lower front slide rule dial, recessed checkered grill, 2 knobs, BC, AC/DC $45.00

250, Portable, 1946, Metal, inner slide rule dial, 4 knobs, flip back top, horizontal louvers, BC, AC/bat $35.00

254, Portable, 1948, Cloth covered, inner dial, 2 knobs, fold-down front, handle, BC, AC/DC/bat $30.00

260, Portable, 1947, Metal, inner slide rule dial, 1 knob, 12 pushbuttons, flip-up lid, BC, 5SW, AC/bat $75.00

280, Table, 1947, Wood, right dial, 2 horizontal grill bars, 4 knobs, rounded corners, BC, SW, bat $30.00

303, Table-R/P, 1947, Wood, front slide rule dial, lower horizontal louvers, 4 knobs, lift top, BC, AC $30.00

304, Table-R/P, 1948, Wood, front slide rule dial, lower horizontal grill bars, 4 knobs, lift top, BC, AC $30.00

321, Table, 1946, Wood, center slide rule dial, upper metal grill, 2 knobs, 5 pushbuttons, BC, AC/DC $40.00

324, Console-R/P, 1949, Mahogany, right tilt-out slide rule dial, 4 knobs, pull-out phono, BC, FM, AC $70.00

328, Console-R/P, 1949, Blonde wood, right tilt-out slide rule dial, 4 knobs, pull-out phono, BC, FM, AC $70.00

356, Table, 1948, Plastic, step-down top, slide rule dial, horizontal louvers, 2 knobs, BC, FM, AC/DC $45.00

376, Console-R/P, 1948, Wood, left tilt-out slide rule dial & 4 knobs, pull-out phono drawer, BC, FM, AC $70.00

400, Table, 1950, Plastic, slanted right front dial, left vertical grill bars, 2 knobs, BC, AC/DC $35.00

408, Table, 1950, Plastic, center front raised half-moon dial, 2 thumbwheel knobs, AM, FM, AC/DC $45.00

409, Table, 1952, Plastic, front half-moon dial over horiz bars, thumbwheel knobs, BC, FM, AC/DC $45.00

410, Table, 1951, Wood, lower front slide rule dial, large upper grill, 2 knobs, feet, AM, AC/DC $30.00

411, Table, 1950, Plastic, large right front dial, left vertical grill bars, 2 knobs $35.00

412, Table, 1952, Plastic, right front rectangular dial, left vertical grill bars, 2 knobs, BC, AC/DC **$35.00**

414, Table, 1950, Plastic, large right front dial, left horizontal grill bars, 2 knobs, BC, AC/DC $35.00
415F, Table, 1953, Plastic, right rectangular dial, left horizontal grill bars, 2 knobs, BC, AC/DC **$35.00**
417, Console-R/P, 1947, Wood, slide rule dial, 5 knobs, 12 pushbuttons, fold-down front, BC, 2SW, 2FM, AC **$95.00**
417A, Console-R/P, 1947, Wood, slide rule dial, 5 knobs, 12 pushbuttons, fold-down front, BC, 2SW, 2FM, AC **$95.00**
419, Table, 1954, Plastic, right front round dial over checkered grill, 2 knobs, feet, BC, AC/DC **$20.00**
422, Table, 1951, Plastic, lower slide rule dial, upper recessed vertical bars, 2 knobs, BC, AC/DC **$40.00**
423, Table, 1951, Plastic, lower front slide rule dial, upper vertical grill bars, 2 knobs, BC, AC/DC **$40.00**

428, Table, 1955, Plastic, raised center w/half-moon dial, vert front bars, feet, 2 knobs, BC, AC/DC $30.00
430, Table, 1952, Wood, right front rectangular dial, left horizontal grill bars, 2 knobs, BC **$35.00**

440, Table, 1954, Plastic, center semi-circular dial over horizontal front bars, 3 knobs, AM, FM, AC $45.00
442, Table, 1951, Plastic, lower slide rule dial, upper recessed grill w/vert bars, 2 knobs, BC, AC **$40.00**

453, Table, 1956, Plastic, right front round dial over checkered grill, 2 knobs, feet, BC, AC/DC **$20.00**
500, Table-C, 1950, Plastic, upper thumbwheel dial, left round clock, horizontal grill bars, BC, AC **$45.00**
502, Console-R/P, 1948, Wood, inner dial, 5 knobs, 9 pushbuttons, pull-out phono drawer, BC, 3SW, 2FM, AC **$100.00**
507, Table-C, 1950, Plastic, front thumbwheel dial, left alarm clock, right vertical grill bars, BC, AC **$45.00**
510, Table-C, 1951, Plastic, upper front thumbwheel dial, left alarm clock, right circular louvers, AC **$45.00**
511, Table-C, 1951, Plastic, upper thumbwheel dial, right circular louvers, left alarm clock, BC, AC **$45.00**

515, Table-C, 1951, Plastic, upper thumbwheel dial, left alarm clock, right vertical grill bars, BC, AC $45.00
515F, Table-C, 1951, Plastic, upper thumbwheel dial, left alarm clock, right vertical grill bars, BC, AC **$45.00**
516F, Table-C, 1951, Plastic, upper thumbwheel dial, left alarm clock, right vertical grill bars, AC **$45.00**

517F, Table-C, 1951, Plastic, upper thumbwheel dial, left alarm clock, right vertical grill bars, AC $45.00
518F, Table-C, 1951, Plastic, upper thumbwheel dial, left alarm clock, right vertical grill bars, AC **$45.00**
521, Table-C, 1950, Upper front thumbwheel dial, right horiz grill bars, left alarm clock, BC, AC **$50.00**
535, Table-C, 1951, Plastic, right front dial, left alarm clock, center vertical bars, 5 knobs, AC **$30.00**
542, Table-C, 1953, Plastic, upper thumbwheel dial, right vert grill bars, left alarm clock, BC, AC **$45.00**
543, Table-C, 1953, Plastic, upper thumbwheel dial, right vert grill bars, left alarm clock, BC, AC **$45.00**
546, Table-C, 1953, Plastic, slide rule dial, right & left vert grill bars, center clock, 5 knobs, AC **$30.00**
548, Table-C, 1953, Plastic, slide rule dial, right & left vert grill bars, center clock, 5 knobs, AC **$30.00**
549, Table-C, 1953, Plastic, lower slide rule dial, center front alarm clock, vertical grill bars, BC, AC **$30.00**
551, Table-C, 1953, Plastic, lower slide rule dial, center clock, vertical grill bars, 4 knobs, BC, AC **$30.00**

590, Table-C, 1955, Plastic, right front dial, left clock, raised center lattice grill, 5 knobs, BC, AC $25.00

600, Portable, 1950, Plastic, slide rule dial, recessed horizontal grill bars, handle, 2 knobs, BC, bat $45.00

601, Portable, 1950, Maroon plastic, lower slide rule dial, 2 thumbwheel knobs, handle, BC, AC/DC/bat $45.00

603, Portable, 1950, Tan plastic, lower slide rule dial, 2 thumbwheel knobs, handle, BC, AC/DC/bat $45.00

604, Portable, 1950, Green plastic, lower slide rule dial, 2 thumbwheel knobs, handle, BC, AC/DC/bat $45.00

610, Portable, 1951, Plastic, front half-moon dial, vertical grill bars, handle, 2 knobs, BC, AC/DC/bat $35.00

650, Portable, 1950, Maroon plastic, slide rule dial, horiz grill bars, handle, 2 knobs, BC, AC/DC/bat $35.00

741, Console-R/P, 1952, Wood, inner right dial, 3 knobs, lower pull-out phono drawer, double doors, BC, AC $55.00

752, Console-R/P, 1951, Wood, inner right dial, 4 knobs, lower pull-out phono, double doors, BC, FM, AC $60.00

755, Console-R/P, 1951, Wood, inner right slide rule dial, 2 knobs, lower pull-out phono, BC, FM, AC $70.00

895, Table-C, 1956, Plastic, right round dial, left square clock, checkered front panel, feet, BC, AC $25.00

A-52, Tombstone, Wood, center front dial w/oval escutcheon, upper grill w/cut-outs, 4 knobs $115.00

A-53, Tombstone, 1935, Walnut finish, center front window dial, upper cloth grill w/cut-outs, BC, SW, AC $115.00

A-63, Tombstone, 1935, Walnut veneer, Deco, center window dial, cloth grill w/vertical bars, BC, SW, AC $145.00

A-64, Tombstone, 1935, Wood, center horizontal pointer dial, upper grill w/cut-outs, arched top, 4 knobs $140.00

A-65, Console, 1935, Walnut-veneer, upper window dial, lower cloth grill w/vertical bars, BC, SW, AC $120.00

A-67, Console, 1935, Wood, upper horiz pointer dial, lower grill w/scrolled cut-outs, 4 knobs, BC, SW $150.00

A-70, Tombstone, 1935, Wood, center horiz pointer dial, upper & lower vert cut-outs, fluting, BC, SW, AC $140.00

A-75, Console, 1935, Walnut-veneer, upper front dial, lower cloth grill w/3 vertical bars, BC, SW, AC $145.00

A-87, Console, 1935, Wood, upper front dial, vertical front lines & grill bars, 3 knobs, feet, 4 band, AC $150.00

A-125, Console, 1935, Wood, upper dial, lower grill w/rectangular cut-outs, fluted sides, base, AC $185.00

BX, Table, 1932, Metal case w/lacquer finish, right front dial, center round grill, 2 knobs, AC/DC $75.00

C415, Table-C, 1958, Plastic, lower front slide rule dial, upper clock, right & left vertical bars, BC, AC $30.00

C-415C, Table-C, 1958, Plastic, lower front slide rule dial, upper clock, right & left vert grill bars, AC $30.00

CT455A, Table-C, 1960, Plastic, transistor, right dial knob over perforated grill, left alarm clock, AM, bat $25.00

E-52, Table, 1937, Wood, center front dial surrounded by 4-section cloth grill, 4 knobs, BC, SW, AC $55.00

E-81, Tombstone, 1936, Wood, center front rectangular dial, upper grill w/cut-outs, 4 knobs, BC, SW, AC $125.00

F-51, Table, 1937, Plastic, Deco, right vertical dial, rounded left side, horizontal bars, 4 knobs, AC $125.00

F-53, Table, 1937, Wood, center rectangular dial, upper & lower grills w/horizontal bars, BC, SW, AC $60.00

F-63, Table, 1937, Walnut, left gold dial, streamlined right side w/Deco grill, 4 knobs, BC, SW, AC $95.00

F-65, Console, 1937, Wood, upper rectangular dial, lower cloth grill w/3 vert bars, 6 tubes, BC, SW, AC $125.00

F-66, Console, 1937, Wood, upper rectangular dial, lower cloth grill w/3 vert bars, 6 tubes, BC, SW, AC $125.00

F-70, Table, 1937, Wood, left rectangular dial, right grill w/4 horizontal bars, 7 tubes, BC, SW, AC $65.00

F-74, Table, 1938, Wood, left front dial, tuning eye, right grill w/2 horiz bars, 4 knobs, BC, SW, AC $85.00

F-75, Console, 1937, Wood, upper rectangular dial, lower cloth grill w/vert bars, 7 tubes, BC, SW, AC $125.00

F-81, Tombstone, 1937, Wood, rectangular dial, upper cloth grill w/horizontal bars, 4 knobs, BC, SW, AC $145.00

F-86, Console, 1937, Wood, upper rectangular dial, lower cloth grill w/3 vert bars, 8 tubes, BC, SW, AC $135.00

F-96, Console, 1937, Wood, upper dial, pushbuttons, lower cloth grill w/2 vertical bars, 5 knobs, AC $145.00

F-107, Console, 1937, Wood, rectangular dial, cloth grill w/vert bars, pushbuttons, 10 tubes, BC, SW, AC $165.00

F-109, Console-R/P, 1937, Wood, upper dial, lower cloth grill w/vert bars, inner phono, 10 tubes, BC, SW, AC $165.00

F-135, Console, 1937, Wood, rectangular dial, cloth grill w/vert bars, pushbuttons, 13 tubes, BC, SW, AC $200.00

F-665, Chairside, 1937, Walnut, top dial & knobs, front grill w/vertical bars, step-down top, BC, SW, AC $130.00

FB-52, Cathedral, Wood, center front rectangular dial, upper vertical grill bars, battery $100.00

FB-77, Console, 1938, Wood, upper rectangular dial, lower cloth grill w/3 vertical bars, 4 knobs, bat $115.00

G-50, Table, 1937, Wood, right round telephone dial, left cloth grill w/vertical bars, 2 knobs, AC $65.00

G-53, Table, 1938, Wood, 2 right slide rule dials, pushbuttons, left grill w/vertical bars, BC, SW, AC $60.00

G-55, Console, 1938, Wood, upper round telephone dial, lower cloth grill w/2 vert bars, 5 tubes, BC, AC $120.00

G-56, Console, 1938, Wood, 2 upper dials, lower grill w/vert bars, pushbuttons, 5 tubes, BC, SW, AC $120.00

G-61, Table, 1938, Wood, rounded right w/3 dials, left wrap-around grill, pushbuttons, BC, SW, AC $85.00

G-68, Console-R/P, 1938, Wood, inner dial & phono, front grill w/vert bars, lift top, 6 tubes, BC, SW, AC $130.00

G-76 "Radiogrande", Console, 1938, Wood, upper front dial, pushbuttons, tuning eye, lower grill w/cut-outs, AC $215.00

G-85, Console, 1938, Wood, upper dial, lower cloth grill, tuning eye, 10 pushbuttons, 8 tubes, BC, SW, AC $165.00

G-97, Console, 1938, Wood, upper dial, cloth grill, 7 pushbuttons, top pop-up station timer, BC, SW, AC $185.00

G-105, Console, 1938, Wood, upper dial, grill w/vert bars, tuning eye, 14 pushbuttons, 10 tubes, BC, SW, AC $185.00

GB-400, Portable, 1948, Luggage-style, thumbwheel dial & volume knobs, center grill, handle, BC, battery $35.00

GB-401, Table, 1939, Two-tone wood, slanted front w/vertical bars, 2 thumbwheel knobs, battery $60.00

GD-41, Table, 1938, Two-tone wood, right dial, left cloth grill w/3 section cut-outs, 2 knobs, AC/DC $50.00

GD-52, Table, 1938, Wood, thumbwheel dial & volume knobs, center grill, 6 pushbuttons, BC, AC $70.00

GD-60, Table, 1938, Wood, thumbwheel dial & volume knobs, center grill, 6 pushbuttons, BC, AC $70.00

GD-520, Table, 1939, Plastic, Deco, small case, right metal dial panel, left wrap-around horiz bars, 2 knobs $150.00

GD-600, Table, 1938, Wood, right front rectangular dial, left wrap-over vertical grill bars, 2 knobs $60.00

H-71, Console-R/P, 1931, Wood, lowboy, inner dial & grill, double doors, lift top, inner phono, 9 tubes, AC $285.00

H-87, Console, 1939, Wood, slanted dial, pushbuttons, cloth grill w/ bars, rounded sides, BC, SW, AC $165.00

H-500, Table, 1939, Plastic, Deco, step-down top, center thumbwheel dial, left front lattice grill, AC $115.00

H-502, Table, 1939, Plastic, Deco, step-down top, center thumbwheel dial, left front lattice grill, AC $115.00

H-520, Table, 1939, Plastic, Deco, step-down top, center thumbwheel dial, left front lattice grill, AC $115.00

H-530, Tombstone, 1939, Wood, small case, lower right dial, upper 2-section cloth grill, sideways chassis $95.00

H-600, Table, 1939, Swirled plastic, right square dial, left lattice grill, horiz case lines, 2 knobs, AC $100.00

H-610, Table, 1939, Plastic, right front square dial, 4 pushbuttons, left lattice grill, 2 knobs, AC $65.00

H-620, Table, 1939, Plastic, right front square dial, left horizontal louvers, 3 knobs, BC, SW, AC $60.00

H-620U, Table, 1939, Plastic, right front square dial, left horizontal louvers, 3 knobs, BC, SW, AC $60.00

H-630, Table, 1939, Wood, right dial, 4 pushbuttons, left cloth wrap-over grill, 3 knobs, BC, SW, AC $50.00

H-632, Table, 1939, Wood, center dial, pushbuttons, side wrap-around grills, 3 knobs, BC, SW, AC $60.00

H-634, Table, 1939, Wood, right front rectangular dial, pushbuttons, left wrap-around grill, BC, 2SW $60.00

H-639, Table-R/P, Deco, right front dial, pushbuttons, rounded left side, horiz grill louvers, lift top $50.00

HB-412, Portable, 1939, Leatherette, inner right front dial, left grill, fold-up front, handle, AC/DC/bat **$30.00**

HJ-624, Table, 1939, Wood, right front dial, pushbuttons, left wrap-around grill, thumbwheel knobs **$85.00**

J-54W, Table, Plastic, upper front slide rule dial, lower horizontal louvers, 2 knobs, BC **$45.00**

J-62, **Table, 1941, Mahogany, upper slide rule dial, lower criss-cross grill, side handles, 3 knobs, AC** **$80.00**

J-71, Table, Wood, 3 center dials, right & left wrap-around horizontal grill bars, pushbuttons **$85.00**

J-80, **Cathedral, 1932, Wood, center window dial, cloth grill w/cut-outs, fluted columns, 3 knobs, AC** **$295.00**

J-82, Cathedral, 1932, Wood, center window dial, cloth grill w/cut-outs, fluted columns, 3 knobs, AC **$295.00**

J-86, Console, 1932, Wood, lowboy, upper window dial, lower grill w/cut-outs, stretcher base, AC **$140.00**

J-88, Console-R/P, 1932, Wood, lower window dial, upper gothic grill, 3 knobs, inner phono, 8 tubes, AC **$140.00**

J-100, Cathedral, 1932, Wood, front window dial, upper grill w/gothic cut-outs, 3 knobs, 10 tubes, AC **$395.00**

J-501W, Table, 1941, Ivory plastic, upper slide rule dial, lower horizontal louvers, 2 knobs, AC/DC **$45.00**

J-620, Table, 1941, Blond wood, slide rule dial, lower criss-cross grill, side handles, 3 knobs, AC **$80.00**

J-644, Table, 1941, Plastic, right front dial, left horizontal louvers, rounded top, 2 knobs, AC **$40.00**

J-805, Console, 1940, Wood, slide rule dial, cloth grill w/bars, 6 pushbuttons, 4 knobs, BC, SW, AC **$145.00**

JCP-562, Table, 1942, Wood, right front dial, left cloth grill, 2 knobs **$35.00**

JFM-90, Table, Wood, FM only, center black dial, 6 pushbuttons, rounded front corners, 2 knobs **$45.00**

K-40-A, Table, 1933, Walnut, right dial, center cloth grill w/cut-outs, lower molding, 2 knobs, AC/DC **$85.00**

K-41, Table, 1933, Metal, right front dial, center round cloth grill, decorative case lines, 2 knobs, AC **$75.00**

K-43, **Cathedral, 1933, Wood, low case, center dial surrounded by cloth grill w/cut-outs, BC, SW, AC** **$165.00**

K-50, Cathedral, 1933, Wood, lower right front window dial, upper cloth grill w/cut-outs, 3 knobs, AC **$200.00**

K-50-P, Cathedral, 1933, Wood, lower right front window dial, upper cloth grill w/cut-outs, 4 knobs, AC **$200.00**

K-51, Table, 1932, Wood, inner window dial, grill w/gothic cut-outs, double sliding carved doors, AC **$175.00**

K-51-P, Table, 1932, Wood, inner window dial, grill w/gothic cut-outs, double sliding carved doors, AC **$175.00**

K-52, Cathedral, 1933, Wood, lower right front window dial, upper grill w/cut-outs, 4 knobs, BC, SW, AC **$225.00**

K-53-M, Table, 1933, Wood, right window dial, left & right quarter-round grill cut-outs, 4 knobs, AC **$90.00**

K-54 "Music Box", Table-R/P, 1932, Ornate case, right side controls, front grill w/cut-outs, lift top, inner phono, AC **$95.00**

K-60, Cathedral, 1933, Wood, mantle-clock design, lower window dial, upper round cloth grill, handle, AC **$250.00**

K-60-P, Cathedral, 1933, Wood, mantle-clock design, lower window dial, upper round grill, handle, 4 knobs **$250.00**

K-62, Console, 1931, Wood, lowboy, center window dial, lower grill w/cut-outs, 3 knobs, 6 legs, AC **$150.00**

K-63, Cathedral, 1933, Wood, mantle-clock design, lower window dial, upper grill, 4 knobs, BC, SW, AC **$295.00**

K-64, Cathedral, 1933, Wood, mantle-clock design, round dial, upper grill, handle, 4 knobs, BC, SW, AC **$265.00**

K-65, Console, 1933, Walnut, window dial, lower grill w/cut-outs, 6 legs, stretcher base, 6 tubes, AC **$145.00**

K-70, Cathedral, Wood, lower window dial w/escutcheon, upper 3-section cloth grill, 3 knobs **$200.00**

K-80, Tombstone, 1933, Wood, center round dial, upper grill w/cut-outs, angled top, 8 tubes, 4 knobs, AC **$300.00**

K-82, Grandfather Clock, 1931, Wood, Colonial design, inner dial & controls, front door, upper clock face, AC **$400.00**

K-85, Console, 1933, Wood, upper round dial, lower cloth grill w/cut-outs, 6 legs, 4 knobs, BC, SW, AC **$165.00**

K-106, Console, 1933, Wood, upper window dial, lower grill w/cut-outs, 6 legs, 20 tubes, 5 knobs, AC **$275.00**

K-126, Console, 1933, Wood, upper window dial, lower grill w/ cut-outs, 6 legs, 12 tubes, BC, SW, AC **$175.00**

L-50, Table, 1932, Ornate "carved" front & sides, right dial, center grill, 2 knobs, handle, AC/DC **$125.00**

L-570, Table, 1941, Catalin, upper slide rule dial, lower horizontal louvers, handle, 2 ribbed knobs .... **$550.00+**

L-604, Table, 1941, Two-tone walnut, upper slanted slide rule dial, lower horizontal louvers, AC **$50.00**

L-613, Table, Two-tone wood, upper slanted slide rule dial, lower horizontal louvers, 2 knobs **$45.00**

L-624, Table, 1941, Ivory plastic, upper slanted slide rule dial, lower horizontal louvers, 2 knobs, AC **$45.00**

L-630, Table, 1940, Two-tone wood, slide rule dial, cloth grill w/ center horiz bar, 4 knobs, BC, SW **$45.00**

L-632, Table, Wood, upper curved slide rule dial, lower horizontal louvers, 2 knobs, BC, SW **$45.00**

L-633, Table, 1941, Walnut, upper slide rule dial, lower horizontal grill bars, rounded front sides, AC **$55.00**

L-640, Table, 1941, Wood, upper front slide rule dial, lower grill, pushbuttons, 4 knobs, BC, 2SW, AC **$60.00**

L-641, Table, 1942, Wood, upper slide rule dial, lower grill w/ horizontal bars & oval, 4 knobs, BC, SW **$45.00**

L-652, Table, Wood, large center front slide rule dial, 5 pushbuttons, top louvers, 2 knobs **$50.00**

L-660, Table, 1941, Wood, recessed slide rule dial, chrome escutcheon, 5 pushbuttons, 2 knobs, BC, AC **$60.00**

L-678, Table-R/P, 1941, Wood, center front dial, right & left wrap-around grills, lift top, inner phono, AC **$35.00**

L-740, Table, 1941, Walnut, center front slide rule dial, 5 pushbuttons, top grill, 4 knobs, BC, SW, AC **$75.00**

L-915, Console, 1940, Walnut, upper slide rule dial, pushbuttons, lower grill w/vertical bars, BC, SW, AC **$125.00**

L-916, Console, 1941, Wood, slanted slide rule dial, pushbuttons, lower grill w/vertical bars, BC, 2SW **$120.00**

LB-530X, Portable, 1940, Leatherette, inner top dial & knobs, front louvers, fold-open top, handle, AC/bat **$40.00**

LC-619, Console-R/P, 1941, Two-tone walnut, inner right slide rule dial, left lift top, inner phono, BC, SW **$130.00**

LC-638, Table-R/P, 1941, Walnut, front slide rule dial, horizontal louvers, 4 knobs, lift top, inner phono, AC **$35.00**

LC-648, Console-R/P, 1941, Two-tone walnut, inner right slide rule dial, door, left lift top, inner phono, BC **$120.00**

LC-658, Table-R/P, 1941, Mahogany veneer, outer front dial & grill, 4 knobs, lift top, inner phono, BC, AC **$30.00**

LF-115, Console, 1940, Walnut, upper slide rule dial, pushbuttons, lower grill w/vert bars, BC, FM, SW **$135.00**

LF-116, Console, 1940, Walnut, upper slide rule dial, pushbuttons, lower grill w/vert bars, BC, FM, SW **$125.00**

LFC-1118, Console-R/P, 1940, Walnut, inner slide rule dial, pushbuttons, left lift top, inner phono, BC, FM, SW **$150.00**

LFC-1128, Console-R/P, 1940, Walnut, inner slide rule dial, pushbuttons, left pull-out phono drawer, BC, FM, SW **$140.00**

Longfellow, Grandfather Clock, 1931, Mahogany, Colonial design, inner dial & knobs, front door, upper clock face, AC ... **$500.00**

M-51A, Tombstone, 1935, Two-tone wood, lower square airplane dial, cloth grill w/vertical bars, 4 knobs **$125.00**

M-61, Tombstone, 1934, Wood, shouldered, center square dial, upper grill w/vertical bars, BC, SW, AC **$160.00**

M-81, Tombstone, 1934, Wood, center front square dial, upper grill w/vertical bars, 5 knobs, BC, SW, AC **$145.00**

M-103, Table, Plastic, lower front curved slide rule dial, upper horizontal louvers, 2 knobs **$45.00**

P-672, Portable, 1958, Plastic, side dial knob, horizontal front grill bars with GE logo, handle **$20.00**

P710A, Portable, 1958, Transistor, right front dial, left lattice grill, center knob, AM, battery **$30.00**

P715-D, Portable, 1958, Metal/leatherette, transistor, upper dial, perforated grill, pull-up handle, AM, bat **$30.00**

P725, Portable, 1958, Transistor, right side dial knob, perforated & crimped front grill, handle, AM, bat **$35.00**

P725A, Portable, 1958, Transistor, right side dial knob, perforated & crimped front grill, handle, AM, bat **$35.00**

P726, Portable, 1958, Transistor, right side dial knob, perforated & crimped front grill, handle, AM, bat **$35.00**

P-735A, Portable, 1958, Plastic, right side dial knob, large grill, handle, left side knob, BC, AC/DC/bat **$25.00**

P-746A, Portable, 1958, Two-tone plastic, transistor, right round dial, left vert grill bars, AM, battery **$30.00**

P750A, Portable, 1958, Leather case, transistor, right side dial knob, front lattice grill, handle, AM, bat **$40.00**

P760A, Portable, 1959, Transistor, right side dial knob, front lattice grill, handle, AM, battery **$30.00**

P761A, Portable, 1959, Transistor, right side dial knob, front lattice grill, handle, AM, battery **$30.00**

P765A, Portable, 1958, Transistor, upper round dial, perforated plaid grill, rechargeable, AM, battery **$35.00**

P766A, Portable, 1958, Transistor, upper round dial, perforated plaid grill, rechargeable, AM, battery **$35.00**

P776A, Portable, 1959, Leather case, transistor, right round dial, horizontal grill bars, handle, AM, bat **$25.00**

P776B, Portable, 1959, Leather case, transistor, right round dial, horizontal grill bars, handle, AM, bat $25.00
P780A, Portable, 1960, Transistor, upper slide rule dial, lower lattice grill, handle, 2 knobs, AM, battery $25.00
P780B, Portable, 1960, Transistor, upper slide rule dial, lower lattice grill, handle, 2 knobs, AM, battery $25.00
P787A, Portable, 1960, Transistor, right slide rule dial, large perforated grill area, side knobs, AM, bat $30.00
P790B, Portable, 1960, White & black, transistor, right dial knob, left grill w/circular cut-outs, AM, bat $30.00
P791A, Portable, 1960, Transistor, right front dial knob, left grill w/circular cut-outs, AM, battery $30.00

P791B, Portable, 1960, Turquoise & white, transistor, right dial, left grill w/circular cut-outs, AM, bat $30.00
P-795A, Portable, 1958, Black leatherette, transistor, right side dial knob, front lattice grill, AM, battery $30.00

P-796A, Portable, 1958, Blue leatherette, transistor, right side dial knob, front lattice grill, AM, battery $35.00
P-797A, Portable, 1958, Beige leatherette, transistor, right side dial knob, front lattice grill, AM, battery $30.00

P800A, Portable, 1959, Plastic, transistor, right dial w/magnifier, left lattice grill, center knob, AM, bat $35.00
P805, Portable, 1959, Plastic, transistor, right front round dial knob, left perforated grill, handle, bat $30.00

P805A, Portable, 1959, Plastic, transistor, right front round dial knob, left perforated grill, handle, bat $30.00
P806A, Portable, 1959, Plastic, transistor, right front round dial knob, left perforated grill, handle, bat $30.00
P830A, Portable, 1960, Charcoal plastic, transistor, upper slide rule dial, lower perforated grill, AM, bat $25.00
P831A, Portable, 1960, Blue plastic, transistor, upper slide rule dial, lower perforated grill, AM, bat $30.00

S-22, Tombstone/Stand, 1931, Wood, lower front dial, upper floral cloth grill, brass handle, columns, 3 knobs, AC $250.00
S-22 "Junior", Console, 1931, Wood, lowboy, upper window dial, lower cloth grill w/cut-outs, 8 tubes, 3 knobs $125.00
S-22A, Tombstone, 1932, Wood, lower front window dial, upper cloth floral grill, side columns, 3 knobs $215.00
S-22C, Tombstone, 1934, Wood, lower front window dial, upper cloth floral grill, side columns, 3 knobs, AC $215.00
S-22X "Junior", Tombstone-C, 1931, Wood, lower window dial, upper grill w/center clock, side columns, 3 knobs, AC $275.00
T-12, Cathedral, 1931, Wood, lower window dial w/escutcheon, upper cloth grill w/cut-outs, 3 knobs, AC $200.00
T-115A, Table, 1958, Plastic, raised upper slide rule dial, large front lattice grill, 2 knobs, BC, AC/DC $25.00
T125A, Table, 1959, Pink plastic, right front dial knob, center lattice grill, left knob, BC, AC/DC $25.00
T126A, Table, 1959, Beige plastic, right front dial knob, center lattice grill, left knob, BC, AC/DC $25.00
T127A, Table, 1959, White plastic, right front dial knob, center lattice grill, left knob, BC, AC/DC $25.00
T-125C, Table, 1959, Pink plastic, right front round dial, center lattice grill, left knob, feet, AC/DC $25.00
T-126C, Table, 1959, Beige plastic, right front round dial, center lattice grill, left knob, feet, AC/DC $25.00

T-127C, Table, 1959, Off-white plastic, right round dial, center lattice grill, left knob, feet, AC/DC $25.00
T-128C, Table, 1959, Yellow plastic, right front round dial, center lattice grill, left knob, feet, AC/DC $25.00

T-129C, Table, 1959, Turquoise plastic, right round dial, center lattice grill, left knob, feet, AC/DC $25.00
YRB-60-2, Table, 1948, Plastic, right dial printed on grill, horizontal wrap-around bands, BC, AC/DC $55.00
YRB 79-2, Table, 1948, Plastic, lower slide rule dial, upper recessed horiz louvers, 2 knobs, BC, AC/DC $50.00
YRB 83-1, Table, 1948, Wood, lower front slide rule dial, upper horizontal louvers, 2 knobs, BC, AC/DC $50.00
X-415, Table, 1948, Wood, front multi-band dial, upper step-back top w/horiz louvers, 5 knobs, AC $75.00

GENERAL IMPLEMENT
General Implement Corp.,
Terminal Tower, Cleveland, Ohio

9A5, Table, 1948, Plastic, right square dial, left horizontal grill bars, handle, 2 knobs, BC, AC/DC $85.00

GENERAL MOTORS

201 "Pioneer", Console, 1931, Wood, lowboy, inner dial & knobs, double doors, lower grill, 7 tubes, battery $110.00
250 "Little General", Cathedral, 1931, Walnut, lower window dial, scalloped grill w/cut-outs, 7 tubes, 3 knobs, AC $250.00

250A "Little General", Cathedral, 1931, Wood, lower window dial, upper cloth grill w/cut-outs, 7 tubes, 3 knobs, AC $250.00

251 "Valere", Console, 1931, Wood, lowboy, upper window dial, lower cloth grill w/cut-outs, 3 knobs, AC $130.00
252 "Cosmopolitan", Console, 1931, Wood, inner dial, 3 knobs, small sliding door, lower vertical grill bars, 10 tubes, AC $135.00

281, Ashtray/Remote Control, Metal, looks like ashtray on floor stand, upper window dial, 2 knobs $185.00

GENERAL TELEVISION
General Radio & Television Corp.,
Chicago, Illinois

4B5, Table, 1947, Plastic, right square front dial, left horizontal louvers, 2 knobs, BC, AC/DC $35.00
6C5, Table, 1948, Plastic, right front square dial, left cloth grill w/3 vertical bars, 2 knobs $45.00
9B6P, Table, 1948, Plastic, right square dial over metal perforated grill, handle, 2 knobs, BC, AC/DC $40.00
14A4F, Table, 1946, Plastic, right front square dial, left horizontal louvers, 2 knobs, BC, battery $35.00
17A5, Table, 1946, Wood, square dial, plastic grill louvers, zig-zag strip on base, 2 knobs, BC, AC/DC $45.00
19A5, Table, 1946, Wood, right square dial outlined in plastic, plastic louvers, 2 knobs, BC, AC/DC $40.00
21A4, Portable, 1947, Luggage-style, cloth w/lower stripes, right square dial, handle, BC, battery $25.00
22A5C, Table-R/P, 1947, Wood, right front square dial, 2 knobs, left grill, inner phono, lift top, BC, AC $30.00
23A6, Portable, 1947, Luggage-style, cloth covered, right square dial, 2 knobs, handle, BC, AC/DC/bat $25.00
24B6, Table, 1948, Wood, right square dial, left horizontal grill bars, handle, 2 knobs, BC, AC/DC $50.00
25B5, Portable, 1947, Two-tone, right square dial, perforated grill, 2 knobs, handle, BC, AC/DC/battery $30.00

26B5, Portable, 1947, Two-tone leatherette, right half-moon dial, handle, 2 knobs, BC, AC/DC/bat $40.00
27C5, Table, 1948, Plastic, right square dial, left cloth grill w/3 vertical bars, 2 knobs, BC, AC/DC $45.00
27C5L, Table, 1948, Plastic, right square dial, left cloth grill w/3 vertical bars, 2 knobs, BC, AC/DC $45.00
49, Table, Wood, right front dial, left grill, plastic grill bars/bezel/handle/knobs, AC/DC $60.00
526, Table, 1940, Wood, right dial, left grill w/horiz bars, plastic bezel/handle/knobs, BC, AC/DC $50.00

534, Table-N, Wood, looks like grand piano, inner dial, lift top, 5 tubes, 2 knobs, BC, AC/DC $350.00
591, Table, 1940, Catalin w/plastic trim, right dial, left horizontal grill bars, handle, 2 knobs ....$650.00+

GILFILLAN
Gilfillan Bros. Inc.,
1815 Venice Boulevard,
Los Angeles, California

The Gilfillan company, originally formed as a smelting and refining outfit, began to manufacture and sell radio parts in 1922 and by 1924 they were advertising complete radios and soon grew to be one of the largest radio manufacturers on the West coast. The last radios made by Gilfillan were produced in 1948.

5, Cathedral, 1932, Wood, lower window dial, upper cloth grill w/cut-outs, scrolled top, 3 knobs, AC $275.00
5-F, Table, 1941, Plastic, right front dial, left horizontal louvers, 5 tubes, 2 knobs $45.00
5-L, Portable, 1941, Striped, right front square dial over horizontal grill bars, handle, 2 knobs, AC/DC $35.00
10, Table, 1926, Carved walnut, rectangular case, slanted front panel w/center dial, 5 tubes, bat $150.00
20, Console, 1926, Wood, lowboy, upper slanted panel w/center dial, lower built-in speaker, battery $195.00
56B, Table, 1946, Wood, right front square dial, left metal perforated grill, 2 knobs, BC, AC/DC $35.00
58M, Table, 1948, Plastic, Deco, rounded right w/half-moon dial, left wrap-around louvers, AC/DC $125.00
58W, Table, 1948, Plastic, Deco, rounded right w/half-moon dial, left wrap-around louvers, AC/DC $125.00
63X, Tombstone, 1935, Wood, center front round dial, upper grill w/cut-outs, 4 knobs, 3 band, AC $125.00
66AM, Table, 1946, Wood, upper slanted slide rule dial, lower cloth grill, 2 knobs, BC, AC $35.00

66B **"The Overland"**, Portable, 1946, Leatherette, copper grill, double front doors, handle, 2 knobs, BC, AC/DC/battery $50.00
66PM **"The El Dorado"**, Table-R/P, 1946, Wood, inner right vertical dial, 4 knobs, lift top, inner phono, BC, AC $25.00
66S, Table, 1939, Wood, upper front dial, right & left side wrap-around grills w/horizontal bars, AC $65.00
68-48, Console-R/P/Rec, 1949, Wood, inner right slide rule dial, 4 knobs, left pull-out phono drawer, BC, AC $70.00
68B-D, Portable, 1948, "Alligator", square dial, metal perforated grill, handle, 2 knobs, BC, AC/DC/bat $35.00
68F, Table, 1948, Wood, right front square dial, left metal perforated grill, 3 knobs, AM, FM, AC/DC $35.00
80, Table, 1927, Wood, low rectangular case, center front window dial, 6 tubes, 3 knobs, AC $150.00
86U, Table, 1947, Wood, center slide rule dial, upper grill, 4 knobs, 6 pushbuttons, BC, 2SW, AC $70.00
100, Console, 1929, Wood, lowboy, inner front dial, lower cloth grill w/cut-outs, doors, 8 tubes, AC $165.00
105, Console, 1929, Wood, lowboy, upper front window dial, lower grill, fluted columns, 3 knobs, AC $130.00
108-48, Console-R/P/Rec, 1949, Wood, inner right dial, 5 knobs, left pull-out phono & recorder drawer, BC, FM, AC $70.00
119, Console, 1940, Wood, lowboy, half-round Hepplewhite-style, upper slide rule dial, AC $300.00
GN-1 **"Neutrodyne"**, Table, 1924, Wood, low rectangular case, inner panel, fold-down front, doors, 5 tubes, battery ..$165.00
GN-2 **"Neutrodyne"**, Table, 1924, Wood, low rectangular case, 3 dial front panel, 5 tubes, battery $130.00
GN-3 **"Neutrodyne"**, Table, 1925, Wood, high rectangular case, 2 dial slant front panel with 4 exposed tubes, bat $195.00
GN-5 **"Neutrodyne"**, Table, 1925, Mahogany, low rectangular case, 3 dial front panel, 5 tubes, battery $140.00
GN-6, Table, 1925, Mahogany, tall case, slanted front panel w/2 dials and 4 exposed tubes, battery $200.00

GLOBE
Globe Electronics, Inc.,
225 West 17th Street,
New York, New York

5BP1, Portable, 1947, Luggage-style, inner left dial, 3 knobs, drop-front, handle, BC, AC/DC/battery $35.00
6P1, Table-R/P, 1947, Wood, front slanted slide rule dial, 2 knobs, 3/4 lift top, inner phono, BC, AC $35.00
6U1, Table, 1947, Wood, upper slanted slide rule dial, lower cloth grill, 2 knobs, BC, AC/DC $35.00
7CP-1, Console-R/P, 1947, Wood, center front square dial, 2 knobs, lift top, lower record storage, BC, AC $50.00
51, Table, 1947, Swirled plastic, round dial, horizontal wrap-around louvers, 2 knobs, BC, AC/DC $85.00
62C, Table-R/P, 1947, Wood, front slanted slide rule dial, 4 knobs, 3/4 lift top, inner phono, BC, SW, AC $35.00
85, Table-C, 1948, Wood, large electric clock, right thumbwheel dial, left thumbwheel knob, BC, AC $65.00
454, Portable, 1948, Inner dial, geometric grill, 2 knobs, flip-open front, handle, BC, AC, DC, battery $45.00
456, Portable, 1948, Leatherette, lower front slide rule dial, upper grill, handle, 2 knobs, BC, battery $20.00
457, Table, 1948, Plastic, right half-moon dial, horizontal wrap-around louvers, 2 knobs, BC, AC/DC $75.00
500, Portable, 1947, Leatherette, inner top slide rule dial, 2 knobs, louvers, fold-up lid, BC, AC/DC $30.00
517, Table-R/P, 1947, Leatherette, left front dial, perforated grill, 2 knobs, open top, BC, AC/DC $15.00
552, Table, 1947, Plastic, streamline, right dial, wrap-around louvers on both sides, BC, AC/DC $115.00
553, Table, 1947, Plastic, half-moon dial, horizontal wrap-around louvers, 2 knobs, BC, AC/DC $75.00

558, Table-C, 1948, Wood, large electric clock, right thumbwheel dial, left thumbwheel knob, BC, AC $65.00
559, Table-N, 1948, Horse stands on wood base, 2 thumbwheel knobs, horizontal grill slats, BC, AC/DC $200.00

————————————

GLOBE TROTTER
Globe Trotter Radio Co.

Globe Trotter, Table-N, 1936, Looks like a world globe on a stand, tunes by turning globe, 6-color maps, AC/DC $450.00

————————————

GLORITONE
U. S. Radio & Television Corp.

9B, Console, 1932, Wood, lowboy, upper quarter-moon dial, lower grill w/cut-outs, 9 tubes, 6 legs $140.00
24, Cathedral, 1933, Wood, center window dial, upper cloth grill w/cut-outs, fluted columns, 2 knobs $225.00
25A, Table, 1932, Wood, rectangular case, front window dial, center cloth grill w/cut-outs $65.00
26, Cathedral, 1931, Wood, center front window dial, upper grill w/cut-outs, side columns, 3 knobs $195.00

26-P, Cathedral, 1929, Wood, center window dial w/escutcheon, upper cloth grill w/cut-outs, 3 knobs $200.00
27, Cathedral, 1930, Wood, modern, thumbwheel dial, upper grill w/cut-outs, w or w/o columns, AC $195.00
27S, Console, 1930, Wood, lowboy, upper front dial, lower "clover-leaf" grill cut-outs, rounded top $200.00
99, Cathedral, 1931, Wood, front half-round dial w/escutcheon, upper 3 section cloth grill, 3 knobs $225.00
99A, Cathedral, 1931, Wood, front half-round dial w/escutcheon, upper 3 section cloth grill, 3 knobs $225.00

————————————

GRANCO
Granco Products, Inc.,
36-17 20th Avenue, Long Island City, New York

601, Table, 1959, Plastic, upper right front window dial, lower lattice panel, 2 knobs, HiFi, FM, AC $20.00

610, Table, 1955, Plastic, center front slide rule dial, horizontal grill bars, 2 knobs, FM, AC/DC $20.00
611, Table, 1959, Plastic, center front slide rule dial, lattice grill, 2 knobs, Hi-Fi, FM, AC/DC $20.00
720, Table, 1955, Plastic, center front slide rule dial, horizontal grill bars, 2 knobs, AM, FM, AC/DC $20.00
750, Table-R/P, 1956, Wood, front slide rule dial, 4 knobs, 3/4 lift top, inner 3 speed phono, BC, FM, AC $25.00
770, Table-C, 1957, Plastic, left dial over horiz bars, right round alarm clock, 2 knobs, BC, FM, AC $25.00
AT-130, Table, 1959, Wood, two slide rule dials, tuning eye, 5 knobs, high fidelity, 13 tubes, AM, FM, AC $60.00
RP-1220, Console-R/P, 1958, Wood, inner right dial, 5 knobs, left lift top, inner phono, hi-fi, 12 tubes, BC, FM, AC $75.00

————————————

GRANTLINE
W. T. Grant Co.,
1441 Broadway, New York, New York

500, Table, 1946, Plastic, small right front dial, horizontal grill bars, 2 knobs, BC, AC/DC $45.00
501, Table, 1946, Plastic, small right front dial, horizontal grill bars, 5 tubes, 2 knobs, BC, AC/DC $45.00
501-7, Table, 1948, Plastic, dial at top of semi-circular front design, 2 knobs, BC, AC/DC $50.00

502, Series A, Table, Plastic, Deco, right dial, left rounded grill, horiz bars, 4 pushbuttons, side knob $120.00
504-7, Table, 1947, Plastic, right front round dial over horizontal louvers, 2 knobs, BC, AC/DC $30.00
508-7, Portable, 1948, Plastic, dial at top of semi-circular grill bars, handle, 2 knobs, BC, AC/DC/bat $50.00
510A, Portable, 1947, Upper slide rule dial, metal grill, 2 knobs on top of case, handle, BC, AC/DC/bat $30.00
605, Table, 1946, Plastic, streamline, right half-moon dial, 2 knobs, 6 pushbuttons, BC, AC/DC $165.00
651, Table, 1947, Plastic, upper slanted slide rule dial, lower horiz louvers, 2 knobs, BC, AC/DC $45.00
5610, End Table, 1948, Wood, front slide rule dial, 2 knobs, grill cut-outs, raised top edge, BC, AC/DC $125.00
6541, Table-R/P, Wood, slanted slide rule dial, horiz grill bars, 3 knobs, lift top, inner phono, AC $30.00
6547, Table-R/P, 1947, Front slanted slide rule dial, lower horizontal louvers, 4 knobs, lift top, BC, AC $30.00

————————————

GRAYBAR
Graybar Electric Co.

310, Table, 1928, Wood, low rectangular case, center dial, lift top, 7 tubes, 2 knobs, 1 switch, AC $95.00

330, Table, 1929, Wood, low rectangular case, center window dial, lift top, 2 knobs, 1 switch, AC $95.00
GB-4, Cathedral, 1931, Wood, lower window dial w/escutcheon, upper 3-section cloth grill, 3 knobs, AC $150.00

GREBE
A. H. Grebe,
10 Van Wyck Avenue,
Richmond Hill, New York

A. H. Grebe began as a young man to produce crystal sets and by 1920 had formed A. H. Grebe & Company. The popular Grebe Synchrophase, introduced in the mid-twenties, was the highlight of the company's career. The ever-present Doctor Mu, the oriental "sage of radio", and his advice on life was a regular feature of Grebe advertising. By 1932, the Grebe company was out of business.

60, Cathedral, 1933, Walnut, center window dial w/escutcheon, upper grill w/cut-outs, 6 tubes, 3 knobs $250.00
80, Tombstone, 1933, Walnut, center window dial w/escutcheon, upper grill w/cut-outs, 8 tubes, 3 knobs $160.00
84, Console, 1933, Walnut, lowboy, inner window dial, grill cut-outs, sliding doors, 8 tubes, 6 legs $175.00
160, Console, 1930, Wood, lowboy, upper window dial & carvings, lower cloth grill w/cut-outs, AC $175.00
270, Console, 1929, Wood, lowboy, upper window dial, lower cloth grill w/cut-outs, stretcher base $150.00
285, Console, 1929, Wood, upper front window dial, lower cloth grill w/cut-outs, sliding doors, AC $175.00
450, Console-R/P, 1929, Wood, window dial, lower cloth grill w/cut-outs, stretcher base, inner phono, AC $160.00
AC-Six, Table, 1928, Wood, low rectangular case, front window dial w/escutcheon, 7 tubes, AC $125.00
Challenger 5, Table, 1938, Plastic, center front square dial, right & left wrap-around grills, 5 tubes, AC/DC $50.00

CR-3, Table, 1920, Wood, 2 styles, low rectangular case, 2 dial black front panel, lift top, battery $400.00
CR-3A, Table, 1920, Wood, low rectangular case, 2 dial front panel, 1 exposed tube, very rare $2,000.00

CR-5, Table, 1921, Wood, low rectangular case, 2 dial black bakelite front panel, battery $450.00

CR-8, Table, 1921, Wood, low rectangular case, 3 dial black bakelite front panel, battery $500.00

CR-9, Table, 1921, Wood, low rectangular case, 2 dial black bakelite panel, lift top, 3 tubes, battery $400.00
CR-12, Table, 1923, Wood, low rectangular case, 2 dial black bakelite front panel, 4 tubes, battery $425.00
CR-13, Table, 1923, Wood, low rectangular case, 3 dial black bakelite front panel, lift top, battery $525.00
CR-14, Table, 1923, Wood, low rectangular case, 2 dial black bakelite panel, lift top, 3 tubes, battery $550.00
CR-18, Table, 1926, Wood, rectangular case, 2 front thumbwheel dials, top exposed meter coils, bat $750.00

MU-1 w/chain, Table, 1925, Wood, rectangular case, 3 diamond-shaped thumbwheel dials, with dial chain, bat $250.00
MU-1 w/o chain, Table, 1925, Wood, rectangular case, 3 diamond-shaped thumbwheel dials, w/o dial chain, bat $250.00

SK-4, Console, 1930, Wood, lowboy, large dial escutcheon w/ thumbwheel tuning, lower round grill, AC $170.00

Synchronette, Table, 1933, Wood, Deco, right dial, center grill w/ cut-outs, stepped top, 5 tubes, 2 knobs, AC **$85.00**

Synchrophase Seven, Table, 1927, Two-tone wood, low rectangular case, center thumbwheel dial, front pillars, bat **$175.00**

DAVID GRIMES

M3XP "Inverse Duplex", Table, 1925, Wood, rectangular case, slanted 3 dial front panel, 3 tubes, battery **$175.00**
4DL, Table, 1924, Wood, 3 inner pointer dials, fold-down front, columns, 4 tubes, battery **$225.00**

5B "Baby Grand Duplex", Table, 1925, Wood, low rectangular case, metal front panel w/ 2 pointer dials, 5 tubes, bat **$200.00**
6D, Table, Wood, low rectangular case, inner panel w/3 pointer dials, fold-down front, bat **$250.00**

GRUNOW
General Household Utilities Co.,
Chicago, Illinois

7C, Console, Wood, upper front round dial, lower shield-shaped grill w/2 vert bars, 5 knobs **$125.00**
450, Tombstone, 1934, Wood, small case, lower front dial, large grill w/chrome cut-outs, BC, SW, AC **$185.00**
470, Tombstone, 1935, Walnut, rounded shoulders, lower front dial, upper grill w/cut-outs, 4 tubes, AC **$110.00**
500, Tombstone, 1933, Wood, front window dial, upper cloth grill w/ chrome cut-outs, step-down top, AC **$200.00**
501, Table, 1933, Two-tone wood, right window dial, center grill w/ chrome cut-outs, 2 knobs, AC **$150.00**

520, Table, 1935, Right & left windows w/escutcheons, center grill w/ chrome bars, 5 tubes, AC/DC **$125.00**

570, Tombstone, Two-tone wood, lower round dial, upper cloth grill w/cut-outs, 5 tubes, 3 knobs **$120.00**
580, Tombstone, 1935, Walnut, center front round dial, upper grill w/ cut-outs, 5 tubes, 4 knobs, AC **$125.00**
581, Console, 1935, Wood, upper front round dial, lower cloth grill w/ cut-outs, 4 knobs, AC **$140.00**

588, Table, 1937, Wood, right Tele-dial, left wrap-around grill w/ horizontal bar, 5 tubes, 3 knobs **$85.00**
594, Table, 1937, Ivory finish, "violin-shape", center chrome airplane dial, 5 tubes, 2 knobs **$120.00**
640, Tombstone, 1935, Wood, center front round dial, upper cloth grill w/cut-outs, 6 tubes, 4 knobs, AC **$110.00**
641, Console, 1935, Two-tone wood, upper round dial, lower cloth grill with cut-outs, 6 tubes, AC **$125.00**
653, Console, 1937, Walnut, upper slanted Tele-dial, lower grill w/ vertical bars, 6 tubes, 4 knobs, AC **$200.00**
654, Upright Table, 1937, Wood, upper round airplane dial, lower horiz grill bars, 6 tubes, 4 knobs, BC, SW **$100.00**
660, Tombstone, 1934, Wood, lower round dial, upper grill w/cut-outs, fluting, 6 tubes, 4 knobs, BC, SW **$140.00**
671, Console, 1934, Wood, upper round dial, lower grill with cut-outs, 6 tubes, 4 knobs, BC, SW, AC **$130.00**
681, Console, 1935, Wood, upper round dial, lower cloth grill with vertical bars, 6 tubes, 4 knobs, AC **$130.00**
700, Tombstone, 1933, Wood, window dial, chrome grill cut-outs, step-down top, 7 tubes, 5 knobs, AC **$200.00**

750, Tombstone, 1934, Wood, center round dial, upper grill w/cut-outs, step-down top, BC, 3SW, AC **$125.00**
761, Console, 1935, Wood, upper round dial, lower grill w/2 vert bars, 7 tubes, 5 knobs, all-wave, AC **$125.00**
871, Console, 1935, Wood, upper round dial, lower cloth grill with 2 vertical bars, 8 tubes, 5 knobs, AC **$130.00**
1081, Console, 1937, Two-tone walnut, upper Tele-dial, lower grill w/ vert bars, 10 tubes, BC, SW, AC **$250.00**
1171, Console, 1935, Wood, upper round dial, lower grill w/2 vert bars, 11 tubes, 5 knobs, all-wave, AC **$165.00**
1183, Console, 1937, Wood, upper Tele-dial, lower grill w/6 vert bars, 11 tubes, 4 knobs, BC, SW, AC **$275.00**
1191, Console, 1936, Wood, upper round dial, tuning eye, lower grill w/horiz bars, 11 tubes, BC, SW, AC **$180.00**
1241, Console, 1935, Wood, upper round dial, lower cloth grill with 2 vert bars, 12 tubes, all-wave, AC **$170.00**
1291, Console, 1936, Wood, upper Tele-dial, lower grill w/5 vert bars, 12 tubes, 4 knobs, BC, 3SW, AC **$300.00**
1541, Console, 1936, Wood, upper Tele-dial, lower grill w/7 vert bars, 15 tubes, 5 knobs, BC, SW, AC **$350.00**

GUILD
**Guild Radio & Television Co.,
460 North Eucalyptus Avenue,
Inglewood, California**

380T "Town Crier", Table-N, Wood, looks like old lantern, inner vertical slide rule dial, 4 knobs, front door **$150.00**

484, Table-N, 1956, Wood, spice chest design, inner dial, 3 knobs, double doors, 2 drawers, BC, AC/DC $125.00
556 "Country Belle", Wall-N, 1956, Wood, looks like old wall telephone, side crank tunes stations, 2 side knobs, AC **$95.00**
785 "Grafonola", Table-R/P-N, 1959, Wood, looks like crank phonograph complete w/horn, top phono, side louvers **$175.00**
T/K 1577 "The Teakettle", Table-N, Wood, china & brass, looks like tea kettle, top lifts for controls, bottom speaker **$125.00**

GULBRANSEN
**Gulbransen Co.,
816 North Kedzie Avenue,
Chicago, Illinois**

130, Cathedral, 1931, Wood, window dial, upper cloth grill w/lyre cut-out, fluted columns, 3 knobs, AC **$400.00**
135, Console, 1931, Wood, lowboy, upper window dial, lower grill w/ cut-outs, stretcher base, AC **$130.00**

235, Console, 1931,Wood, upper front window dial, lower cloth grill w/cut-outs, stretcher base, AC **$130.00**
9950, Console, 1930, Wood, highboy, upper front window dial, speaker grill underneath cabinet, AC **$150.00**

HALLDORSON
**Halldorson Co.,
1772 Wilson Ave., Chicago, Illinois**

RD-400, Table, 1925, Wood, low rectangular case, 2 dial front panel, 4 tubes, battery **$95.00**
RF-500, Table, 1925, Wood, low rectangular case, 3 dial front panel, 5 tubes, battery **$100.00**

HALLICRAFTERS
**The Hallicrafters Co.,
5th & Kostner Avenues, Chicago, Illinois**

The Hallicrafters Company was formed by Bill Halligan in 1933 for the manufacture of receivers and phonos for other companies. By 1935, Hallicrafters was producing its own ham radio equipment. During WW II, the company had many military contracts. Hallicrafters enjoyed a reputation for reasonable prices which they were able to offer due to their large volume of business.

5R10, Table, 1951, Metal cabinet, large center front slide rule dial, 4 knobs, BC, 3SW, AC/DC **$50.00**
5R10A, Table, 1952, Metal, large slide rule dial, top left perforated grill, 4 knobs, multi-band, AC/DC **$50.00**
5R14, Table, 1951, Plastic, center front round metal dial over dotted grill, 2 knobs, BC, AC/DC **$40.00**
5R24, Portable, 1952, Center front round dial over woven grill, handle, 2 knobs, BC, AC/DC/battery **$35.00**
5R30A "Continental", Table, 1952, Right front slide rule dial, checkered perforated grill, 3 knobs, BC, SW, AC/DC ... **$40.00**
5R50, Table-C, 1952, Right front slide rule dial, left square alarm clock, top grill, 6 knobs, BC, SW, AC **$45.00**

5R60, Table, 1955, Mahogany plastic, left front round dial, right checkered grill, 2 center knobs, feet $35.00
5R61, Table, 1955, Ivory plastic, right front round dial, left checkered grill, 2 center knobs, feet **$35.00**
EC-306, Table-R/P, 1948, Wood, center front dial, right & left grills, lift top, inner phono, BC, SW, AC **$40.00**
EX-306, Table-R/P, 1948, Wood, center front dial, right & left grills, lift top, inner phono, BC, SW, AC **$40.00**

S-53, Table, 1948, Metal, center front slide rule dial, 4 switches, 5 knobs, BC, 4SW, AC $50.00

S-55, Table, 1949, Metal, center front slide rule dial, top perforated grill, 4 knobs, BC, FM, AC/DC $40.00

S-58, Table, 1949, Metal, left rectangular slide rule dial, right perforated grill, 4 knobs, BC, FM, AC $55.00

S-80, Table, 1952, Upper slanted slide rule dial, lower perforated grill w/"h" logo, 3 knobs, BC, SW $45.00

TR-88, Portable, 1958, Leather case, transistor, left dial knob, lattice grill, "h" logo, handle, AM, bat $30.00

TW-25, Portable, 1953, Leatherette case, large center front round dial w/inner perforated grill, handle $35.00

TW-500, Portable, 1954, Leatherette, flip-up front w/map, telescope antenna, handle, AC/DC/battery $75.00

TW-1000 "World Wide" , Portable, 1953, Leatherette, flip-up front w/map, telescope antenna, handle, 8 bands, AC/DC /bat $125.00

TW-2000 "World Wide", Portable, 1955, Leatherette, flip-up front w/map, telescope antenna, handle, 8 bands, AC/DC/bat**$125.00**

HALSON
Halson Radio Mfg. Co.,
New York City, New York

10, Table, 1938, Two-tone wood, rectangular dial, left wrap-around grill, 3 knobs, BC, SW, AC/DC $55.00

43B-A, Table, 1933, Walnut, right dial, center round grill w/cut-outs, 2 knobs, step-down top, AC/DC $85.00

610, Tombstone, 1934, Wood, center front round dial, upper cloth grill w/cut-outs, 3 knobs, BC, SW, AC $125.00

A5, Table, 1938, Catalin, small case, right front round dial, left grill w/3 horiz sections, 2 knobs ....$700.00+

T10, Table, 1937, Wood, right front dial, left cloth grill w/Deco cut-outs, 2 knobs, BC, AC $45.00

HANSEN
Hansen Storage Co.,
120 Jefferson St., Milwaukee, Wisconsin

American Crest, Table, 1925, Wood, low rectangular case, 2 dial front panel, 6 tubes, battery $125.00

Bluebird, Table, 1925, Wood, step-down case, slanted two dial panel, 4 tubes, battery $125.00

Gold Finch, Table, 1925, Wood, low rectangular case, 2 dial front panel, 5 tubes, battery $110.00

HARMONY
Harmony Mfg. Co.,
2812 Griffith Ave., Cincinnati, Ohio

5, Table, 1925, Wood, low rectangular case, front panel w/left dial, 5 tubes, battery $110.00

HARPERS

547F, Portable, Plastic, transistor, right front round dial, left horizontal grill bars, AM, battery $35.00

HARTMAN
Hartman Electrical Mfg. Co.,
31 E. Fifth St.,
Mansfield, Ohio

12-A, Console, 1925, Wood, lowboy, inner left 3 dial panel, right speaker, fold-down front, 5 tubes, bat $180.00

12-B, Table, 1925, Wood, low rectangular case, left 3 dial panel, right built-in speaker, 5 tubes, bat $150.00

12-C, Table, 1925, Wood, low rectangular case, center 3 dial panel, 5 tubes, battery $125.00

HETEROPLEX
Heteroplex Mfg. Co.,
423 Market St.,
Philadelphia, Pennsylvania

De Luxe, Table, 1925, Wood, low rectangular case, 2 dial front panel, top door, 3 tubes, battery $115.00

HITACHI
Electronic Utilities Co., Div. of the Sampson Company, 2244 South Western Avenue, Chicago, Illinois

TH-621, Portable, 1959, Plastic, transistor, round dial, lattice grill, thumbwheel knob, strap, AM, battery $25.00

TH-666, Portable, 1959, Transistor, upper right dial, lower perforated grill, thumbwheel knob, AM, battery $30.00

TH-667, Portable, 1960, Transistor, right front thumbwheel dial, large checkered grill area, AM, battery $25.00

TH-862R "Marie", Portable, 1960, Plastic, transistor, right window dial, large metal perforated grill, AM, battery ..$15.00

WH-822M, Portable, 1960, Transistor, top dial, large front grill area, thumbwheel knobs, BC, Marine, battery $20.00

HOFFMAN
Hoffman Radio Corp., 3430 South Hill Street, Los Angeles, California

A-200, Table, 1946, Plastic, upper front slide rule dial, lower horizontal louvers, 2 knobs, BC, AC/DC $45.00

A-300, Table, 1946, Wood, upper slide rule dial, lower horiz louvers, rounded front, 3 knobs, BC, AC $30.00

A-301, Table, Wood, upper slanted slide rule dial, pushbuttons, lower horiz louvers, 3 knobs $40.00

A-309, Table, 1947, Wood, upper slide rule dial, lower cloth grill w/ metal strips, 2 knobs, BC, AC/DC $45.00

A-401, Table-R/P, 1947, Wood, upper slide rule dial, lower grill, 3 knobs, lift top, inner phono, BC, AC $30.00

A-500, Console-R/P, 1946, Wood, lowboy style, inner slide rule dial, 3 knobs, 6 pushbuttons, lift top, BC, AC $65.00

A-501, Console-R/P, 1946, Wood, inner right slide rule dial, 5 knobs, pushbuttons, doors, BC, 2SW, AC $100.00

A-700, Portable, 1947, Slide rule dial, 2 knobs, horizontal louvers w/ vert. bars, handle, BC, AC/DC/bat $40.00

B-400, Table-R/P, 1947, Wood, upper front dial, lower grill, 2 knobs, 3/4 lift top, inner phono, BC, AC $30.00

B-1000, Console-R/P, 1947, Wood, modern, tilt-out front, inner dial, knobs, pushbuttons & phono, BC, 2SW, AC $90.00

B-8002, Console-R/P, 1958, Wood, inner left dial, right phono, lift top, large front grill, feet, BC, FM, AC $50.00

C-501, Console-R/P, 1948, Wood, modern, inner slide rule dial, 5 knobs, lift top, inner phono, BC, 2SW, AC $85.00

C-502, Console-R/P, 1948, Wood, modern, inner slide rule dial, 5 knobs, 6 pushbuttons, lift top, 15 tubes, BC, FM, AC ... $95.00

C-503, Console-R/P, 1948, Wood, modern, inner slide rule dial, 5 knobs, 6 pushbuttons, lift top, 14 tubes, BC, 2SW, AC $95.00

C-507, Console-R/P, 1948, Wood, slide rule dial, 4 knobs, pull-out phono, front doors fold open, 10 tubes, BC, FM, AC $75.00

C-514, Console-R/P, 1948, Wood, inner right slide rule dial, 5 knobs, pushbuttons, left lift top, 18 tubes, BC, FM, AC $125.00

C-518, Console-R/P/Rec, 1949, Wood, inner right slide rule dial, 5 knobs, left pull-out phono drawer, 21 tubes, BC, FM, AC $150.00

C-1007, Console-R/P/Rec, 1949, Wood, inner right slide rule dial,

5 knobs, pushbuttons, pull-out phono, 23 tubes, BC, FM, AC $180.00

HORN
Herbert H. Horn Radio Manufacturing Co., 1625-29 South Hill Street, Los Angeles, California

156A "Tiffany Tone", Table, 1934, Wood, Deco, front dial, center cloth grill w/cut-outs, raised side panels, 2 knobs $95.00

Riviera PR, Console-R/P, 1935, Wood, Deco, tubular cabinet, inner front dial, double doors, lift top, inner phono $500.00

Riviera R, Console, 1935, Wood, Deco, tubular cabinet, inner front dial & knobs, lower grill, double doors $500.00

HOWARD
Howard Radio Company, 451-469 East Ohio Street, Chicago, Illinois

Howard began selling radio parts in 1922 and complete sets in 1924. By 1949, the company was out of business.

4BT, Table, 1939, Wood, upper front dial, lower cloth grill w/cut-outs, 3 knobs, battery $35.00

20, Cathedral, 1931, Wood, center front window dial, upper grill w/ cut-outs, 7 tubes, 3 knobs, AC $285.00

40, Console, 1931, Wood, lowboy, upper front window dial, lower grill w/cut-outs, stretcher base $135.00

60, Console-R/P, 1931, Wood, window dial, grill cut-outs, lift top, inner phono, 9 tubes, 3 knobs, AC $145.00

150, Table, 1925, Wood, low rectangular case, slanted 3 dial front panel, 5 tubes, battery $95.00

200, Table, 1925, Wood, low rectangular case, 3 dial front panel, 6 tubes, battery $110.00

225, Table, 1937, Wood, right front oblong dial, left grill w/Deco lines, 4 knobs, BC, SW, AC $65.00

250, Console, 1925, Wood, lowboy, inner 3 dial panel, fold-down front, built-in speaker, 5 tubes, bat $160.00

250, Table, 1937, Two-tone wood, right front dial, left vertical grill cut-outs, 4 knobs, BC, SW, AC $60.00

303, Console, 1939, Walnut, rectangular dial, 4 pushbuttons, lower grill w/vertical bar, 6 tubes, AC $125.00

305, Table, 1939, Two-tone wood, lower slide rule dial, pushbuttons, upper grill, 4 knobs, BC, SW, AC $55.00

307, Table, 1940, Two-tone wood, lower front rectangular slide rule dial, upper grill, BC, SW, AC $55.00

307-TP, Table-R/P, 1941, Wood, lower right front slide rule dial, left grill, lift top, inner phono, AC $30.00

308-C, Console, 1939, Wood, upper slide rule dial, pushbuttons, lower cloth grill w/vert bar, BC, SW, AC $135.00

308-TT, Console, 1939, Walnut, slanted top, pushbuttons, tuning eye, horiz grill bars, 4 knobs, BC, SW, AC $165.00

368, Table, 1937, Wood, right oblong dial, left grill w/vertical wrap-over bars, 4 knobs, BC, SW, AC $60.00

400, Console, 1937, Wood, upper oblong dial, lower grill w/horiz bars, 12 tubes, 4 knobs, BC, SW, AC $140.00

425, Console, 1937, Wood, upper oblong dial, lower grill w/horiz bars, 14 tubes, 4 knobs, BC, SW, AC $150.00

472AC, Chairside-R/P, 1948, Wood, top dial & knobs, front grill & pull-out phono drawer, storage, BC, FM, AC $75.00

472AF, Console-R/P, 1948, Wood, pull-out right dial & left phono drawers, criss-cross grill, BC, FM, AC $85.00

472C, Chairside-R/P, 1948, Wood, top dial & knobs, front grill & pull-out phono drawer, storage, BC, FM, AC $75.00

472F, Console-R/P, 1948, Wood, pull-out right dial & left phono drawers, criss-cross grill, BC, FM, AC $85.00

474, Table, 1948, Plastic, right square dial, horizontal wrap-around bars, 3 knobs, BC, FM, AC/DC $40.00

518-S, Console, 1940, Walnut, rectangular dial, pushbuttons, lower horiz grill bars, 4 knobs, BC, SW, AC $150.00

575, Table, 1939, Two-tone walnut, slide rule dial, pushbuttons, upper grill, 4 knobs, BC, SW, AC $55.00

580, Table, 1940, Wood, center front slide rule dial, pushbuttons, tuning eye, 4 knobs, BC, SW, AC $65.00

700, Table, 1940, Plastic, right front dial, horizontal louvers wrap around case, 2 knobs, BC, AC $40.00

718C, Console, 1941, Walnut, upper rectangular dial, 6 pushbuttons, lower vert grill bars, BC, SW, AC $130.00

780, Table, 1941, Two-tone wood, slanted slide rule dial, criss-cross grill, 6 knobs, BC, SW, AC $50.00

808CH, Chairside-R/P, 1941, Wood, top dial & knobs, front grill w/ cut-outs, lift top, inner phono, BC, SW, AC $120.00

901A, Table, 1946, Two-tone wood, right front dial, left horizontal louvers, 2 knobs $30.00

901A-I, Table, 1946, Plastic, right square dial, left horiz louvers & side horiz bars, 2 knobs, BC, AC/DC $50.00

901AP, Table-R/P, 1946, Wood, front square dial, 3 knobs, horizontal louvers, lift top, inner phono, BC, AC $30.00

906, Table, 1947, Wood, lower front slide rule dial, upper criss-cross grill w/center bar, BC, AC $45.00

906C, Chairside-R/P, 1947, Wood, dial/4 knobs on top of case, pull-out phono drawer, record storage, BC, AC $75.00

906-S, Table, 1948, Wood, lower slide rule dial, upper criss-cross grill w/center bar, BC, SW, AC $50.00

906-SB, Table, 1948, Wood, lower slide rule dial, upper criss-cross grill w/center bar, BC, AC, AC $50.00

909M, Console-R/P, 1947, Wood, pull-out radio drawer, pull-out phono drawer, criss-cross grill, BC, SW, AC $75.00

920, Table, 1946, Plastic, right front square dial, left horizontal grill bars, 3 knobs, BC, battery $35.00

A, Table, 1925, Wood, low rectangular case, 3 dial front panel, 5 tubes, battery $135.00

F-17, Console, 1934, Wood, lowboy, upper round dial, lower cloth grill w/cut-outs, 6 legs, BC, SW, AC $125.00

HUDSON-ROSS

Three Little Pigs, Cathedral, 1933, Ivory or green wood, center dial, upper grill w/hand-painted 3 Little Pigs cut-out $425.00

HYMAN
Henry Hyman & Co., Inc.,
476 Broadway, New York, New York

V-60 "Bestone", Table, 1925, Wood, high rectangular case, 2 dial front panel, 4 tubes, battery $130.00

JACKSON
Jackson Ind., Inc.,
58E Cullerton Street, Chicago, Illinois

150, Console-R/P, 1951, Wood, inner slide rule dial, 4 knobs, lower pull-out phono, double doors, BC, AC $50.00

254, Console-R/P, 1952, Wood, inner right vertical dial, left pull-out phono, double front doors, BC, AC $50.00

255, Console-R/P, 1952, Wood, inner slide rule dial, 4 knobs, left pull-out phono, double doors, AM, FM, AC $50.00

350, Console-R/P, 1951, Wood, inner front slide rule dial, 4 knobs, pull-out phono, double doors, BC, FM, AC $55.00

DP-51, Table-R/P, 1952, Luggage-style, inner right dial, lattice grill, left phono, lift top, handle, BC, AC $20.00

JP-50, Table-R/P, 1952, Inner right dial, lattice grill, 3 knobs, left phono, lift top, handle, BC, AC $20.00

JP-200, Table-R/P, 1952, Leatherette, inner right half-moon dial, left phono, lift top, handle, BC, AC $20.00

JACKSON-BELL
Jackson-Bell Company,
Los Angeles, California

The Jackson-Bell Company began business in 1926 selling radios. The company is best known for its midget radios manufactured from 1930 to 1932 with which they grew to be one of the largest radio producers on the West coast. Jackson-Bell was out of business by 1935.

4, Cathedral, Wood, front window dial w/escutcheon, upper cloth grill w/cut-outs, 3 knobs $200.00

4 "Peter Pan", Cathedral, 1935, Wood, window dial, upper grill w/ Peter Pan cut-out, pointed top, 3 knobs, AC $295.00

6, Table, 1930, Wood, center front window dial, right & left scenic tapestry grills, 3 knobs, AC $195.00

25 "Peter Pan", Tombstone, 1932, Wood, shouldered, window dial, upper grill w/Peter Pan cut-out, 3 knobs, AC $295.00

25AV "Pandora", Table, 1932, Wood, chest-style, carved case, lift top, inner window dial, 3 knobs, storage, AC $275.00

26-SW, Tombstone, 1932, Wood, shouldered, lower window dial, upper grill w/gothic cut-outs, 3 knobs $150.00

34, Table, C1933, Blue mirror glass case, center round dial surrounded by cut-outs, 2 knobs, AC $1,200.00

38, Console, 1932, Wood, lowboy, upper window dial, lower cloth grill w/cut-outs, 8 tubes, 6 legs $140.00

54 "Peter Pan", Tombstone, 1935, Wood, shouldered, window dial, upper cloth grill w/Peter Pan cut-out, 3 knobs $295.00

60, Cathedral, 1929, Wood, right front thumbwheel dial, center Deco "sun-burst" grill, pointed top, AC $175.00

62, Cathedral, 1930, Wood, lower front window dial, upper cloth grill w/swan cut-outs, 3 knobs, AC $300.00

87, Cathedral, 1931, Wood, half-round dial, upper grill w/tulip cut-outs, side spindles, 3 knobs, AC $250.00

JEFFERSON-TRAVIS
Jefferson-Travis, Inc.,
380 Second Avenue, New York, New York

MR-3, Portable, 1947, Leatherette, right slide rule dial, round grill, 3 knobs, handle, BC, SW, AC/DC/bat $30.00

JEWEL
Jewel Radio Corp.,
583 Sixth Avenue, New York, New York

300, Table, 1947, Plastic, right round dial, left horizontal wrap-around louvers, 2 knobs, BC, AC/DC $45.00

304 "Pixie", Portable, 1948, "Alligator", inner right dial, flip-open lid, outer perforated grill, strap, BC, bat..............................$50.00

501, Table, 1947, Wood, upper slanted slide rule dial, horizontal grill openings, 2 knobs, BC, AC/DC $35.00

502, Table, 1947, Wood, slanted slide rule dial, horiz louvers, 2 knobs, decorative veneer, BC, AC/DC $40.00

504, Table, 1947, Wood, slanted slide rule dial, grill w/large block openings, 2 knobs, BC, AC/DC $40.00

505 "Pin Up", Wall-C, 1947, Plastic, lower right window dial, 2 side knobs, vert bars, center clock, BC, AC $85.00

801 "Trixie", Portable, 1948, "Alligator", inner right dial, 3 knobs, fold-open door, strap, BC, AC/DC/battery $50.00

814, Portable, 1948, Plastic, left front recessed dial, 2 knobs, top grill bars, strap, BC, battery $40.00

915, Table-C, 1950, Plastic, right round dial, left clock, center perforated grill, 4 knobs, BC, AC $35.00

920, Table-C, 1949, Plastic, right round dial, left electric clock, horizontal grill bars, 4 knobs, BC, AC $35.00

940, Table-C, Plastic, right front dial, left clock, center lattice grill, 6 knobs, BC, AC $30.00

949, Portable, 1950, Leatherette, center front round dial, cloth grill, handle, 2 knobs, BC, AC/DC/bat $25.00

955, Table, 1950, Plastic, right dial, left lattice grill, step-down right top, 2 knobs, BC, AC/DC $50.00

956, Table, 1951, Plastic, right front clear dial, left checkered grill, center tuning knob, BC, AC/DC $30.00

960, Table, 1950, Plastic, dial markings over checkered grill, large dial pointer, 2 knobs, BC, AC/DC $30.00

5007, Table, 1952, Plastic, right front dial, left checkered grill, large center dial knob, BC, AC $30.00

5010, Portable, 1950, Leatherette, center front dial, cloth grill, handle, 2 knobs, BC, AC/DC/battery $25.00

5020, Table-R/P, 1951, Wood, outer front half-moon dial, 3 knobs, lift top, inner phono, BC, AC $35.00

5040, Table-C, 1952, Plastic, center round alarm clock, right & left upper thumbwheel knobs, BC, AC $50.00

5050, Portable, 1951, Plastic, dial & checkered grill on top of case, handle, 3 knobs, BC, AC/DC/bat $45.00

5057U "Wakemaster", Table-C, 1950, Plastic, right slide rule dial, center clock, checkered grills, 5 knobs, BC, AC $45.00

5125U, Table-C, 1953, Plastic, right side dial knob, center front alarm clock, checkered grill, BC, AC $30.00

5200, Table, 1953, Plastic, right half-moon dial, left diagonal grill bars, 2 knobs, feet, BC, AC/DC $35.00

5205, Table/Lamp, 1953, Plastic, right front dial knob, center vertical bars, upper slanted lamp, BC, AC $60.00

5250, Table-C, 1953, Plastic, 2 upper front thumbwheel knobs, center alarm clock, side bars, BC, AC $50.00

5310, Portable, 1953, Plastic, center front round dial over horizontal grill bars, handle, BC, battery $30.00

KADETTE
International Radio Corporation,
Ann Arbor, Michigan

24 "Clockette", "Futura", Table, 1937, Wood, clock-style, round dial with inner metal screen grill, feet, 2 knobs, AC $125.00

35, Table, 1936, Wood, lower front telephone dial, upper grill w/ horizontal bars, 2 knobs, BC, AC $75.00

36, Table, 1935, Wood, right front square dial, left grill w/horizontal bars, 4 knobs, BC, SW, AC $45.00

41 "Jewel", Table, 1935, Walnut bakelite, right front dial, center grill w/plastic cut-outs, 2 knobs, AC/DC $175.00

43 "Jewel", Table, 1935, Ivory plastic, right front dial, center grill w/ plastic cut-outs, 2 knobs, AC/DC $200.00

44 "Jewel", Table, 1935, Red plastic, right front dial, center grill w/ plastic cut-outs, 2 knobs, AC/DC $300.00

47 "Jewel", Table, 1935, Black plastic, right front dial, center grill w/ plastic cut-outs, 2 knobs, AC/DC $250.00

48 "Jewel", Table, 1935, Marble plastic, right front dial, center grill w/plastic cut-outs, 2 knobs, AC/DC $250.00

66X, Table, 1936, Wood, lower front dial, upper horizontal grill bars, rounded front sides, 3 knobs $70.00

77, Tombstone, 1936, Wood, lower front airplane dial, upper grill w/ horiz & vert bars, BC, SW, AC/DC **$90.00**

87, Table, 1936, Wood, rectangular case, right round dial over horiz grill bars, 3 knobs, BC, SW, AC **$50.00**

96, Table, Two-tone wood, right round dial, left round grill w/3 vertical bars, 3 knobs **$55.00**

1019, Table, Wood, right front gold dial, left cloth grill w/3 horizontal bars, 3 knobs, BC, SW **$50.00**

K-25 "Clockette", Table, 1937, Catalin, clock-style, center front round dial w/inner metal screen grill, 2 knobs .... **$1,000.00+**

K-150, Table, 1937, Walnut plastic, lower dial, upper grill w/vertical bars, oval medallions, AC/DC **$100.00**

K-151, Table, 1937, Ivory plastic, lower dial, upper grill w/vertical bars, oval medallions, AC/DC **$120.00**

K-1024, Table, 1937, Two-tone walnut & maple, right front dial, left rectangular grill, 3 knobs, AC/DC **$75.00**

L "Classic", Table, 1936, 3 plastics, lower dial, back/front grills w/ horiz bars, 2 glass knobs, AC/DC **$800.00**

S947 "Kadette Jr.", Table, 1933, Plastic, pocket-size, Deco, thumbwheel dial, grill cut-outs, 2 tubes, BC, AC/DC **$300.00**

The Colonial "Clockette", Table, 1937, Wood, clock-style, large center front round dial, side fluting, 2 knobs, AC **$125.00**

The Modern "Clockette", Table, 1937, Wood, Deco, clock-style, large center front round dial, 2 knobs, AC **$125.00**

Topper, Table, 1940, Plastic, upper front slide rule dial, top speaker "dome" grill & 2 knobs, AC/DC **$195.00**

KELLOGG
Kellogg Switchboard & Supply Company,
1066 West Adams Street,
Chicago, Illinois

The Kellogg Switchboard & Supply Company began in 1897 producing telephone equipment and by 1922 was manufacturing radio parts. Experiencing some financial difficulty in the late 1920's, Kellogg discontinued radio production in 1930.

504 "Wavemaster", Table, 1925, Wood, low rectangular case, 1 dial front panel, lift top, feet, battery **$245.00**

507 "Wavemaster", Table, 1926, Wood, low rectangular case, center escutcheon with 4 knobs, 6 tubes, battery **$135.00**

521, Console, 1928, Wood, lowboy, upper dial & escutcheon, lower grill w/cut-outs, stretcher base **$175.00**

523, Console, 1929, Wood, lowboy, inner window dial & knobs, double doors, stretcher base, 9 tubes, AC **$225.00**

KENNEDY
The Colin B. Kennedy Company, Inc.,
Rialto Building,
San Francisco, California

The Colin B. Kennedy Company was formed in 1919 for the production of radios. Business boomed until the mid-twenties when sales began to decline due to lower priced competitive models and, by 1926, the company had declared bankruptcy.

20, Table, 1925, Mahogany, slanted front panel, center large round dial, 5 tubes, 4 knobs, battery **$200.00**

42 "Coronet", Upright Table, 1931, Walnut, right window dial, upper grill w/"tulip" cut-outs, flared base, 3 knobs **$325.00**

52A, Cathedral, Wood, center front window dial, upper cloth grill w/ cut-outs, 3 knobs **$325.00**

110 "Universal", Table, 1922, Wood, low rectangular case, black bakelite front panel, 3 tubes, battery **$900.00**

110/525, Table, 1922, Wood, 2 units, receiver & two-stage amp, black front panels, battery **$1,300.00**

164, Console, 1932, Wood, lowboy, upper quarter-round dial, lower grill w/cut-outs, 6 legs, 10 tubes **$175.00**

220, Table, 1921, Wood, square case, black bakelite front panel, lift top, 1 tube, battery **$600.00**

220/525, Table, 1922, Wood, 2 units - receiver & 2 stage amp, black front panels, lift tops, battery **$1,000.00**

266, Console, 1932, Wood, lowboy, upper quarter-round dial, lower grill w/cut-outs, 6 legs, 12 tubes **$200.00**

281/521, Table, 1921, Mahogany, 2 units - receiver & 2 stage amp, 3 dial receiver panel, lift tops, bat **$850.00**

311, Portable, 1923, Oak, inner left control panel w/1 exposed tube, right storage, lift top, handle, bat **$450.00**

366A, Console, 1932, Wood, lowboy, inner quarter-round dial, double front doors, 6 legs, 12 tubes **$200.00**

525, Table/Amp, 1921, Wood, tall rectangular case, 2 stage amp **$400.00**

III, Portable, 1923, Leatherette, inner black panel w/3 exposed tubes, lift-off cover w/storage, bat **$475.00**

V, Table, 1923, Wood, high rectangular case, 2 dial slant front panel, 3 exposed tubes, battery **$375.00**

VI, Table, 1924, Wood, tall rectangular case, 2 dial slant front panel, 4 exposed tubes, battery **$425.00**

X, Table, 1923, Mahogany, inner 2 dial slanted panel, 3 exposed tubes, speaker, fold-up top, bat **$515.00**

XI, Table, 1924, Mahogany, inner 2 dial slanted panel, 4 exposed tubes, speaker, fold-up top, bat **$525.00**

XV, Table, 1924, Wood, tall rectangular case, 2 dial slant front panel, 5 exposed tubes, battery **$425.00**

XXX, Table, 1925, Wood, rectangular case, 2 dial slant front panel, battery **$375.00**

KENT

422, Table, Metal, midget, right dial over wrap-around horizontal bars, handle, BC, AC/DC $85.00

KING

King Quality Products, Inc., Buffalo, New York
King-Buffalo, Inc., Buffalo, New York
King-Hinners Radio Co., Inc., Buffalo, New York
King Manufacturing Corporation, Buffalo, New York

King Quality Products, Inc., began business in 1924. The company was owned by Sears, Roebuck and in the mid-twenties produced radios with both the Sears Silvertone as well as the King label. By the early 1930's, King was out of business.

4, Upright Table, Wood, small case, lower half-round dial, upper round grill w/cut-outs, 2 knobs $195.00
10, Table, Wood, low rectangular case, 3 dial black front panel, 5 tubes, battery $125.00
25, Table, 1925, Wood, low rectangular case, 3 dial front panel, 5 tubes, battery $115.00
25-C, Console, 1925, Wood, inner 3 dial panel & speaker grill, fold-down front, lower storage, battery $195.00
30, Table, 1925, Wood, low rectangular case, slanted 3 dial front panel, 5 tubes, battery $100.00
30-S, Table, 1925, Wood, low rectangular case, slanted 3 dial front panel, speaker grill, battery $140.00
61, Table, 1926, Two-tone wood, low rectangular case, 3 dial front panel, 6 tubes, battery $110.00
61-H, Console, 1926, Wood, inner 3 dial panel & speaker grill, fold-down front, lower storage, battery $180.00
62, Table, 1926, Two-tone wood, rectangular case, oval front panel w/window dial, 3 knobs, bat $100.00
71 "Commander", Table, 1926, Wood, low rectangular case, center front window dial, loop antenna, feet, battery $175.00
80 "Baronet", Table, 1927, Wood, low rectangular case, center front dial, 3 knobs, fluted columns, 6 tubes, bat $120.00
80-H "Viking", Console, 1927, Wood, inner 1 dial panel & speaker grill, fold-down front, lower storage, battery $200.00
81 "Crusader", Table, 1927, Two-tone wood, low rectangular case, oval panel w/1 dial, 3 knobs, battery $100.00

81-H "Chevalier", Console, 1927, Walnut, highboy, inner 1 dial oval panel, upper round grill, double doors, battery $185.00
FF, Table, 1929, Two-tone metal, low rectangular case, center front window dial, 3 knobs, AC $85.00

KINGSTON
Kingston Radio, Kokomo, Indiana

Founded in 1933, Kingston manufactured many farm battery radios which were sold under other labels, such as Airline and Truetone. The company was out of business by 1951.

55, Table, 1934, Wood, right dial, center cloth grill w/cut-outs, step-down sides, 2 knobs, AC/DC $80.00
600A, Tombstone, 1934, Wood, Deco, center front window dial, upper grill w/cut-outs, 6 tubes, 3 knobs $125.00
600B, Console, 1934, Wood, Deco, upper front dial, lower arrow-shaped grill, 3 knobs, 6 tubes $150.00

KITCHENAIRE
The Radio Craftsmen, Inc.,
1341 South Michigan Avenue,
Chicago, Illinois

No #, Wall, 1946, Metal, kitchen wall set w/ side shelves, 2 knobs, herringbone grill, BC, AC/DC $150.00

KNIGHT
Allied Radio Corp.,
833 West Jackson Boulevard,
Chicago, Illinois

4D-450, Portable, 1948, Leatherette & plastic, upper slide rule dial, lattice grill, handle, 2 knobs, BC, bat $30.00
4G-420, Table-R/P, 1950, Front slide rule dial, 3 knobs, lift top, inner phono, BC, AC $25.00
5A-152, Table, 1947, Plastic, upper slanted slide rule dial, lower horiz louvers, 2 knobs, BC, AC/DC $35.00
5A-154, Table, 1947, Two-tone wood, slanted slide rule dial, horizontal louvers, 2 knobs, BC, AC/DC $45.00
5A-190, Table, 1947, Plastic, right dial, horizontal grill louvers, 2 knobs, step-down top, BC, AC/DC $60.00
5B-160, Table-R/P, 1947, Leatherette, open top, slide rule dial, horizontal louvers, 3 knobs, BC, AC $20.00
5B-175, Table, 1947, Plastic, right square dial, left vertical wrap-over louvers, 2 knobs, BC, AC/DC $45.00
5B-185, Table-R/P, 1947, Wood, front dial, criss-cross grill, 2 knobs, lift top, inner phono, BC, AC $35.00
5C-290, Portable, 1947, Leatherette, top dial, lower horizontal louvers, 3 knobs, handle, BC, AC/DC/bat $20.00
5D-250, Table, 1949, Plastic, upper raised slanted slide rule dial, horiz louvers, 2 knobs, BC, AC/DC $50.00
5D-455, Portable, 1948, Leatherette/plastic, slide rule dial, lattice grill, handle, 2 knobs, BC, AC/DC/bat $30.00
5F-525, Table, 1949, Plastic, right round dial, checkerboard grill, ridged base, 2 knobs, BC, AC/DC $40.00
5F-565, Portable, 1949, Leatherette, right square dial, horizontal louvers, handle, 3 knobs, BC, AC/DC/bat $40.00
5H-570, Table, 1951, Plastic, lower right round dial over graduated vertical bars, 2 knobs, BC, AC/DC $25.00
5H-605, Table-C, 1951, Plastic, lower right dial knob over horizontal bars, upper square clock, BC, AC $45.00

5H-700, Table-R/P, 1951, Right front square dial, 3 knobs, switch, 3/4 lift top, inner phono, BC, AC **$20.00**

5J-705, Table, 1952, Plastic, right front square dial, left vert wrap-over grill bars, 2 knobs, BC, AC **$45.00**

5K-715, Table-C, 1953, Wood, right dial, left alarm clock, center vertical wrap-over grill bars, BC, AC **$45.00**

6A-122, Table, 1946, Wood, upper slanted slide rule dial, lower cloth grill, 4 knobs, BC, SW, AC/DC **$40.00**

6A-127, Portable, 1946, Leatherette, slide rule dial, criss-cross grill, 3 knobs, handle, BC, AC/DC/bat **$35.00**

6A-195, Table, 1947, Plastic, right square dial, vertical wrap-over grill bars, 2 knobs, BC, SW, AC/DC **$45.00**

6C-225, Table, 1947, Plastic, upper slide rule dial, horiz louvers w/center bar, 3 knobs, BC, AC/DC **$50.00**

6D-235, Table, 1949, Wood, right dial, left vertical grill bars, curved sides, 3 knobs, BC, AC/DC **$60.00**

6D-360, Console-R/P, 1948, Two-tone wood, upper slide rule dial, 5 knobs, pull-out phono drawer, BC, SW, AC **$110.00**

6H-580, Table, 1951, Plastic, large front dial w/horizontal design lines, 2 knobs, BC, AC/DC **$25.00**

6K-718, Portable, 1953, Two-tone leatherette, center front round dial, handle, 2 knobs, BC, AC/DC/bat **$35.00**

7B-220, Table, 1947, Wood, upper slanted slide rule dial, criss-cross grill, 4 knobs, BC, FM, AC/DC **$45.00**

7D-405, Table-R/P, 1948, Wood, upper front slide rule dial, 4 knobs, lift top, inner phono, BC, AC **$30.00**

8B-210, Table-R/P, 1947, Wood, front dial, 4 knobs, 3/4 lift top, inner phono, BC, AC **$30.00**

11C-300, Console-R/P, 1947, Wood, inner slide rule dial, 4 knobs, left pull-out phono drawer, BC, FM, AC **$75.00**

11D-302, Console-R/P, 1949, Wood, inner right slide rule dial, 4 knobs, left fold-down phono, BC, FM, AC **$75.00**

68B-151K, Table, Wood, right oval black dial, left cloth grill w/cut-outs, tuning eye, 4 knobs, bat **$60.00**

94S-445, Table-C, 1955, Wood, right square dial, left alarm clock, center horizontal grill bars, BC, AC **$35.00**

96-326, Table-R/P/Rec, 1951, Leatherette, inner left dial, phono, recorder, lift top, handle, BC, AC **$30.00**

449, Portable, 1950, Leatherette, slide rule dial, horizontal grill bars, handle, 2 knobs, BC, AC/DC/bat **$25.00**

KODEL
The Kodel Manufacturing Company,
118 Third Street, West, Cincinnati, Ohio
The Kodel Radio Corporation,
503 East Pearl Street, Cincinnati, Ohio

The Kodel Manufacturing Company was formed in 1924 by Clarence Ogden, an inventor of battery chargers. The company's radio business boomed along with their production of battery eliminators. The advent of AC radio, however, was the beginning of the end for Kodel and they were out of business by the 1930's.

Big Five "Logodyne", Console, 1925, Mahogany, desk-style, inner 3 dial panel, fold-down front, built-in speaker, bat **$175.00**

Big Five "Logodyne", Table, 1925, Mahogany, low rectangular case, slanted 3 dial front panel, 5 tubes, battery **$120.00**

C-11 "Gold Star", Table, 1924, Leatherette, tall rectangular case, 1 dial front panel, 1 tube, battery **$165.00**

C-12 "Gold Star", Table, 1924, Wood, rectangular case, center front dial, 2 tubes, battery **$140.00**

C-13 "Gold Star", Table, 1924, Wood, low rectangular case, 2 dial front panel, 3 tubes, battery **$155.00**

C-14, Table, 1924, Wood, low rectangular case, 2 dial front panel, 4 tubes, battery **$130.00**

P-12, Portable, 1924, Leatherette, inner 1 dial panel, fold-down front, 2 tubes, handle, battery **$160.00**

Standard Five "Logodyne", Console, 1925, Mahogany, slanted 3 dial center panel, right built-in speaker, 5 tubes, battery **$175.00**

Standard Five "Logodyne", Table, 1925, Mahogany, low rectangular case, 3 dial front panel, 5 tubes, battery **$120.00**

KOLSTER
Federal Telegraph Company,
Woolworth Building, New York City, New York
Federal-Brandes, Inc., Newark, New Jersey

To avoid any confusion with the rival Federal Telephone & Telegraph Company, the Federal Telegraph Company used the Kolster name for its radio line which began in 1925. Plagued with internal problems from the beginning and after many reorganization attempts, Kolster was out of business by 1930.

6D, Table, 1926, Wood, rectangular case, center dial w/escutcheon, lift top, 6 tubes, 3 knobs, bat **$90.00**

6E, Console, 1926, Wood, lowboy, upper front dial, center fleur-de-lis grill, lower storage, battery **$165.00**

6G, Console, 1926, Wood, inner front dial, 3 knobs, lower speaker, double doors, storage, battery **$185.00**

6H, Console, 1926, Wood, lowboy, inner front dial, 3 knobs, lower speaker, double doors, storage, bat **$185.00**

6J, Table, 1927, Wood, low rectangular case, center window dial w/large escutcheon, 3 knobs, AC **$100.00**

7A, Table, 1926, Two-tone wood, low rectangular case, center front dial, 3 knobs, battery **$100.00**

7B, Table, 1926, Two-tone wood, center front dial, upper built-in speaker, 3 knobs, battery **$130.00**

8A, Table, 1926, Wood, center front window dial w/escutcheon, 8 tubes, 3 knobs, battery **$110.00**

8B, Console, 1926, Wood, center window dial, upper built-in speaker, left side loop antenna, battery **$225.00**

8C, Console, 1926, Wood, inner dial, fold-down front, upper speaker, built-in loop antenna, battery **$250.00**

K-20, Table, 1928, Wood & metal, rectangular case, center window dial, 3 knobs, carvings, paw feet, AC **$150.00**

K-21, Table, 1928, Wood, rectangular case, center window dial, front carvings, paw feet, 3 knobs **$130.00**
K-45, Console, 1928, Wood, lowboy, right side window dial & knobs, front floral grill cloth, 11 tubes, AC **$400.00**
K-48, Console, Wood, lowboy, upper window dial w/escutcheon, lower grill w/cut-outs, 3 knobs **$135.00**

K-60, Tombstone, 1931, Wood, center window dial, upper cloth grill w/repwood cut-outs, 3 knobs, AC **$145.00**
K-80, Console, Wood, carved front panel w/upper window dial, lower 3-section grill, 3 knobs **$195.00**
K-90, Console, Wood, inner window dial, 3-section cloth grill, 3 knobs, double sliding doors **$150.00**
K-110, Tombstone, 1932, Wood, shouldered, center window dial, upper cloth grill w/cut-outs, 8 tubes, AC **$120.00**
K-114, Tombstone, 1932, Wood, shouldered, center window dial, upper cloth grill w/cut-outs, 9 tubes, bat **$100.00**
K-120, Console, 1932, Wood, lowboy, upper front window dial, lower cloth grill w/cut-outs, 8 tubes, bat **$100.00**
K-130, Console, 1932, Wood, lowboy, upper window dial, lower grill w/cut-outs, 6 legs, 9 tubes, AC **$150.00**
K-140, Console, 1932, Wood, lowboy, upper dial, lower grill w/cut-outs, doors, 6 legs, 10 tubes, AC **$185.00**

———————————

KOWA
Lafayette Radio,
165-08 Liberty Ave., Jamaica, New York

KT-31, Portable, 1960, Transistor, upper diamond-shaped window dial, lower perforated grill, AM, bat **$25.00**

———————————

KRAFT
Kraft Mfg. & Dist. Co.

Puppytune, Table-Lamp, 1949, Radio/lamp shaped like puppy, right square dial, left grill, 2 knobs, BC, AC/DC **$175.00**

———————————

LAFAYETTE
Radio Wire Television, Inc.,
100 Sixth Avenue,
New York, New York

60, Tombstone, 1935, Wood, center front round dial, upper grill w/ horizontal bars, 4 knobs, BC, SW **$135.00**
BB-27, Table, 1939, Wood, right dial, pushbuttons, tuning eye, curved left wrap-around grill, AC/DC **$65.00**
C-119, Table, Wood, right front rectangular slide rule dial, left horizontal louvers, 4 knobs **$45.00**
CC-55, Portable, 1939, Striped cloth covered, right front dial, left grill, handle, 2 knobs, AC/DC/battery **$30.00**
D-68, Table, 1938, Wood, right dial, pushbuttons, tuning eye, left grill w/horiz bars, 3 knobs, AC/DC **$65.00**
D-73, Table, 1939, Plastic, streamline, right dial, curved left wrap-around louvers, 3 knobs, AC/DC **$100.00**
FA-15W, Table, 1947, Wood, small right rectangular dial, horizontal louvers, 2 knobs, BC, AC/DC **$35.00**
FS-112, Portable, 1959, Transistor, upper front thumbwheel dial, lower perforated grill, AM, battery **$25.00**
FS-200, Portable, 1960, Transistor, right thumbwheel dial, left perforated grill area w/logo, AM, battery **$20.00**
J62C, Table-R/P, 1947, Wood, front slanted slide rule dial, 4 knobs, 3/4 lift top, inner phono, BC, SW, AC **$35.00**
MC10B, Table, 1947, Plastic, right front square dial, left horizontal louvers, 2 knobs, BC, AC/DC **$45.00**
MC11, Table, 1947, Two-tone, upper slanted slide rule dial, rounded top, 3 knobs, BC, AC/DC **$40.00**
MC12, Table, 1947, Plastic, upper slanted slide rule dial, horiz louvers, 3 knobs, BC, 2SW, AC/DC **$40.00**
MC13, Table-R/P, 1947, Wood, front square dial, cloth grills, 2 knobs, lift top, inner phono, BC, AC **$25.00**
MC16, Table, 1947, Plastic, right dial, horiz wrap-around louvers, Deco base, 2 knobs, BC, AC/DC **$70.00**

———————————

LAMCO
La Magna Mfg. Co., Inc.,
51 Clinton Place,
East Rutherford, New Jersey

1000, Table, 1947, Plastic, right round dial, left horizontal wrap-around louvers, 2 knobs, BC, AC/DC **$45.00**
3000, Portable, 1948, Two-tone leatherette, right front dial, left cloth grill, handle, 2 knobs **$35.00**

———————————

LASALLE
LaSalle Radio Products Co.,
140 Washington Street,
New York, New York

LTUC, Tombstone, 1935, Wood, center octagonal dial, upper cloth grill w/cut-outs, 3 knobs, BC, SW, AC **$115.00**

———————————

LEARADIO
Lear Incorporated,
110 Ionia Avenue, N. W., Grand Rapids, Michigan

561, Table, 1946, Wood, small upper slide rule dial, center cloth grill, 2 knobs, BC, AC/DC **$35.00**

567, Table, 1946, Plastic, upper slide rule dial, horiz louvers w/ stylized "X", 2 knobs, BC, AC/DC **$65.00**

1281-PC, Console-R/P, 1948, Wood, right tilt-out slide rule dial, 4 knobs, pull-out phono drawer, BC, FM, AC **$100.00**

6611PC, Console-R/P, 1946, Wood, dial & 4 knobs in right drawer, phono in left drawer, BC, 2SW, AC **$80.00**

6615, Table, 1946, Wood, upper slanted slide rule dial, lower cloth grill w/horizontal bars, 3 knobs **$40.00**

6616, Table, 1946, Plastic, upper slide rule dial, lower horizontal louvers, 3 knobs, BC, AC/DC **$35.00**

6617PC, Table-R/P, 1947, Wood, chest-type, inner slide rule dial, 3 knobs, phono, lift top, BC, AC **$35.00**

RM-402C "Learavian", Portable, 1948, Striped cloth, top slide rule dial, 3 knobs, handle, BC, SW, LW, AC/DC/battery **$40.00**

LEE
John Meck Industries,
Plymouth, Indiana

400, Table, 1948, Flocked case, right half-moon dial printed on flocking, 2 plastic knobs **$150.00**

LEWOL

4L "Best", Table, 1934, Embossed leather, right dial, center "leaf" cut-out, angled top, 2 knobs, AC/DC **$175.00**

LEWYT
Lewyt Corp.,
60 Broadway, Brooklyn, New York

615A, Table-R/P, 1947, Wood, inner slide rule dial, 4 knobs, phono, lift atop, front grill, BC, AC **$25.00**

711, Portable, 1948, Lower front dial, "dotted" grill, molded handle, 3 recessed knobs, BC, AC/DC/bat **$40.00**

LEXINGTON
Bloomingdale Bros.,
60th Street & Lexington Avenue.,
New York, New York

6545, Table-R/P, 1947, Wood, upper slide rule dial, lower criss-cross grill, 3 knobs, open top, BC, AC **$20.00**

LIBERTY
American Communications Co.,
306 Broadway, New York, New York

507A, Table, 1947, Plastic, right square dial, horizontal wrap-around louvers, 2 knobs, BC, AC/DC **$45.00**

A6P, Table, 1947, Plastic, right square dial, horizontal wrap-around louvers, 2 knobs, BC, AC/DC **$45.00**

LIBERTY TRANSFORMER
Liberty Transformer Co.,
555 N. Parkside Ave.,
Chicago, Illinois

Sealed Five, Table, 1925, Wood, low rectangular case, 3 dial front panel, 5 tubes, battery **$120.00**

LINCOLN
Allied Radio Corp.,
833 West Jackson Boulevard,
Chicago, Illinois

5A-110, Table, 1946, Wood, lower slide rule dial, large upper cloth grill, 2 knobs, BC, AC/DC **$30.00**

S13L-B, Table, 1946, Wood, upper slanted slide rule dial, lower cloth grill, 2 knobs, BC, SW, AC/DC **$45.00**

LINMARK
Shriro, Inc.,
276 Fourth Ave., New York, New York

T-40, Portable, 1960, Transistor, left front thumbwheel window dial, lower perforated grill, AM, bat **$15.00**

T-61, Portable, 1959, Transistor, left front thumbwheel window dial, round perforated grill, AM, bat **$25.00**

T-62, Portable, 1960, Transistor, upper wedge-shaped thumbwheel dial, perforated grill, AM, battery **$25.00**

T-63, Portable, 1960, Transistor, right thumbwheel window dial, large perforated grill area, AM, bat **$25.00**

T-80, Portable, 1960, Transistor, right thumbwheel dial, lower perforated grill, swing handle, AM, bat **$25.00**

LOG CABIN

Log Cabin, Table-N, C1935, Wood, looks like log cabin, knobs in windows, speaker grill in door **$200.00**

LYRIC
The Rauland Corp.,
4245 Knox Avenue, Chicago, Illinois

546T, Table, 1946, Plastic, right square dial, cloth grill w/ criss-cross center, 2 knobs, BC, AC/DC **$45.00**

MACO
Maco Electric Corp.
1776 Broadway, New York, New York

AB-100, Portable, 1960, Transistor, right thumbwheel window dial, lower perforated grill, AM, battery **$15.00**

T-16, Portable, 1960,Transistor, right front round dial, left perforated grill area, AM, battery **$15.00**

MAGIC-TONE
Radio Developement & Research Corp.,
233 West 54th Street,
New York, New York

501, Table, 1946, Wood, upper slide rule dial, lower grill w/4 horizontal bars, 2 knobs, BC, AC/DC **$40.00**

504, Table-N, 1947, Liquor bottle shape, dial on neck, controls in cap, base, BC, AC/DC **$350.00**

510, Portable, 1948, "Snakeskin", purse-type, dial on top of case, strap, side knob, BC, battery **$60.00**

900, Table, 1948, Keg lamp, keg front "spigot" is dial, keg top knob is volume, base, BC, AC/DC **$250.00**

MAGNAVOX
The Magnavox Company,
Oakland, California

The Magnavox Company was formed in 1917 for the manufacture of microphones and loudspeakers and, during WW I, the company did much business with the US government. Magnavox soon began to produce radios but lost money throughout the 1920's due to many internal problems.

10, Table, 1925, Wood, low rectangular case, center front dial, 5 tubes, battery **$165.00**

25, Table, 1925, Wood, high rectangular case, lower 1 dial panel, upper grill w/spindles, battery **$210.00**

28M, End Table-R/P, 1941, Wood, top dial & knobs, front grill w/ diamond cut-out, lift top, inner phono, AC **$115.00**

75, Console, 1925, Mahogany, inner 1 dial panel, fold-down front, upper grill w/spindles, battery **$275.00**

154B, Console-R/P, 1947, Wood, inner right dial, 5 knobs, 6 pushbuttons, pull-out phono drawer, BC, SW, AC **$85.00**

155B "Regency Symphony", Console-R/P, 1947, Wood, inner right dial, 5 knobs, 8 pushbuttons, left lift top, BC, SW, AC **$95.00**

AM-2, Portable, 1957, Transistor, right front round dial knob, large perforated grill area, AM, battery **$45.00**

AM-20, Table-C, Plastic, right side dial knob, left front clock over horizontal bars, feet, BC, AC **$25.00**

AM-22, Portable, 1960, Plastic, transistor, right window dial, left metal perforated grill, AM, battery **$15.00**

AM-23, Portable, 1960, Plastic, transistor, right dial, perforated grill, telescoping antenna, AM, battery **$25.00**

AW-24, Portable, 1960, Transistor, slide rule dial, perforated grill, telescoping antenna, BC, SW, battery **$20.00**

AW-100 "Intercontinental", Portable, 1958, Transistor, inner 4 band dial, fold-up front w/map, telescoping antenna, battery **$100.00**

J "Junior", Table, 1925, Wood, square case, center front round dial, lift top, 5 tubes, 3 knobs, battery **$175.00**

T, Table, 1925, Wood, inner right "works in a drawer", left speaker, fold-down front, 5 tubes **$145.00**

TRF-5, Table, 1924, Wood, low rectangular case, center round dial, side cut-outs, feet, 5 tubes, bat **$160.00**

TRF-50, Table, 1924, Wood, inner center dial, double doors w/ carvings, built-in speaker, 5 tubes, bat **$230.00**

MAGUIRE
Maguire Industries, Inc.,
West Putnam,
Greenwich, Connecticut

6L, Table, Plastic, step-back right w/slide rule dial, left horiz wraparound louvers, 3 knobs **$60.00**

500BW, Table, 1946, Plastic, left round dial, horizontal louvers, 2 knobs, BC, AC/DC **$35.00**

561DW, Table, 194 , Plastic, left round dial, horizontal grill bars, 2 knobs, BC, AC/DC **$40.00**

571, Table, 1948, Plastic, left round dial, right horizontal louvers, 2 knobs, BC, AC/DC **$40.00**

661, Table, 1947, Plastic, left round dial, horizontal wrap-around bars, 2 knobs, handle, BC, AC/DC **$55.00**

700A, Table-R/P, 1946, Wood, lower front slide rule dial, horizontal louvers, 3 knobs, 3/4 lift top, BC, AC $35.00

700E, Table-R/P, 1947, Wood, lower front slide rule dial, criss-cross grill, 3 knobs, 3/4 lift top, BC, AC $30.00

MAJESTIC
Grigsby-Grunow Company,
5801 Dickens Avenue, Chicago, Illinois
Majestic Radio & Television Corporation,
St. Charles, Illinois

The Grigsby-Grunow Company was formed in 1928. The company's radio sales were extraordinary due to the superiority of their speakers over that of others on the market. However, the Depression soon caught up with Grigsby-Grunow, and by 1933 the company was bankrupt. The business was re-formed into the Majestic Radio & Television Corporation which made the Majestic line and General Household Utilities which made the Grunow line until 1937.

1 "Charlie McCarthy", Table-N, 1938, Right front slide rule dial, rounded left w/Charlie McCarthy figure, 2 knobs, AC..$1,000.00

1A-59, Table, 1939, Two-tone wood, right front slide rule dial, left wrap-around louvers, 2 knobs, AC $50.00

1BR50-B, Portable, 1939, Luggage-type, striped cloth covered, front dial & knobs, handle, AC/DC/battery $30.00

2C60-P, Console-R/P, 1939, Wood, Deco, slide rule dial, pushbuttons, fold-down front, inner phono, BC, SW, AC $165.00

3BC90-B, Console, 1939, Walnut, Deco, slide rule dial, pushbuttons, tuning eye, vertical grill bars, BC, SW $200.00

3-C-80, Console, 1939, Wood, upper slide rule dial, pushbuttons, lower grill & horiz bars, BC, SW, AC $165.00

4L1, Portable, 1955, Plastic, center front round dial w/eagle, top thumbwheel knob, handle, BC, bat $35.00

5A410, Table, 1946, Plastic, raised slide rule dial, lower horizontal wrap-around louvers, 2 knobs $100.00

5A430, Table, 1946, Wood, small upper slide rule dial, lower cloth grill, 2 knobs, BC, AC/DC $35.00

5A445, Table-R/P, 1947, Wood, right front dial, criss-cross grill, 3 knobs, lift top, inner phono, BC, AC..........................$30.00

5A445R, Table-R/P, 1947, Wood, right dial, 3 knobs, left grill, lift top, inner phono, BC, AC $30.00

5AK711, Table, 1947, Plastic, low case, slide rule dial, vertical grill bars on top, 2 knobs, BC, AC/DC $75.00

5AK780, End Table-R/P, 1947, Step down end table, top dial & 3 knobs, phono in base w/lift top, BC, AC $75.00

5C-2, Table-C, 1952, Plastic, upper front dial, center clock, right & left vertical bars, 5 knobs, BC, AC $45.00

5LA5, Table, 1951, Plastic, raised upper slide rule dial, wrap-around louvers, 2 knobs, BC, AC/DC $100.00

5LA7, Table, 1951, Plastic, raised upper slide rule dial, wrap-around louvers, 2 knobs, BC, AC/DC $60.00

5M1, Portable, 1955, Plastic, center front round dial w/eagle, top switch, handle, BC, AC/DC/battery $35.00

5T, Table-C, 1939, Plastic, Deco, center round dial surrounds clock, rear grill, 2 knobs, BC, AC/DC $100.00

6FM714, Table, 1948, Plastic, raised slide rule dial, horiz wrap-around louvers, 2 knobs, BC, SW, AC/DC $60.00

6FM773, Console-R/P, 1949, Wood, inner right dial, 2 knobs, left phono, lift top, criss-cross grill, BC, FM, AC $55.00

6G780, Portable, Plastic, transistor, left round dial, lower metal perforated grill w/logo, AM, bat $35.00

7C432, Table, 1947, Wood, upper slide rule dial, lower horiz louvers, 4 knobs, tubular feet, BC, AC $40.00

7C447, Table-R/P, 1947, Wood, inner vertical dial, 4 knobs, lift top, criss-cross grill, feet, BC, AC $30.00

7FM888, Console-R/P, 1949, Wood, inner right slide rule dial, 2 knobs, pull-out phono drawer, BC, FM, AC/DC $70.00

7JK777R, Console-R/P, 1947, Wood, inner dial, 4 knobs, phono, lift top, outer criss-cross grill, BC, AC $65.00

7JL866, End Table-R/P, 1949, Wood, step-down top, slide rule dial, 4 knobs, front pull-out phono drawer, BC, AC $90.00

7P420, Portable, 1947, Leatherette, grill w/eagle, drop front, handle, telescope antenna, BC, AC/DC/bat $45.00

7YR752, Table-R/P/Rec, 1947, Wood, outer dial, cloth grill, 3 knobs, lift top, inner phono/recorder, BC, AC $25.00

8FM744, Table, 1947, Wood, slide rule dial, large cloth grill, curved top, 2 knobs, BC, FM, AC/DC $50.00

8FM775, Console-R/P, 1947, Wood, upper front slide rule dial, 3 knobs, pull-out phono drawer, BC, FM, AC $55.00

8FM776, Console-R/P, 1947, Wood, inner slide rule dial, 3 knobs, pull-out phono drawer, doors, BC, FM, AC $65.00

8FM889, Console-R/P, 1949, Wood, inner right slide rule dial, 3 knobs, left pull-out phono drawer, BC, FM, AC $75.00

8JL885, Console-R/P, 1948, Wood, upper slide rule dial, 5 knobs, lower pull-out phono drawer, BC, SW, AC $85.00

8S452, Table-R/P, 1946, Wood, inner dial & knobs, lift top, horizontal wrap-around louvers, BC, SW, AC $40.00

8S473, Console-R/P, 1946, Wood, inner right dial & knobs, lift-top, criss-cross grill, BC, SW, AC $85.00

10FM891, Console-R/P, 1949, Wood, inner right slide rule dial, 4 knobs, left pull-out phono drawer, BC, SW, FM, AC $75.00

12FM475, Console-R/P, 1947, Wood, inner right dial & left phono, lift top, criss-cross grill, BC, SW, FM, AC $80.00

12FM895, Console-R/P, 1949, Wood, inner right slide rule dial, 4 knobs, pull-out phono drawer, BC, SW, FM, AC $80.00

15A, Tombstone, 1932, Wood, shouldered, center front window dial, upper grill w/cut-outs, 3 knobs $120.00

31, Cathedral, 1931, Wood, shouldered, right dial, cloth grill with cut-outs, small finials, 2 knobs $225.00

44 "Duo-Chief", Table, 1933, Wood, lower left dial, upper grill w/ aluminum cut-outs, 2 knobs, BC, SW, AC $150.00
49 "Duo-Modern", Table, 1933, Two-tone wood, lower left dial, grill w/aluminum cut-outs, 2 knobs, BC, SW, AC $150.00
50, Cathedral, 1931, Wood, lower right front window dial, upper grill w/cut-outs, top spires, 2 knobs $225.00
50, Cathedral/Stand, 1931, Wood, lower right front window dial, upper grill w/cut-outs, top spires, 2 knobs $250.00
52, Cathedral, 1930, Wood, lower right window dial, upper grill w/cut-outs, top spires, 8 tubes, 2 knobs $225.00
52, Table, 1938, Upper right front dial, left grill w/2 horizontal bars, 2 knobs, BC, AC $60.00
55 "Duette", Table, 1933, Two-tone wood, left dial & knobs, grill w/ aluminum lyre cut-out, BC, SW, AC $225.00
59 "Studio", Tombstone, 1933, Two-tone wood, Deco, lower dial & knobs, upper aluminum grill, BC, SW, AC $190.00
60, Table, Two-tone wood, Deco, right front oval dial, left cloth grill w/cut-outs, 4 knobs $85.00
71, Console, 1928, Walnut, upper window dial, lower oval grill w/cut-outs, 7 tubes, 3 knobs, BC, AC $215.00

71B, Console, 1929, Wood, lowboy, upper window dial w/oval escutcheon, round grill w/cut-outs, AC $175.00
72, Console, 1928, Walnut, highboy, inner window dial/grill/3 knobs, double front doors, 7 tubes, AC $195.00
75 "Queen Anne", Console, 1933, Walnut w/oriental wood front panel, window dial, grill cut-outs, Queen Anne legs ...$165.00
76, Table, 1936, Wood, front oval dial w/escutcheon, tuning eye, step-down left grill, BC, SW, AC $80.00
77, Console, 1933, Wood, upper front window dial, lower grill w/cut-outs, 3 knobs, Queen Anne legs $150.00
80FMP2, Console-R/P, 1951, Wood, inner right slide rule dial, 2 knobs, left pull-out phono drawer, BC, FM, AC $75.00
86 "Hyde Park", Console, 1931, Walnut, upper window dial, lower cloth grill w/cut-outs, 8 tubes, 3 knobs, AC $140.00
90, Console, 1929, Wood, lowboy, upper front window dial, lower cloth grill, 3 knobs, stretcher base $135.00
90-B, Console, 1929, Wood, lowboy, upper front window dial, lower cloth grill, 3 knobs, stretcher base $135.00
91, Console, 1929, Walnut, lowboy, upper window dial, lower grill w/ cut-outs, silver knobs, AC $130.00

92, Console, 1929, Highboy, walnut, inner window dial, oval grill w/cut-outs, doors, 3 knobs, AC $140.00
102, Console-R/P, 1930, Wood, lowboy, front window dial, grill w/cut-outs, lift top, inner phono, storage $150.00
103, Console-R/P, 1930, Wood, lowboy, inner window dial, grill w/ cut-outs, doors, lift top, inner phono $150.00
130, Console, 1930, Walnut, lowboy, upper front window dial, lower cloth grill w/cut-outs, AC $130.00
130, Portable, 1939, Leatherette, small case, upper front round dial, angled top, handle, 2 knobs, bat $75.00
131, Console, 1930, Wood, lowboy, upper front window dial, lower cloth grill w/cut-outs, AC $130.00
132, Console, 1930, Wood, highboy, inner window dial/grill w/cut-outs, double doors, stretcher base $150.00
161, Tombstone, Wood, right front dial, center grill w/chrome cut-outs, 2 knobs, step-down top $195.00
167, Tombstone, 1939, Wood, lower front round dial, upper cloth grill w/5 vertical bars, BC, SW, AC $95.00

181, Console-R/P, 1929, Walnut, lowboy, inner dial, round grill w/cut-outs, 3 knobs, doors, phono, AC **$150.00**

194 "Gothic", Cathedral, 1933, Walnut finish, lower front dial, upper cloth grill w/cut-outs, 4 tubes, BC, SW, AC **$195.00**

195 "Gothic", Cathedral, 1933, Walnut finish, window dial, upper grill w/cut-outs, 5 tubes, 2 knobs, BC, SW, AC **$250.00**

196 "Gothic", Cathedral, 1933, Wood, front window dial, upper grill w/cut-outs, fluted columns, 6 tubes, BC, AC **$250.00**

201 "Sheffield", Tombstone, 1932, Wood, Deco, lower quarter-round dial, upper grill w/cut-outs, 8 tubes, 3 knobs **$130.00**

203 "Fairfax", Console, 1932, Wood, Early English design, quarter round dial, lower grill w/cut-outs, 8 tubes **$150.00**

211 "Whitehall", Console, 1932, Wood, Jacobean design, lower quarter-round dial, upper grill w/cut-outs, 10 tubes **$175.00**

214 "Stratford", Console, 1932, Wood, Art-Modern design, upper quarter-round dial, Deco grill cut-outs, 10 tubes **$175.00**

215 "Croydon", Console, 1932, Wood, Early English design, quarter-round dial, upper grill w/cut-outs, 10 tubes **$200.00**

233, Console-R/P, 1930, Wood, lowboy, inner window dial/grill, double doors, lift top, inner phono, AC **$175.00**

250-Ml "Zephyr", Table, 1939, Plastic, upper slide rule dial, top pushbuttons, horiz wrap-around louvers, 2 knobs **$100.00**

259-EB, Table, 1939, Walnut, slide rule dial, pushbuttons, curved left grill w/cut-outs, BC, SW, AC **$85.00**

310A, Tombstone, Wood, center window dial w/escutcheon, large cloth grill w/cut-outs, 3 knobs $135.00

331, Cathedral, 1933, Wood, shouldered, center window dial, upper grill w/gothic cut-outs, 7 tubes, AC **$265.00**

336, Console, 1933, Wood, lowboy, upper window dial, lower grill w/cut-outs, 8 tubes, 3 knobs, AC **$135.00**

344, Console, 1933, Wood, lowboy, upper doors, criss-cross grill, twin speakers, 11 tubes, 6 legs, AC **$175.00**

351 "Collingwood", Console, 1932, Wood, Tudor design, inner quarter-round dial, grill cut-outs, front doors, 10 tubes**$160.00**

353 "Abbeywood", Console-R/P, 1932, Wood, ornate Charles II design, front dial & knobs, lift top, inner phono, 10 tubes **$235.00**

363, Console, 1933, Wood, upper quarter-round dial, lower grill w/ gothic cut-outs, 11 tubes, 6 legs, AC **$155.00**

370, Table, 1933, Wood, right window dial, center cloth grill w/cut-outs, rounded top, 2 knobs, AC **$90.00**

371, Table, 1933, Wood, right window dial, center grill w/gothic cut-outs, rounded top, 2 knobs **$90.00**

373, Table, 1933, Wood, right window dial, upper grill w/glass insert, rounded top, 2 knobs, AC **$175.00**

381 "Treasure Chest", Table-N, 1933, Wood w/repwood trim, looks like treasure chest, inner dial, speaker in lid, AC $225.00

393, Console, 1933, Wood, lowboy, window dial, lower grill w/ "peacock" cut-out, 6 legs, 3 knobs, AC **$185.00**

400, Table, 1941, Plastic, upper slide rule dial, lower wrap-around horiz louvers, 2 knobs, AC/DC **$100.00**

411 "DeLuxe", Table, 1933, Wood w/inlay, Deco, front dial & knobs, center grill w/aluminum cut-out, AC/DC **$125.00**

448-2, Table, 1940, Wood, right front slide rule dial, left vertical wrap-over bars, battery **$45.00**

461 "Master Six", Tombstone, 1933, Wood, window dial, upper chrome grill, vertical fluted front bar, top & side fluting, AC **$150.00**

463 "Century Six", Table, 1933, Walnut, right front window dial, left grill w/Deco chrome cut-out, side knob, AC **$175.00**

560 "Chatham", Console, 1933, Birch & walnut, upper window dial, lower grill with vertical bars, stretcher base **$130.00**

566 "Tudor", Console, 1933, Oak, antique finish, upper window dial, geometric grill cut-outs, stretcher base **$135.00**

599 "Radiograph", Table-R/P, 1931, Walnut, front dial & knobs, center grill w/cut-outs, lift top, inner phono **$75.00**

651, Table, 1937, Plastic, round dial, curved wrap-around grill w/ horiz bars, 3 knobs, BC, AC/DC **$100.00**

666 "Ritz", Console, 1933, Two-tone wood, Deco, upper window dial, "V" shaped grill with vert cut-outs **$150.00**

672, Chairside, 1937, Wood, top dial & knobs, lift top, inner bar & glassware, front grill, storage, AC **$195.00**

776 "Lido", Console, 1933, Five-tone wood, Deco, step-down top, upper window dial, center vert grill bars **$185.00**

886 "Park Avenue", Console, 1933, Two-tone wood, ultra-modern, right window dial, left horiz rectangular sections **$500.00**

906 "Riviera", Console, 1933, Four-tone wood, modern, upper window dial, vertical grill bars, right shelves **$350.00**

921 "Melody Cruiser", Table-N, 1946, Wood radio shaped like sailing ship, horizontal louvers, chrome sails, AC **$375.00**

P1A50, Table-R/P, 1939, Two-tone wood, right front slide rule dial, left horiz louvers, lift top, AC **$40.00**

T081A, Table, 1940, Wood, right front slide rule dial, left wrap-over vertical bars, 2 knobs **$45.00**

TP221-A, Table-R/P, 1941, Wood, right front dial, left horizontal louvers, 3 knobs, lift top, inner phono, AC **$30.00**

MANTOLA
The B. F. Goodrich Co.,
500 South Main Street,
Akron, Ohio

24B6, Table, 1947, Wood, right square dial, left horizontal grill bars, handle, 2 knobs, BC, AC/DC **$45.00**

92-521, Table, 1949, Plastic, right round dial, horizontal grill bars, 2 knobs, BC, AC/DC **$30.00**

92-522, Table, 1949, Plastic, right round dial, horizontal grill bars, 2 knobs, BC, AC/DC $30.00

92-529, Table, 1951, Wood, lower front slide rule dial, upper cloth grill, 2 knobs, BC, FM, AC $35.00

92-752, Portable, 1949, Slide rule dial at top of recessed horizontal louvers, handle, 2 knobs, AC/DC/bat..........................$35.00

477 5QL, Table, Plastic, Deco, right dial, curved left, step-down top w/pushbuttons, 2 knobs, AC/DC $125.00

R-450A, Table, Plastic, Deco, right front dial, curved left w/horiz wrap-around louvers, 2 knobs $80.00

R-643W, Table, 1946, Wood, upper slanted slide rule dial, cloth grill w/4 bars, 2 knobs, BC, battery $30.00

R-652, Portable, 1946, Luggage-style, inner dial & 2 knobs, fold-down front, handle, BC, AC/DC/bat $30.00

R-654-PM, Table, 1946, Plastic, upper slanted slide rule dial, lower horiz louvers, 2 knobs, BC, AC/DC $40.00

R-654-PV, Table, 1946, Plastic, upper slide rule dial, lower horizontal louvers, 2 knobs, BC, AC/DC $40.00

R-655-W, Table-R/P, 1946, Wood, upper front slide rule dial, 2 horizontal grill bars, 2 knobs, lift top, BC, AC $30.00

R-662, Portable, 1946, Suitcase style, inner right dial, 2 knobs, fold down front, handle, BC, AC/DC/bat $30.00

R-664-PV, Table, 1947, Plastic, upper slide rule dial, lower horizontal louvers, 2 knobs, BC, AC/DC $40.00

R-7543, Table, 1947, Plastic, right square dial, left horizontal louvers, 2 knobs, handle, BC, AC/DC $65.00

R-75143, Table, 1948, Wood, right square dial, left horizontal plastic louvers, 2 knobs, BC, AC/DC $40.00

R-75152, Table-R/P, 1948, Wood, lower front slide rule dial, criss-cross grill, 4 knobs, lift top, BC, AC $30.00

R-76162, Console-R/P, 1948, Wood, front tilt-out dial, pull-out phono drawer, 2 vertical grill bars, BC, AC $65.00

R-76262, Chairside, 1948, Wood, slide-out front radio unit, 4 knobs, horizontal louvers, inner phono, BC, AC $90.00

R-78162, Console-R/P, 1948, Wood, inner right slide rule dial, 4 knobs, left pull-out phono drawer, BC, FM, AC $75.00

MARTIAN
Martian Manufacturing Co, Inc.,
New Jersey

Big 4, Table-Crystal, 1923, Crystal set on tripod base, 4 headphone connections, top crystal & detector $325.00

MARWOL

A-1, Table, 1925, Wood, low rectangular case, 3 dial front panel, 5 tubes, battery $150.00

Baby Grand, Table, 1925, Low rectangular case, slanted 3 dial front panel with 5 exposed tubes, battery $250.00

MASON
Mason Radio Products Co.,
Kingston, New York

45-1A, Table, 1947, Plastic, right square dial, left horizontal louvers, 2 knobs, BC, AC/DC $45.00

MASTER

70, Cathedral, 1930, Wood, right front window dial, upper cloth grill w/cut-outs, side spindles, AC $210.00

MAYFAIR
Radiaphone Corporation,
1142 South Wall Street,
Los Angeles, California

355, Table, Black plastic, right front half-moon dial, left grill w/ horizontal louvers, 2 knobs $40.00

393, Table, White plastic, right front half-moon dial, left grill w/ horizontal louvers, 2 knobs $40.00

510, Table, 1947, Plastic, step-down top, right dial, horizontal grill bars, 2 knobs, BC, AC/DC $40.00

520, Table, 1947, Plastic, upper slide rule dial, lower wrap-around louvers, 3 knobs, BC, AC/DC $45.00

530, Table, 1947, Two-tone wood, upper slide rule dial, lower vert grill bars, 2 knobs, BC, AC/DC $55.00

550, Table-R/P, 1947, Wood, front slide rule dial, vertical grill bars, 3 knobs, lift top, BC, AC...$40.00

MAZDA
Mazda Radio Manufacturing Co.,
3405 Perkins Ave., Cleveland, Ohio

Consomello Grand, Table, 1925, Wood, high rectangular case, 3 dial panel, upper built-in speaker, 6 tubes, battery $180.00

Consomello, Jr., Table, 1926, Wood, small low rectangular case, 2 dial front panel, battery $100.00

MECK
John Meck Industries,
Plymouth, Indiana

4C7, Table, 1948, Plastic, right square dial, left & top horizontal louvers, 2 knobs, BC, AC/DC $50.00

5A7-P11 "Trail Blazer", Table, 1948, Plastic, Deco, right square dial, left wrap-around louvers, 2 knobs, BC, AC/DC $60.00

5A7-PB11 "Trail Blazer, Table, 1948, Plastic, Deco, right square dial, left wrap-around louvers, 2 knobs, BC, AC/DC $60.00

5D7-WL18, Portable, 1947, Two-tone, left round dial, horizontal louvers, 2 knobs, handle, BC, AC/DC/battery $30.00

6A6-W4, Table, 1947, Two-tone wood, slanted slide rule dial, large cloth grill, 2 knobs, BC, AC/DC $40.00

CD-500, Table-R/P, 1948, Wood, right front square dial, horizontal grill bars, 3 knobs, lift top, BC, AC $30.00

CE-500, Table, 1948, Plastic, right dial, left & top horizontal louvers, handle, 2 knobs, BC, AC/DC $50.00

CM-500, Portable, 1948, Leatherette, left round dial, horizontal louvers, handle, 2 knobs, BC, AC/DC/bat $30.00

CR-500, Table, 1948, Wood, lower slide rule dial, large lattice grill, 4 knobs, BC, FM, AC/DC $40.00

CW-500, Table, 1948, Plastic, right front dial, vertical wrap-over grill bars, 2 knobs, BC, AC/DC $55.00

DA-601, Table, 1950, Plastic, right front dial, left vertical wrap-over grill bars, 2 knobs, BC, AC/DC $55.00

DB-602I, Table, 1950, Plastic, right front dial, left vertical wrap-over grill bars, 2 knobs, BC, AC/DC $55.00

EC-720, Table, 1950, Plastic, right square dial, left horiz wrap-around louvers, 2 knobs, BC, AC/DC $60.00

EF-730, Table, 1950, Plastic, Deco, right "sun-rise" dial, horiz wrap-around louvers, 2 knobs, BC, AC/DC $65.00

EV-760, Portable, 1950, Leatherette, left front dial, perforated grill, handle, 2 knobs, BC, AC/DC/battery $20.00

PM-5C5-DW10 "Trail Blazer", Table-R/P, 1946, Wood, right front square dial, left cloth grill, 2 knobs, lift top, BC, AC $30.00

PM-5C5-PW-10, Table-R/P, 1947, Wood, right square dial, cloth grill w/horizontal bars, 2 knobs, lift top, BC, AC $30.00

RC-5C5-P "Trail Blazer", Table, 1946, Wood, right front dial, left wrap around grill, 2 knobs, BC, AC/DC $40.00

MEDCO
Telesonic Corp. of America,
5 West 45th Street, New York, New York

1635, Table, 1947, Wood, right square dial over horizontal louvers, 2 knobs, BC, AC/DC $35.00
1636, Table, 1947, Wood, upper slide rule dial, large lower cloth grill, 3 knobs, BC, SW, AC/DC $40.00
1642, Table, 1947, Plastic, upper slide rule dial, lower horizontal louvers, 2 knobs, BC, AC/DC $40.00
1643, Table, 1947, Plastic, upper slide rule dial, lower horizontal louvers, 3 knobs, feet, BC, AC/DC $40.00

MEISSNER
Meissner Mfg. Co.,
Div. of Maguire Industries, Inc.,
Mt Carmel, Illinois

9-1065, Portable-R/P/Rec/PA, 1946, Suitcase style, inner radio, phono, recorder, PA system, BC, AC $35.00
16A, Console-R/P, 1950, Wood, inner right slide rule dial, 5 knobs, left phono, hinged lift top, BC, FM, AC $65.00
2961, Console-R/P, 1947, Wood, inner right dial, left pull-out phono drawer, double doors, storage, AM, FM, AC $80.00

MELODY BEER

Melody Beer Bottle, Table-N, Plastic, looks like large beer bottle, base, "The beer with a melody" slogan $350.00

METEOR
Sears, Roebuck & Co.,
925 South Homan, Chicago, Illinois

7001, Table, 1956, Plastic, right front round dial, left geometric recessed grill, feet, BC, AC/DC $25.00
7016A, Table-C, 1958, Plastic, right side dial, front oval clock, horizontal bars on three sides, BC, AC $20.00

METRO
Metro Electrical Co., Newark, New Jersey

A3 "Little Gem", Table-Crystal, 1925, Crystal set on tripod base, headphone connections, top crystal & detector $325.00

METRODYNE
Metro Electric Company,
2161-71 N. California Avenue, Chicago, Illinois

Super-Five, Table, 1926, Mahogany, low rectangular case, 3 dial black bakelite front panel, 5 tubes, bat $100.00

Super-Six, Table, 1926, Walnut, low rectangular case, front panel w/ 3 pointer dials, lift top, 6 tubes $130.00
Super-Seven, Table, 1926, Walnut, low rectangular case, 1 dial etched bakelite front panel, 7 tubes, battery $215.00
Super-Eight, Console, 1928, Two-tone wood, inner dial, lower cloth grill w/cut-outs, double doors, 8 tubes, AC $250.00

Super-Eight, Table, 1928, Two-tone walnut w/carvings, low rectangular case, center front dial, 8 tubes, AC $185.00

MICHIGAN
Michigan Radio Corporation,
32 Pearl Street, Grand Rapids, Michigan

Junior, Table, 1923, Wood, low rectangular case, 2 dial black front panel, right lift top, 1 tube, bat $150.00
MRC-2, Table, 1924, Wood, high rectangular case, 2 dial slanted front panel, 2 tubes, battery $200.00
MRC-3, Table, 1923, Wood w/inlay, low rectangular case, 2 dial front panel, storage, 3 tubes, battery $250.00
MRC-4, Table, 1923, Mahogany, low rectangular case, inner 2 dial panel, fold-down front, 4 tubes, bat $285.00

MRC-12, Table, 1924, Wood, detector/amplifier, low rectangular case, slant front black panel, 3 tubes $300.00
Senior, Table, 1922, Mahogany, rectangular case, 2 dial black front panel, right lift top, battery $400.00

MICROPHONE RADIO

Novelty radios in the shape of microphones were produced for many different radio stations. Each radio features the station call letters and usually tunes in only that one station.

Various Styles, Table-N, Usually plastic, upper "microphone" w/ metal grill, lower base with louvers $150.00

MIDLAND
Midland Mfg. Co., Decorah, Iowa

M6B, Table, 1946, Wood, right front round dial, left wrap around grill bars, 2 knobs, BC, AC/DC $45.00

MIDWEST
The Midwest Radio Company,
808 Main Street, Cincinnati, Ohio
Midwest Radio Corporation,
406-D East Eight Street, Cincinnati, Ohio

Midwest Radio Company was founded by A. G. Hoffman as a mail-order radio business. Many early sets were sold with the "Miraco" name – short for Midwest Radio Company.

16-37, Console, 1937, Wood, Deco case, upper front round dial, lower cloth grill w/cut-outs, 16 tubes **$250.00**
18-35, Console, 1935, Wood, Deco case, upper round dial, lower cloth grill w/vertical bars, 18 tubes **$275.00**
18-36, Console, 1936, Wood, Deco case, upper round dial, lower cloth grill w/cut-outs, 18 tubes, 6 band **$300.00**
20-38, Console, 1938, Wood, large case, upper white dial, lower recessed grill w/Deco cut-outs, 20 tubes **$375.00**
M-32, Tombstone, 1936, Wood, lower round airplane dial, upper "herringbone" grill, step-down top, 4 knobs **$165.00**
MW "Miraco", Table, 1923, Mahogany, low rectangular case, 2 dial black front panel, lift top, 4 tubes, bat **$175.00**
P-6, Portable, 1947, Luggage-style, upper slanted slide rule dial, 3 knobs, handle, BC, AC/DC/battery **$35.00**
PB-6, Portable, 1947, Luggage-style, upper slanted slide rule dial, 3 knobs, handle, BC, AC/DC/battery **$35.00**
RG-16, Console-R/P, 1948, Wood, inner slide rule dial, pushbuttons, fold-out phono, 16 tubes, BC, 3SW, FM, AC **$145.00**
ST-16, Console-R/P, 1947, Wood, top slide rule dial, pushbuttons, tilt-out phono, 16 tubes, BC, 3SW, FM, AC **$145.00**
TM-8, Table, 1947, Wood, large slide rule dial, right grill w/2 vertical bars, 4 knobs, 5 bands, AC **$75.00**

Ultra 5 "Miraco", Table, 1924, Wood, low rectangular case, 3 dial front panel (several versions), lift top, bat $150.00
Unitune 5, Table, 1926, Walnut, low rectangular case, slanted front bakelite 1 dial panel, battery **$110.00**
Unitune 8, Table, 1928, Wood, low rectangular case, center front window dial w/escutcheon **$115.00**

MINERVA
Minerva Corp. of America,
238 William Street, New York, New York

411, Table, 1948, Plastic, upper slanted slide rule dial, lower lattice grill, 2 knobs, BC, AC/DC **$35.00**
702H, Table, 1947, Plastic, upper slanted slide rule dial, lower lattice grill, 3 knobs, BC, AC/DC **$35.00**
729 "Portapal", Portable, 1947, Right front slide rule dial, lattice grill, 3 knobs, handle, BC, AC/DC/battery **$35.00**
L-728, Table, 1947, Wood, rounded corners, upper slanted slide rule dial, lower grill, 2 knobs, BC **$40.00**
W117 "Tropic Master", Table, 1946, Metal, inner right square dial, "M" grill, fold down front, handle, BC, SW, AC/DC **$35.00**
W-117-3, Table, 1947, Wood, right dial, horizontal grill bars, 4 knobs, rounded corners, BC, SW, AC **$40.00**
W-702, Table, 1947, Plastic, upper slanted slide rule dial, grill w/ geometric pattern, 3 knobs, BC **$35.00**

W-702B, Table, 1947, Plastic, slanted slide rule dial, metal wrap-around grill, 2 knobs, BC, AC/DC **$45.00**
W710A, Table, 1946, Wood, upper slide rule dial, lower horizontal wooden louvers, 3 knobs, BC, AC/DC **$30.00**

MINUTE MAN
Intercontinental Industries, Inc.,
555 W. Adams St., Chicago, Illinois

6T-170, Portable, 1960, Transistor, top dial & volume knobs, 2 front half-moon perforated grills, AM, bat **$35.00**

MIRRORTONE
John Meck Industries,
Plymouth, Indiana

850 "Deluxe", Table, Plastic, Deco, right dial, left horizontal grill bars, decorative case lines, 2 knobs **$60.00**
RC-6A7-P6, Table, 1948, Plastic, upper slanted slide rule dial, large grill area, 3 knobs, BC, AC/DC **$35.00**

MISSION BELL
Mission Bell Radio Mfg. & Distr. Co.,
2117 West Pico, Los Angeles, California

Mission Bell began business in 1927.

400, Portable, 1939, Cloth covered, right dial, left grill w/3 horizontal sections, handle, 2 knobs, bat **$30.00**
500, Portable, 1939, Right front slide rule dial, left grill w/3 horizontal sections, handle, 2 knobs, bat **$35.00**

MITCHELL
Mitchell Mfg. Co.,
2525 North Clybourn Avenue, Chicago, Illinois

1101, Portable, 1956, Transistor, right front dial knob over large patterned grill, AM, battery **$35.00**
1250 "Lullaby", Radio/Bed Lamp, 1949, Plastic, round front dial knob, vertical grill bars, built-in bed lamp, BC, AC/DC .**$70.00**

1251 "Lullaby", Radio/Bed Lamp, 1949, Plastic, round front dial knob, vertical grill bars, built-in bed lamp, BC, AC/DC $70.00

1252, Table, 1952, Plastic, right front round dial, left grill, 2 knobs, BC, AC/DC **$25.00**
1254 "Madrigal", Table, 1952, Plastic, round front dial ring w/inner pointer over large grill, 2 knobs, BC, AC/DC **$40.00**
1256, Portable, 1952, Plastic, left side dial knob, front concentric rectangles, flex handle, AC/DC/bat **$30.00**

1260 "Lumitone", Table-Lamp, 1940, Plastic, lamp/radio, lower thumbwheel dial, upper horizontal grill bars, AC ..$185.00
1267, Table-C, 1952, Plastic, lower right round dial knob over horizontal grill bars, upper clock, BC, AC **$40.00**

1268R, Table-C, 1951, Plastic, lower right round dial knob over horizontal grill bars, upper clock, BC, AC $40.00
1287, Portable, 1955, Right front slide rule dial, telescope antenna, handle, 4 knobs, BC, SW, AC/DC/bat **$40.00**

_____

MOHAWK
Mohawk Corporation,
2222 Diversy Parkway, Chicago, Illinios
Mohawk Corporation of Illinois

The Mohawk Corporation began business in 1920 as Electrical Dealers Supply House. The first radio was produced in 1924. Most of Mohawks radio cabinets were made by Wurlitzer. In 1928 the Mohawk company was bought out by All-American and the company name changed to All-American Mohawk.

100, Table, 1923, Wood, high rectangular case, 1 dial slant front panel, 5 tubes, battery **$120.00**
105, Portable, 1925, Leatherette, inner bakelite panel w/2 dials, flip-up front w/antenna, handle, bat **$200.00**
110, Table, 1925, Wood, high domed top case, 1 dial front panel, door, battery...**$185.00**
115, Console, 1925, Wood, highboy, domed top case, inner 1 dial front panel, door, battery.....................................**$215.00**
A5, Table, 1925, Wood, high rectangular case, 1 dial slant front panel, 5 tubes, battery **$120.00**
Cherokee, Table, 1926, Walnut, front window dial w/large escutcheon, 3 knobs, lift top, 6 tubes, battery **$95.00**
Chippewa, Console, 1926, Two-tone walnut, inner dial & knobs, fold-down front, lift top, battery **$145.00**
Geneva, Console, 1926, Walnut, desk-style, inner dial, fold-down front, domed top speaker grill, bat **$210.00**
Iroquois, Console, 1927, Wood, inner window dial w/escutcheon, fold-down front, fancy upper grill, bat **$225.00**
Navajo, Table, 1926, Wood, front window dial w/escutcheon, lift top, burl panels, 6 tubes, 3 knobs, bat **$100.00**
Pocahontas, Console, 1926, Two-tone inlaid walnut, double front doors, inner dial & knobs, feet, battery **$195.00**
Pontiac, Console, 1926, Walnut w/burl front, inner dial & knobs, fold-down front door, upper grill, bat **$165.00**
Winona, Table, 1926, Walnut, low rectangular case, window dial w/escutcheon, 3 knobs, lift top, bat **$110.00**

_____

MONITOR
Monitor Equipment Co.,
110 East 42nd Street, New York, New York

M-403, Table-R/P, 1947, Wood, right front dial, left grill w/picture, 3 knobs, open top, BC, AC **$40.00**
M-500, Table, 1947, Plastic, right front dial, left horizontal louvers, 2 knobs, BC, AC/DC **$40.00**
M-510, Portable, 1947, Leatherette, right dial, horizontal grill bars, 2 knobs, handle, BC, AC/DC/battery **$25.00**
M-3070, Console-R/P, 1947, Wood, inner slide rule dial, 4 knobs, pull-out phono drawer, doors, BC, FM, AC **$80.00**
RA-50, Table-R/P, 1947, Wood, front slide rule dial, horizontal grill bars, 4 knobs, lift top, BC, AC **$30.00**
TA56M, Table, 1946, Plastic, upper slide rule dial, lower horizontal louvers, 2 knobs, BC, AC/DC **$45.00**

_____

MOTOROLA
Galvin Mfg. Corporation,
847 West Harrison Street, Chicago, Illinois

The Galvin Manufacturing Company was founded in 1928 by Paul Galvin in the business of producing radio power supplies. Soon branching out into the production of auto radios, the company grew to become one of the largest producers of radios and television under the Motorola brand name.

3A5 "Playboy", Portable, 1941, Maroon metal & chrome, set plays when lid opens, handle, BC, AC/DC/battery **$40.00**
5A1, Portable, 1946, Inner right round dial, set plays when flip-up lid opens, handle, BC, battery **$40.00**

5A5, Portable, 1946, Inner right round dial, set plays when flip up lid opens, handle, BC, AC/DC/bat **$40.00**

5A7, Portable, 1947, Metal, inner dial, horiz louvers, set plays when flip-up lid opens, BC, AC/DC/bat **$45.00**

5A7A "Playmate Jr", Portable, 1948, Metal, inner dial, horiz louvers, set plays when flip-up lid opens, BC, AC/DC/bat **$45.00**

5C1 "Radio-Larm", Table-C, 1950, Plastic, right round dial, center vert grill bars, left alarm clock, 5 knobs, BC, AC **$30.00**

5C11E, Table-C, 1959, Plastic, lower right dial knob, left clock, right vertical grill bars, feet, BC, AC **$15.00**

5C13M, Table-C, 1959, Plastic, off-center front dial, left clock, right vertical grill bars, feet, BC, AC **$15.00**

5C13P, Table-C, 1959, Plastic, off-center dial, left alarm clock, right vertical grill bars, feet, BC, AC **$15.00**

5C22M, Table-C, 1958, Plastic, off-center front dial, left clock, right vertical grill bars, feet, BC, AC **$15.00**

5C23GW, Table-C, 1958, Plastic, left dial & clock panel over front horizontal grill bars, "M" logo, BC, AC **$15.00**

5H1, Table, 1950, Plastic, large center front round dial w/stylized "M" over dotted grill, 2 knobs **$30.00**

5H11, Table, 1950, Plastic, center front round dial w/stylized "M", dotted grill, 2 knobs, BC, AC/DC **$30.00**

5H11U, Table, 1950, Plastic, center front round dial w/stylized "M", dotted grill, 2 knobs, BC, AC/DC **$30.00**

5H12, Table, 1952, Plastic, center front round dial w/stylized "M", dotted grill, 2 knobs, BC, AC/DC **$30.00**

5J1 "Jewel Box", Portable, 1950, Plastic, inner dial, random pattern grill, flip-up front, handle, BC, AC/DC/battery **$85.00**

5L1 "Music Box", Portable, 1950, Two-tone plastic, right dial, checkered grill, handle, 2 knobs, BC, AC/DC/battery ... **$30.00**

5L1U, Portable, 1950, Two-tone plastic, right front dial, center checkered grill panel, handle, 2 knobs **$30.00**

5M1 "Playmate Jr", Portable, 1950, Metal, small case, inner dial, 2 knobs, flip-up front, handle, BC, AC/DC/battery **$45.00**

5M1U, Portable, 1950, Metal, small case, inner dial, 2 knobs, flip-up front, handle, BC, AC/DC/battery **$45.00**

5P22RW-1, Portable, Leatherette, right front round dial knob, lower horizontal bars, handle, AC/DC/bat **$35.00**

5R1, Table, 1951, Plastic, right round dial w/"M" logo, lower high-low checkered panel, 2 knobs **$40.00**

5R11U, Table, 1950, Plastic, right round dial, lower front hi-low checkered panel, 2 knobs, BC, AC/DC **$40.00**

5R23G, Table-R/P, 1958, Front dial over large perforated grill, slanted lift top, inner phono, BC, AC **$35.00**

5T, Tombstone, 1937, Wood, lower front dial w/escutcheon, upper grill w/bars, 5 tubes, 4 knobs, BC, SW **$100.00**

5T11M, Table, 1959, Plastic, right front round dial, left lattice grill, 2 knobs, BC, AC/DC **$20.00**

5T13P, Table, 1959, Plastic, lower right front dial knob, upper & lower vertical bars, feet, BC, AC/DC **$30.00**

5T22Y, Table, 1957, Plastic, large lower right dial, horizontal front grill bars, left knob, BC, AC/DC **$30.00**

5X11U, Table, 1950, Plastic, center round dial w/inner perforated grill, stand, 2 knobs, BC, AC/DC **$40.00**

5X12U, Table, 1950, Plastic, center round dial w/inner perforated grill, stand, 2 knobs, BC, AC/DC **$40.00**

5X21U, Table, 1951, Plastic, center round dial w/inner perforated grill, stand, 2 knobs, BC, AC/DC **$40.00**

6F11, Console-R/P, 1950, Wood, upper front half-moon dial, 4 knobs, lower pull-out phono drawer, BC, AC **$50.00**

6L1 "Town & Country", Portable, 1950, Plastic, uniquely shaped dial, perforated grill, handle, 2 knobs, BC, AC/DC/bat .. **$35.00**

6P34E "700 Ranger", Portable, 1957, Leatherette, right front dial, lower metal perforated grill, handle, AC/DC/bat **$35.00**

6-T, Table, 1937, Wood, right dial w/large escutcheon, left grill w/3 vert bars, 4 knobs, BC, SW, AC **$85.00**

6T15S "Custom 6", Table, 1959, Plastic, lower right front dial knob, upper vertical grill bars, BC, AC/DC **$30.00**

6X11U, Table, 1950, Plastic, center front round dial w/inner perforations, 2 knobs, BC, AC/DC **$40.00**

6X28B, Portable, 1959, Blue plastic, transistor, window dial, vert grill bars, thumbwheel knobs, AM, bat **$35.00**

6X28N, Portable, 1959, Mocha plastic, transistor, window dial, vert grill bars, thumbwheel knobs, AM, bat **$30.00**

6X28P, Portable, 1959, Pink plastic, transistor, window dial, vert grill bars, thumbwheel knobs, AM, bat **$35.00**

6X28W, Portable, 1959, White plastic, transistor, window dial, vert grill bars, thumbwheel knobs, AM, bat **$30.00**

6X39A-2 "Weatherama", Portable, 1958, Transistor, right front round dial, perforated grill, base, 2 band, battery **$30.00**

7F11, Console-R/P, 1950, Wood, inner right half-moon dial, 4 knobs, pull-out phono, double doors, BC, AC **$55.00**

7X23E, Portable, 1959, Transistor, modern, "jet plane" grill w/ window dial, swing handle, AM, battery $60.00

7X24S, Portable, 1959, Transistor, modern, "jet plane" grill w/window dial, thumbwheel knob, AM, bat **$60.00**

7X24W, Portable, 1959, Transistor, modern, "jet plane" grill w/ window dial, thumbwheel knob, AM, bat **$60.00**

7X25P, Portable, 1959, Salmon, transistor, upper round dial knob over horiz bars, swing handle, AM, bat **$20.00**

7X25W, Portable, 1959, White, transistor, upper round dial knob over horiz bars, swing handle, AM, bat **$20.00**

8FM21, Console-R/P, 1951, Wood, upper half-moon dial, 4 knobs, lower pull-out phono drawer, BC, FM, AC **$50.00**

8X26E, Portable, 1959, Transistor, upper front round dial, lower horizontal bars, swing handle, AM, bat **$40.00**

8X26S, Portable, 1959, Transistor, upper front round dial, lower horizontal bars, swing handle, AM, bat **$40.00**

9A, Chairside, 1937, Two-tone wood, oval Deco case, top dial & knobs, tuning eye, front vertical bars **$150.00**

9FM21, Console-R/P, 1950, Wood, inner half-moon dial, 4 knobs, pull-out phono, open storage, BC, FM, AC **$50.00**

10-Y-1, Console, 1937, Wood, upper round dial, tuning eye, lower grill w/vert bars, 10 tubes, BC, SW, AC **$160.00**

12-Y-1, Console, 1937, Walnut, upper round dial, tuning eye, lower grill w/vert bars, 12 tubes, BC, SW, AC **$175.00**

17-FM-41, Console, 1941, Wood, slide rule dial, tuning eye, pushbuttons, vertical grill bars, BC, SW, AC **$145.00**

40-60W, Table, 1940, Wood, right front square dial, left cloth grill, wooden handle, 3 knobs, AC/DC **$50.00**

40-65BP "Headliner", Portable, 1940, Cloth covered, right front dial, left cloth grill, handle, 2 knobs, BC, AC/DC/bat **$25.00**

40BW, Table, 1940, Two-tone wood, right front square dial, raised left grill w/cut-outs, 3 knobs **$65.00**

41A, Table, 1939, Plastic, Deco, right dial, raised left grill w/graduated horiz louvers, 3 knobs, bat **$75.00**

41B, Table, 1939, Plastic, right front dial, curved left grill w/horiz louvers, 3 knobs, BC, battery **$35.00**

41D-1, Portable, 1939, London Tan leatherette, right front dial, left grill, handle, 2 knobs, BC, battery **$25.00**

41D-2, Portable, 1939, Tan striped cloth, right front dial, left grill, handle, 2 knobs, BC, battery **$25.00**

41E, Table, 1939, Two-tone wood, right front dial, left grill w/cut-outs, 3 knobs, BC, battery **$40.00**

41F, Table, 1939, Two-tone wood, right front dial, left grill w/cut-outs, 3 knobs, BC, battery **$40.00**

41S "Sporter", Portable, 1939, Leatherette, "camera-case" style, top dial & knobs, front grill, carrying strap, bat **$40.00**

42B1, Portable, 1953, Two-tone, right dial and left knob on case top, vertical grill bars, handle, BC, bat **$25.00**

45B12, Table, 1946, Wood, right front dial, cloth grill, 3 knobs, contrasting veneer, BC, battery **$40.00**

45P2 "Pixie", Portable, 1956, Plastic, right metal dial plate, vertical grill bars, handle, 2 thumbwheel knobs $40.00

47B11, Table, 1947, Wood, right half-moon dial, left cloth grill w/ "Motorola", 3 knobs, BC, battery **$40.00**

48L11, Portable, 1948, Plastic, right dial, graduated horizontal louvers, handle, 2 knobs, BC, battery **$35.00**

49L11Q, Portable, 1949, Plastic, right front dial/left volume knobs over center grill, flex handle, BC, bat **$35.00**

50XC, Table, 1940, Catalin, Deco, right square dial, left round grill, handle, 2 hexagonal knobs, BC .... **$2,000.00+**

50XH1, Table, Plastic, right front dial, left horizontal louvers, rounded corners, 2 knobs, BC $40.00

51A, Table, 1939, Plastic, Deco, right dial, raised left grill w/graduated horiz louvers, 2 knobs, BC, AC/DC **$85.00**

51C, Table, 1939, Plastic, Deco, right dial, raised left grill w/graduated horiz louvers, 2 knobs, BC, AC/DC **$85.00**

51D, Portable, 1941, Cloth covered, inner right dial, left grill, fold-down front, handle, AC/DC/battery **$30.00**

51X16, Table, 1939, Catalin, right square dial, left cloth grill w/curved bars, handle, 2 knobs, BC, AC .... **$3,000.00+**

51X17, Table, 1941, Two-tone leatherette, right square dial, left horizontal louvers, 2 knobs, AC/DC **$40.00**

51X18, Table, 1941, Ivory finish, upper slanted slide rule dial, lower grill, side bars, 2 knobs, AC **$40.00**

51X19, Table, 1942, Wood, right front square dial, left horizontal plastic louvers, 2 knobs, BC, AC/DC **$40.00**

51X20, Table, Two-tone wood, upper front slanted slide rule dial, lower woven grill, 2 knobs **$45.00**

52, Table, 1939, Catalin, right front dial, left grill w/7 vertical louvers, 2 hexagonal knobs .... **$800.00+**

52B1U, Portable, 1953, Plastic, top left dial knob, front vertical grill bars, handle, BC, AC/DC/battery **$30.00**

52C1, Table-C, 1953, Plastic, right front dial, left alarm clock, center vertical bars, 4 knobs, BC, AC $20.00

52C6, Table-C, 1952, Plastic, right dial, left alarm clock, center vertical grill bars, 5 knobs, BC, AC $20.00

52CW1, Table/Wall-C, 1953, Plastic, side dial knob, large upper front alarm clock, lower checkered panel, AC $25.00

52H, Table, 1952, Plastic, upper dial w/stick pointer & center knob, lower vertical bars, BC, AC/DC $30.00

52H11U, Table, 1952, Plastic, upper dial w/stick pointer & center knob, lower vertical bars, BC, AC/DC $30.00

52L1, Portable, 1953, Plastic, top right dial, front lattice grill & stylized "M", handle, BC, AC/DC/bat $30.00

52M1U, Portable, 1952, Small case, inner right round dial knob, lattice grill, flip-up front, BC, AC/DC/bat $40.00

52R12A, Table, 1952, Plastic, right front round dial knob, left round disc w/stylized "M", BC, AC/DC $25.00

52R12U, Table, 1952, Plastic, right front round dial knob, left round disc w/stylized "M", BC, AC/DC $25.00

52R14, Table, 1952, Plastic, right front round dial knob, left round disc w/stylized "M", BC, AC/DC $25.00

52X110, Table, Plastic, upper front slide rule dial, large lower perforated grill area, 2 knobs, BC $30.00

53A, Table, 1939, Two-tone wood, right dial, left wrap-around horiz grill bars, 2 knobs, BC, AC/DC $60.00

53C, Table, 1939, Plastic, Deco, right dial, raised left grill w/graduated horiz louvers, 2 knobs, BC, AC/DC $85.00

53C1, Table-C, 1954, Plastic, right front dial, left alarm clock, side horizontal bars, 3 knobs, BC, AC $30.00

53C1B, Table-C, 1954, Plastic, right front dial, left alarm clock, side horizontal bars, feet, 3 knobs $30.00

53D1, Table-C-N, 1954, Clock/radio/pen & pencil set, center radio w/ round alarm clock, 5 knobs, BC, AC $50.00

53F2, Table-R/P, 1954, Plastic, top round dial, vertical front grill bars, lift top, inner phono, BC, AC $40.00

53H1, Table, 1954, Plastic, modern, lower slide rule dial, large perforated grill area, 2 knobs, BC, AC $65.00

53LC, Portable-C, 1953, Plastic, right dial, left clock, large lower grill, handle, side knobs, AC/DC/bat $40.00

53LC1, Portable-C, 1953, Plastic, right dial, left clock, large lower grill, handle, side knobs, AC/DC/bat $40.00

53X, Table, 1954, Plastic, half-moon dial on case top, front horizontal louvers w/logo, BC, AC/DC $50.00

53X1, Table, 1954, Plastic, half-moon dial on case top, front horizontal grill bars, BC, AC/DC $50.00

55F11, Table-R/P, 1946, Wood, upper slide rule dial & 2 knobs, lower grill, lift top, inner phono, BC, AC $30.00

55L2, Portable, 1956, Leatherette, right front round dial knob, "V" shaped checkered grill, handle, bat $55.00

55L2U, Portable, 1956, Leatherette, right round dial knob, "V" shaped checkered grill, handle, AC/DC/bat $55.00

55L4, Portable, 1956, Leatherette, right front round dial knob, "V" shaped checkered grill, handle, bat $55.00

55M2, Portable, 1956, Leatherette, right dial knob, plastic lattice trapezoid grill, handle, AC/DC/bat $35.00

55X13A, Table, 1946, Two-tone wood, upper front dial, lower cloth grill, 2 knobs, BC, AC/DC $40.00

56B1, Portable, 1959, Leatherette, right front round dial, center plastic grill w/"V" louvers, handle $35.00

56CJ, Table-C, 1956, Plastic, left front dial, right alarm clock, center panel with "M" logo, 5 knobs, AC $25.00

56M1, Portable, 1959, Leatherette, right front chrome dial plate, lattice grill, handle, BC, AC/DC/bat $50.00

56T1, Portable, 1956, Transistor, lower right front round dial, left perforated grill, stand, AM, battery $50.00

56X11, Table, 1947, Walnut plastic, right front square dial, horizontal louvers, 2 knobs, BC, AC/DC $35.00

56X12, Table, 1947, Plastic, upper slide rule dial, horiz louvers, slanted sides, 2 knobs, BC, AC/DC $45.00

57A, Table, Plastic, lower right round dial knob over horizontal front bars, center "M" logo $25.00

57CE, Table-C, 1957, Plastic, lower right front dial, left alarm clock, vertical grill bars, feet, BC, AC $20.00

57X11, Table, 1947, Plastic, upper slide rule dial, horiz louvers, slanted sides, 2 knobs, BC, AC/DC $45.00

57X12, Table, 1947, Ivory plastic, upper slide rule dial, lower horiz louvers, slanted sides, 2 knobs $45.00

58A11, Table, 1948, Plastic, right front square dial, left horizontal louvers, 2 knobs, BC, AC/DC $35.00

58G11, Table, 1949, Plastic, right front square dial, left horizontal louvers, 2 bullet knobs, BC, AC/DC $30.00

58L11, Portable, 1948, Plastic, right front dial knob, graduated horiz louvers, handle, BC, AC/DC/bat $35.00

58R11, Table, 1948, Plastic, raised center panel w/round dial & left lattice grill, 2 knobs, BC, AC/DC $35.00

58R11A, Table, 1949, Plastic, raised center panel w/right round dial & lattice grill, 2 knobs, BC, AC/DC $35.00

58X, Table, 1949, Plastic, upper slide rule dial, lower trapezoid grill w/3 vertical bars, 2 knobs $45.00

58X11, Table, 1949, Plastic, slide rule dial, trapezoid grill w/3 vertical bars, 2 knobs, BC, AC/DC $45.00

59F11, Portable-R/P, 1949, Leatherette, inner dial, 3 knobs, phono, record storage, lift top, handle, BC, AC $20.00

59H11U, Table, 1950, Plastic, dial numerals on horizontal grill bars, stick pointer, 2 knobs, BC, AC/DC $30.00

59L12Q, Portable, 1949, Plastic, right dial/left volume knobs, wedged grill, flex handle, BC, AC/DC/bat $35.00

59R11, Table, 1949, Plastic, right round dial over checkered front, recessed base, 2 knobs, BC, AC/DC $25.00

59T-4, Table, 1938, Wood, 2 top thumbwheel knobs & 4 pushbuttons, front cloth grill w/cut-outs, AC $65.00

59T-5, Table, 1938, Wood, right slide rule dial, pushbuttons, left grill w/Deco cut-outs, 4 knobs, AC $75.00

59X11, Table, 1950, Plastic, large half-moon dial over horizontal grill bars, 2 knobs, BC, AC/DC $40.00

59X21U, Table, 1950, Plastic, half-moon dial over horizontal grill bars, 2 knobs, BC, SW, AC/DC $40.00

61A, Table, 1939, Plastic, right dial, left wrap-around horiz louvers, 5 pushbuttons, 2 knobs, BC, AC/DC $65.00

61B, Table, 1939, Plastic, right dial, left wrap-around horiz louvers, 5 pushbuttons, 3 knobs, BC, SW, AC/DC $65.00

61C, Table, 1939, Wood, right slide rule dial, left grill w/horiz bars, pushbuttons, 4 knobs, BC, SW, AC $60.00

61-CA, Table, 1940, Wood, right slide rule dial, left grill w/horizontal bars, pushbuttons, 4 knobs $60.00

61D, Console, 1939, Wood, upper slide rule dial, lower grill w/3 vert bars, 6 pushbuttons, BC, SW, AC $125.00

61E, Table, 1939, Plastic, right dial, left wrap-around horiz louvers, 5 pushbuttons, 2 knobs, BC, AC/DC $65.00

61F, Table-R/P, 1939, Wood, right front slide rule dial, left grill w/cut-outs, lift top, inner phono, AC $45.00

61F23, Console-R/P, Wood, inner right dial, left phono, 4 knobs, double lift top, front grill w/vert bars $85.00

62B, Table, 1939, Plastic, right dial, left wrap-around horiz louvers, 5 pushbuttons, 3 knobs, BC, SW, AC/DC $65.00

62C1, Table-C, 1952, Plastic, right front dial, center divider panel, left alarm clock, 5 knobs, BC, AC $20.00

62CW1, Table-C, 1953, Wood, lower front slide rule dial, large clock face, side knobs & grill, BC, AC $45.00

62E, Table, 1939, Plastic, right dial, left wrap-around horiz louvers, 5 pushbuttons, 2 knobs, BC, AC/DC $65.00

62X21, Table, 1954, Plastic, upper dial, lower vertical grill bars, stylized "M", 2 knobs, BC, SW, AC/DC $40.00

63E, Table, 1939, Wood, right front dial, left grill w/vertical bars, 5 pushbuttons, 2 knobs, BC, AC $60.00

63X21, Table, 1954, Plastic, lower front slide rule dial, large upper grill, 2 knobs, BC, SW, AC/DC $25.00

65F11, Table-R/P, 1946, Wood, outer front dial & grill, 2 knobs, lift top, inner phono, BC, AC $30.00

65F21, Console-R/P, 1946, Wood, inner slide rule dial, phono, 4 knobs, 6 pushbuttons, lift top, BC, SW, AC $85.00

65L11, Portable, 1946, Cloth covered, inner right dial, cloth grill, 2 knobs, flip-up lid, BC, AC/DC/bat $35.00

65T21, Table, 1946, Wood, right front slide rule dial over large cloth grill, 4 knobs, BC, SW, AC $45.00

66T1, Portable, 1958, Transistor, right front round dial, perforated grill, large handle, AM, battery $30.00

67F11, Table-R/P, 1948, Plastic, outer vertical dial, 2 knobs, pushbuttons, lift top, inner phono, BC, AC $45.00

67F12, Table-R/P, 1948, Walnut, outer vertical dial, 2 knobs, pushbuttons, lift top, inner phono, BC, AC $30.00

67F12B, Table-R/P, 1948, Blonde, outer vertical dial, 2 knobs, pushbuttons, lift top, inner phono, BC, AC $35.00

67F14, Console-R/P, 1949, Wood, step-down top, slide rule dial, 4 knobs, pull-out phono drawer, BC, AC $90.00

67F61BN, Table-R/P, 1948, Wood, front slide rule dial, grill, 4 knobs, lift top, inner phono, BC, 5SW, AC $45.00

67L11, Portable, 1948, "Alligator", inner slide rule dial, 2 square knobs, lift-up front, BC, AC/DC/bat $40.00

67X11, Table, 1947, Walnut plastic, slide rule dial, horiz wrap-around louvers, 2 knobs, BC, AC/DC $50.00

67X12, Table, 1947, Ivory plastic, slide rule dial, horiz wrap-around louvers, 2 knobs, BC, AC/DC $50.00

67X13, Table, 1947, Wood, upper front slide rule dial, lower cloth grill, 2 knobs, BC, AC/DC $35.00

67XM21, Table, 1948, Plastic, slide rule dial, horizontal wrap-around louvers, 2 knobs, BC, FM, AC/DC $50.00

68F11, Table-R/P, 1949, Plastic, outer vertical slide rule dial, lift top, inner phono, BC, AC $45.00

68F12, Table-R/P, 1949, Wood, outer front vertical slide rule dial, lift top, inner phono, BC, AC $35.00

68F14, Console-R/P, 1949, Wood, vertical slide rule dial, criss-cross grill, lift top, lower storage, BC, AC $65.00

68L11, Portable, 1948, Plastic coated cloth, dial moves inside handle, thumbwheel knobs, BC, AC/DC/bat $70.00

68T11, Table, 1949, Plastic, upper slanted slide rule dial, large center cloth grill, 2 knobs, AM, AC $40.00

68X11, Table, 1949, Two-tone plastic, slanted slide rule dial, geometric grill, 2 knobs, BC, AC/DC $65.00

68X11Q, Table, 1949, Two-tone plastic, slanted slide rule dial, geometric grill, 2 knobs, BC, AC/DC $65.00

69L11, Portable, 1949, Plastic over cloth, suitcase-style, dial moves inside clear handle, BC, AC/DC/bat $70.00

69X11, Table, 1950, Plastic, large half-moon dial w/stick pointer, woven grill, 2 knobs, BC, AC/DC $40.00

71-A, Table, 1940, Walnut, right slide rule dial, pushbuttons,wrap-around grill bars, 4 knobs, BC, SW $60.00

72XM21, Table, 1952, Plastic, half-moon dial, lower half-moon checkered grill, 2 knobs, BC, AC/DC $25.00

75F21, Console-R/P, 1947, Wood, upper slanted slide rule dial, 4 knobs, pull-out phono drawer, BC, SW, AC $100.00

75F31, Console-R/P, 1947, Wood, inner slide rule dial, 4 knobs, 6 pushbuttons, 2 lift tops, BC, SW, FM, AC $80.00

76F31, Console-R/P, 1948, Wood, inner right slide rule dial & knobs, left phono, fold-down front, BC, FM, AC $75.00

76T1, Portable, 1957, Charcoal, transistor, right round dial, large metal perforated grill, handle, AM, bat $35.00

76T2, Portable, 1957, Brown, transistor, right round dial, large metal perforated grill, handle, AM, bat $35.00

77FM21, Console-R/P, 1948, Wood, step-down top, wrap-over dial, 4 knobs, pull-out phono drawer, BC, FM, AC $90.00

77FM22, Console-R/P, 1949, Wood, outer right vertical dial, 4 knobs, left lift top, inner phono, BC, FM, AC $65.00

77FM22M, Console-R/P, 1948, Wood, step-down top w/slide rule dial, grill & 4 knobs, pull-out phono, BC, FM, AC $90.00

77FM22WM, Console-R/P, 1948, Wood, step-down top w/slide rule dial, grill & 4 knobs, pull-out phono, BC, FM, AC $90.00

77FM23, Console-R/P, 1948, Wood, step-down top w/slide rule dial, grill & 4 knobs, pull-out phono, BC, FM, AC $90.00

77XM21, Table, 1948, Walnut plastic, slide rule dial, horiz wrap-around louvers, 2 knobs, BC, FM, AC/DC $50.00

77XM22, Table, 1948, Walnut, top wrap-over dial, cloth wrap-around grill, 2 knobs, BC, FM, AC/DC $40.00

77XM22B, Table, 1948, Blonde, top wrap-over dial, cloth wrap-around grill, 2 knobs, BC, FM, AC/DC $40.00

78F11, Console-R/P, 1949, Wood, step-down top, slide rule dial, 4 knobs, front pull-out phono drawer, BC, AC $90.00

78F12M, Console-R/P, 1949, Wood, right front slide rule dial, 4 knobs, lift top, inner phono, storage, BC, AC $65.00

78FM21, Console-R/P, 1949, Wood, step-down top, slide rule dial, 4 knobs, front pull-out phono drawer, BC, AC $90.00

78FM22, Console-R/P, 1948, Wood, right vertical dial, left pull-out phono, lower grill, storage, BC, FM, AC $75.00

79FM21R, Console-R/P, 1950, Wood, upper front half-moon dial, 4 knobs, pull-out phono drawer, BC, FM, AC $80.00

79XM21, Table, 1950, Plastic, large half-moon dial w/ stick pointer over grill, 2 knobs, AM, FM, AC/DC $50.00

81C, Console, 1939, Wood, slide rule dial, lower grill w/vert bars, 6 pushbuttons, 8 tubes, BC, SW, AC $135.00

81F21, Console-R/P, 1941, Wood, upper slide rule dial, 5 pushbuttons, pull-out phono drawer, BC, SW, AC $120.00

82A, Console, 1939, Wood, slide rule dial, automatic tuning clock, vert grill bars, 4 knobs, 8 tubes, BC, SW, AC $185.00

85F21, Console-R/P, 1946, Wood, slanted slide rule dial, 4 knobs, 6 pushbuttons, pull-out phono, BC, SW, AC $100.00

85K21, Console, 1946, Wood, slanted slide rule dial, cloth grill, 4 knobs, contrasting veneer, BC, SW, AC $100.00

88FM21, Console-R/P, 1949, Wood, inner right slide rule dial, 4 knobs, left pull-out phono drawer, BC, FM, AC $65.00

95F31, Console-R/P, 1947, Wood, right tilt-out slide rule dial, 4 knobs, inner left phono, BC, SW, FM, AC $75.00

99FM21R, Console-R/P, 1949, Wood, inner right slide rule dial, 4 knobs, left pull-out phono, storage, BC, FM, AC $60.00

107F31, Console-R/P, 1948, Wood, tilt-out dial, 4 knobs, pushbuttons, pull-out phono drawer, BC, SW, FM, AC $80.00

107F31B, Console-R/P, 1948, Wood, tilt-out dial, 4 knobs, pushbuttons, pull-out phono drawer, BC, SW, FM, AC $80.00

496BT-1, Table, 1939, Wood, right front dial, left grill w/horizontal cut-outs, 4 tubes, 3 knobs, battery $40.00

A-1, Portable, 1941, Maroon metal & chrome, set turns on when lid is opened, handle, BC, battery $50.00

B-150, Bike Radio, 1940, Mounts on bike handlebars, horizontal grill bars, separate battery pack $85.00

C5G, Table-C, 1959, Plastic, center round dial, pushbuttons, left alarm clock, right vertical bars, AM $15.00

HS-678, Portable, 1959, Plastic, transistor, upper round dial knob, wrap-around horiz bars, handle, bat $25.00

L12N "Power 8", Portable, 1960, Plastic, transistor, right dial knob, lower horiz bars, left knob, handle, AM, bat..................$20.00

L13S, Portable, 1960, Plastic, transistor, right dial knob, lower horiz bars, left knob, handle, AM, bat $25.00

L14E, Portable, 1960, Plastic, transistor, right dial knob, lower horiz bars, left knob, handle, AM, bat $25.00

X11B, Portable, 1960, Transistor, upper dial, perforated grill w/"M" logo, rear swing-out stand, AM, bat $30.00

X11E, Portable, 1960, Transistor, upper dial, perforated grill w/ "M" logo, rear swing-out stand, AM, bat $30.00

X12A "Power Eight", Portable, 1960, Plastic, transistor, lower right dial, vert grill bars, thumbwheel knobs, AM, bat $15.00

X12A-1, Portable, 1960, Plastic, transistor, lower right dial, vert grill bars, thumbwheel knobs, AM, bat $15.00

X12E-1, Portable, 1960, Plastic, transistor, lower right dial, vert grill bars, thumbwheel knobs, AM, bat $15.00

X14B, Portable, 1960, Blue plastic, transistor, upper left window dial, circular perforated grill, AM, bat $35.00

X14E, Portable, 1960, Black plastic, transistor, upper left window dial, circular perforated grill, AM, bat $30.00

X14R, Portable, 1960, Red plastic, transistor, upper left window dial, circular perforated grill, AM, bat $35.00

X14W, Portable, 1960, White plastic, transistor, upper left window dial, circular perforated grill, AM, bat $30.00

MURDOCK

Wm. J. Murdock Company,
430 Washington Avenue,
Chelsea, Massachusetts

The Wm. J. Murdock Company began business making radio parts and related items and the company produced its first complete radio in 1924. The Murdock model 200, circa 1925, featured a very unusual loud speaker which screwed into the top of the set. By 1928, the company was out of the radio production business.

100, Table, 1924, Wood, low rectangular case, 3 dial front panel, top screw-in loud speaker, bat $175.00

101, Table, 1925, Wood, low rectangular case, 3 dial front panel, 5 tubes, battery $125.00

200 "Neutrodyne", Table, 1925, Wood, low rectangular case, 3 dial panel, top screw-in loud speaker horn, battery $175.00

201, Table, 1925, Wood, rectangular case, 3 dial front panel, 5 tubes, battery $125.00

203, Table, 1925, Wood, low rectangular case, 3 dial front panel, 6 tubes, battery $135.00

204, Console, 1925, Wood, inner front 2 dial panel, fold-down front, upper speaker grill, storage, bat $250.00

CS-32, Table, 1923, Wood, low rectangular case, 3 dial black bakelite front panel, 5 tubes, battery $175.00

MURPHY

G. C. Murphy Co.,
531 5th Ave., McKeesport, Pennsylvania

113, Table, 1946, Wood, slant front, lower dial, upper horizontal louvers, BC, AC/DC $35.00

MUSIC MASTER

Music Master Corporation,
Tenth and Cherry Streets,
Philadelphia, Pennsylvania

The Music Master Corporation was formed in 1924 in the business of manufacturing horns for radios. The company began to produce radios in 1925 but, due to mismanagement and financial difficulties, was out of business by 1926.

60, Table, 1925, Mahogany, low rectangular case, 3 dial front panel, 5 tubes, battery $125.00

100, Table, 1925, Mahogany, low rectangular case, slanted 3 dial front panel, 5 tubes, feet, battery $135.00

140, Table, 1925, Wood, high rectangular case, center front dial over stylized "M" logo, lift top, bat $400.00

175, Table, 1925, Mahogany, rectangular case, 2 dial slanted panel, lower speaker grill, 6 tubes, bat $295.00

215, Console, 1925, Mahogany, upper slanted 2 dial panel, lower grill w/cut-outs, spinet legs, bat $350.00

250, Table, 1925, Mahogany, low rectangular case, front window dial, 7 tubes, battery $195.00

460 Console, 1925, Mahogany, inner slanted 1 dial panel, fold-down front, bowed legs, 7 tubes, bat $265.00

NANOLA

Leopold Sales Corp.,
P.O. Box 276, Highland Park, Illinois

6TP-106, Portable, 1960, Transistor, upper thumbwheel dial knob, lower perforated grill, AM, battery $20.00

NATIONAL UNION

National Union Radio Corp.,
Newark, New Jersey

571, Table, 1947, Wood, upper right slide rule dial, large lattice grill, 2 knobs, BC, AC/DC $30.00

G-613, Portable, 1947, Leatherette, slide rule dial, 2 knobs, 1 switch, cloth grill, handle, BC, AC/DC/bat $30.00

G-619 "Presentation", Table, 1947, Wood, upper slanted slide rule dial, lower cloth grill, 2 knobs, BC, AC/DC $35.00

NAYLOR

Sterling Five, Table, Wood, low rectangular case, 2 dial black bakelite front panel, 5 tubes, battery $125.00

NEC

Kanematsu New York, Inc.,
Import & Export,
150 Broadway, New York, New York

NT-61, Portable, 1960, Transistor, right thumbwheel dial knob under clear plastic, left grill, AM, bat $30.00

NT-620, Portable, 1960, Transistor, right wedge-shaped thumbwheel dial, left perforated grill, AM, bat $30.00

NEUTROWOUND
**Neutrowound Radio Mfg. Co.,
Homewood, Illinois**

1926, Table, 1926, Metal, very low rectangular case w/3 top raised dial areas, 6 tubes w/top caps $300.00

1927, Table, 1927, Metal, very low rectangular case w/3 top raised dial areas, 6 tubes w/top caps $300.00

NORDEN-HAUCK
**Norden-Hauck, Inc.,
1617 Chestnut Street,
Philadelphia, Pennsylvania**

Super-10 "Admiralty", Table, 1926, Rectangular case 36" long, 2 dial front panel, 10 tubes, battery $1,000.00

NORELCO
**North American Phillips Co., Inc.,
230 Duffy Ave.,
Hicksville, New York**

L2X97T, Portable-C, 1960, Transistor, right front round dial, left round clock face, AM, battery $30.00
L3X86T, Portable, 1960, Transistor, slide rule dial, perforated grill, pushbuttons, handle, AM, SW, battery $45.00
L3X88T, Portable, 1960, Transistor, slide rule dial, perforated grill, pushbuttons, handle, AM, SW, battery $45.00
L4X95T, Portable, 1960, Transistor, slide rule dial, perforated grill, pushbuttons, handle, AM, SW, battery $45.00

OLSON
**Olson Radio Corp.,
260 S. Forge St., Akron, Ohio**

RA-315, Portable, 1960, Plastic, transistor, left front dial w/decorative lines, lattice grill, AM, battery $30.00

OLYMPIC
**Olympic Radio & TV/Hamilton Radio,
510 Avenue of the Americas, New York, New York**

6-501, Table, 1946, Plastic, right round slanted dial, left horizontal louvers, 2 knobs, BC, AC/DC $65.00
6-501W, Table, 1946, Plastic, right round slanted dial, left horizontal louvers, 2 knobs, BC, AC/DC $65.00
6-502, Table, 1946, Wood, right slanted panel w/aqua dial, horizontal grill bars, 2 knobs, BC, AC/DC $60.00
6-502P, Table, 1946, Wood, right slanted round black dial, vertical grill bars, 2 knobs, BC, AC/DC $60.00
6-504L, Table-R/P, 1946, Right front round dial, 3 knobs, flip up top covers phono only, BC, AC $20.00
6-601W, Table, 1946, Plastic, upper slide rule dial, lower horizontal grill bars, 4 knobs, BC, SW, AC $40.00
6-604W, Table, 1947, Plastic, slide rule dial, horizontal wrap-around louvers, 4 knobs, BC, SW, AC/DC $40.00
6-606, Portable, 1946, Luggage style, slide rule dial, plastic escutcheon/grill, 2 knobs, BC, AC/DC/bat $35.00

6-606-A, Portable, 1947, Luggage style, slide rule dial, plastic escutcheon/grill, 2 knobs, BC, AC/DC/bat $35.00
6-606-U, Portable, 1947, Luggage style, slide rule dial, plastic escutcheon/grill, 2 knobs, BC, AC/DC/bat $35.00
6-617, Table-R/P, 1946, Wood, front slanted slide rule dial, 4 knobs, horizontal louvers, lift top, BC, AC $35.00
7-421W, Table, 1949, Plastic, slanted round dial, horizontal wrap-around louvers, 2 knobs, BC, AC/DC $65.00
7-435V, Table, 1948, Plastic, slanted right round dial, wrap-around louvers, 3 knobs, BC, SW, AC/DC $70.00
7-526, Portable, 1947, Leatherette, slide rule dial, plastic escutcheon, handle, 2 knobs, BC, AC/DC/bat $35.00
7-532W, Table, 1948, Plastic, slide rule dial, horizontal wrap-around louvers, 4 knobs, BC, FM, AC/DC $45.00
7-537, Table, 1948, Plastic, slanted right round dial, wrap-around louvers, 3 knobs, BC, FM, AC/DC $65.00
7-622, Table-R/P, 1948, Wood, upper front slide rule dial, 4 knobs, lift top, inner phono, BC, AC $30.00
7-638, Console-R/P, 1948, Wood, front slide rule dial/grill/4 knobs, lift top, inner phono, lower storage, AC $60.00
7-724, Console-R/P, 1947, Wood, inner right slide rule dial, 4 knobs, storage, door, left lift top, BC, SW, AC $70.00
7-925, Console-R/P, 1948, Wood, inner right slide rule dial, 4 knobs, pull-out left phono drawer, BC, FM, AC $70.00

8-451, Portable, 1948, Inner right slide rule dial, lattice grill, 2 knobs, flip-up top, handle, BC, battery **$45.00**

8-533W, Table, 1949, Plastic, slide rule dial, horizontal wrap-around louvers, 4 knobs, BC, FM, AC/DC **$45.00**

8-618, Table-R/P, 1948, Wood, upper slanted slide rule dial, 4 knobs, lift top, inner phono, BC, SW, AC **$30.00**

8-934, Console-R/P, 1948, Wood, inner right slide rule dial, 4 knobs, lift top for inner left phono, BC, FM, AC **$70.00**

402, Table-C, 1955, Plastic, right side dial knob over louvers, center front alarm clock, feet, BC, AC **$40.00**

412, Table-R/P, 1959, Two-tone, front dial, 3 knobs, lift top, inner phono, high-fidelity, BC, AC/DC **$20.00**

447, Portable, 1958, Transistor, top dial & knob, front checkerboard grill, handle, AM, battery **$40.00**

450-V, Portable, 1957, Plastic, right front round dial, horizontal grill bars, handle, BC, battery **$30.00**

461, Portable, 1959, Plastic, front dial, perforated grill, handle, telescope antenna, BC, AC/DC/bat **$35.00**

465, Table-C, 1959, Plastic, upper right front round dial, left alarm clock, lattice grill, feet, BC, AC **$20.00**

489, Portable, 1951, Inner slide rule dial, lattice grill, set plays when flip-up front opens, BC, battery **$45.00**

501, Table-R/P, Two-tone wood, right square dial, left grill, 3 knobs, lift top, inner phono, AC **$35.00**

552, Table, 1960, Plastic, raised top w/slide rule dial, lower checkered grill area, 2 knobs, AM **$25.00**

555, Table-C, 1959, Plastic, lower front slide rule dial, upper alarm clock, 4 knobs, feet, BC, AC **$25.00**

666, Portable, 1959, Plastic, transistor, off-center round dial, left diamond pattern grill, AM, battery **$30.00**

689M, Console-R/P, 1959, Wood, inner right dial, left 4 speed phono, lift top, high fidelity, BC, FM, AC **$30.00**

766, Portable, 1959, Leather case, transistor, upper round dial, perforated grill, strap, AM, battery **$25.00**

768, Portable, 1959, Plastic, transistor, right round dial, left checkered grill, handle, AM, battery **$30.00**

770, Portable, 1959, Plastic, transistor, right front round dial, Olympic torch on grill, BC, battery **$35.00**

771, Portable, 1959, Transistor, right round dial, "wishbone" grill, thumbwheel knob, handle, AM, bat **$35.00**

777, Portable, 1960, Transistor, small round window dial, lower perforated grill area, AM, battery **$30.00**

808, Portable, 1960, Transistor, right front dial, large lattice grill area, handle, 2 knobs, AM, battery **$25.00**

FM-15, Table, 1960, Plastic, trapezoid case, slide rule dial over checkered grill, 3 knobs, AM/FM **$25.00**

LP-163, Table, Plastic, right front slanted round dial, horizontal wrap-around louvers, 2 knobs **$65.00**

LP-3244, Table, Plastic, center half-round dial w/inner grill, lower "Olympic" logo, feet, 2 knobs **$40.00**

OPERADIO
The Operadio Corporation,
Chicago, Illinois

1925, Portable, 1925, Leatherette, inner 2 dial panel, removable cover contains antenna, 6 tubes, bat **$250.00**

OZARKA
Ozarka, Inc.,
804 Washington Boulevard,
Chicago, Illinois

Ozarka, Incorporated began manufacturing radios in 1922. Despite the stock market crash, Ozarka's business remained strong, however the company was out of business by 1932.

78 "Armada", Table, Wood, looks like treasure chest, center window dial, 3 knobs, 6 tubes, handles **$325.00**

89, Table, 1928, Metal, low rectangular case, center window dial w/thumbwheel tuning, 9 tubes **$175.00**

95, Cathedral, 1932, Wood, fancy scrolled cabinet, center window dial, upper grill w/cut-outs, feet **$500.00**

Corona, Table, Wood, low case w/curved sides & carvings, center front dial, 6 tubes, 3 knobs, bat **$275.00**

J-1 "Junior", Table, 1925, Wood, slanted top with 5 exposed tubes, lower built-in speaker grill w/cut-outs **$325.00**

Minuet, Console, Wood, upper front dial, large speaker grill w/cut-outs, 6 tubes, 3 knobs, battery **$250.00**

S-1 "Senior", Table, 1925, Wood, high rectangular case, 3 dial slanted front panel, battery **$200.00**

S-5 "Senior", Table, Wood, high rectangular case, slanted 3 dial panel, lift top, 5 tubes, battery $215.00
S-7 "Senior", Table, Wood, high rectangular case, slanted 3 dial panel, lift top, 7 tubes, battery $250.00
V-16, Console, 1932, Wood, large Gothic cabinet w/carvings, upper window dial, 16 tubes, 2 speakers, 4 knobs, AC $395.00

PACKARD-BELL
Packard-Bell Company,
1115 South Hope Street, Los Angeles, California

Packard-Bell began business in 1933, during the Depression and always enjoyed a reputation for good quality radios. Packard-Bell had many government contracts during WW II for electrical equipment. The company was sold in 1971.

5D8, Table, 1948, Plastic, right dial, left horizontal louvers, handle, 2 bullet knobs, BC, AC/DC $45.00
5DA, Table, 1947, Bakelite, right black dial, left horizontal louvers, 2 knobs, handle, BC, AC/DC $45.00
5FP, Table, 1946, Plastic, right front black dial, left horizontal louvers, 2 knobs, BC, AC $40.00
5R, Table, 1957, Plastic, dial numerals over large front checkered grill area, 2 knobs, BC, AC/DC $35.00
5R1, Table, 1957, Plastic, dial numerals over large front checkered grill area, 2 knobs, BC, AC/DC $35.00
5R5, Table, 1959, Plastic, dial numerals over front horizontal bars, large dial pointer w/logo, feet $25.00
6RC1, Table-C, 1958, Plastic, lower front slide rule dial, upper alarm clock, horizontal louvers, BC, AC $20.00
6RT1, Portable, 1959, Leather case, transistor, right dial, left lattice cut-outs, handle, AM, battery $30.00
47, Table, 1935, Wood, center front round dial, right & left wrap-around louvers, 5 knobs, BC, SW $80.00
65A, Table, 1940, Wood, right front square dial, left cloth grill w/ horizontal bars, handle, 3 knobs $50.00
100, Table, 1949, Plastic, right dial, Deco grill w/ horizontal bars, handle, 2 knobs, BC, AC/DC $45.00
100A, Table, 1949, Plastic, right front dial, left grill w/horizontal bars, handle, 2 knobs, BC, AC/DC $45.00
471, Portable, 1947, Luggage-style, leatherette, right dial, 3 knobs, handle, BC, AC/DC/battery $25.00
531, Table, 1954, Plastic, large center front dial over perforated grill, 2 knobs, BC, AC/DC $25.00
551, Table, 1946, Plastic, upper curved slide rule dial, lower vertical grill bars, handle, BC, AC $65.00
561, Table-R/P, 1946, Leatherette, left front dial, right grill, 3 knobs, fold-up lid, BC, AC $25.00
568, Portable-R/P, 1947, Leatherette, inner square black dial, grill, 3 knobs, phono, fold-back top, BC, AC $25.00
572, Table, 1947, Two-tone wood, upper curved dial, horiz louvers, 2 knobs under handle, BC, AC/DC $65.00
621, Table-C, 1952, Plastic, right front square dial, left alarm clock, top grill bars, 5 knobs, BC, AC $25.00
631, Table, 1954, Plastic, large perforated metal front grill w/raised dial numerals, 2 knobs $25.00
651, Table, 1946, Wood, curved slide rule dial, horizontal louvers, 2 knobs under handle, BC, SW, AC $65.00
661, Table-R/P, 1946, Wood, upper front dial, lower grill, 4 knobs, inner phono, 3/4 lift top, BC, AC $25.00
662, Console-R/P, 1947, Wood, desk-style, inner slide rule dial, 4 knobs, phono, fold-up top, BC, AC $100.00
682, Table, 1949, Right front square dial, left criss-cross grill, handle, 3 knobs, BC, AC/DC $45.00
771, Table, 1948, Wood, upper slanted slide rule dial, lower criss-cross grill, 4 knobs, BC, SW, AC $40.00
861 "Phonocord", Console-R/P/Rec/PA, 1947, Wood, inner slide rule dial, 4 knobs, phono, lift top, criss-cross grill, BC, AC $90.00

880, Console-R/P, 1948, Wood, inner right slanted slide rule dial, 4 knobs, left phono, lift-up front, BC, AC $90.00
880-A, Chairside-R/P, 1948, Wood, slide rule dial, 4 knobs, lift top, inner phono, lower storage area, BC, AC $85.00
881-A, Console-R/P/Rec, 1948, Wood, slanted inner right slide rule dial, 4 knobs, left phono, lift-up front, BC, AC $90.00
881-B, Chairside-R/P/Rec, 1948, Wood, slide rule dial, 4 knobs, lift top, inner phono, lower storage area, BC, AC $85.00
884, Table-R/P, 1949, Upper slide rule dial, criss-cross grill, 4 knobs, lift top, inner phono, BC, FM, AC $30.00
892, Console-R/P, 1949, Wood, slide rule dial, 4 knobs, front fold-out phono, criss-cross grill, BC, FM, AC $55.00
1052A, Table-R/P/Rec/PA, 1946, Left front dial, right grill, fold-up top, inner phono, BC, AC $30.00
1054-B, Console-R/P/Rec/PA, 1947, Wood, top slanted slide rule dial, lower pull-out phono drawer, BC, SW, AC $65.00
1063, Console-R/P/PA, 1947, Wood, inner dial, 4 knobs, 6 pushbuttons, phono, lift top, drop front, BC, SW, AC ...$90.00
1181, Console-R/P/Rec, 1949, Wood, inner right slide rule dial, 5 knobs, left phono, hinged lift top, BC, FM, AC $90.00
1273, Console-R/P, 1948, Wood, center front pull-out drawer w/slide rule dial, 5 knobs, phono, BC, FM, AC $80.00
1472, Console-R/P/Rec, 1948, Wood, slide rule dials, 4 knobs, 6 pushbuttons, pull-out phono drawer, BC, FM, AC $80.00

PAN AMERICAN
Pan American Electric Co.,
132 Front Street, New York, New York

"Clock", Table-N, 1946, Wood, arched case looks like mantle clock, round plastic dial, 2 knobs, AC/DC $85.00

PARAGON
Adams-Morgan Co., 16 Alvin Place,
Upper Montclair, New Jersey
The Paragon Electric Corporation,
Upper Montclair, New Jersey

The Adams-Morgan Company began business in 1910 selling wireless parts. By 1916 their first Paragon radio was marketed. After much internal trouble between the company executives leading to the demise of the Adams-Morgan Company and the creation of the Paragon Electric Corporation in 1926, the company finally went out of business completely in 1928.

DA-2, Table, 1921, Wood, detector/2-stage amplifier, black front panel $450.00
RA-10, Table, 1921, Wood, low rectangular case, 2 dial black front panel, battery $550.00

RA10/DA2, Table, 1921, Two rectangular case units, RA 10-receiver & DA 2-two stage amplifier, battery $1,000.00
RB-2, Table, 1923, Mahogany, low rectangular case, 3 dial black front panel, lift top, battery $300.00
RD-5, Table, 1922, Wood, low rectangular case, battery ..$600.00
RD5/A2, Table, 1922, Two rectangular case units, RD5-receiver & A2-two stage amplifier, battery $1,000.00

Two, Table, 1924, Mahogany finish, low rectangular case, 1 center dial, 2 tubes, battery **$300.00**

IIIA, Table, 1924, Wood, inner black bakelite panel, double front doors, lift top, 3 tubes, battery **$325.00**

PARSON'S
Parson's Laboratories, Inc.,
1471 Selby Ave., St Paul, Minnesota

66A, Table, 1932, Wood, right front half-moon dial w/escutcheon, left scalloped grill, 6 tubes, bat **$65.00**

PATHE
Pathe Phonograph & Radio Corp.,
20 Grand Ave., Brooklyn, New York

B-5, Table, 1925, Wood, low rectangular case, slanted 3 dial front panel, 5 tubes, battery **$125.00**

Minute Man, Table, 1925, Wood, low rectangular case, 3 dial front panel, 5 tubes, battery **$125.00**

PEPSI

Pepsi Bottle, Table-N, Plastic, looks like large Pepsi bottle, base, AC **$750.00**

PERRY

3856, Cathedral, 1932, Wood, quarter-round dial w/escutcheon, upper cloth grill w/cut-outs, 3 knobs, AC **$300.00**

PERWAL
Perwal Radio & Television Co.,
140 North Dearborn Street,
Chicago, Illinois

51, Table, 1937, Wood, right front 3 color airplane dial, left cloth grill with Deco cut-outs, AC/DC **$55.00**

52, Table, 1937, Two-tone wood, center front airplane dial, top grill w/cut-outs, 3 knobs, AC/DC **$70.00**

581, Table, 1937, Center front round airplane dial, top grill with cut-outs, 2 knobs, AC/DC **$70.00**

741, Table, 1937, Wood, round dial, tuning eye, vertical grill bars, 7 tubes, 3 knobs, BC, SW, AC/DC **$75.00**

PFANSTIEHL
Pfanstiehl Radio Co.,
11 South La Salle Street, Chicago, Illinois
Fansteel Products Company, Inc.,
North Chicago, Illinois

The Pfanstiehl Radio Company was formed in 1924 and in that same year marketed its first radio. During the late 1920's business gradu-

ally declined and by 1930, the company was out of the radio business.

7 "Overtone", Table, 1924, Wood, low rectangular case, 3 dial front panel, 5 tubes, battery **$110.00**

8, Table, 1924, Wood, low rectangular case, slanted 2 dial front panel, 5 tubes, battery **$100.00**

10 "Overtone", Table, 1924, Wood, low rectangular case, 1 dial front panel, 6 tubes, battery **$95.00**

10-C, Console, 1924, Wood, highboy, inner 1 dial panel, fold-down front, upper speaker, 6 tubes, bat **$185.00**

10-S "Overtone", Console, 1924, Wood, 1 dial front panel & built-in speaker on detachable stand, 6 tubes, battery **$135.00**

18, Table, 1926, Wood, front panel w/half-round double-pointer dial, 3 knobs, 5 tubes, battery **$100.00**

20, Table, 1926, Wood, rectangular case, 1 dial front panel, 6 tubes, battery **$95.00**

50, Table, 1927, Wood, low rectangular case, center window dial w/ escutcheon, 7 tubes, AC **$100.00**

PHILCO
Philadelphia Storage Battery Company,
Ontario & C Streets,
Philadelphia, Pennsylvania
Philco Corp.,
Tioga & C Streets,
Philadelphia, Pennsylvania

Philco began business in 1906 as the Philadelphia Storage Battery Company, a maker of batteries and power supplies. In 1927 the company produced its first radio and grew to be one of the most prolific of all radio manfacturers.

14-X, Console, 1933, Walnut, upper dial, inclined grill w/3 vertical bars, fluted columns, BC, SW, AC **$135.00**

15DX, Console, 1932, Wood, inner dial & controls, bowed tambour doors, lower inclined grill, 11 tubes **$225.00**

15X, Console, 1932, Walnut, upper window dial, lower inclined grill w/vert columns, 11 tubes, AC **$200.00**

16B, Cathedral, 1933, Two-tone wood, center front window dial, upper grill w/cut-outs, BC, SW, AC **$300.00**

16B, Tombstone, 1933, Two-tone wood, center window dial, upper grill w/cut-outs, 11 tubes, BC, SW, AC **$285.00**

16-L, Console, 1933, Wood, lowboy, upper window dial, lower grill w/ cut-outs, 11 tubes, BC, SW, AC **$145.00**

16RX, Chairside-2 piece, 1933, Wood, small table w/dial & knobs, separate speaker w/inclined grill, BC, SW, AC **$300.00**

16-X, Console, 1933, Wood, upper window dial, lower grill w/vert bars, 11 tubes, 4 knobs, BC, SW, AC **$145.00**

17, Cathedral, Wood, center front window dial, upper cloth grill w/cut-outs, 11 tubes, 4 knobs **$275.00**

17-D, Console, 1933, Wood, highboy, inner front dial & controls, double doors, 6 legs, BC, SW, AC **$165.00**

18-B, Cathedral, 1933, Two-tone wood, center window dial, 3-section grill, 8 tubes, 4 knobs, BC, AC **$325.00**

18-H, Console, 1933, Wood, lowboy, upper window dial, lower grill w/cut-outs, 8 tubes, 6 legs, BC, AC **$150.00**

18-L, Console, 1933, Wood, lowboy, upper front window dial, lower cloth grill w/cut-outs, BC, SW, AC **$130.00**

18MX, Console, 1934, Mahogany w/black, upper window dial, lower grill w/2 vert bars, 8 tubes, BC, AC **$135.00**

19, Cathedral, Wood, recessed panel w/window dial, upper grill w/cut-outs, 6 tubes, 4 knobs **$225.00**

19-H, Console, 1933, Wood, lowboy, upper front window dial, lower cloth grill w/cut-outs, 6 legs, AC **$145.00**

20, Cathedral, 1930, Wood, center window dial, upper grill w/ scrolled cut-outs, 7 tubes, 3 knobs, AC **$300.00**

20, Console, 1930, Wood, small case, lowboy, upper window dial, lower cloth grill, 7 tubes, 3 knobs **$125.00**

20 Deluxe, Cathedral, 1931, Two-tone wood, window dial, upper cloth grill w/scrolled cut-outs, 3 knobs **$300.00**

20B, Cathedral, 1930, Wood, window dial, upper cloth grill w/scrolled cut-outs, 7 tubes, 3 knobs, AC **$275.00**

21B, Cathedral, 1930, Wood, center window dial, upper grill w/cut-outs, fluted columns, 7 tubes, 3 knobs **$325.00**

22L, Console-R/P, 1932, Wood, lowboy, window dial, grill cut-outs, lift top, inner phono, 7 tubes, 6 legs, AC **$145.00**

28C, Table, 1934, Wood, window dial, right & left "butterfly" grills, 6 tubes, 4 knobs, BC, SW, AC/DC **$150.00**

28L, Console, 1934, Wood, lowboy, window dial, grill w/cut-outs, 6 tubes, 4 knobs, BC, SW, AC/DC **$140.00**

29TX, Table-2 piece, 1934, Wood, table set w/dial & knobs, separate speaker cabinet w/cut-outs, BC, SW, AC **$200.00**

29X, Console, 1934, Wood, window dial, inclined grill w/3 vert columns, 6 tubes, 4 knobs, BC, SW, AC **$150.00**

32L, Console, 1934, Wood, lowboy, upper window dial, grill w/cut-outs, 6 tubes, 4 knobs, BC, battery **$100.00**

34L, Console, 1934, Wood, lowboy, window dial, lower grill w/cut-outs, 7 tubes, 4 knobs, BC, SW, bat **$100.00**

37-9, Console, 1937, Two-tone wood, round automatic tuning dial, grill w/3 vertical bars, 9 tubes **$175.00**

37-10, Console, 1937, Two-tone wood, round automatic tuning dial, grill w/3 vertical bars, 9 tubes **$175.00**

37-11, Console, 1937, Two-tone wood, round automatic tuning dial, grill w/4 vertical bars, 10 tubes **$175.00**

37-33, Cathedral, 1937, Two-tone wood, lower round dial, upper grill w/vertical cut-outs, 5 tubes, bat **$100.00**

37-33, Console, 1937, Wood, upper round dial, lower cloth grill w/cut-outs, 5 tubes, BC, battery **$100.00**

37-34, Cathedral, 1937, Wood, center front round dial, upper grill w/cut-outs, 5 tubes, 4 knobs, BC, bat **$100.00**

37-34, Console, 1937, Wood, upper round dial, lower cloth grill w/cut-outs, 5 tubes, BC, battery **$100.00**

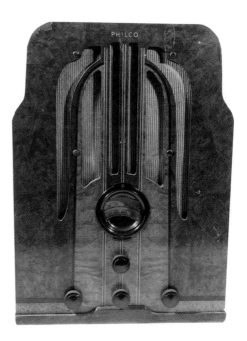

37-38, Tombstone, 1937, Wood, shouldered, center round dial, upper grill w/cut-outs, 4 knobs, battery **$80.00**

37-38, Console, 1937, Wood, upper round dial, lower cloth grill w/3 vert bars, 6 tubes, BC, SW, battery **$100.00**

37-60, Cathedral, 1937, Wood, center front round dial, upper grill w/cut-outs, 5 tubes, 4 knobs, BC, SW, AC **$160.00**

37-60, Console, 1937, Wood, upper round dial, lower cloth grill w/cut-outs, 5 tubes, BC, SW, AC **$130.00**

37-61, Tombstone, 1937, Wood, center front round dial, upper cloth grill w/cut-outs, BC, SW, AC **$130.00**

37-61, Cathedral, 1937, Wood, center front round dial, upper grill w/ cut-outs, 5 tubes, 4 knobs, BC, SW, AC **$160.00**
37-61, Console, 1937, Wood, upper round dial, lower cloth grill w/cut-outs, 5 tubes, BC, SW, AC **$130.00**

37-62, Table, 1937, Two-tone wood, right front round dial, left cloth grill w/Deco cut-outs, 3 knobs $75.00
37-63, Table, 1937, Wood, right front round dial, left cloth grill w/Deco cut-outs, 3 knobs **$80.00**
37-84, Cathedral, 1937, Wood, lower round dial, upper grill w/vertical cut-outs, 4 tubes, 2 knobs, AC **$145.00**
37-89, Cathedral, 1937, Wood, center front round dial, upper grill w/ cut-outs, 6 tubes, 4 knobs, BC, AC **$150.00**
37-89, Console, 1937, Wood, upper round dial, lower cloth grill w/cut-outs, 6 tubes, BC, AC **$130.00**
37-93, Cathedral, 1937, Wood, lower round dial, upper cloth grill w/ vertical cut-outs, 5 tubes, 2 knobs **$135.00**
37-116, Console, 1937, Wood, automatic tuning dial, 5 vert grill bars, lower front cut-outs, 15 tubes, BC, SW **$250.00**
37-600, Table, 1937, Wood, round dial, Deco grill cut-outs, finished front & back, 2 knobs, top grill bars **$75.00**

37-602, Table, 1937, Wood, round dial, Deco grill cut-outs, finished front & back, 2 knobs, top grill bars...........$75.00
37-604, Table, 1937, Wood, dial & knobs on top of case, front round grill w/horizontal bars, AC/DC **$110.00**
37-610, Tombstone, 1937, Wood, 2 styles, center round dial, upper grill w/cut-outs, 4 knobs, BC, SW, AC **$135.00**
37-610, Console, 1937, Wood, upper round dial, lower grill w/3 vert bars, 5 tubes, 4 knobs, BC, SW, AC **$125.00**
37-610, Table, 1937, Wood, Deco, off-center round dial, rounded left grill w/Deco cut-outs, BC, SW, AC **$125.00**
37-611, Tombstone, 1937, Wood, center round dial, upper grill w/ Deco cut-outs, 4 knobs, BC, SW, AC/DC **$135.00**

37-611, Console, 1937, Wood, upper round dial, lower cloth grill w/ cut-outs, 5 tubes, BC, SW, AC/DC **$130.00**
37-611, Table, 1937, Wood, off-center round dial, rounded left grill w/ Deco cut-outs, BC, SW, AC/DC **$125.00**

37-620, Tombstone, 1937, Two-tone wood, various styles, center front dial, grill cut-outs, 4 knobs, BC, 2SW, AC $130.00
37-620, Console, 1937, Wood, various styles, upper round dial, lower cloth grill, 6 tubes, BC, SW, AC **$130.00**
37-623, Tombstone, 1937, Two-tone wood, rounded shoulders, round dial, grill cut-outs, 4 knobs, BC, SW, bat **$85.00**
37-623, Console, 1937, Wood, upper round dial, lower cloth grill w/ 3 vert bars, 6 tubes, BC, SW, battery **$100.00**
37-624, Tombstone, 1937, Two-tone wood, rounded shoulders, round dial, grill cut-outs, 4 knobs, BC, SW, bat **$85.00**
37-624, Console, 1937, Wood, upper round dial, lower cloth grill w/ 3 vert bars, 6 tubes, BC, SW, battery **$100.00**

37-630, Console, 1937, Wood, recessed front, upper round dial, lower cloth grill w/3 vert bars, 4 knobs, BC, SW, AC $135.00

37-630, Table, 1937, Wood, Deco, right round dial, left cloth grill w/ Deco cut-outs, 6 tubes, 4 knobs, BC, SW, AC **$90.00**

37-640, Tombstone, 1937, Wood, rounded shoulders, round dial, vert grill cut-outs, 7 tubes, 4 knobs, BC, SW, AC **$125.00**

37-640, Console, 1937, Wood, upper round dial, lower grill w/vertical cut-outs, 7 tubes, BC, SW, AC **$135.00**

37-641, Tombstone, 1937, Wood, rounded shoulders, round dial, vert grill cut-outs, 4 knobs, BC, SW, AC/DC **$125.00**

37-641, Console, 1937, Wood, round dial, lower grill w/vert cut-outs, 6 tubes, 4 knobs, BC, SW, AC/DC **$125.00**

37-643, Tombstone, 1937, Wood, rounded shoulders, center front round dial, grill cut-outs, 4 knobs, battery **$95.00**

37-650, Tombstone, 1937, Wood, various styles, center dial, grill cut-outs, 9 tubes, 4 knobs, BC, SW, AC **$135.00**

37-650, Console, 1937, Wood, upper round dial, lower cloth grill w/ 3 vertical bars, 8 tubes, BC, SW, AC **$130.00**

37-660, Console, 1937, Wood, upper round dial, lower cloth grill w/ cut-outs, 9 tubes, 6 feet, BC, SW, AC **$150.00**

37-660, Tombstone, 1937, Wood, rounded shoulders, round dial, grill cut-outs, 9 tubes, 4 knobs, BC, SW, AC **$135.00**

37-665, Tombstone, 1937, Wood, rounded shoulders, center round dial, grill cut-outs, 9 tubes, 4 knobs, BC, 2SW, AC **$135.00**

37-670, Tombstone, 1937, Wood, center front round dial, upper grill w/vertical bars, 11 tubes, BC, SW, AC **$150.00**

37-670, Console, 1937, Wood, upper round dial, lower cloth grill w/ 4 vertical bars, 11 tubes, BC, SW, AC **$175.00**

37-675, Console, 1937, Wood, round automatic tuning dial, grill cut-outs, 12 tubes, 5 knobs, BC, SW, AC **$200.00**

37-690, Console, 1937, Wood, inner automatic dial, grill cut-outs, doors, 20 tubes, 3 speakers, BC, SW, AC **$500.00**

37-2620, Console, 1937, Wood, upper round dial, lower cloth grill w/ 3 vert bars, 6 tubes, BC, SW, LW **$125.00**

37-2650, Tombstone, 1937, Wood, rounded shoulders, center round dial, grill cut-outs, 4 knobs, BC, SW, LW, AC **$125.00**

37-2650, Console, 1937, Wood, upper round dial, lower cloth grill w/ 3 vert bars, 8 tubes, BC, SW, LW, AC **$135.00**

37-2670, Tombstone, 1937, Wood, center front round dial, upper grill w/vert bars, 11 tubes, BC, SW, LW, AC **$150.00**

37-2670, Console, 1937, Wood, upper round dial, lower cloth grill w/ 4 vert bars, 11 tubes, BC, SW, LW, AC **$175.00**

38-1, Console, 1938, Wood, slanted front w/automatic tuning dial, grill w/5 vertical bars, 12 tubes, BC, SW, AC **$190.00**

38-2, Console, 1938, Wood, slanted front w/automatic tuning dial, lower grill w/cut-outs, 11 tubes, BC, SW, AC **$175.00**

38-3, Console, 1938, Wood, slanted front w/automatic tuning dial, lower grill w/4 vert bars, 9 tubes, BC, SW, AC **$150.00**

38-4, Console, 1938, Wood, slanted front w/automatic tuning dial, lower vert grill bars, 8 tubes, BC, SW, AC **$150.00**

38-5, Tombstone, 1938, Two-tone wood, rounded shoulders, center round dial, grill w/3 vertical bars, 8 tubes, BC, SW, AC **$145.00**

38-5, Console, 1938, Wood, upper round dial, lower grill w/3 vertical bars, 8 tubes, 4 knobs, BC, SW, AC **$135.00**

38-7, Chairside, 1938, Wood, Deco, top dial & knobs, lower round grill w/horiz bars, storage, BC, SW, AC **$110.00**

38-7, Table, 1938, Wood, right round dial, rounded left side, cloth grill w/Deco cut-outs, BC, SW, AC **$85.00**

38-7, Console, 1938, Wood, slanted front w/automatic tuning dial, lower grill w/vert bars, BC, SW, AC **$125.00**

38-8, Console, 1938, Wood, upper round dial, lower cloth grill w/3 vertical bars, 4 knobs, BC, SW, AC **$125.00**

38-9, Table, 1938, Wood, right round dial, rounded left side, cloth grill w/Deco cut-outs, 4 knobs, BC, SW, AC $80.00

38-9, Console, 1938, Two-tone wood, upper dial, lower grill w/cut-outs, 6 tubes, 4 knobs, BC, SW, AC **$135.00**

38-10, Table, 1938, Wood, right round dial, rounded left side, cloth grill w/Deco cut-outs, 4 knobs, BC, SW, AC **$100.00**

38-10, Console, 1938, Wood, upper round dial, lower grill w/2 vertical bars, 5 tubes, 4 knobs, BC, SW, AC **$130.00**

38-12, Table, 1938, Wood, 2 versions-one w/left wrap-around grill; one w/left front grill, BC, AC/DC **$50.00**

38, Cathedral, 1930, Two-tone wood, center front window dial, upper grill w/cut-outs, 4 knobs, bat $110.00

38, Console, 1933, Wood, lowboy, upper window dial, lower cloth grill w/cut-outs, 4 knobs, battery **$120.00**

38-14, Table, 1938, Plastic, right front dial, left wrap-around cloth grill w/2 horiz bars, 3 knobs, AC $45.00

38-15, Table, 1938, Wood, right dial, left wrap-around grill w/Deco cut-outs, 3 knobs, BC, SW, AC $50.00

38-22, Chairside, 1938, Wood, Deco, top dial & knobs, lower front round grill w/horiz bars, BC, SW, AC/DC $110.00

38-22, Table, 1938, Wood, right round dial, rounded left side, grill w/ Deco cut-outs, BC, SW, AC/DC $75.00

38-22, Console, 1938, Wood, slanted front w/automatic tuning dial, lower grill w/vert bars, BC, SW, AC/DC $125.00

38-23, Console, 1938, Two-tone wood, 2 styles, upper round dial, lower grill, 4 knobs, BC, SW, AC/DC $135.00

38-23, Table, 1938, Wood, round dial, rounded left side, Deco grill cut-outs, 4 knobs, BC, SW, AC/DC $95.00

38-31, Console, 1938, Wood, slanted slide rule dial, pushbuttons, lower grill w/3 vertical bars, BC, SW $130.00

38-33, Tombstone, 1938, Wood, lower rectangular dial, upper grill cut-outs, 5 tubes, 2 knobs, BC, battery $80.00

38-33, Console, 1938, Wood, upper dial, lower cloth grill w/2 vertical bars, 5 tubes, 2 knobs, BC, bat $95.00

38-34, Tombstone, 1938, Wood, center front round dial, grill cut-outs, 5 tubes, 3 knobs, BC, battery $80.00

38-34, Console, 1938, Wood, upper dial, lower cloth grill w/2 vertical bars, 5 tubes, 2 knobs, BC, bat $95.00

38-35, Tombstone, 1938, Two-tone wood, lower dial, upper cloth grill w/cut-outs, 2 knobs, bat/AC $80.00

38-38, Console, 1938, Wood, 2 styles, upper round dial, lower grill w/ bars, 6 tubes, BC, SW, battery $100.00

38-38, Table, 1938, Wood, right round dial, rounded left w/Deco grill cut-outs, 4 knobs, BC, SW, bat....................................$60.00

38-39, Table, 1938, Wood, right round dial, rounded left w/Deco grill cut-outs, 4 knobs, BC, SW, bat....................................$50.00

38-40, Table, 1938, Wood, Deco, right front round dial, left cloth grill w/Deco cut-outs, 4 knobs $65.00

38-60, Tombstone, 1938, Two-tone wood, center front round dial, upper 2-section grill, 4 knobs, BC, SW $130.00

38-60, Console, 1938, Wood, upper round dial, lower cloth grill w/2 vertical bars, 4 knobs, BC, SW, AC $130.00

38-62, Console, 1938, Wood, upper front round dial, lower cloth grill w/3 vertical bars, 3 knobs, BC, AC $130.00

38-62, Table, 1938, Two-tone wood, right dial, left cloth grill with Deco cut-outs, 3 knobs, BC, AC $65.00

38-89, Tombstone, 1938, Two-tone wood, center round dial, upper 2-section grill, 6 tubes, 4 knobs, BC $130.00

38-89, Console, 1938, Two-tone wood, upper round dial, lower grill w/cut-outs, 4 knobs, BC, SW, AC $135.00

38-90, Table, 1938, Wood, Deco, right front round dial, left cloth grill with Deco cut-outs, 4 knobs $75.00

38-93, Tombstone, 1938, Two-tone wood, rounded shoulders, round dial, grill cut-outs, 2 knobs, BC, AC $110.00

38-116, Console, 1938, Wood, slanted front w/automatic tuning dial, vert grill bars, 15 tubes, BC, SW, AC $200.00

38-620, Table, 1938, Wood, Deco, right front round dial, rounded left w/Deco grill cut-outs, 4 knobs $100.00

38-623, Console, 1938, Wood, upper round dial, lower cloth grill w/ cut-outs, 6 tubes, 4 knobs, BC, SW, bat $110.00

38-623, Table, 1938, Wood, Deco, right round dial, rounded left w/ Deco grill cut-outs, BC, SW, battery $75.00

38-643, Tombstone, 1938, Wood, rounded shoulders, round dial, upper grill cut-outs, 4 knobs, BC, SW, bat $100.00

38-643, Console, 1938, Wood, upper round dial, lower grill w/3 vertical bars, 7 tubes, BC, SW, battery $110.00

38-690, Console, 1938, Wood, automatic tuning, lower grill w/cut-outs, vert front bars, 20 tubes, 3 speakers, BC, SW, AC $450.00

38-2670, Tombstone, 1938, Wood, center round dial, upper cloth grill w/vertical cut-outs, 11 tubes, 4 knobs $150.00

38L, Console, 1933, Wood, lowboy, upper window dial, lower grill w/ cut-outs, 5 tubes, 4 knobs, BC, bat............................$125.00

38-TP51 "Transitone", Table, 1938, Plastic, right front dial, upper pushbuttons, left grill w/horiz bars, 2 knobs, BC $60.00

39-6, Table, 1939, Walnut w/inlay, right front dial, left cloth grill w/ Deco cut-outs, 2 knobs, BC $40.00

39-7, Table, 1939, Walnut, 2 styles, right dial, pushbuttons, left grill w/Deco bars, 2 knobs, BC $50.00

39-17, Console, 1939, Wood, upper dial, 6 pushbuttons, large lower cloth grill w/3 vertical bars, BC, AC $135.00

39-18, Console, 1939, Wood, upper dial, 6 pushbuttons, lower cloth grill w/3 vertical bars, BC, AC/DC $135.00

39-19, Console, 1939, Wood, upper dial, 6 pushbuttons, lower cloth grill w/3 vertical bars, BC, SW, AC $135.00

39-25, Table, 1939, Wood, slant front, right slide rule dial, pushbuttons, left Deco grill, BC, SW, AC $90.00

39-25, Console, 1939, Wood, slanted slide rule dial, pushbuttons, lower grill w/vert bars, BC, SW, AC $120.00

39-30, Table, 1939, Wood, slant front, right slide rule dial, pushbuttons, left Deco grill, BC, SW, AC $90.00

39-31, Console, 1939, Walnut, 2 styles, slanted slide rule dial, pushbuttons, lower grill w/vertical bars, BC, SW $145.00

39-35, Console, 1939, Wood, slanted slide rule dial, 4 vert grill bars, pushbuttons, 6 tubes, BC, SW, AC $145.00

39-36, Console, 1939, Walnut, slanted slide rule dial, pushbuttons, vertical grill bars, 6 tubes, BC, SW $135.00

39-40, Console, 1939, Walnut, slanted slide rule dial, pushbuttons, vert grill bars, 8 tubes, BC, SW, AC $150.00

39-45, Console, 1939, Walnut, inner dial, pushbuttons, fold-up front, vert grill bars, 9 tubes, BC, SW, AC $155.00

39-55, Console, 1939, Wood, inner slide rule dial, fold-up front, vert grill bars, remote control, 10 tubes, BC, SW, AC $195.00

39-70, Tombstone, 1939, Wood, rounded shoulders, lower front square dial, grill cut-outs, 2 knobs, battery **$100.00**

39-71, Portable, 1939, Striped cloth covered, right front dial, left grill, leather handle, 2 knobs, bat **$25.00**

39-72, Portable, 1939, Striped cloth covered, right front dial, left grill, 2 knobs, handle, battery **$25.00**

40-130, Table, 1940, Wood, upper slide rule dial, lower grill w/ diagonal cut-outs, 4 knobs, BC, AC $50.00

40-135, Table, 1940, Wood, slide rule dial, lower grill w/diagonal cut-outs, 6 pushbuttons, 4 knobs, BC, AC **$60.00**

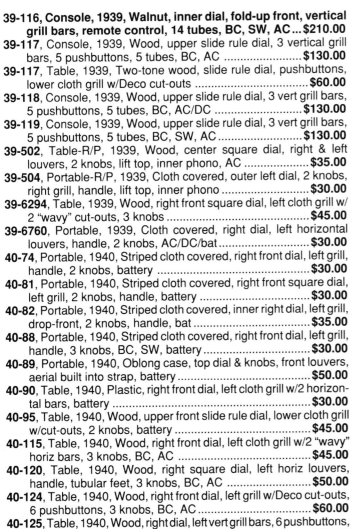

39-116, Console, 1939, Walnut, inner dial, fold-up front, vertical grill bars, remote control, 14 tubes, BC, SW, AC ... $210.00

39-117, Console, 1939, Wood, upper slide rule dial, 3 vertical grill bars, 5 pushbuttons, 5 tubes, BC, AC **$130.00**

39-117, Table, 1939, Two-tone wood, slide rule dial, pushbuttons, lower cloth grill w/Deco cut-outs **$60.00**

39-118, Console, 1939, Wood, upper slide rule dial, 3 vert grill bars, 5 pushbuttons, 5 tubes, BC, AC/DC **$130.00**

39-119, Console, 1939, Wood, upper slide rule dial, 3 vert grill bars, 5 pushbuttons, 5 tubes, BC, SW, AC **$130.00**

39-502, Table-R/P, 1939, Wood, center square dial, right & left louvers, 2 knobs, lift top, inner phono, AC **$35.00**

39-504, Portable-R/P, 1939, Cloth covered, outer left dial, 2 knobs, right grill, handle, lift top, inner phono **$30.00**

39-6294, Table, 1939, Wood, right front square dial, left cloth grill w/ 2 "wavy" cut-outs, 3 knobs **$45.00**

39-6760, Portable, 1939, Cloth covered, right dial, left horizontal louvers, handle, 2 knobs, AC/DC/bat............................**$30.00**

40-74, Portable, 1940, Striped cloth covered, right front dial, left grill, handle, 2 knobs, battery **$30.00**

40-81, Portable, 1940, Striped cloth covered, right front square dial, left grill, 2 knobs, handle, battery **$30.00**

40-82, Portable, 1940, Striped cloth covered, inner right dial, left grill, drop-front, 2 knobs, handle, bat **$35.00**

40-88, Portable, 1940, Striped cloth covered, right front dial, left grill, handle, 3 knobs, BC, SW, battery................................**$30.00**

40-89, Portable, 1940, Oblong case, top dial & knobs, front louvers, aerial built into strap, battery **$50.00**

40-90, Table, 1940, Plastic, right front dial, left cloth grill w/2 horizontal bars, battery **$30.00**

40-95, Table, 1940, Wood, upper front slide rule dial, lower cloth grill w/cut-outs, 2 knobs, battery **$45.00**

40-115, Table, 1940, Wood, right front dial, left cloth grill w/2 "wavy" horiz bars, 3 knobs, BC, AC **$45.00**

40-120, Table, 1940, Wood, right square dial, left horiz louvers, handle, tubular feet, 3 knobs, BC, AC **$50.00**

40-124, Table, 1940, Wood, right front dial, left grill w/Deco cut-outs, 6 pushbuttons, 3 knobs, BC, AC **$60.00**

40-125, Table, 1940, Wood, right dial, left vert grill bars, 6 pushbuttons, handle, 3 knobs, BC, AC/DC **$65.00**

40-140, Table, 1940, Wood, upper slide rule dial, lower grill w/ Deco cut-outs, 4 knobs, BC, SW, AC $50.00

40-145, Table, 1940, Wood, slide rule dial, grill w/Deco cut-outs, 6 pushbuttons, 4 knobs, BC, SW, AC $60.00

40-150, Table, 1940, Wood, Deco, slanted front, slide rule dial, grill cut-outs, 8 pushbuttons, BC, SW, AC **$90.00**

40-155, Table, 1940, Wood, slant front, slide rule dial, Deco grill, 8 pushbuttons, 4 knobs, BC, SW, AC **$90.00**

40-158, Console, 1940, Wood, upper dial, lower grill w/vertical bars, 6 pushbuttons, 6 tubes, 4 knobs, BC **$135.00**

40-160, Console, 1940, Wood, upper dial, lower grill w/vertical bars, 6 pushbuttons, 6 tubes, 4 knobs, BC **$135.00**

40-170, Chairside, 1940, Wood, all sides finished, top dial/knobs/6 pushbuttons, horiz grill bars, BC, AC **$145.00**

40-180, Console, 1940, Wood, slanted slide rule dial, 8 pushbuttons, vert grill bars, 7 tubes, BC, SW, AC .$140.00

40-185, Console, 1940, Wood, slanted slide rule dial, 4 vert grill bars, 8 pushbuttons, 8 tubes, BC, SW, AC **$140.00**

40-190, Console, 1940, Wood, slanted slide rule dial, vert grill bars, 8 pushbuttons, 8 tubes, BC, SW, AC **$145.00**

40-195, Console, 1940, Wood, slanted slide rule dial, thumbwheel knobs, 8 pushbuttons, 10 tubes, BC, SW, AC **$160.00**

40-200, Console, 1940, Wood, inner dial/pushbuttons, fold-up front, vert grill bars, 11 tubes, BC, SW, AC **$175.00**

40-201, Console, 1940, Wood, inner slide rule dial, pushbuttons, fold-up front, front cut-outs, BC, SW, AC **$200.00**

40-205, Console, 1940, Walnut, inner dial, fold-up front, vert grill bars, remote control, 12 tubes, BC, AC **$200.00**

40-215, Console, 1940, Walnut, inner dial, fold-up front, vertical grill bars, remote control, BC, SW, AC **$200.00**

40-216, Console, 1940, Walnut, inner dial, fold-up front, vertical grill bars, remote control, 13 tubes, BC, SW, AC **$210.00**

40-300, Console, 1940, Wood, inner slide rule dial, 7 pushbuttons, fold-up front, 12 tubes, BC, SW, AC **$175.00**

40-501, Table-R/P, 1940, Two-tone wood, right front dial, left grill w/ cut-outs, open top phono, BC, AC **$40.00**

40-504, Portable-R/P, 1940, Cloth covered, left front dial, right grill, lift top, inner crank phono, battery **$35.00**

40-525, Console-R/P, 1940, Wood, front dial, pushbuttons, lower grill w/3 vertical bars, lift top, inner phono...................**$120.00**

41-65, Console, 1941, Wood, upper slide rule dial, 6 pushbuttons, lower grill w/3 vert bars, 4 knobs **$135.00**

41-87PT, Portable, 1941, Right front dial, left horizontal louvers, 2 knobs, handle, AC/DC/battery **$25.00**

41-95, Table, 1941, Two-tone wood, slanted slide rule dial, lower vertical grill bars, 2 knobs, bat **$45.00**

41-100, Table, 1941, Wood, upper slide rule dial, lower 3 section grill, 6 pushbuttons, 2 knobs, BC, bat **$45.00**

41-220, Table, 1941, Wood, right front dial, left grill, handle, feet, 6 tubes, 3 knobs, BC, AC/DC **$45.00**

41-221, Table, 1941, Wood w/ivory trim, right dial, left horiz louvers, handle, feet, 3 knobs, BC, SW, AC/DC **$45.00**

41-225, Table, 1941, Two-tone wood, right square dial, 6 pushbuttons, left grill, 3 knobs, BC, AC/DC **$55.00**

41-226, Table, 1941, Wood, Deco, right dial, upper pushbuttons, left louvers, fluted left side, BC, SW, AC $125.00

41-230, Table, 1941, Brown plastic, upper slanted slide rule dial, lower cloth grill, 4 knobs, BC, AC **$50.00**

41-231, Table, 1941, Wood, Deco, right dial, pushbuttons, horiz louvers, round left side, BC, SW, AC/DC $75.00

41-235, Table, 1941, Wood, upper slanted slide rule dial, lower cloth grill, plastic trim, 4 knobs, BC **$45.00**

41-240, Table, 1941, Wood, upper slanted slide rule dial, lower grill w/horiz bars, 4 knobs, BC, SW **$50.00**

41-245, Table, 1941, Wood, slanted slide rule dial, 6 pushbuttons, grill w/horiz bars, 4 knobs, BC, SW **$60.00**

41-246, Table, 1941, Wood, slanted slide rule dial, 6 pushbuttons, grill w/horiz bars, 4 knobs, BC, SW **$65.00**

41-250, Table, 1941, Wood, Deco, slanted front, slide rule dial, 8 pushbuttons, horiz louvers, BC, SW **$100.00**

41-255, Table, 1941, Wood, Deco, slanted front, slide rule dial, pushbuttons, horiz louvers, BC, SW **$100.00**

41-256, Table, 1941, Wood, Deco, slanted front, slide rule dial, pushbuttons, horiz louvers, BC, SW **$100.00**

41-258, Console, 1941, Wood, upper front dial, lower cloth grill with 4 vertical bars, 6 tubes, BC, AC **$125.00**

41-260, Console, 1941, Wood, upper slide rule dial, pushbuttons, vertical grill bars, 7 tubes, BC, SW, AC **$130.00**

41-265, Console, 1941, Wood, upper slide rule dial, pushbuttons, vertical grill bars, 7 tubes, BC, SW **$125.00**

41-280, Console, 1941, Wood, slanted slide rule dial, 8 pushbuttons, vertical grill bars, 8 tubes, BC, SW **$140.00**

41-285, Console, 1941, Wood, slanted slide rule dial, 8 pushbuttons, vertical grill bars, 9 tubes, BC, SW ..$135.00

41-287, Console, 1941, Wood, inner slide rule dial, pushbuttons, fold-up front, grill cut-outs, 9 tubes, BC, SW, AC **$150.00**

41-290, Console, 1941, Wood, slanted slide rule dial, 8 pushbuttons, vertical grill bars, 10 tubes, BC, SW. $140.00

41-295, Console, 1941, Walnut, slanted slide rule dial, 8 pushbuttons, vert grill bars, 11 tubes, BC, SW **$165.00**

41-296, Console, 1941, Walnut, slanted slide rule dial, 8 pushbuttons, vertical grill bars, 9 tubes, BC, SW **$150.00**

41-300, Console, 1941, Walnut, slide rule dial, pushbuttons, fold-up front, vert front panels, 12 tubes, BC, SW, AC **$185.00**

41-315, Console, 1941, Walnut, inner slide rule dial, pushbuttons, fold-up front, lower vert bars, 12 tubes, BC, SW **$185.00**

41-316, Console, 1941, Wood, inner slide rule dial, fold-up front, vert grill bars, remote control, 15 tubes, BC, SW **$215.00**

41-601, Table-R/P, 1941, Wood, front slide rule dial, horizontal louvers, lift top, inner phono, BC, AC **$35.00**

41-603, Table-R/P, 1941, Wood, center front square dial, right & left horiz grill bars, lift top, inner phono **$35.00**

41-608, Console-R/P/Rec, 1941, Wood, slide rule dial, 6 pushbuttons, tilt-out front, inner phono/recorder, BC, SW, AC **$135.00**

41-625, Console-R/P, 1941, Wood, rectangular dial, tilt-out front w/ vertical bars, inner phono, BC, SW, AC **$140.00**

41-842, Portable, 1941, Leatherette, right front dial, left horizontal louvers, handle, 2 knobs, AC/DC/bat **$25.00**

41-843, Portable, 1941, Leatherette, inner right dial, horiz louvers, fold-down front, handle, AC/DC/bat **$35.00**

41-844, Portable, 1941, Inner right dial, left horiz louvers, 2 knobs, tambour door, handle, AC/DC/bat **$40.00**

41-851, Portable, 1941, Leatherette, right front dial, left horizontal louvers, 3 knobs, BC, SW, AC/DC/bat **$25.00**

41-854, Portable, 1941, Leatherette, inner dial, horiz louvers, tambour door, handle, BC, SW, AC/DC/bat **$40.00**

42-1-T-96 "Transitone" , Table, 1942, Wood, right front dial w/red pointer, left square cloth grill, 2 knobs $45.00

42-100, Console, 1942, Wood, inner slide rule dial, pushbuttons, slide-down cover, lower vert grill bars $165.00

42-321, Table, 1942, Wood, upper slanted slide rule dial, lower rectangular cloth grill, 2 knobs, feet **$40.00**

42-322, Table, 1942, Wood, upper front slanted slide rule dial, lower horiz louvers, 3 knobs, BC, SW **$45.00**

42-327, Table, 1942, Two-tone wood, slide rule dial, 2 horiz grill bars, 6 pushbuttons, 3 knobs, BC, SW, AC/DC **$60.00**

42-340, Table, 1942, Two-tone wood, upper slanted slide rule dial, lower rectangular grill, 4 knobs **$50.00**
42-345, Table, 1942, Wood, upper slide rule dial, lower cloth grill, 6 pushbuttons, 4 knobs, BC, SW, AC **$50.00**

42-350, Table, 1942, Wood, slide rule dial, pushbuttons, wrap-around grill bars, 4 knobs, BC, SW, FM, AC **$75.00**
42-355, Table, 1942, Walnut, right slide rule dial, pushbuttons, left grill w/horiz bars, BC, SW, FM, AC **$70.00**
42-380, Console, 1942, Wood, slanted slide rule dial, pushbuttons, lower vert bars, 8 tubes, BC, SW, AC **$135.00**
42-390, Console, 1942, Wood, slide rule dial w/plastic escutcheon, pushbuttons, vert grill bars, 4 knobs **$145.00**

42-842, Portable, 1942, Leatherette, plastic grill/dial escutcheon, horizontal louvers, handle, 2 knobs **$35.00**
42-1002, Table-R/P, 1942, Wood, front slanted slide rule dial, horiz bars, 2 knobs, lift top, inner phono, AC **$35.00**
42-1003, Table-R/P, 1942, Wood, front slide rule dial, right & left wrap-around louvers, lift top, inner phono, AC **$45.00**
42-1005, Console-R/P, 1942, Wood, slide rule dial, lower tilt-out front w/vertical bars & inner phono, BC, SW **$135.00**
42-1008, Console-R/P, 1942, Wood, upper slanted slide rule dial, tilt-out front w/vertical bars, inner phono **$135.00**
42-1013, Console-R/P, 1942, Walnut, inner dial, pushbuttons, fold-up cover, tilt-out front, BC, SW, FM, AC **$200.00**
42-1015, Console-R/P, 1942, Wood, inner slide rule dial, pushbuttons, fold-up cover, tilt-out front, BC, FM, SW **$200.00**

42-KR3, Table, 1942, Wood, rounded case, front recessed right dial & left horiz louvers, base, 2 knobs **$55.00**
42-KR5, Table/Refrigerator-C, 1942, Wood, rounded sides, right dial, left clock, curved base to fit top of refrigerator **$60.00**

42-PT93 "Transitone", Table, 1942, Walnut w/black, right dial, left cloth grill, center divider, handle, feet, 2 knobs **$50.00**

42-PT94 "Transitone", Table, 1942, Wood, right dial, left horiz louvers, plastic escutcheon, handle, feet, 2 knobs .$50.00
42-PT95 "Transitone", Table, 1942, Wood, right dial, left horiz louvers, plastic escutcheon, handle, feet, 2 knobs **$50.00**
43, Cathedral, 1934, Wood, center front window dial, upper grill w/ cut-outs, 8 tubes, 4 knobs, BC, SW **$250.00**
43B, Cathedral, 1934, Wood, center front window dial, upper grill w/ cut-outs, 8 tubes, 4 knobs, BC, SW **$250.00**

44, Cathedral, 1934, Wood, center front window dial, upper cloth grill w/cut-outs, 4 knobs, AC **$225.00**
45, Table, 1934, Wood, Deco, center window dial, right & left "butterfly" grills, top louvers, AC **$150.00**
45C, Table, 1934, Wood, Deco, center front window dial, right & left "butterfly" grills, top louvers **$150.00**
45L, Console, 1934, Wood, lowboy, upper window dial, lower grill w/ cut-outs, 6 tubes, 4 knobs, BC, SW, AC **$150.00**
46-131, Table, 1946, Plastic, upper front slide rule dial, lower horizontal louvers, 2 knobs, BC, battery **$30.00**
46-132, Table, 1946, Upper slide rule dial, lower cloth grill w/ horizontal center bar, 2 knobs, BC, bat **$35.00**
46-142, Table, 1946, Plastic, upper slanted slide rule dial, lower horizontal louvers, 2 knobs, BC, bat **$40.00**

46-200 "Transitone", Table, 1946, Plastic, right front dial over horizontal wrap-around louvers, 2 knobs, BC, AC/DC ..**$40.00**
46-250 "Transitone", Table, 1946, Plastic, upper slide rule dial, lower horizontal louvers, 2 knobs, BC, AC/DC **$50.00**
46-350, Portable, 1946, Wood & leatherette, upper slide rule dial, tambour cover, handle, BC, AC/DC/bat **$45.00**

46-420, Table, 1946, Plastic, curved dial & 2 knobs on top of case, horizontal louvers, BC, AC/DC **$60.00**

46-420-I, Table, 1946, Plastic, curved dial & 2 knobs on top of case, horizontal louvers, BC, AC/DC **$60.00**
46-421, Table, 1946, Walnut, upper front slide rule dial, center oblong cloth grill, 2 knobs, BC, AC/DC **$50.00**
46-421-I, Table, 1946, Ivory, upper front slide rule dial, center oblong cloth grill, 2 knobs, BC, AC/DC **$50.00**
46-427, Table, 1946, Wood, upper slide rule dial, lower horiz louvers, 6 tubes, 3 knobs, BC, SW, AC/DC **$45.00**
46-480, Console, 1946, Wood, slanted slide rule dial, cloth grill w/ vertical bars, 4 knobs, BC, SW, FM, AC **$135.00**
46-1201, Table-R/P, 1946, Wood, top slide rule dial, 2 knobs, lower slide-in record slot, cloth grill, BC, AC **$75.00**
46-1203, Table-R/P, 1946, Wood, right front vertical dial, 3 knobs, horizontal louvers, lift top, BC, AC **$40.00**
46-1209, Console-R/P, 1946, Wood, 2 slide rule dials, pushbuttons, 4 knobs, lower tilt-out phono, BC, SW, AC **$130.00**
46-1213, Console-R/P, 1946, Wood, inner slide rule dial, 10 pushbuttons, fold-out phono door, BC, SW, FM, AC .**$100.00**
46-1217, Console-R/P, 1946, Wood, inner left dial & knobs, lift cover, right pull-out phono, grill cut-outs, AC **$150.00**

46-1226, Console-R/P, 1946, Wood, upper slanted slide rule dial, tilt-out phono door, 4 knobs, BC, SW, AC **$135.00**

47, Cathedral, 1932, Two-tone wood, center window dial, upper cloth grill w/cut-outs, 4 knobs, bat **$120.00**

47-204, Table, 1947, Leatherette, plastic panel w/right dial & horizontal louvers, 2 knobs, BC, AC/DC **$40.00**

47-205, Table, 1947, Leatherette & plastic, right front dial, left horizontal louvers, 2 knobs, AC/DC **$40.00**

47-1227, Console-R/P, 1947, Wood, upper slanted slide rule dial, 4 knobs, lower tilt-out phono, BC, FM, AC **$135.00**

47-1230, Console-R/P, 1947, Wood, upper slide rule dial, 4 knobs, pushbuttons, tilt-out phono, BC, SW, FM, AC **$135.00**

48-141, Table, 1948, Walnut plastic, upper slide rule dial, lower horizontal louvers, 2 knobs, BC, bat **$35.00**

48-145, Table, 1948, Ivory plastic, upper slide rule dial, lower horizontal louvers, 2 knobs, BC, bat **$35.00**

48-150, Table, 1948, Wood, upper slanted slide rule dial, large perforated grill, 2 knobs, BC, battery **$35.00**

48-200 "Transitone", Table, 1948, Walnut plastic, right dial over horizontal wrap-around louvers, 2 knobs, BC, AC/DC $45.00

48-200-I "Transitone", Table, 1948, Ivory plastic, right dial over horizontal wrap-around louvers, 2 knobs, BC, AC/DC .. **$45.00**

48-206, Table, 1948, Leatherette, plastic escutcheon/louvers, rounded corners, 2 knobs, BC, AC/DC $45.00

48-214 "Transitone", Table, 1948, Wood, right front dial, left two-tone lattice grill, 5 tubes, 2 knobs, BC, AC/DC **$40.00**

48-225 "Transitone", Table, 1948, Plastic, right square dial, left perforated grill, rounded top, 2 knobs, BC, AC/DC **$35.00**

48-230 "Transitone", Table, 1948, Plastic, modernistic dial & grill, rounded top, slanted sides, 2 knobs, BC, AC/DC **$75.00**

48-250 "Transitone", Table, 1948, Walnut plastic, slide rule dial, horiz wrap-around louvers, 2 knobs, BC, AC/DC **$40.00**

48-250-I "Transitone", Table, 1948, Ivory plastic, slide rule dial, horiz wrap-around louvers, 2 knobs, BC, AC/DC **$40.00**

48-300 "Transitone", Portable, 1948, Leatherette, slide rule dial, horizontal louvers, 2 knobs, handle, BC, AC/DC/bat **$30.00**

48-360, Portable, 1948, "Alligator", upper slide rule dial, lower wood louvers, tambour lid, handle, AC, bat **$40.00**

48-460, Table, 1948, Plastic, curved dial & 2 knobs on top of case, horizontal louvers, BC, AC/DC $60.00

48-460-I, Table, 1948, Ivory plastic, curved dial & 2 knobs on top of case, horizontal louvers, BC, AC/DC **$60.00**

48-461, Table, 1948, Wood, upper slanted slide rule dial, lower recessed grill, 2 knobs, BC, AC/DC **$40.00**

48-464, Table, 1948, Brown plastic, slanted slide rule dial, horizontal louvers, 4 knobs, BC, SW, AC/DC **$45.00**

48-472, Table, 1948, Walnut plastic, slanted slide rule dial, horizontal louvers, 4 knobs, BC, FM, AC/DC $40.00

48-472-I, Table, 1948, Ivory plastic, slanted slide rule dial, horizontal louvers, 4 knobs, BC, FM, AC/DC **$40.00**

48-475, Table, 1948, Wood, raised top, upper slanted slide rule dial, 4 knobs, 6 pushbuttons, BC, FM, AC **$60.00**

48-482, Table, 1948, Wood, lower slanted slide rule dial, 10 pushbuttons, upper grill, BC, SW, FM, AC **$55.00**

48-1201, Table-R/P, 1948, Wood, curved front, top slide rule dial & knobs, lower slide-in record slot, BC, AC **$75.00**

48-1253, Table-R/P, 1948, Walnut, front slide rule dial, 2 knobs, horiz louvers, 3/4 lift top, inner phono, BC, AC **$40.00**

48-1256, Table-R/P, 1948, Wood, right vertical slide rule dial, 3 knobs, horizontal grill bars, lift top, BC, AC **$35.00**

48-1260, Console-R/P, 1948, Wood, inner right slide rule dial, inner left "slide-in" phono, lower grill, BC, AC **$65.00**

48-1262, Console-R/P, 1948, Two-tone wood, slanted slide rule dial, 4 knobs, lower tilt-out phono, BC, AC **$110.00**

48-1263, Console-R/P, 1948, Walnut, upper slanted slide rule dial, 4 knobs, front tilt-out phono, BC, SW, AC **$110.00**

48-1264, Console-R/P, 1948, Walnut, slide rule dial, 4 knobs, tilt-out phono, criss-cross grill, BC, FM, AC **$75.00**

48-1266, Console-R/P, 1948, Wood, inner dial, 4 knobs, 6 pushbuttons, fold-down phono door, BC, SW, FM, AC **$80.00**

48-1270, Console-R/P, 1948, Wood, slide rule dials, 4 knobs, pushbuttons, fold-down phono, 13 tubes, BC, SW, FM, AC **$95.00**

48-1274 "Hepplewhite", Console-R/P, 1948, Mahogany, inner dial & knobs, fold-up cover, tilt-out front phono, 15 tubes, BC, SW, FM, AC **$110.00**

48-1276 "Sheraton", Console-R/P, 1948, Wood, inner slide rule dial, 4 knobs, pushbuttons, tilt-out phono, 15 tubes, BC, SW, FM, AC **$125.00**

48-1284, Console-R/P, 1948, Wood, inner right slide rule dial, 4 knobs, left pull-down phono door, BC, SW, AC **$75.00**

48-1286, Console-R/P, 1948, Wood, inner slide rule dial, 4 knobs, lift-up front panel, tilt-out phono, BC, FM, AC **$95.00**

48-1290, Console-R/P, 1948, Wood, vertical dial, 4 knobs, pushbuttons, fold-down phono door, BC, SW, FM, AC **$100.00**

48-1401, Table-R/P, 1948, Wood & plastic, modern, top slide rule dial, left arched grill, slide-in phono, AC **$125.00**

49-100, Table, 1949, Brown plastic, upper slide rule dial, lower horizontal louvers, 4 tubes, battery **$35.00**

49-101, Table, 1949, Plastic, slide rule dial, vertical wrap-over grill bars, 2 knobs, BC, AC/DC/bat **$35.00**

49-472, Table, 1949, Plastic, upper slanted slide rule dial, lower horizontal louvers, 4 knobs, BC, FM **$45.00**

49-500 "Transitone", Table, 1949, Walnut plastic, right dial over horiz wrap-around louvers, 2 knobs, BC, AC/DC **$40.00**

49-500-I "Transitone", Table, 1949, Ivory plastic, right dial over horizontal wrap-around louvers, 2 knobs, BC, AC/DC .. **$40.00**

49-501 "Transitone", Table, 1949, Brown plastic, modern, round dial, left arched perforated grill, 2 knobs, BC, AC/DC . **$275.00**

49-501-I "Transitone", Table, 1949, Ivory plastic, modern, round dial, left arched perforated grill, 2 knobs, BC, AC/DC . **$275.00**

49-503 "Transitone", Table, 1949, Plastic, modern checkerboard grill design, curved sides, 2 knobs, BC, AC/DC **$75.00**

49-504 "Transitone", Table, 1949, Walnut plastic, upper slide rule dial, lower horizontal louvers, 2 knobs, BC, AC/DC **$40.00**

49-504-I "Transitone", Table, 1949, Ivory plastic, upper slide rule dial, lower horizontal louvers, 2 knobs, BC, AC/DC **$40.00**

49-505 "Transitone", Table, 1949, Plastic, center slide rule dial, curved-in checkerboard front, 2 knobs, BC, AC/DC **$50.00**

49-506 "Transitone", Table, 1949, Wood, right dial/center decorative "wedge" over plastic checkered grill, 2 knobs **$45.00**

49-601, **Portable, 1949, Plastic, lower slide rule dial, horizontal grill bars, 2 knobs, handle, BC, battery** **$35.00**

49-602 "Transitone", Portable, 1949, Plastic, lower slide rule dial, horizontal louvers, handle, 2 knobs, BC, AC/DC/bat **$35.00**

49-603, Table, 1949, Leather & plastic, folds open, tiny window dial, 2 thumbwheel knobs, BC, AC/DC **$55.00**

49-605, Portable, 1949, Plastic, upper slide rule dial, large lower grill, handle, 6 tubes, BC, AC/DC/bat **$50.00**

49-607, Portable, 1949, "Alligator", upper slide rule dial, wooden louvers, handle, 2 knobs, BC, AC/DC/bat **$55.00**

49-900-E, Table, 1949, Ebony plastic, step-down top w/ curved dial & 2 knobs, horiz louvers, BC, AC/DC **$60.00**

49-900-I, Table, 1949, Ivory plastic, step-down top w/ curved dial & 2 knobs, horiz louvers, BC, AC/DC **$60.00**

49-901, Table, 1949, Plastic, modern, right thumbwheel dial/switch, raised dotted grill, BC, AC/DC **$95.00**

49-902, Table, 1949, Plastic, recessed front, right square dial, horizontal louvers, 2 knobs, BC, AC/DC **$40.00**

49-904, Table, 1949, Plastic, large right slide rule dial on clear plastic, feet, 4 knobs, BC, SW, AC/DC **$40.00**

49-905, Table, 1949, Plastic, slanted front design, left horizontal louvers, 3 knobs, BC, FM, AC/DC **$35.00**

49-906, Table, 1949, Plastic, upper slide rule dial, lower horizontal grill bars, 4 knobs, BC, FM, AC/DC **$35.00**

49-909, Table, 1949, Wood, center front slide rule dial, large upper woven grill, 4 knobs, BC, FM, AC **$35.00**

49-1100, Console-R/P, 1949, Wood, front scrolled dial, 4 knobs, 3 vertical grill bars, inner phono, BC, AC **$75.00**

49-1101, Console, 1949, Wood, upper dial, large lower cloth grill w/ 3 vertical bars, 9 tubes, BC, FM, AC **$75.00**

49-1401, Table-R/P, 1949, Wood & plastic, top slide rule dial, left rear arched grill, slide-in phono, BC, AC **$125.00**

49-1405, Table-R/P, 1949, Wood, right front square dial, left horizontal louvers, 3 knobs, lift top, BC, AC **$35.00**

49-1600, Console-R/P, 1949, Wood, slide rule dial, 4 knobs, fold-down phono door, criss-cross grill, BC, AC $60.00

49-1602, Console-R/P, 1949, Wood, upper slanted slide rule dial, 4 knobs, front tilt-out phono door, BC, AC $75.00

49-1604, Console-R/P, 1949, Wood, upper front slide rule dial, lower pull-out phono, record storage, AC/DC $70.00

49-1606, Console-R/P, 1949, Wood, slide rule dial, 4 knobs, fold-down phono door, lower grill, storage, BC, AC $70.00

49-1613, Console-R/P, 1949, Wood, inner right slide rule dial, storage, left pull-down phono door, BC, FM, AC $80.00

49-1615, Console-R/P, 1949, Wood, inner left slide rule dial, pushbuttons, 5 knobs, fold-out phono, BC, FM, AC $85.00

49B, Cathedral, 1934, Two-tone wood, center front window dial, 3 section grill, 4 knobs, BC, SW, DC $200.00

49H, Console, 1934, Wood, lowboy, upper window dial, lower grill w/ cut-outs, 4 knobs, BC, SW, DC $130.00

49X, Console, 1934, Wood, upper window dial, lower cloth grill w/ vertical bars, 4 knobs, BC, SW, DC $130.00

50, Cathedral, 1931, Wood, center front window dial, upper cloth grill w/cut-outs, 3 knobs, battery $150.00

50-520 "Transitone", Table, 1950, Brown plastic, lower slide rule dial, upper horiz louvers, 2 knobs, feet, BC, AC/DC $30.00

50-520-I "Transitone", Table, 1950, Ivory plastic, lower slide rule dial, upper horiz louvers, 2 knobs, feet, BC, AC/DC $30.00

50-522 "Transitone", Table, 1950, Brown plastic, lower slide rule dial, upper horizontal louvers, 2 knobs, BC, AC/DC $35.00

50-522-I "Transitone", Table, 1950, Ivory plastic, lower slide rule dial, upper horizontal louvers, 2 knobs, BC, AC/DC $35.00

50-524, Table, 1950, Mahogany, lower slide rule dial, large upper grill, 4 tubes, 2 knobs, BC, AC/DC $30.00

50-526 "Transitone", Table, 1950, Maroon plastic, lower slide rule dial, half-moon lattice grill, 2 knobs, BC, AC/DC $45.00

50-527 "Transitone", Table-C, 1950, Brown plastic, thumbwheel dial on case top, left clock, horiz grill bars, BC, AC $45.00

50-527-I "Transitone", Table-C, 1950, Ivory plastic, thumbwheel dial on case top, left clock, horiz grill bars, BC, AC $45.00

50-620 "Transitone", Portable, 1950, Plastic, lower slide rule dial, horizontal louvers, handle, 2 knobs, BC, AC/DC/bat $35.00

50-621, Portable, 1950, Plastic, lower slide rule dial, horizontal louvers, handle, 2 knobs, BC, AC/DC/bat $35.00

50-920, Table, 1950, Brown plastic, top slanted half-round dial, wrap-around louvers, 2 knobs, AC/DC $50.00

50-921, Table, 1950, Ivory plastic, half-moon dial on case top, front horiz louvers, 2 knobs, BC, AC/DC $45.00

50-922, Table, 1950, Maroon plastic, half-moon dial on case top, front horiz louvers, 2 knobs, BC, AC/DC $45.00

50-925, Table, 1950, Plastic, diagonal front, right dial, left horizontal louvers, 3 knobs, BC, FM, AC/DC $30.00

50-926, Table, 1950, Wood & leatherette, right front dial, left grill, 6 tubes, 3 knobs, BC, FM, AC/DC $25.00

50-1421, Table-R/P, 1950, Plastic, top right dial, 3 knobs, left lift top, inner phono, front vert bars, BC, AC $65.00

50-1422, Table-R/P, 1950, Plastic, top right dial, 3 knobs, left lift top, inner phono, front vert bars, BC, AC $65.00

50-1423, Table-R/P, 1950, Wood, right front dial, left horiz louvers, lift top, inner phono, 3 knobs, BC, AC $40.00

50-1720, Console-R/P, 1950, Wood, upper front slide rule dial, 4 knobs, lower fold-down phono door, BC, FM, AC $65.00

50-1721, Console-R/P, 1950, Wood, upper front slide rule dial, 4 knobs, lower fold-down phono door, BC, FM, AC $65.00

50-1723, Console-R/P, 1950, Wood, upper slide rule dial, lower left pull-out phono/right storage, BC, FM, AC $70.00

50-1724, Console-R/P, 1950, Wood, inner right slide rule dial, 4 knobs, left phono, double doors, BC, FM, AC $70.00

50-1726, Console-R/P, 1950, Wood, inner right slide rule dial/ storage, door, left pull-out phono, BC, FM, AC $80.00

50-1727, Console-R/P, 1950, Wood, inner left slide rule dial, pushbuttons, 5 knobs, right phono, BC, FM, AC $80.00

51-530 "Transitone", Table, 1951, Plastic, lower front slide rule dial, upper horizontal louvers, 2 knobs, BC, AC/DC $30.00

51-537 "Transitone", Table-C, 1951, Plastic, left top thumbwheel dial, alarm clock, right horizontal grill bars, BC, AC $45.00

51-538 "Transitone", Table-C, 1951, Plastic, thumbwheel dial on case top, left clock, horiz wrap-around grill bars $45.00

51-631, Portable, 1951, Plastic, slide rule dial, wrap-over vertical bars, 2 side knobs, metal handle, BC, AC/DC/bat ...$25.00

51-930, Table, 1951, Plastic, raised top w/half-round dial, wrap-around louvers, 2 knobs, BC, AC/DC $50.00

51-934, Table, 1951, Plastic, slanted slide rule dial, horizontal louvers, 3 knobs, AM, FM, AC/DC $35.00

51-1330, Table-R/P, 1951, Plastic, slide rule dial, horizontal louvers, 2 knobs, lift top, inner phono, BC, AC $40.00

51-1730, Console-R/P, 1951, Wood, upper slide rule dial, center fold-down phono door, lower grill, 4 knobs $60.00

51-1732, Console-R/P, 1951, Wood, inner right slide rule dial, 4 knobs, pull-out phono, double doors, BC, FM, AC $60.00

51B, Cathedral, 1932, Wood, center window dial, upper grill w/cut-outs, scrolled base, 5 tubes, 3 knobs $250.00

52, Cathedral, 1932, Wood, center window dial w/escutcheon, upper cloth grill w/cut-outs, 3 knobs $250.00

52-540 "Transitone" , Table, 1952, Plastic, lower front slide rule dial, upper horizontal grill bars, 2 knobs, BC $25.00

52-541 "Transitone", Table, 1952, Plastic, lower front slide rule dial, upper horizontal grill bars, 2 knobs, BC **$25.00**

52-542 "Transitone", Table, 1952, Plastic, lower slide rule dial, upper semi-circular horiz grill bars, 2 knobs, BC **$35.00**

52-542I "Transitone", Table, 1952, Plastic, lower slide rule dial, upper semi-circular horiz grill bars, 2 knobs, BC **$35.00**

52-544 "Transitone", Table-C, 1952, Plastic, raised left top thumbwheel dial, horizontal grill bars, left clock, BC, AC **$45.00**

52-544-I "Transitone", Table-C, 1952, Plastic, raised left top thumbwheel dial, horizontal grill bars, left clock, BC, AC $45.00

52-643, Portable, 1952, Plastic, lower slide rule dial, lattice grill, flex handle, 2 knobs, BC, AC/DC/bat **$30.00**

52-942, Table, 1952, Plastic, half-moon dial on case top, front horizontal louvers, 2 knobs, BC, AC/DC $45.00

52-944, Table, 1952, Plastic, upper slanted slide rule dial, lower horiz louvers, 3 knobs, BC, FM, AC/DC **$30.00**

52-1340, Table-R/P, 1952, Plastic, lower front dial, upper lattice grill, 2 knobs, lift top, inner phono, BC, AC **$40.00**

52C, Table, 1932, Wood, rectangular case, right window dial, left grill w/cut-outs, 3 knobs, AC **$60.00**

52L, Console, 1932, Wood, lowboy, upper window dial, lower cloth grill w/cut-outs, 4 knobs, AC **$135.00**

53-559 "Transitone", Table, 1953, Plastic, right front square dial over horizontal grill bars, 2 knobs, 2 band, AC/DC **$20.00**

53-560 "Transitone", Table, 1953, Plastic, right front square dial over horizontal grill bars, 2 knobs, BC, AC/DC **$25.00**

53-561 "Transitone", Table, 1953, Plastic, right front round dial, left horizontal grill bars, 2 knobs, feet **$35.00**

53-562 "Transitone", Table, 1953, Plastic, right front dial, left horizontal grill bars, 2 knobs, BC, SW, AC/DC **$30.00**

53-563 "Transitone", Table, 1953, Plastic, modern, right slide rule dial, raised left checkered grill, 2 band, AC/DC . $50.00

53-564 "Transitone", Table, 1953, Plastic, right front dial w/ stylized "P", left horizontal grill bars, feet, 2 knobs . $35.00

53-566 "Transitone", Table, 1953, Plastic, modern, slide rule dial, raised top w/vertical grill bars, 2 band, AC/DC **$65.00**

53-656, Portable, 1953, Plastic, lower slide rule dial, lattice grill, handle, side knob, BC, SW, AC/DC/bat **$30.00**

53-701 "Transitone", Table-C, 1953, Plastic, right front square dial, left alarm clock, center lattice grill, BC, AC **$30.00**

53-702 "Transitone", Table-C, 1953, Plastic, right front round dial over horizontal grill bars, left clock, 2 band, AC **$30.00**

53-706, Lamp/Radio/Clock, 1953, Wood, tall rectangular case, lower front dial, center grill, upper alarm clock, AC **$120.00**

53-804, Table-C, 1953, Plastic, right round dial, left alarm clock, center checkered grill, 2 band, AC **$30.00**

53-958, Table, 1953, Wood, right & left front vertical dials, center cloth grill, 3 knobs, BC, FM, AC/DC **$35.00**

53-960, Table, 1953, Wood, right front recessed 9 band vertical slide rule dial, left grill, 5 knobs, AC **$50.00**

53-1750, End Table-R/P, 1953, Wood, drop-leaf, top dial, 2 knobs, lower front pull-out phono drawer, 2 band, AC **$90.00**

53-1754, Console-R/P, 1953, Wood, inner front slide rule dial, 4 knobs, lower phono, double doors, 2 band, AC **$50.00**

54, Table, 1933, Walnut w/inlay, right front dial, center grill w/cut-outs, top louvers, BC, AC/DC **$75.00**

54S, Table, 1934, Two-tone wood, right dial, center grill w/cut-outs, top louvers, 2 knobs, AC/DC **$75.00**

57C, Table, 1933, Wood, right front dial, center grill w/vertical cut-outs, top louvers, 2 knobs, AC **$65.00**

59C, Table, 1936, Walnut w/inlay, right front dial, center grill w/cut-outs, top louvers, BC, AC **$65.00**

59S, Table, 1934, Two-tone wood, right dial, center grill w/cut-outs, top louvers, 2 knobs, AC **$75.00**

60, Cathedral/Various Styles, 1936, Wood, center window dial w/ escutcheon, upper cloth grill w/cut-outs, 4 knobs **$190.00**

60, Console, C1935, Wood, upper window dial with escutcheon, lower cloth grill w/cut-outs, 4 knobs **$130.00**

60, Tombstone, 1934, Wood, round center front window dial, upper cloth grill w/Deco cut-outs, 4 knobs $125.00
60B, Cathedral/Various Styles, 1935, Wood, center window dial, upper cloth grill with cut-outs, 4 knobs, BC, SW, AC ..**$175.00**
60B, Tombstone, 1934, Wood, round center front window dial, upper cloth grill w/Deco cut-outs, 4 knobs **$125.00**
60L, Console, 1933, Wood, lowboy, upper window dial, lower cloth grill w/cut-outs, 4 knobs, BC, AC **$130.00**
65, Console, 1930, Wood, lowboy, upper window dial, lower cloth grill w/"oyster-shell" cut-out, AC **$135.00**
65, Table, 1929, Metal, low rectangular case, center front window dial w/escutcheon, 4 knobs, AC **$125.00**
66, Tombstone, 1933, Two-tone wood, center window dial, shield-shaped grill w/vertical bars, 4 knobs **$100.00**
66B, Tombstone, 1935, Wood, center round window dial, upper grill w/Deco cut-outs, 4 knobs, BC, SW, AC **$110.00**

69, Table, Wood w/scrolled inlay, right dial, center cloth grill w/ cut-outs, top louvers $75.00
70, Cathedral, 1931, Walnut, center window dial, upper grill w/cut-outs, fluted columns, 7 tubes, AC **$350.00**
70, Console, 1931, Walnut, lowboy, window dial, lower grill w/cut-outs, stretcher base, 4 knobs **$160.00**
70, Grandfather Clock, 1932, Wood, Colonial-style, front window dial, cloth grill w/cut-outs, upper clock, AC **$350.00**

70B, Cathedral, 1931, Wood, center window dial, upper cloth grill w/ cut-outs, fluted columns, 7 tubes **$350.00**
70B, Tombstone, 1938, Two-tone wood, rounded shoulders, center square dial, grill cut-outs, 2 knobs **$100.00**
70L, Console, 1931, Wood, lowboy, upper front window dial, lower cloth grill w/shield cut-out, AC **$140.00**

71, Cathedral, 1932, Two-tone wood, center window dial, upper cloth grill w/cut-outs, 4 knobs, AC $250.00
71B, Cathedral, 1932, Wood, center window dial w/escutcheon, upper grill w/cut-outs, 7 tubes, AC **$250.00**
71H, Console, 1932, Wood, lowboy, upper window dial, lower grill w/ cut-outs, 7 tubes, 6 legs, AC.....................................**$140.00**
71L, Console, 1932, Wood, lowboy, upper window dial, lower cloth grill w/shield cut-out, 4 knobs, AC **$140.00**
77, Console, 1930, Wood, lowboy, upper window dial, lower grill w/ vert bars, 7 tubes, 4 knobs, AC **$140.00**

80 "Junior", Cathedral, 1933, Two-tone wood, center window dial, upper cloth grill w/cut-outs, 2 knobs, AC.......$150.00
80B, Cathedral, 1932, Two-tone wood, center window dial, upper cloth grill w/cut-outs, 2 knobs, AC **$150.00**

81 "Junior", Cathedral, 1933, Wood, center front window dial, upper cloth grill with cut-outs, 3 knobs $150.00

81B, Cathedral, 1933, Two-tone wood, center window dial, upper cloth grill w/cut-outs, 3 knobs, AC $150.00

84, Cathedral, 1936, Two-tone wood, center dial, upper cloth grill with Deco cut-outs, 2 knobs $140.00

84B, Cathedral/Various Styles, 1934, Two-tone wood, lower dial, upper cloth grill w/cut-outs, 4 tubes, 2 knobs, AC $150.00

89, Cathedral, 1934, Two-tone wood, center dial, upper cloth grill with cut-outs, 4 knobs, BC, SW, AC $200.00

89B, Cathedral/Various Styles, 1936, Two-tone wood, center dial, upper cloth grill with cut-outs, 4 knobs, BC, AC $175.00

89L, Console, 1935, Wood, upper window dial, lower cloth grill w/cut-outs, 6 tubes, 4 knobs, BC, AC $140.00

90, Cathedral, 1931, Wood, center front window dial, upper cloth grill w/cut-outs, fluted columns, AC $450.00

90, Console, 1931, Walnut, lowboy, inner dial & 4 knobs, double front doors, stretcher base, AC $195.00

90B, Cathedral, 1931, Wood, center front window dial, upper cloth grill w/cut-outs, fluted columns, AC $400.00

90H, Console, 1931, Wood, lowboy, inner window dial, cloth grill w/cut-outs, double front doors, AC $195.00

91, Cathedral, 1934, Two-tone wood, center front window dial, upper cloth grill w/cut-outs, 4 knobs $325.00

91L, Console, 1932, Wood, lowboy, upper front window dial, lower cloth grill w/cut-outs, 4 knobs, AC $160.00

91X, Console, 1933, Wood, upper window dial w/escutcheon, lower inclined grill, 9 tubes, 4 knobs, AC $175.00

93B, Tombstone, 1937, Two-tone wood, rounded shoulders, round dial, grill w/vert cut-outs, 2 knobs $125.00

95, Console, 1930, Wood, highboy, lower window dial, tapestry grill, doors, side fleur-de-lis, AC $295.00

96, Console, 1930, Wood, lowboy, upper window dial w/escutchon, lower 3 section cloth grill, AC $140.00

96H, Console, 1930, Wood, upper front window dial, lower cloth grill w/cut-outs, stretcher base $150.00

112L, Console, 1931, Wood, lowboy, recessed window dial, lower grill w/cut-outs, ridged top, 11 tubes $200.00

116, Tombstone, 1935, Wood, center oval dial, upper cloth grill w/cut-outs, 10 tubes, 4 knobs, BC, SW $135.00

116B, Tombstone, 1936, Wood, center front oval dial, upper cloth grill w/cut-outs, 4 knobs, BC, SW, AC $135.00

116X, Console, 1936, Walnut, upper oval dial, lower grill w/vert bars, right & left cut-outs, 11 tubes, BC, SW, AC $250.00

116PX, Console-R/P, 1935, Wood, front oval dial, lower grill w/cut-outs, lift top, inner phono, BC, SW, AC $250.00

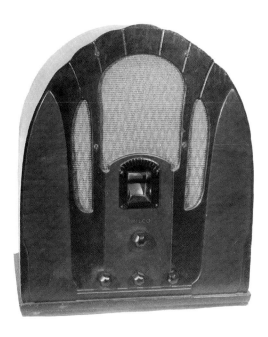

118, Cathedral, 1934, Two-tone wood, center front window dial, upper cloth grill w/cut-outs, 4 knobs $300.00

118, Tombstone, Two-tone wood, shouldered, center dial, upper cloth grill w/cut-outs, 4 knobs $135.00

118B, Cathedral, 1934, Two-tone wood, center window dial, upper 3-section grill, 4 knobs, BC, AC $300.00

118H, Console, 1935, Wood, upper window dial, lower grill w/cut-outs, 8 tubes, 4 knobs, BC, SW, AC $150.00

118MX, Console, 1935, Wood, upper window dial, lower grill w/2 vert bars, 8 tubes, 4 knobs, BC, SW, AC $150.00

118RX, Chairside-2 piece, 1935, Wood, chairside unit w/top dial, separate speaker w/Deco grill, BC, SW, AC $200.00

118X, Console, 1935, Wood, upper window dial, lower grill w/3 vert bars, 8 tubes, 4 knobs, BC, SW, AC $150.00

144, Cathedral, 1934, Wood, center front window dial, 3-section grill, 6 tubes, 4 knobs, BC, SW, AC $225.00

144B, Cathedral, 1934, Wood, center front window dial, 3-section grill, 6 tubes, 4 knobs, BC, SW, AC **$225.00**
144H, Console, 1935, Wood, upper window dial, lower grill w/cut-outs, 6 tubes, 4 knobs, BC, SW, AC **$140.00**
144X, Console, 1935, Wood, upper window dial, lower grill w/3 vert bars, 6 tubes, 4 knobs, BC, SW, AC **$140.00**
200X, Console, 1934, Wood, upper window dial, lower grill w/vertical bars, 10 tubes, 4 knobs, BC, AC **$165.00**
270, Grandfather Clock, Wood, Colonial-style, front window dial, cloth grill w/cut-outs, upper clock face **$350.00**
280 Jr - Mantle Clock, Table-N, 1932, Wood, Colonial mantle clock-style, lower knobs, upper clock face, top finials **$275.00**
500X, Console-R/P, 1935, Wood, window dial, inclined grill w/2 vert bars, lift top, inner phono, 11 tubes, BC, SW, AC **$170.00**
501X, Console-R/P, 1935, Wood, window dial, inclined grill w/2 vert bars, lift top, inner phono, 11 tubes, BC, SW, AC **$170.00**
505L, Console-R/P, 1935, Wood, lowboy, window dial, grill cut-outs, lift top, inner phono, 5 tubes, 6 legs, BC, AC **$125.00**
511, Table, 1928, Metal, rectangular case, center window dial, hand-painted flowers, 3 knobs, AC **$115.00**
512, Table, 1928, Metal, rectangular case, center window dial, hand-painted flowers, 3 knobs, AC **$115.00**
513, Table, 1928, Metal, rectangular case, center window dial, hand-painted flowers, 3 knobs, AC **$115.00**

514, Table, 1928, Metal, rectangular case, center window dial, hand-painted flowers, 3 knobs, AC **$115.00**
515, Table, 1928, Metal, rectangular case, center window dial, hand-painted flowers, 3 knobs, AC **$115.00**
610, Tombstone, 1936, Wood, shouldered, center front oval dial, upper cloth grill w/cut-outs, 4 knobs **$115.00**
610, Console, 1936, Wood, upper front oval dial, lower cloth grill w/cut-outs, fluting, 4 knobs, BC, SW **$130.00**
620, Tombstone, 1934, Wood, shouldered, center front oval dial, upper grill w/cut-outs, 4 knobs, BC, SW **$115.00**
620F, Console, 1936, Wood, upper front oval dial, lower cloth grill w/cut-outs, 4 knobs, BC, SW **$135.00**
623, Tombstone, 1935, Wood, shouldered, center front oval dial, upper cloth grill w/cut-outs, battery **$85.00**
623K, Console, 1935, Wood, upper front oval dial, lower cloth grill w/cut-outs, 4 knobs, BC, SW, bat **$100.00**
624, Console, 1935, Wood, upper front oval dial, lower cloth grill w/cut-outs, 4 knobs, BC, SW, bat **$100.00**
630, Chairside, 1936, Two-tone wood, top oval dial & knobs, front grill w/3 vertical bars, BC, SW, AC **$135.00**
635B, Tombstone, 1935, Wood, shouldered, center oval dial, upper grill w/cut-outs, 6 tubes, 4 knobs, BC, SW **$120.00**
640, Tombstone, 1936, Wood, shouldered, center oval dial, upper grill w/cut-outs, 4 knobs, BC, SW **$125.00**
640X, Console, 1936, Wood, upper front oval dial, lower cloth grill w/cut-outs, 4 knobs, BC, SW **$135.00**

645K, Console, 1936, Wood, upper oval dial, lower cloth grill w/cut-outs, 7 tubes, 4 knobs, BC, SW **$135.00**
650H, Console, 1936, Wood, lowboy, inner top dial & controls, lift lid, front lyre grill, 8 tubes, BC, SW, AC **$140.00**
650MX, Console, 1936, Wood, inner top dial & controls, lift lid, front cloth grill w/4 vert bars, BC, SW **$150.00**
650PX, Console-R/P, 1936, Two-tone wood, upper oval dial, lower grill w/vert cut-outs, 4 knobs, BC, SW **$150.00**
650RX, Chairside - 2 piece, 1936, Wood, small table w/top dial & knobs, separate speaker w/inclined grill, BC, SW **$225.00**
650X, Console, 1936, Two-tone wood, upper oval dial, lower cloth grill w/3 vert bars, 4 knobs, BC, SW **$130.00**
660X, Console, 1936, Two-tone wood, oval dial, cloth grill w/3 vert bars, 10 tubes, 4 knobs, BC, SW **$140.00**
680X, Console, 1935, Wood, inner dial, lift top, vertical grill bars, front cut-outs, 15 tubes, BC, SW, AC **$175.00**
B569 "Transitone", Table, 1954, Plastic, right front half-moon dial, left horizontal grill bars, 2 knobs, BC, AC/DC **$25.00**

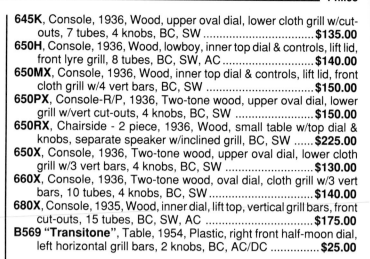

B570 "Transitone", Table, 1954, Plastic, right round dial over horiz grill bars, arched base, 2 knobs, BC, AC/DC ..$30.00
B574 "Multiwave", Table, 1954, Plastic, right front dial, left horizontal grill bars, 2 knobs, feet, 2 band, AC/DC **$30.00**

B578 "Transitone", "Multiwave", Table, Plastic, right dial w/center "P" logo, left horizontal grill bars, feet, 2 knobs **$30.00**
B650, Portable, 1954, Plastic, top thumbwheel knobs, front dotted grill, fold-back handle, BC, battery **$30.00**
B710 "Transitone", Table-C, 1953, Plastic, right front dial, left alarm clock, center lattice grill, 5 knobs, BC, AC **$30.00**
B-956, Table, 1953, Plastic, off-center dial w/stylized "P", front lattice grill, 3 knobs, feet, BC, FM **$35.00**
B1349, Table-R/P, 1954, Plastic, front slide rule dial, lattice grill, 2 knobs, lift top, inner phono, AC **$45.00**
B-1352, Table-R/P, 1954, Lower front slide rule dial, upper grill, 2 knobs, lift top, inner phono, 2 band, AC **$25.00**
C580 "Transitone", Table, 1955, Plastic, right round dial, lower metal "houndstooth" perforated grill, 2 knobs **$30.00**
C-660, Portable, 1955, Plastic, two top thumbwheel knobs, front dotted grill panel, handle, BC, battery **$30.00**

C-663, Portable, 1955, Plastic, top right dial knob, front vertical grill bars, handle, BC, AC/DC/battery **$35.00**

C-666, Portable, 1955, Plastic, lower front slide rule dial, upper lattice grill, handle, BC, AC/DC/bat **$30.00**

C-667, Portable-N, 1955, Plastic, 2 slide rule dials, lattice grill, flashlight, handle, 2 bands, AC/DC/bat **$60.00**

D-665-124, Portable, 1956, Vanity-case style, inner controls, lift-up mirrored lid, front grill, BC, AC/DC/bat **$60.00**

D-1345, Table-R/P-C, 1957, Inner right dial, outer front clock & perforated grill, lift top, handle, BC, AC **$40.00**

E-810-124, Table, 1957, Plastic, right dial over checkered panel, left metal perforated grill, BC, AC/DC $20.00

E-812-124, Table, 1957, Plastic, large right round dial over perforated panel & horiz grill bars, BC, AC/DC **$20.00**

E-818, Table, 1958, Bowed front, center slide rule dial, twin speakers, "P" logo, 2 knobs, BC, AC/DC **$35.00**

E-1370, Table-R/P, 1957, Right side dial, large front perforated grill w/crest, lift top, handle, BC, AC **$25.00**

F-673-124, Portable, 1957, Leather case, right half-moon dial, metal perforated grill, handle, left side knob $25.00

F675-124, Portable, 1957, Leather case, right front dial, large center perforated grill, handle, AC/DC/bat **$25.00**

F752-124, Table-C, 1958, Plastic, lower right dial, upper alarm clock, left lattice grill, feet, BC, AC **$20.00**

F-809, Table, 1958, Plastic, center front round dial, horizontal wrap-around grill bars, feet, BC, AC **$20.00**

F-974, Table, 1959, Plastic, center front round dial over large lattice grill, 2 knobs, AM, FM, AC/DC **$20.00**

G-681-124, Portable, 1959, Plastic, left front dial, right checkered grill, "Scantenna" handle, BC, AC/DC/bat **$30.00**

G-822 "Deluxe", Table, 1959, Plastic, lower off-center front dial over vertical grill bars, 2 knobs, BC, AC/DC **$20.00**

PT-4 "Transitone", Table, 1941, Plastic, upper slide rule dial, horizontal wrap-around louvers, 2 knobs, AC/DC **$40.00**

PT-6, Table, 1940, Two-tone wood, slide rule dial, cloth grill w/3 horizontal bars, 2 knobs, AC **$50.00**

PT-12 "Transitone", Table, 1940, Two-tone wood, upper slanted slide rule dial, lower horizontal louvers, 2 knobs $40.00

PT-25 "Transitone", Table, 1939, Brown plastic, right front square dial, left & right horiz louvers, 2 knobs, BC . $40.00

PT-26 "Transitone", Table, 1940, Brown plastic, right front square dial, left & right horiz louvers, 2 knobs, BC **$40.00**

PT-27 "Transitone", Table, 1939, Ivory plastic, right front dial over horiz wrap-around louvers, 2 knobs, BC, AC/DC **$40.00**

PT-33 "Transitone", Table, 1940, Plastic, right front dial, horizontal front bars & louvers, handle, 2 knobs **$40.00**

PT-36 "Transitone", Table, 1939, Ivory & black plastic, modern, right dial, horizontal wrap-around grill bars, BC **$90.00**

PT-38 "Transitone", Table, 1939, Walnut, right front dial, left cloth grill w/Deco cut-outs, 2 knobs, BC, SW **$45.00**

PT-41 "Transitone", Table, 1939, Walnut, right front dial, left cloth grill w/Deco cut-outs, 2 knobs, BC, SW **$45.00**

PT-42 "Transitone", Table, 1939, Two-tone wood, right front square dial, left cloth grill, tapered sides, 2 knobs **$45.00**

PT-43 "Transitone", Table, 1939, Wood & plastic, right dial, left wrap-around grill bars, handle, 2 knobs, BC, AC **$65.00**

PT-46 "Transitone", Table, 1939, Brown plastic, right front dial, 6 pushbuttons, left mitered louvers, 2 knobs, BC **$65.00**

PT-48 "Transitone", Table, 1939, Ivory plastic, right front dial, 6 pushbuttons, left mitered louvers, 2 knobs, BC **$65.00**

PT-50 "Transitone", Table, 1939, Maple, right front dial, left cloth grill w/3 horizontal bars, 2 knobs, BC **$45.00**

PT-61 "Transitone", Table, 1940, Two-tone wood, Deco, right dial, left horizontal louvers, curved base, BC, AC/DC **$95.00**

PT-63 "Transitone", Portable, 1940, Striped cloth covered, right front square dial, left cloth grill, handle, 2 knobs **$25.00**
PT-66 "Transitone", Table, 1939, Wood & plastic, right dial, left grill, 6 pushbuttons, base, handle, 2 knobs, BC **$60.00**
PT-69 "Transitone", Table-C, 1939, Wood, trapezoid-shaped, lower right dial, left grill, upper clock, 2 knobs, BC, AC **$85.00**
PT-87 "Transitone", Portable, 1942, Right front dial, left horizontal louvers, handle, 2 knobs, 5 tubes, battery **$30.00**
PT-88 "Transitone", Portable, 1942, Inner right dial, left horizontal louvers, 2 knobs, fold-down front, handle, 5 tubes **$35.00**

T1000-124, Table-C, 1960, Metal & plastic, transistor, 3 modules-center clock, right & left speakers, base **$85.00**
TC-47, Portable-C, 1960, Transistor, right round dial, center round grill, left round clock, AM, battery **$40.00**
TH-1 "Transitone", Table, 1939, Wood, right front gold dial, left grill with horizontal cut-outs, 2 knobs **$45.00**

PT-91 "Transitone", Table, 1941, Plastic, right front dial over horizontal wrap-around louvers, feet, 2 knobs, AC . $40.00
T-4, Portable, 1959, Transistor, right front dial, left horizontal bars, small center knob, feet, AM, bat **$30.00**
T-6, Portable, 1959, Leather case, transistor, right front round dial, handle, left side knob, AM, bat **$25.00**

PHILHARMONIC
Espey Mfg. Co., Inc.,
528 East 72nd Street, New York, New York

100C, Console-R/P, 1948, Wood, inner slide rule dial, 4 knobs, phono, lift top, criss-cross grill, BC, AC **$60.00**
100T "Minuet", Table-R/P, 1948, Wood, upper front dial, upper horizontal grill bars, 4 knobs, lift top, BC, AC **$35.00**
149-C, Console-R/P, 1949, Wood, inner right dial, left phono, lift top, front criss-cross grill, BC, AC **$55.00**
249-C, Console-R/P, 1949, Wood, inner right dial, lift top, pull-out phono drawer, criss-cross grill, BC, AC **$60.00**
349-C, Console-R/P, 1949, Wood, inner slide rule dial, pull-out phono drawer, criss-cross grill, BC, FM, AC **$75.00**
8712, Console-R/P, 1947, Wood, inner slide rule dial, 4 knobs, record storage, lift top phono, BC, 2 SW, AC **$70.00**

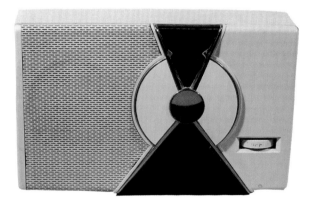

T-7-126, Portable, 1957, Plastic, transistor, modern, center front round dial, left perforated grill, AM, bat $50.00
T-9 "Trans-World", Portable, 1959, Leather case, transistor, inner slide rule dial, fold-up front w/map, 7 bands, bat **$60.00**
T-50, Portable, 1959, Plastic, transistor, upper front round dial, lower horizontal grill bars, battery **$25.00**
T-60, Portable, 1959, Plastic, transistor, upper thumbwheel dial, horiz grill bars, swing handle, AM, bat **$30.00**
T-65, Portable, 1959, Plastic, transistor, right round dial, left horizontal louvers, handle, AM, battery **$25.00**
T-75, Portable, 1959, Leather case, transistor, off-center round dial, perforated grill, strap, AM, bat **$30.00**
T-78, Portable, 1960, Leather case, transistor, small right window dial, round grill, handle, AM, bat **$25.00**
T-700, Portable, 1957, Leather case, transistor, right front dial knob, metal grill, handle, battery **$40.00**

PHILLIPS 66
Phillips Petroleum Co.,
Bartlesville, Oklahoma

3-12A, Portable, 1948, Upper slanted slide rule dial, lower horizontal louvers, handle, 2 knobs, battery **$35.00**
3-20A, Table-R/P, 1948, Wood, slanted slide rule dial, horizontal louvers, 3/4 lift top, inner phono, BC, AC **$35.00**
3-81A, Console-R/P, 1948, Wood, inner right slide rule dial, 4 knobs, pull-out phono drawer, 14 tubes, BC, SW, FM, AC **$85.00**

PILOT
Pilot Electrical Manufacturing Company,
323 Berry Street, Brooklyn, New York
Pilot Radio & Television Corporation
Pilot Radio and Tube Corporation,
Lawrence, Massachusetts
Pilot Radio Corporation,
37-06 36th Street, Long Island City, New York

The Pilot Electrical Manufacturing Company began business in 1922 making batteries and parts and eventually expanded into the production of radios.

53, Tombstone, 1934, Wood, center round dial, upper grill w/cut-outs, 5 tubes, 4 knobs, BC, SW, AC **$125.00**

63, Tombstone, 1934, Wood, center round "compass" dial, upper grill w/cut-outs, 6 tubes, BC, SW, AC **$125.00**

93, Table, 1934, Walnut w/inlay, center front grill w/cut-outs, 5 tubes, 3 knobs, BC, SW, AC/DC **$75.00**

114, Tombstone, 1934, Wood, center front round dial, upper cloth grill, BC, SW, AC **$120.00**

133, Table, 1940, Wood, right front dial, tuning eye, left horizontal wrap-around louvers, 4 knobs **$65.00**

B-2, Table, 1934, Walnut, Deco, right window dial, center grill cut-outs, 4 tubes, 2 knobs, AC/DC **$90.00**

B1151 "Lone Ranger", Table, 1939, Plastic, right dial w/Lone Ranger & Silver, left horiz louvers, handle, 2 knobs **$85.00**

C-63, Console, 1934, Wood, upper front round dial, lower cloth grill, 6 tubes, 4 knobs, feet, BC, SW, AC **$125.00**

E-20, Table, 1934, Burl walnut, Deco, right front window dial, center grill cut-outs, 3 knobs, AC/DC **$90.00**

L-8, Cathedral, 1933, Wood, center half-moon dial w/escutcheon, upper grill w/cut-outs, 4 knobs, AC **$235.00**

T-3, Table, Wood, upper slide rule dial, lower horizontal louvers, 3 knobs, BC, SW **$50.00**

T-301, Table, 1941, Wood, upper slanted slide rule dial, lower cloth grill w/2 horiz bars, AC, BC, FM...................................**$50.00**

T-341, Table, 1941, Walnut, right slide rule dial, pushbuttons, left vertical grill bars, 8 band, AC/DC **$85.00**

T-411-U, Table, 1947, Wood, center vertical slide rule dial, right & left grill areas, 4 knobs, BC, SW, AC **$40.00**

T-500U, Table, 1947, Two-tone plastic, large dial, wrap-around grill bars, 3 knobs, BC, SW, AC/DC **$85.00**

T-511, Table, 1946, Wood, center vert slide rule dial, right/left grill areas, 4 knobs, BC, SW, AC/DC **$40.00**

T-521, Table, 1947, Wood, lower front slide rule dial, large upper grill area, 4 knobs, BC, FM, AC/DC **$40.00**

T-531AB, Table, 1947, Wood, lower slide rule dial, upper horiz louvers, 4 knobs, BC, 3 SW, FM, AC/DC **$45.00**

T-601 "Pilotuner", Table, 1947, Wood, FM Tuner, center front FM dial, 2 knobs **$25.00**

T-741, Table, 1948, Wood, lower front slide rule dial, upper horizontal louvers, 4 knobs, BC, AC/DC **$40.00**

TP-32, Table-R/P, 1941, Walnut, right front slide rule dial, left wrap-around louvers, 4 knobs, BC, SW, AC **$30.00**

TP-1062, Table-R/P, 1941, Wood, recessed right dial, left wrap-around horiz louvers, 3 knobs, lift top, AC **$30.00**

TX-42, Table-R/P, 1941, Wood, right front slide rule dial, left criss-cross grill, 4 knobs, inner phono, AC **$30.00**

X3, Table, Wood, upper slanted slide rule dial, lower horiz louvers, 3 knobs, BC, SW, AC/DC **$40.00**

Super Wasp - AC, Table, 1928, Kit, SW receiver, metal front panel w/two window dials, plug-in coils, AC **$375.00**

Super Wasp - DC, Table, 1928, Kit, SW receiver, metal front panel w/two window dials, plug-in coils, DC **$375.00**

Wasp, Table, 1928, Kit, SW receiver, metal front panel w/left window dial, 4 knobs, plug-in coils **$275.00**

PIONEER

578, Tombstone, Wood, lower round dial, upper grill w/cut-outs, 3 knobs, "Pioneer" nameplate **$100.00**

PORTO BARADIO
Porto Products,
412 North Orleans Street, Chicago, Illinois

PA-510, Table-N, 1949, Plastic bar/radio, slide rule dial, horizontal louvers, handles, 2 knobs, BC, AC/DC .**$200.00 w/glassware**

PB-520, Table-N, 1948, Plastic bar/radio, slide rule dial, horizontal louvers, handles, 2 knobs, BC, AC/DC .**$200.00 w/glassware**

PRECEL
Precel Radio Mfg. Co.,
227 Erie St., Toledo, Ohio

Superfive, Table, 1925, Wood, tall rectangular case, slanted 3 dial panel, lower storage, 5 tubes, battery **$150.00**

PREMIER
Premier Crystal Laboratories, Inc.,
63 Park Row, New York, New York

15LW, Table, 1946, Wood, upper slide rule dial, lower cloth grill, rounded corners, 2 knobs, BC, AC/DC **$35.00**

PRIESS
Priess Radio Corporation,
693 Broadway, New York, New York

The Priess Radio Corporation began business in1924. Beset by financial difficulties, the company was bankrupt by 1927.

PR3, Table, 1925, Mahogany, low rectangular case, 2 dial front panel, top loop antenna, 5 tubes, bat **$195.00**

PR4 "Straight 8", Table, 1925, Wood, low rectangular case, center front dial, top loop antenna, 8 tubes, battery $220.00
Straight Eight, Console, 1925, Two-tone walnut, inner panel and speaker grill, double front doors, 8 tubes, bat **$285.00**
Straight Eight, Table, 1925, Two-tone mahogany w/inlay, low rectangular case, top loop antenna, 8 tubes, bat **$230.00**
Straight Nine, Table, 1926, Mahogany, low rectangular case, center front dial, top loop antenna, 9 tubes, bat **$245.00**

PURITAN
Pure Oil Co.,
35 Wacker Drive,
Chicago, Illinois

501, Table, 1946, Plastic, upper front slide rule dial, lower criss-cross grill, 2 knobs, BC, AC/DC **$35.00**
502X, Table, 1946, Plastic, upper front slide rule dial, lower criss-cross grill, 2 knobs, BC, AC/DC **$35.00**
503, Table-R/P, 1946, Wood, top vertical dial, 3 knobs, front cloth grill, lift top, inner phono, BC, AC **$30.00**
504, Table, 1946, Wood, upper slanted slide rule dial, lower criss-cross grill, 4 knobs, BC, SW, AC **$40.00**
506, Table, 1946, Plastic, upper slide rule dial, horiz louvers, step-down top, 3 knobs, BC, AC/DC **$50.00**
508, Table, 1946, Wood, upper slanted slide rule dial, lower cloth grill, 4 knobs, BC, 2SW, AC **$55.00**
509, Portable, 1947, Upper slanted slide rule dial, horiz grill bars, 2 knobs, handle, BC, AC/DC/battery **$25.00**
515, Table, 1947, Wood, upper slanted slide rule dial, two-tone cloth grill, 2 knobs, BC, AC/DC **$35.00**

R. K. RADIO LABORATORIES
Chicago, Illinois

S-4233 "Radio Keg", Table-N, 1933, Looks like keg w/copper hoops, front window dial & knobs, rear grill, AC/DC **$300.00**

RADIO VISION

414, Tombstone, Wood, lower dial & grill w/shield-shaped cut-out, upper scene w/battleships, planes and Statue of Liberty, water "moves" when radio plays **$200.00**

RADIOBAR
Radiobar Co. of America,
296 Broadway, New York, New York

200-RBP, Console-R/P/Bar, 1940, Wood, inner left dial, right pull-out phono, upper bar unit with doors & lift top **$1,000.00/**
with glassware
508, Console/Bar, 1933, Wood, lower front dial, knobs & grill, upper bar unit with folding doors & lift top **$1,200.00/**
with glassware
510, Console/Bar, 1933, Walnut, lower inside dial & knobs, doors, upper bar with folding doors & lift top **$1,200.00/**
with glassware
528, Console/Bar, 1933, Wood, Deco, lower inside dial & knobs, upper bar with folding doors & lift top **$1,200.00/**
with glassware

No #/Philco chassis, Console/Bar, 1936, Wood, lower oval dial, 4 knobs & grill, upper bar w/folding doors and lift top .... $800.00/with glassware

RADIOETTE
Alamo Electronics Corp.,
San Antonio, Texas

PR-2, Portable, 1948, Trapeziod shape, front dial, circular grill bars, handle, 3 knobs, BC, AC/DC/bat $40.00

RADIOLA
Radio Corp. of America,
Home Instrument Division,
Camden, New Jersey

61-1, Table, 1947, Plastic, upper front slide rule dial, lower horizontal louvers, 3 knobs, BC, AC/DC $50.00
61-5, Table, 1947, Upper slanted slide rule dial w/stars, horizontal louvers, 3 knobs, BC, SW, AC/DC $45.00

61-8, Table, 1947, Plastic, upper slanted slide rule dial, lower horiz grill bars, 2 knobs, BC, AC/DC $45.00

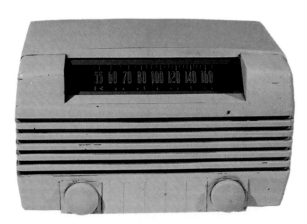

61-9, Table, 1947, Plastic, upper slanted slide rule dial, lower horiz grill bars, 2 knobs, BC, AC/DC $45.00
75ZU, Table-R/P, 1948, Wood, front dial, horizontal grill bars, 2 knobs, lift top, inner phono, BC, AC $30.00
76ZX11, Table, 1948, Plastic, upper slanted slide rule dial, lower horiz grill bars, 2 knobs, BC, AC/DC $45.00
76ZX12, Table, 1948, Plastic, upper slanted slide rule dial, lower horiz grill bars, 2 knobs, BC, AC/DC $45.00
500, Table, 1941, Plastic, midget, right front dial, left horizontal wrap-around louvers, 2 knobs, BC $85.00

512, Table, 1941, Wood, right front dial, left square cloth grill, flared base, 2 knobs $60.00
515, Table, 1941, Wood, upper slanted slide rule dial, lower horizontal louvers, 3 knobs, BC, SW, AC $45.00

517, Table, 1942, Two-tone wood, upper slanted slide rule dial, lower horizontal louvers, 2 knobs $50.00
520, Table, 1942, Wood, right front dial, left horizontal grill bars, 2 knobs $35.00

RAYENERGY
RayEnergy Radio & Television Corp.,
32 West 22nd Street,
New York, New York

AD, Table, 1946, Wood, right front dial, left cloth grill w/ cut-outs, 2 knobs, BC, AC/DC $40.00
AD4, Table, 1946, Wood, right front dial, left contrasting horizontal louvers, 2 knobs, BC, AC/DC $45.00
SRB-1X, Table, 1947, Plastic, upper front slide rule dial, lower horizontal louvers, 2 knobs, BC, AC/DC $45.00

RAYTHEON
Raytheon Mfg. Co., TV & Radio Div.,
5921 West Dickens Avenue,
Chicago, Illinois

8TP1, Portable, 1955, Tan leatherette, transistor, top knobs, metal perforated grill, handle, AM, bat $125.00
8TP2, Portable, 1955, Brown leatherette, transistor, top knobs, metal perforated grill, handle, AM, bat $125.00
8TP3, Portable, 1955, Beige leatherette, transistor, top knobs, metal perforated grill, handle, AM, bat $125.00
8TP4, Portable, 1955, Red leatherette, transistor, top knobs, metal perforated grill, handle, AM, bat $140.00
T-100-1, Portable, 1955, Black & yellow plastic, transistor, right dial, center grill, wrist chain, AM, bat $100.00
T-100-2, Portable, 1955, Ivory & yellow plastic, transistor, right dial, center grill, wrist chain, AM, bat $100.00
T-100-3, Portable, 1955, Black & red plastic, transistor, right dial, center grill, wrist chain, AM, bat $100.00
T-100-4, Portable, 1955, Ivory & red plastic, transistor, right dial, center grill, wrist chain, AM, bat $100.00
T-100-5, Portable, 1955, Ivory & grey plastic, transistor, right dial, center grill, wrist chain, AM, bat $100.00

T2500, Portable, 1956, Leatherette, transistor, 2 top knobs, front grill w/crest, handle, AM, battery $110.00

RCA
Radio Corporation of America,
233 Broadway, New York, New York

RCA was formed in 1919 and soon became one of the largest distributors of radios. The company was one of the pioneers of early radio & broadcasting and began the National Broadcasting Company (NBC) in 1926. As well as being one of the most prolific of radio manufacturers, RCA also made vacuum tubes, Victrolas, marine apparatus, transmitters and other broadcasting equipment. Their mascot, "Nipper", was featured in many company logos listening at the horn for "his master's voice".

1-BT-29 "Transicharg Super", Portable, 1959, Transistor, right front dial, left horizontal bars, thumbwheel knob, AM, battery $35.00
1-BT-32 "Transicharg Deluxe", Portable, 1959, White & pink, transistor, right dial, left horizontal bars, swing handle, AM, bat $35.00
1-BT-34 "Transicharg Deluxe", Portable, 1959, White & green, transistor, right dial, left horizontal bars, swing handle, AM, bat $35.00
1-BT-36 "Transicharg Deluxe", Portable, 1959, Grey & white, transistor, right dial, left horizontal bars, swing handle, AM, bat $35.00
1-BT-41, Portable, 1958, Leather case, transistor, front round dial, perforated grill, handle, AM, battery $25.00
1-BT-46, Portable, 1958, Leather case, transistor, front round dial, perforated grill, handle, AM, battery $25.00
1-BT-48, Portable, 1958, Leather case, transistor, front round dial, perforated grill, handle, AM, battery $25.00
1BT58 "Globe Trotter", Portable, 1959, Leather case, transistor, slide rule dial, perforated grill, side knobs, handle, bat .$35.00
1-MBT-6 "Strato-World", Portable, 1959, Leather, transistor, flip-up front w/map, telescope antenna, handle, 7 band, bat $75.00
1R81, Table, 1952, Plastic, center front round dial over horizontal bars, right side knobs, AM, FM, AC $35.00
1-T-4H, Portable, 1960, Transistor, upper round dial, lower perforated grill, swing handle, AM, battery $25.00
1-T-4J, Portable, 1960, Transistor, upper round dial, lower perforated grill, swing handle, AM, battery $25.00

1-T-5J, Portable, 1960, Transistor, rounded top w/dial & 2 thumbwheel knobs, horiz bars, handle, AM, bat $35.00

1X, Table, Plastic, right front dial, left vertical grill bars, decorative case lines, 2 knobs $50.00
1X2, Table, Plastic, right front dial, left vertical grill bars, decorative case lines, 2 knobs $50.00
1X51, Table, 1952, Plastic, front half-moon dial, lower horizontal grill bars, side knob, BC, AC/DC $40.00
1X54, Table, 1952, Plastic, front half-moon dial, lower horizontal grill bars, side knob, BC, AC/DC $40.00
1X55, Table, 1952, Plastic, front half-moon dial, lower horizontal grill bars, side knob, BC, AC/DC $40.00

1X56, Table, 1952, Plastic, front half-moon dial, lower horizontal grill bars, side knob, BC, AC/DC $40.00

1X591, Table, 1952, Plastic, lower thin slide rule dial, upper pleated gold grill, side knobs, BC, AC/DC $50.00

1X592, Table, 1952, Plastic, lower thin slide rule dial, upper pleated gold grill, side knobs, BC, AC/DC $50.00
2B400, Portable, 1952, Plastic, center front round dial, lower metal perforated grill, handle, BC, battery $35.00
2B401, Portable, 1952, Plastic, center front round dial, lower metal perforated grill, handle, BC, battery $35.00
2BX63, Portable, 1953, Plastic, upper slide rule dial, lower horiz bars, handle, side knobs, BC, AC/DC/bat $25.00

2-C-521, Table-C, 1953, Plastic, right round dial, left alarm clock, center horiz bars, 5 knobs, BC, AC $20.00
2-S-7, Console-R/P, 1953, Wood, inner right slide rule dial, 2 knobs, pull-out phono, double doors, BC, AC $45.00
2-S-10, Console-R/P, 1953, Wood, inner right dial, 5 knobs, upper phono, double doors, BC, FM $60.00
2US7, Table-R/P, 1952, Wood, front slide rule dial, lower grill, 2 knobs, 3/4 lift top, inner phono, BC, AC $25.00
2-X-52, Table, 1952, Plastic, dial on top of case, center front horizontal louvers, 2 knobs, AC/DC $40.00
2-X-61, Table, 1953, Plastic, dial on top of case, center front horizontal louvers, 2 knobs, AC/DC $40.00
2-XF-91, Table, 1953, Plastic, right front FM dial, left AM dial, center horizontal bars, AM, FM, AC/DC $35.00
3-BX-671 "Strato-World", Portable, 1954, Leather, inner dial, flip-up front w/map, telescope antenna, 7 bands, AC/DC/bat $100.00

3-RF-91, Table, 1952, Plastic, large center round dial over horizontal front lines, side knobs, AM, FM $45.00
3-X-521, Table, 1954, Plastic, right front dial, large left checkered grill, left side knob, BC $20.00
3-X-532, Table, 1954, Plastic, right front dial, large left checkered grill, left side knob, BC $20.00
4T, Cathedral, 1935, Wood, center front window dial, grill cut-outs, 5 tubes, 2 knobs, BC, AC $140.00
4-X-552, Table, 1955, Plastic, top right thumbwheel dial, left horizontal metal grill bars, BC, AC/DC $35.00

4-Y-511, Table-R/P, 1954, Plastic, front round dial, large grill, 3 knobs, lift top, inner 45 phono, BC, AC $50.00
5BT, Tombstone, 1936, Wood, center front dial, vertical grill cut-outs, 5 tubes, 2 knobs, DC $125.00
5-BX-41, Portable, 1955, Plastic, upper front flip-up dial, horiz grill bars, 2 side knobs, BC, AC/DC/bat $45.00
5-C-581, Table-C, 1955, Lower base with slide rule dial, center front round alarm clock, top knob, BC, AC $60.00
5Q55, Table, 1939, Plastic, lower slanted slide rule dial, upper cloth grill, 2 knobs, BC, 2SW, AC $60.00
5T, Tombstone, 1936, Wood, center rectangular dial, upper cloth grill w/cut-outs, 5 tubes, BC, SW, AC $140.00

5T1, Tombstone, 1936, Wood, center dial, upper cloth grill w/ cut-outs, tuning eye, 4 knobs, BC, SW, AC $150.00
5T4, Table, 1936, Wood, rectangular case, right front dial, left cloth grill w/wrap-around bars $70.00

5X, Table, 1936, Wood, center dial, rounded sides, finished front & back, top louvers, 3 knobs, AC $100.00
5X4, Table, 1936, White lacquer/black base, center dial, 3 knobs, made for kitchen use, BC, SW, AC $60.00
6-BX-6B, Portable, 1955, Plastic, side dial knob, large front grill area w/horiz bars, handle, AC/DC/bat $30.00
6K2, Console, Wood, upper rectangular dial, lower cloth grill w/2 vert bars, 6 tubes, 4 knobs $125.00

6T, Tombstone, 1936, Wood, center front black dial, upper grill with cut-outs, 4 knobs, BC, SW, AC **$125.00**

6T2, Tombstone, 1936, Wood, center front rectangular dial, upper cloth grill w/cut-outs, 4 knobs **$125.00**
6T10, Tombstone, 1936, Black lacquer/chrome frame, center dial, vert chrome grill bar, 4 knobs, BC, SW **$475.00**
6-X-5A, Table, 1956, Plastic, right front dial knob over lattice grill, left Nipper logo, feet, AC/DC **$25.00**

6-X-7, Table, 1956, Plastic, right front dial knob over lattice grill, left Nipper logo, feet, BC, AC/DC **$25.00**
6-XD-5A, Table, 1956, Plastic, lower dial, large upper lattice grill w/ center Nipper, 2 knobs, BC, AC/DC **$25.00**
6-XF-9, Table, C1953, Plastic, center front dial, right & left checkered grills, feet, 4 knobs, AM, FM **$35.00**
7-BT-9J, Portable, 1958, Transistor, right front dial, perforated grill, top thumbwheel knob, AM, battery **$30.00**
7-BT-10K, Portable, 1956, Leather, transistor, upper dial, horizontal grill bars, handle, side knobs, AM, bat **$30.00**
7-BX-6L, Portable, 1956, Plastic, right front dial/left volume over lattice grill, handle, BC, AC/DC/battery **$35.00**
7-BX-7L, Portable, 1956, Plastic, right front dial/left volume over lattice grill, handle, BC, AC/DC/battery **$35.00**
7-BX-8J "Globe Trotter", Portable, 1957, Plastic, slide rule dial, horizontal grill bars, handle, top antenna, BC, AC/DC/bat **$25.00**

7-BX-10, Portable, 1954, Leather case, inner dial, flip-up front w/ map, telescoping antenna, BC, SW, AC/DC/bat $100.00
7-HFR-1, Console-R/P/Rec, 1957, Wood, left dial, pull-out phono, inner top right reel-to-reel recorder, BC, FM, AC **$55.00**
7K1, Console, 1937, Wood, upper front dial, tuning eye, lower cloth grill w/3 vertical bars, 4 knobs **$135.00**
7T, Tombstone, 1936, Wood, center front rectangular dial, upper cloth grill w/cut-outs, BC, SW, AC **$160.00**
7T1, Tombstone, 1936, Wood, front rectangular dial, tuning eye, upper cloth grill w/vert bars, 4 knobs **$195.00**
8B41 "Jewel Box", Portable, 1949, Black plastic, inner round dial, horiz grill bars, flip-open lid, handle, BC, battery **$45.00**
8B42 "Jewel Box", Portable, 1949, Brown plastic, inner round dial, horiz grill bars, flip-open lid, handle, BC, battery **$45.00**
8B43 "Jewel Box", Portable, 1949, Red plastic, inner round dial, horiz grill bars, flip-open lid, handle, BC, battery **$50.00**
8B46 "Jewel Box", Portable, 1949, Ivory plastic, inner round dial, horiz grill bars, flip-open lid, handle, BC, battery **$45.00**

8-BT-7LE, Portable, 1957, Plastic, transistor, large right dial knob over horizontal front bars, AM, battery **$35.00**
8-BT-8FE, Portable, 1957, Plastic, transistor, large right dial knob over horizontal front bars, AM, battery **$35.00**
8-BT-10K, Portable, 1957, Leather case, transistor, upper front dial, horizontal grill bars, handle, AM, bat **$30.00**
8BX5, Portable, 1948, "Snakeskin" & plastic, round dial over horiz bars, handle, side knob, BC, AC/DC/bat **$35.00**
8BX6 "Globe Trotter", Portable, 1948, Aluminum & plastic, slide rule dial, thumb-wheel knobs, handle, BC, AC/DC/ bat **$40.00**
8BX54, Portable, 1948, Front round dial over horizontal grill bars, handle, side knob, BC, AC/DC/battery **$35.00**
8BX55, Portable, 1948, Front round dial over horizontal grill bars, handle, side knob, BC, AC/DC/battery **$35.00**
8F43, Table, 1950, Wood, upper slanted slide rule dial, lower horizontal louvers, 2 knobs, BC, battery **$35.00**

8K, Console, 1936, Wood, upper dial, tuning eye, lower cloth grill w/ cut-outs, 8 tubes, BC, SW, AC $150.00

8R71, Table, 1949, Plastic, recessed dial on top of case, large front cloth grill, 4 knobs, BC, FM, AC $45.00

8-RF-13, Console on Legs, 1958, Wood, slide rule dial, horizontal bars over cloth grill, long legs, BC, FM, AC $40.00

8V7, Console-R/P, 1949, Wood, inner right dial/knobs, left phono, 2 lift tops, front horiz grill bars, BC, AC $70.00

8V90, Console-R/P, 1949, Wood, inner right slide rule dial, 4 knobs, phono, criss-cross grill, BC, FM, AC $65.00

8V111, Console-R/P, 1949, Wood, right tilt-out slide rule dial, 4 knobs, left fold-down phono, BC, FM, AC $80.00

8V151, Console-R/P, 1949, Wood, right tilt-out dial, 8 pushbuttons, left fold-down phono, BC, SW, FM, AC $100.00

8X53, Table, 1948, Wood, upper slanted slide rule dial, horizontal grill openings, 2 knobs, BC, AC/DC $40.00

8X71, Table, 1949, Maroon plastic, dotted slide rule dial, horiz grill bars, 3 knobs, AM, FM, AC/DC $40.00

8X72, Table, 1949, Ivory plastic, dotted slide rule dial, horiz grill bars, 3 knobs, AM, FM, AC/DC $40.00

8X521, Table, 1948, Maroon plastic, round dial on case top, horiz louvers, right side knob, BC, AC/DC...................... $55.00

8X522, Table, 1948, Ivory plastic, round dial on case top, horiz louvers, right side knob, BC, AC/DC $55.00

8-X-541, Table, 1949, Maroon plastic, round dial knob, horizontal grill bars, brass top strip, BC, AC/DC $35.00

8-X-542, Table, 1949, Ivory plastic, round dial knob, horizontal grill bars, brass top strip, BC, AC/DC $35.00

8-X-547, Table, 1949, White plastic, round dial knob, horizontal grill bars, brass top strip, BC, AC/DC $35.00

8X681, Table, 1949, Maroon plastic, large round dial/grill, thumbwheel knobs, base, BC, SW, AC/DC $45.00

8X682, Table, 1949, Ivory plastic, large round dial/grill, thumbwheel knobs, base, BC, SW, AC/DC $45.00

9-BT-9E, Portable, 1957, White plastic, transistor, round dial, horiz grill bars, thumbwheel knob, AM, bat $30.00

9-BT-9H, Portable, 1957, Green plastic, transistor, round dial, horiz grill bars, thumbwheel knob, AM, bat $30.00

9-BT-9J, Portable, 1957, Grey plastic, transistor, round dial, horiz grill bars, thumbwheel knob, AM, bat $30.00

9BX5, Portable, 1948, Leatherette & plastic, center front round dial, handle, BC, AC/DC/battery $40.00

9BX56, Portable, 1949, Plastic, upper dial, dotted lower panel, wire stand, 2 side knobs, BC, AC/DC/bat $30.00

9K-2, Console, 1936, Wood, upper dial, tuning eye, lower grill w/ bars, 9 tubes, 5 knobs, BC, SW, AC $150.00

9T, Tombstone, 1935, Wood, lower dial, upper cloth grill w/cut-outs, tuning eye, 9 tubes, BC, SW, AC $210.00

9TX, Table, 1939, Catalin, right front dial knob, wrap-around horizontal louvers, BC .... $700.00+

9TX23 "Little Nipper", Table, 1939, Wood, Deco, right front vertical slide rule "V" dial, left grill, 2 knobs, AC/DC $75.00

9TX31 "Little Nipper", Table, 1939, Plastic, right front dial, left horizontal wrap-around grill bars, 2 knobs, AC $50.00

9W101, Console-R/P, 1949, Wood, inner right slide rule dial, 4 knobs, left pull-out phono drawer, BC, FM, AC $70.00

9W103, Console-R/P, 1949, Wood, inner right slide rule dial, left phono, lower grill, storage, feet, BC, FM, AC $75.00

9W106, Console-R/P, 1950, Wood, inner right slide rule dial, 3 knobs, left pull-out phono drawer, BC, FM, AC $85.00

9X11, Table, 1939, Catalin, right dial, left cloth grill with "W" cut-out, 2 right side knobs, AC/DC .... $700.00 +

9X12, Table, 1939, Catalin, right dial, left cloth grill with "W" cut-out, 2 right side knobs, AC/DC .... $700.00 +

9X13, Table, 1939, Catalin, right dial, left cloth grill with "W" cut-out, 2 right side knobs, AC/DC .... $700.00+

9X14, Table, 1939, Catalin, right dial, left cloth grill with "W" cut-out, 2 right side knobs, AC/DC .... $700.00 +

9-X-561, Table, 1950, Plastic, narrow slide rule dial, circular front grill bars, 2 side knobs, BC, AC/DC $40.00

9-X-562, Table, 1950, Plastic, narrow slide rule dial, circular front grill bars, 2 side knobs, BC, AC/DC $40.00

9X571, Table, 1950, Plastic, thin lower slide rule dial, "bull-horn" louvers, side knobs, BC, AC/DC $55.00

9X572, Table, 1949, Plastic, thin lower front slide rule dial, "bull-horn" louvers, side knobs, BC, AC $55.00

9-X-641, Table, 1950, Plastic, upper slanted slide rule dial, lower horiz grill bars, 2 knobs, BC, AC/DC $45.00

9-X-642, Table, 1950, Plastic, upper slanted slide rule dial, lower horiz grill bars, 2 knobs, BC, AC/DC $45.00

9X652, Table, 1950, Plastic, upper slanted slide rule dial, horizontal louvers, 3 knobs, BC, SW, AC/DC $45.00

9-Y-7, Table-R/P, 1949, Wood, slide rule dial, horiz grill bars, 2 knobs, 3/4 lift top, inner phono, BC, AC $30.00

9Y51, Table-R/P, 1950, Plastic, upper front slide rule dial, side knobs, 3/4 lift top, inner phono, BC, AC $50.00

9Y510, Table-R/P, 1951, Plastic, front slanted slide rule dial, side knobs, 3/4 lift top, inner phono, BC, AC $50.00

10T, Tombstone, 1936, Wood, center dial, tuning eye, upper grill w/ 2 vert bars, 5 knobs, BC, SW, LW, AC $210.00

10T11, Tombstone, 1937, Black lacquer w/chrome supports, center dial, upper grill, tuning eye, 5 knobs $475.00

13K, Console, 1936, Wood, upper dial, tuning eye, lower grill w/vert bars, 13 tubes, 5 knobs, AM, 3SW $215.00

14BT2, Table, 1940, Wood, rectangular case, right slide rule dial, left horizontal louvers, 3 knobs $60.00

14X, Table, 1941, Plastic, upper slanted slide rule dial, lower horiz louvers, 5 tubes, 2 knobs, BC, SW $45.00

15K, Console, 1936, Wood, 2 dials, tuning eye, lower grill w/center vert bars, 15 tubes, 5 knobs, AC $250.00

15X, Table, 1940, Plastic, upper slanted dial w/red dot pointer, lower horiz louvers, 3 knobs, AC $40.00

16T4, Table, 1940, Wood, right slide rule dial, pushbuttons, left curved grill w/horiz bars, 4 knobs $85.00

16X3, Table, 1941, Wood, upper slanted slide rule dial, lower horizontal louvers, 3 knobs, AC $50.00

16X4, Table, 1941, Two-tone wood, upper slide rule dial, large lower grill area, pushbuttons, AC $50.00

16X13, Table, 1940, Upper slanted slide rule dial, lower horizontal louvers, 3 knobs, BC, SW, AC $45.00

18T, Table, 1940, Wood, upper slanted slide rule dial, pushbuttons, lower cloth grill, BC, 2SW, AC $50.00

19K, Console, 1940, Walnut, slide rule dial, 6 pushbuttons, vertical grill bars, 9 tubes, BC, 2SW, AC $195.00

25BP "Pick-Me-Up", Portable, 1941, Two-tone leatherettte, right front dial, left horiz louvers, handle, AC/DC/bat $30.00

25BT2, Table, 1942, Wood, right front slide rule dial, left horizontal grill bars, 3 knobs, battery $40.00

26X1, Table, 1941, Plastic, upper slanted slide rule dial, lower horiz louvers, 3 knobs, BC, SW, AC $40.00

26X3, Table, 1941, Upper slanted slide rule dial, lower horizontal louvers, 3 knobs, BC, SW, AC/DC $40.00

26X4, Table, 1941, Wood, slanted slide rule dial, pushbuttons, horiz louvers, 3 knobs, BC, SW, AC $55.00

28 "Carryette", Table, 1933, Wood, rectangular case, window dial, center 3-section cloth grill, 4 knobs, feet $95.00

28B, Table, 1933, Wood, window dial, center cloth grill with cut-outs, vertical side bands, 3 knobs $75.00

28C "Colonial", Table, 1933, Wood, chest-type, window dial, 3-section grill, rounded top, side handles, AC $95.00

28D, Table, 1933, Wood, window dial, grill cut-outs, 3 knobs, rounded front w/tambour doors, AC $145.00

28E, Table, 1933, Wood, front window dial, cloth grill w/cut-outs, 3 knobs, fluted columns, AC $80.00

28T, Table, 1941, Wood, slanted slide rule dial, pushbuttons, lower cloth grill, 4 knobs, BC, 2SW, AC $55.00

28X, Table, 1941, Wood, upper slanted slide rule dial, lower horizontal louvers, 4 knobs, BC, SW, AC $50.00

28X5, Table, 1941, Wood, slanted slide rule dial, pushbuttons, horiz louvers, 4 knobs, BC, SW, AC $60.00

29K, Console, 1941, Wood, slanted slide rule dial, pushbuttons, vertical grill bars, 9 tubes, BC, SW, AC $150.00

29K2, Console, 1941, Wood, slanted slide rule dial, pushbuttons, vert grill bars, 9 tubes, 4 knobs, BC, 2SW, AC $195.00

36X, Table, 1941, Wood, slanted slide rule dial, horizontal wraparound louvers, 3 knobs, AC/DC $40.00

40X53 "La Siesta", Table, 1939, Wood, painted w/Mexican scene, right dial, left cloth grill, handle, 2 knobs, AC $300.00

40X54 "Treasure Chest", Table, 1939, Wood, nautical-type, left "ship's wheel" grill, 2 knobs, side rope handles, AC ... $200.00

40X56 "1939 World's Fair", Table, 1939, Repwood, Trylon & Perisphere pressed into front, right dial, handle, 2 knobs, AC $900.00

40X57 "San Francisco Expo", Table, 1939, Repwood, Golden Gate bridge pressed in front, right dial, left grill, handle, AC $750.00

45-E, Side Table/Bookcase, 1940, Wood, vertical slide rule dial, right/left horiz bars, 2 shelf book case, BC, AC/DC $55.00

45-EW, Side Table/Bookcase, 1940, Wood, vertical slide rule dial, right/left horiz bars, 2 shelf book case, BC, AC/DC $55.00

45-W-10, Console-R/P, 1951, Wood, inner right pull-out drawer w/ slide rule dial, 5 knobs, phono, BC, FM, AC $65.00

45X1, Table, 1940, Brown plastic, right front dial, left horizontal wraparound louvers, 2 knobs, AC $45.00

45X11, Table, 1940, Brown plastic, right dial, left vert grill bars, decorative case lines, 2 knobs, AC $50.00

45X12, Table, 1940, Ivory plastic, right dial, left vert grill bars, decorative case lines, 2 knobs, AC $50.00

45X13, Table, 1940, Two-tone wood, right front dial, left horizontal louvers, 2 knobs, AC $50.00

45X16, Table, 1940, Two-tone wood, right front dial, left horizontal louvers, 2 knobs, BC, AC/DC $50.00

46X3, Table, 1939, Two-tone wood, right front dial, left horizontal louvers, finished back, AC $45.00

46X11, Table, 1940, Plastic, right front dial, left vertical grill bars, 3 knobs, BC, SW, AC $50.00

46X12, Table, 1939, Ivory plastic, right front dial, left vertical grill bars, 3 knobs, BC, SW, AC $50.00

46X13, Table, 1940, Right front dial, left square checkerboard grill, 3 knobs, BC, SW, AC $45.00

46X23, Table, 1940, Wood, right front airplane dial, left horizontal louvers, 3 knobs, BC, SW $45.00

54B1, Portable, 1946, Inner round dial, perforated grill, thumbwheel knob, flip open lid, handle, BC, bat $50.00

54B2, Portable, 1946, Inner round dial, perforated grill, thumbwheel knob, flip open lid, handle, BC, bat $50.00

54B5, Portable, 1947, Plastic, left round dial, horizontal wrap-around louvers, 3 knobs, handle, BC, bat $45.00

55F, Table, 1946, Wood, right front slide rule dial, left horizontal louvers, 3 knobs, BC, battery $40.00

55U, Table-R/P, 1946, Wood, upper front slide rule dial, 2 knobs, horizontal louvers, lift top, BC, AC $30.00

55X, Table, 1942, Wood, center front dial, right & left grills w/ horizontal louvers, 2 knobs, AC/DC $55.00

56X, Table, 1946, Plastic, upper dial w/red dot pointer, lower horiz louvers, 3 knobs, BC, AC/DC $40.00

56X2, Table, 1946, Plastic, upper dial w/red dot pointer, lower horiz louvers, 3 knobs, BC, AC/DC $40.00

56X5 "The 12,000 Miler", Table, 1946, Wood, upper slanted slide rule dial, lower horizontal louvers, 3 knobs, BC, SW $50.00

56X10, Table, 1946, Plastic, upper slide rule dial, lower horizontal louvers, 3 knobs, BC, SW, AC/DC $40.00

58AV, Console-R/P, 1946, Wood, inner right slide rule dial, 4 knobs, 6 pushbuttons, lift top, BC, SW, AC $50.00

59AV1, Console-R/P, 1946, Wood, slide rule dial, 6 knobs, 6 pushbuttons, doors, pull-out phono, BC, 2 SW, AC $75.00

61-8, Table, 1948, Brown plastic, slanted slide rule dial, horiz wrap-around louvers, 2 knobs, BC, AC/DC $45.00

61-9, Table, 1948, Ivory plastic, slanted slide rule dial, horiz wrap-around louvers, 2 knobs, BC, AC/DC $45.00

64F3, Table, 1946, Wood, upper slide rule dial, lower horiz louvers, 2 knobs, burl graining, BC, bat $50.00

65AU, Table-R/P, 1947, Wood, upper front slide rule dial, horizontal louvers, 2 knobs, lift top, BC, AC $40.00

65BR9, Portable, 1947, Center round dial, lower horizontal grill bars, side switch, handle, BC, AC/bat $45.00

65F, Table, 1948, Wood, right front slide rule dial, left horizontal louvers, 3 knobs, BC, battery $45.00

65U, Table-R/P, 1947, Wood, front slide rule dial, horizontal louvers, 2 knobs, lift top, BC, AC $45.00

65X1, Table, 1948, Plastic, upper slanted slide rule dial, lower horiz louvers, 2 knobs, BC, AC/DC $45.00

65X2, Table, 1946, Plastic, upper slanted slide rule dial, lower horiz louvers, 2 knobs, BC, AC/DC $45.00

66BX "Globe Trotter", Portable, 1947, Aluminum/plastic, flip-up dial lid, horizontal grill bars, handle, BC, AC/DC/bat $40.00

66X1, Table, 1946, Walnut plastic, concave slide rule dial, cloth grill, rear hand-hold, 3 knobs, BC, SW, AC/DC $60.00

66X2, Table, 1948, Plastic, concave slide rule dial, cloth grill, rear hand-hold, 3 knobs, BC, SW, AC/DC $60.00

66X3, Table, 1946, Two-tone wood, concave slide rule dial, "zebra" cloth grill, 3 knobs, AC/DC $55.00

66X7, Table, 1946, Dark blue Catalin, concave slide rule dial, horiz grill bars, 3 knobs, BC, SW, AC ....$500.00+

66X8, Table, 1946, Maroon Catalin, concave slide rule dial, horizontal grill bars, 3 knobs, BC, SW, AC ....$500.00 +

66X9, Table, 1946, Green swirl Catalin, concave slide rule dial, horiz grill bars, 3 knobs, BC, SW, AC ....$500.00+

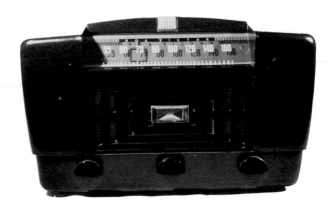

66X11, Table, 1947, Brown plastic, slanted slide rule dial, rectangular grill cut-outs, 3 knobs, BC, AC/DC $50.00

66X12, Table, 1947, Ivory plastic, slanted slide rule dial, rectangular grill cut-outs, 3 knobs, BC, AC/DC **$50.00**

66X13, Table, 1948, Walnut, slanted slide rule dial, horiz louvers, brass trim, 3 knobs, BC, SW, AC/DC $40.00
66X14, Table, 1948, Blonde, slanted slide rule dial, horiz louvers, brass trim, 3 knobs, BC, SW, AC/DC **$40.00**
66X15, Table, 1948, Mahogany, slanted slide rule dial, horiz louvers, brass trim, 3 knobs, BC, SW, AC/DC **$40.00**
67AV1, Console-R/P, 1946, Wood, dial & 4 knobs in right tilt-out drawer, lift top, inner phono, BC, SW, AC **$75.00**
68R1, Table, 1947, Plastic, center front slide rule dial, upper vertical grill bars, 4 knobs, BC, FM **$40.00**
75X11, Table, 1948, Maroon plastic & brass, right dial, left cloth grill w/vert bars, 2 knobs, BC, AC/DC **$70.00**
75X12, Table, 1948, Ivory plastic & brass, right dial, left cloth grill w/vert bars, 2 knobs, BC, AC/DC **$70.00**
75X14, Table, 1948, Mahogany plastic & brass, right dial, left cloth grill w/vert bars, 2 knobs, BC, AC/DC **$70.00**
75X15, Table, 1948, Walnut plastic & brass, right dial, left cloth grill w/vert bars, 2 knobs, BC, AC/DC **$70.00**
75X16, Table, 1948, Blonde plastic & brass, right dial, left cloth grill w/vert bars, 2 knobs, BC, AC/DC **$70.00**
75X17, Table, 1948, Lacquered cabinet w/oriental designs, right dial, left vert grill bars, brass trim **$125.00**
75X18, Table, 1948, Lacquered cabinet w/oriental designs, right dial, left vert bars, brass trim, AC **$125.00**
76ZX11, Table, 1948, Walnut plastic, slanted slide rule dial, horizontal louvers, 2 knobs, BC, AC/DC **$45.00**
76ZX12, Table, 1948, Ivory plastic, slanted slide rule dial, horizontal louvers, 2 knobs, BC, AC/DC **$45.00**
77U, Table-R/P, 1948, Wood, upper front slide rule dial, horizontal louvers, 2 knobs, lift top, BC, AC **$35.00**
77V1, Console-R/P, 1948, Wood, inner right slide rule dial, 3 knobs, left phono, lift top, BC, AC **$65.00**

77V2, Console-R/P, 1948, Wood, right tilt-out slide rule dial, 4 knobs, left lift top, inner phono, BC, SW, AC **$65.00**

79-10, Tombstone, Wood, lower front dial, tuning eye, upper cloth grill w/vertical bars, 5 knobs $145.00
84BT, Tombstone, 1937, Wood, rounded shoulders, center front dial, vertical grill cut-outs, 2 knobs, bat **$100.00**
85E, Chairside, 1937, Walnut finish, Deco, top square dial & knobs, front grill w/horiz bars, storage, AC **$120.00**
85T, Table, 1936, Wood, right front square dial, left cloth grill w/2 horizontal bars, 3 knobs, AC **$40.00**

85-T-1, Table, 1936, Wood, right front dial, left wrap-around grill w/2 horiz bars, 3 knobs, BC, SW, AC $40.00
86T, Table, 1937, Wood, Deco, right dial, rounded left side, horizontal grill bar, 3 knobs, BC, SW, AC **$60.00**
86T1, Table, 1937, Wood, right front dial, left cloth grill w/horizontal bars, 3 knobs, BC, SW, AC **$50.00**
86T2, Table, 1937, Two-tone wood, right slide rule dial, curved left w/curved grill cut-out, 3 knobs **$75.00**
86T3, Tombstone, 1938, Wood, center front slide rule dial, upper cloth grill w/vertical cut-outs, 3 knobs **$115.00**
86T6, Table, 1938, Wood, right slide rule dial, rounded left w/Deco grill cut-outs, pushbuttons, BC, SW **$75.00**
87K-1, Console, 1938, Wood, slanted slide rule dial, vertical grill bars, pushbuttons, tuning eye, BC, SW **$145.00**

87K-2, Console, 1938, Wood, upper slide rule dial, vertical grill bars, pushbuttons, tuning eye, BC, SW **$145.00**

87T, Table, 1937, Wood, right dial, left wrap-around cloth grill w/horiz bars, 3 knobs, BC, SW, AC **$50.00**

87T-2, Table, 1938, Wood, slide rule dial, upper cloth grill, 6 pushbuttons, tuning eye, 2 knobs, BC, SW **$80.00**

94BP1, Portable, 1939, Leatherette, right front square dial, left grill, handle, 2 knobs, battery **$30.00**

94BT1, Tombstone, 1938, Wood, lower airplane dial, upper grill w/splayed vertical bars, 2 knobs, battery **$60.00**

95T5, Table, 1938, Wood, slide rule dial, pushbuttons, upper grill w/vert bars, 2 knobs, BC, SW, AC **$60.00**

96E, Chairside, 1938, Wood, Deco, top dial/knobs/pushbuttons, front grill w/vertical center bar, BC, AC **$100.00**

96E2, Chairside, 1939, Two-tone wood, half-round, top dial, pushbuttons, vertical grill bars, BC, SW, AC **$145.00**

96K, Console, 1938, Wood, slide rule dial, vertical grill bar, 6 pushbuttons, 6 tubes, BC, SW, AC **$135.00**

96K6, Console, 1939, Wood, slide rule dial, cloth grill w/vertical center bar, 6 pushbuttons, BC, SW, AC **$145.00**

96T, Table, 1938, Wood, lower slide rule dial, upper cloth grill w/vert bars, pushbuttons, BC, AC **$55.00**

96T1, Table, 1938, Wood, lower slide rule dial, pushbuttons, upper vert grill bars, 2 knobs, BC, AC **$60.00**

96T2, Table, 1938, Wood, lower slide rule dial, vert grill bars, 6 pushbuttons, 2 knobs, BC, SW, AC **$55.00**

96T3, Table, 1938, Wood, lower front slide rule dial, upper grill, 6 pushbuttons, 2 knobs, BC, SW, AC **$55.00**

96T4, Table, 1939, Wood, lower slide rule dial, pushbuttons, upper grill w/2 chrome bars, 2 knobs **$60.00**

96T6, Table, 1939, Wood, lower slide rule dial, large upper grill, 5 pushbuttons, 2 knobs, BC, SW, AC/DC **$50.00**

96-X-1, Table, 1939, Plastic, Deco, right dial, curved left w/wrap-around louvers, raised top, 3 knobs, AC **$175.00**

97E, Chairside, 1938, Wood, top dial/knobs/pushbuttons/tuning eye, front grill w/horiz bars, BC, SW, AC **$110.00**

97KG, Console, 1938, Wood, slanted slide rule dial, pushbuttons, tuning eye, vert grill bars, BC, SW, AC **$140.00**

97T, Table, 1938, Wood, slide rule dial, grill w/wrap-over vert bars, pushbuttons, tuning eye, BC, SW, AC **$65.00**

97X, Table, 1938, Wood, lower slide rule dial, upper cloth grill w/vert bars, pushbuttons, BC, AC/DC **$50.00**

98K, Console, 1938, Wood, upper dial, vert grill bar, 8 pushbuttons, tuning eye, 8 tubes, BC, SW, AC **$140.00**

98K2, Console, 1939, Wood, slide rule dial, vertical grill bars, pushbuttons, tuning eye, BC, 2SW, AC **$135.00**

98X, Table, 1938, Wood, slide rule dial, tuning eye, pushbuttons, 3-section wrap-over grill, BC, SW **$65.00**

99K, Console, 1938, Wood, upper curved dial, vertical grill bars, 8 pushbuttons, 9 tubes, BC, SW, AC **$145.00**

99T, Table, 1938, Wood, slide rule dial, vertical grill bars, 8 pushbuttons, tuning eye, BC, SW, AC **$75.00**

100, Cathedral, 1933, Wood, center front window dial, cloth grill w/ cut-outs, side moldings, BC, AC **$175.00**

102, Table, 1934, Metal, right front dial, center grill w/Deco cut-outs, left volume knob, BC, AC/DC **$80.00**

103, Tombstone, 1934, Wood, lower front round dial, upper cloth grill w/vertical cut-outs, fluting **$85.00**

110, Cathedral, 1933, Wood, front window dial, upper cloth grill w/ cut-outs, side moldings, BC, SW, AC **$225.00**

110K "Presidential", Console, 1940, Wood, oblong slide rule dial, pushbuttons, vert grill bars, 10 tubes, BC, 3SW, AC ..**$250.00**

111, Table, 1934, Inlaid walnut, right window dial, scalloped grill cut-outs, 4 knobs, BC, SW, AC **$90.00**

111K, Console, 1941, Wood, oblong slide rule dial, pushbuttons, lower grill w/vert bars, BC, 3SW, AC **$150.00**

114, Table, 1934, Two-tone wood, right front dial, center 3-section cloth grill, 2 knobs, BC, SW, AC **$75.00**

118, Tombstone, 1934, Wood, center front round dial, upper cloth grill w/cut-outs, 5 tubes, BC, SW, AC **$125.00**

119, Tombstone, 1935, Wood, center front round dial, upper cloth grill w/vertical bars, 4 knobs, AC **$100.00**

120, Cathedral, 1933, Wood, window dial, upper grill w/cut-outs, side/top moldings, 4 knobs, BC, AC $225.00

121, Cathedral, 1933, Wood, round dial, cloth grill w/cut-outs, side/top moldings, 4 knobs, BC, SW, AC **$240.00**

125, Tombstone, 1934, Wood, center front round dial, upper 4-section grill, fluting, 4 knobs, BC **$115.00**

128, Tombstone, 1934, Wood, shouldered, rounded top, center front round dial, grill cut-outs, BC, 2SW, AC $275.00

140, Tombstone, 1933, Wood, scalloped shoulders, airplane dial, grill w/cut-outs, 4 knobs, BC, SW, AC **$250.00**

142-B, Cathedral, 1933, Wood, window dial, upper grill w/cut-outs, side/top moldings, 4 knobs, BC, bat **$150.00**

143, Tombstone, 1933, Wood, shouldered, rounded top, round dial, grill w/vertical cut-outs, BC, 3SW, AC **$280.00**

211, Console, 1934, Wood, lowboy, upper round dial, lower cloth grill w/cut-outs, 5 tubes, BC, SW, AC **$135.00**

211K, Console, 1941, Wood, slanted slide rule dial, pushbuttons, vertical grill bars, 11 tubes, 2 speakers, BC, 3SW, AC **$165.00**

220, Console, 1933, Wood, lowboy, upper round dial, lower grill w/cut-outs, 4 knobs, 6 legs, BC, AC **$150.00**

221, Console, 1934, Wood, lowboy, upper round dial, lower cloth grill w/cut-outs, 6 tubes, BC, SW, AC **$135.00**

224, Console, 1934, Wood, lowboy, upper round dial, lower cloth grill w/cut-outs, 6 legs, BC, SW, AC **$150.00**

240 **"All-Wave"**, Console, 1933, Wood, lowboy, recessed upper round dial, lower grill w/cut-outs, 6 legs, BC, SW, AC **$150.00**

242, Console, 1934, Wood, lowboy, upper round dial, lower cloth grill w/cut-outs, 6 legs, BC, SW, AC **$155.00**

260, Console, 1933, Wood, lowboy, upper window dial, lower grill w/cut-outs, 10 tubes, 6 legs, BC, AC **$160.00**

262, Console, 1934, Wood, lowboy, round dial, lower grill w/cut-outs, 10 tubes, 6 legs, BC, SW, AC **$160.00**

280, Console, 1933, Wood, lowboy, upper window dial, lower grill w/cut-outs, 12 tubes, 6 legs, BC, AC **$175.00**

281, Console, 1934, Wood, lowboy, inner dial, lower grill w/cut-outs, doors, 12 tubes, 6 legs, BC, SW, AC **$195.00**

300 "Duo", Table-R/P, 1933, Wood, front dial and cloth grill w/fancy cut-outs, lift top, inner phono, BC, AC **$100.00**

301 "Duo", Table-R/P, 1934, Wood, upright style, round dial, upper grill, lift top, inner phono, BC, SW, AC **$100.00**

310 "Duo", Console-R/P, 1933, Wood, highboy, window dial, cloth grill w/cut-outs, stretcher base, BC, SW, AC **$130.00**

340 "All Wave Duo", Console-R/P, 1934, Wood, lowboy, center round dial, lower grill w/cut-outs, 8 tubes, 6 legs, AC .**$150.00**

341, Console-R/P, 1934, Wood, lowboy, round dial, grill w/lyre cut-out, lift top, inner phono, BC, SW, AC **$150.00**

381, Console-R/P, 1934, Wood, lowboy, inner round dial, doors, lift top, inner phono, 12 tubes, BC, SW, AC **$225.00**

551, Table, 1950, Black plastic, center front round dial, curved louvers, top strip, side knob, AC/DC **$45.00**

552, Table, 1950, Ivory plastic, center front round dial, curved louvers, top strip, side knob, AC/DC **$45.00**

610V2, Console-R/P, 1948, Wood, inner right slide rule dial, 4 knobs, 6 pushbuttons, pull-down phono, BC, FM, AC **$70.00**

612V3, Console-R/P, 1947, Wood, slide rule dial, 6 knobs, 8 pushbuttons, pull-down door, BC, SW, FM, AC **$100.00**

710V2, Console-R/P, 1948, Wood, right tilt-out slide rule dial/4 knobs, pull-out phono drawer, BC, FM, AC **$70.00**

711V2, Console-R/P, 1947, Wood, right tilt-out dial, 2 knobs & 6 pushbuttons, fold-out phono, BC, SW, FM, AC **$70.00**

810K1, Console, 1937, Wood, square dial, tuning eye, 3 vertical grill bars, 9 tubes, 3 knobs, BC, SW, AC **$140.00**

811K, Console, 1937, Wood, slide rule dial, pushbuttons, tuning eye, 3 vert grill bars, 11 tubes, BC, SW, AC **$150.00**

812K, Console, 1937, Wood, slide rule dial, pushbuttons, tuning eye, horiz grill bars, 12 tubes, BC, SW, AC **$175.00**

813K, Console, 1937, Wood, curved dial, pushbuttons, tuning eye, horiz grill bars, 13 tubes, BC, SW, AC **$200.00**

816-K , Console, 1937, Wood, curved dial, pushbuttons, tuning eye, horiz grill bars, 16 tubes, BC, SW, AC **$300.00**

910KG, Console, 1938, Wood, curved dial, pushbuttons, cloth grill w/ 2 vert bars, 10 tubes, BC, SW, AC **$175.00**

911K, Console, 1938, Wood, curved dial, pushbuttons, tuning eye, vert grill bars, 11 tubes, BC, SW, AC **$190.00**

A55, Console-R/P, 1950, Wood, upper front slide rule dial, 4 knobs, right pull-out phono drawer, BC, AC **$50.00**

A78, Console-R/P, 1950, Wood, inner right slide rule dial, left pull-out phono, double doors, storage, AC **$75.00**

A-82, Console-R/P, 1951, Wood, inner right slide rule dial w/top phono, left pull-out phono drawer, BC, AC **$80.00**

A-108, Console-R/P, 1951, Wood, inner pull-out dial w/top phono, 5 knobs, pull-out phono, doors, BC, FM, AC **$80.00**

B-411, Portable, 1951, Plastic, center front round dial knob, 1 thumbwheel knob, handle, BC, battery **$35.00**

BP-10, Portable, 1941, Leatherette & plastic, inner dial, 2 thumbwheel knobs, flip-open front, handle, bat **$65.00**

BT7-8, Tombstone, 1935, Wood, rounded shoulders, lower round dial, grill w/cut-outs, 4 knobs, BC, SW, bat **$85.00**

BT42, Table, 1940, Two-tone wood, rectangular case, right front slide rule dial, left square grill, AC **$50.00**

BX6 "Globe Trotter", Portable, 1950, Leatherette & aluminum, slide rule dial, 2 thumbwheel knobs, slide-down cover, BC, AC/DC/bat **$40.00**
BX55, Portable, 1950, Plastic, "alligator" panels, horiz louvers, 2 side knobs, handle, BC, AC/DC/bat **$40.00**

BX-57, Portable, 1950, Plastic, "alligator" panels, horiz louvers, 2 side knobs, handle, BC, AC/DC/bat **$40.00**
C1E, Table-C, 1959, Plastic, lower right round dial, left clock, right vertical grill bars, BC, AC **$20.00**
C13-2, Console, 1935, Wood, inner dial, tuning eye, 5 knobs, doors, criss-cross grill, 13 tubes, BC, SW, AC **$175.00**
D-22, Console-R/P/Rec, 1935, Wood, inner dial, tuning eye, 5 knobs, doors, criss-cross grill, 22 tubes, lift top, AC **$500.00**
HF2, Console, 1938, Wood, curved dial, pushbuttons, cloth grill w/ vertical bars, 12 tubes, BC, SW, AC **$195.00**
HF4, Console, 1938, Wood, 18th Century-style, half-round, inner dial & knobs, 12 tubes, BC, SW, AC **$195.00**
HF6, Console, 1938, Wood, curved dial, pushbuttons, cloth grill w/ vertical bars, 14 tubes, 7 bands, AC **$200.00**
K-60, Console, 1939, Wood, upper slide rule dial, 8 pushbuttons, 2 vert grill bars, 4 knobs, BC/SW, AC **$120.00**
K-80, Console, 1938, Wood, slide rule dial, lower grill w/3 vert bars, pushbuttons, 8 tubes, BC, SW, AC **$140.00**
K81, Console, 1939, Wood, upper slide rule dial, pushbuttons, lower 3-section cloth grill, BC, 2SW, AC **$145.00**
K130, Console, 1939, Wood, inner slide rule dial, pushbuttons, doors, vertical grill bars, BC, 2SW, AC **$175.00**
P31, Portable, 1932, Leatherette, inner wood panel w/window dial, 2 grills, snap-off front, handle, 8 tubes **$200.00**

PX600, Portable, 1952, Plastic, front slide rule dial, perforated grill w/ circles, handle, BC, AC/DC/bat **$35.00**
Q36, Table, 1947, Wood, center front vertical multi-band dial, right & left cloth grills, 4 knobs **$45.00**
QB55X, Table, 1948, Plastic, slanted slide rule dial, horizontal louvers, 3 knobs, BC, SW, LW, battery **$40.00**
R-4 "Superette", Cathedral, 1932, Wood, front window dial, upper 3-section cloth grill, fluted columns, 3 knobs **$300.00**
R5 "Radiolette", Cathedral, 1931, Wood, lower front window dial, upper cloth grill w/cut-outs, 3 knobs, AC **$195.00**

R-7 "Superette", Tombstone, Wood, shouldered, lower window dial, upper 3-section cloth grill, 3 knobs, AC **$130.00**

R-8 "Superette", Tombstone, 1932, Wood, lower front window dial, upper cloth grill w/gothic cut-outs, 3 knobs ..**$145.00**
R-15 "Victor", Console, Wood, lowboy, decorative carving, upper dial, lower cloth grill, 7 tubes, 2 knobs **$165.00**
R22S, Table, 1933, Ornate case, round front dial, center cloth grill w/ cut-outs, metal handle, AC/DC **$110.00**
R-22-W, Table, 1933, Wood, front dial, center cloth grill w/cut-outs, sloped sides, AC/DC **$95.00**

R-27, Table, 1933, Wood, round front dial, center cloth grill w/cut-outs, sloped sides, AC **$75.00**

R28, Cathedral, 1933, Wood, lower front window dial, upper cloth grill w/cut-outs, 4 knobs, AC **$200.00**

R-28B, Table, 1933, Wood, front window dial, center cloth grill w/cut-outs, 4 knobs, AC **$75.00**

R28P, Tombstone, 1933, Wood, lower round dial, upper 4 section cloth grill, 5 tubes, 4 knobs, BC, SW **$125.00**

R-32 "Victor", Console, 1929, Wood, lowboy, upper front sliding dial w/wood escutcheon, large lower cloth grill **$130.00**

R-34 "Victor", Console, 1929, Wood, lowboy, upper sliding dial w/wood escutcheon, lower grill w/2 vert bars **$130.00**

R37, Cathedral, 1933, Wood, lower window dial, upper cloth grill w/cut-outs, side & top moldings, AC $300.00

R38, Console, 1933, Wood, lowboy, window dial, lower grill w/cut-outs, 6 legs, stretcher base, AC..................................**$145.00**

R-70, Cathedral, 1933, Wood, lower front window dial w/escutcheon, upper grill w/cut-outs, 3 knobs **$225.00**

R73, Tombstone, 1933, Wood, shouldered, lower window dial, cloth grill w/cut-outs, 3 knobs, 8 tubes **$185.00**

R-78, Console, 1932, Wood, lowboy, inner dial, double doors, 6 legs, stretcher base, 12 tubes, AC...............................**$185.00**

R110, Tombstone, 1933, Wood, shouldered, lower window dial, cloth grill w/cut-outs, 3 knobs, 4 tubes **$145.00**

Radiola I, Table, 1923, Wood, model ER-753A, box-type, crystal, fold-open front, storage, handle $600.00

Radiola II, Portable, 1923, Mahogany, 2 dial inner panel, removable fold-open front, handle, 2 tubes, battery **$300.00**

Radiola III , Table, 1924, Wood, top black bakelite 1 dial panel with 2 exposed tubes, battery $125.00

Radiola III-A, Table, 1924, Wood, low rectangular case, top 1 dial black panel with 4 exposed tubes, battery $150.00

Radiola IV, Table, 1923, Mahogany, high rectangular case, inner left panel, door, right speaker, 3 tubes, bat **$425.00**

Radiola V , Table, 1923, Metal w/mahogany finish, rectangular case, wood base & top, 2 dials, 3 tubes, bat **$450.00**

Radiola VI, Table, 1923, Metal w/mahogany finish, rectangular case, wood base & top, 2 dials, battery **$450.00**

Radiola VII, Table, 1923, Wood, rectangular case, 2 dial black bakelite front panel, 5 tubes, battery **$1,400.00**

Radiola VIIB, Table, 1924, Wood, inner panel w/5 exposed tubes, lift top, front oval speaker grill, battery **$1,200.00**

Radiola VIII "Super-VIII", Console, 1924, Wood, lowboy, inner 2 dial panel, fold-down front, built-in speaker, storage, 6 tubes, bat **$500.00**

Radiola X , Table, 1924, Wood, tall case, lower 2 dial panel, upper speaker grill w/cut-outs, 4 tubes, bat **$350.00**

Radiola 16 , Table, 1927, Wood, low rectangular case, right front oval dial, 2 knobs, lift top, 6 tubes, bat **$90.00**

Radiola 17, Table, 1927, Wood, low rectangular case, center dial w/escutcheon, 7 tubes, 3 knobs, lift top, AC **$95.00**

Radiola 18, Table, 1927, Wood, low rectangular case, center dial w/escutcheon, lift top, 7 tubes, 2 knobs, AC **$95.00**

Radiola 20 "Floor Model", Console, 1925, Wood, highboy, slanted front panel with 2 thumbwheel dials, 4 knobs, lift top, bat **$185.00**

Radiola 20, Table, 1925, Wood, slanted front panel with 2 thumbwheel dials, 4 knobs, lift top, battery **$150.00**

Radiola 24, Portable, 1925, Leatherette, 2 inner dials, removable front cover stores loop antenna, handle, 6 tubes, bat .**$325.00**

Radiola 25, Table, 1925, Two-tone wood, slanted front w/2 thumbwheel dials, top loop antenna, 6 tubes, battery **$200.00**

Radiola 26, Portable, 1925, Walnut, inner 2 dial panel, grill cut-outs, fold-open front w/antenna, handle, 6 tubes, bat **$350.00**

Radiola 28, Console, 1925, Wood, highboy, slanted front panel, 2 thumbwheel dials, loop antenna, 8 tubes, battery **$225.00**

Radiola 30, Console, 1925, Wood, lower front fold-out contols, upper round grill w/cut-outs, fluted legs, bat **$250.00**

Radiola 30A, Console, 1927, Wood, lowboy, inner panel w/2 thumbwheel dials, double front doors, AC **$200.00**

Radiola 33, Console, 1929, Metal, Deco rectangular case on high legs, center front diamond-shaped dial, AC **$135.00**

Radiola 33, Table, 1929, Metal, Deco rectangular case, center front diamond-shaped dial, 3 knobs, AC **$100.00**

Radiola 43, Table, Two-tone wood, center window dial with pressed wood escutcheon, lift top, switch **$95.00**

Radiola 44, Table, 1929, Wood, low rectangular case, center front window dial w/escutcheon, AC **$100.00**

Radiola 48, Console, 1930, Wood, lowboy, ornate case, upper window dial, lower cloth grill w/cut-outs, AC **$235.00**

Radiola 60, Table, 1927, Two-tone walnut, low rectangular case, center front window dial, feet, AC **$110.00**

Radiola 60, Console/Desk, 1927, Wood, highboy, inner panel w/ window dial & left speaker grill, drop-front, AC **$225.00**

Radiola 62, Console, 1927, Walnut w/maple inlay, inner front dial & knobs, double doors, stretcher base, AC **$235.00**

Radiola 64, Console, 1927, Wood, lowboy, inner right dial, upper snap-out grill, large double doors, AC **$285.00**

Radiola 66, Console, 1929, Wood, lowboy, center front tapestry panel with right window dial, 2 knobs, AC **$185.00**

Radiola 80, Console, 1930, Wood, lowboy, upper front window dial, lower grill w/cut-outs, 3 knobs, AC **$175.00**

Radiola 86, Console-R/P/Rec, 1930, Wood, inner front dial, double doors, lift top, inner phono & recorder, 9 tubes, AC **$400.00**

Radiola AC/Radiola Senior, Table, 1923, Wood, 2 boxes, receiver with 1 exposed tube/amplifier with 2 exposed tubes .. **$400.00**

Radiola AR-812, Portable, 1924, Wood, low rectangular case, 2 dial front panel, handle, storage, 6 tubes, battery **$175.00**

Radiola Grand, Table, 1922, Wood, chest-style, inner 1 dial panel, 4 exposed tubes, lift top, front grill, bat **$500.00**

Radiola RAE-59, Console R/P/Rec, Wood, inner dial/3 knobs, doors, lower grill, lift top, inner phono & recorder, 10 tubes, AC **$300.00**

Radiola Regenoflex, Table, 1924, Wood, low rectangular case, 2 dial front panel, side compartments, 4 tubes, bat . $275.00

Radiola RS, Table, 1923, Wood, square case, lift top, inner 1 dial black panel, 2 exposed tubes, battery **$275.00**

Radiola Special, Table, 1923, Tall rectangular case, inner black panel w/controls, cover, 1 tube **$450.00**

RAE45, Console-R/P, 1931, Wood, lowboy, inner sliding dial w/ wood escutcheon, double doors, lift top, AC **$195.00**

RE45 "Victor", Console-R/P, 1929, Wood, lowboy, inner sliding dial w/wood escutcheon, double doors, lift top, AC **$195.00**

RV151, Console-R/P, 1949, Wood, right front tilt-out dial, left pull-out phono unit, storage, BC, FM, AC **$70.00**

SHF-1, Console/R/P/Rec, 1958, Wood, inner slide rule dial, phono & recorder, two separate speakers, BC, FM, AC **$125.00**

SHF-2, Console-R/P, 1958, Wood, inner left slide rule dial, pull-out phono drawer, right grill, BC, FM, AC **$55.00**

T-1EN, Portable, 1960, Transistor, upper round dial knob, lower horiz grill bars, swing handle, AM, bat **$30.00**

T-1JE, Portable, 1960, Transistor, upper round dial knob, lower horiz grill bars, swing handle, AM, bat **$30.00**

T5-2, Tombstone, 1935, Wood, small case, center front dial, upper grill w/cut-outs, 4 knobs, BC, SW, AC $100.00

T7-5, Tombstone, Wood, lower dial, upper cloth grill w/cut-outs, fluting, 7 tubes, 4 knobs, BC, SW $125.00

T60, Table, 1940, Wood, right slide rule dial, left grill w/horizontal bars, pushbuttons, BC, SW $65.00

T62, Table, 1940, Wood, right front slide rule dial, pushbuttons, left cloth grill, 4 knobs, BC, SW $60.00

T-80, Table, 1939, Wood, lower slide rule dial, 6 pushbuttons, tuning eye, cloth grill, BC, 2SW, AC $75.00

TX-1JE, Table, 1960, Plastic, transistor, center vertical dial, right & left speakers w/horiz bars, AM, bat $20.00

U12, Table-R/P, 1939, Wood, right front slide rule dial, pushbuttons, left grill, 4 knobs, lift top, AC $35.00

U20, Console-R/P, 1939, Wood, slide rule dial, pushbuttons, grill w/ crossed bars, lift top, BC, SW, AC $135.00

U42, Console-R/P, 1939, Walnut, inner dial/knobs/phono, double doors, vertical grill bars, BC, SW, AC $150.00

U46, Console-R/P, 1940, Wood, inner right slide rule dial, pushbuttons, left phono, doors, 13 tubes, BC, 2SW, AC $165.00

U-111, Table-R/P, 1939, Wood, left front dial, right grill w/horizontal bars, lift top, inner phono, BC, AC $35.00

U112, Table-R/P, 1939, Wood, left front dial, right square grill, lift top, inner phono, AC $35.00

U-119, Table-R/P, 1939, Wood, tall case, slide rule dial, pushbuttons, vert grill bars, lift top, BC, SW, AC $40.00

U-125, Console-R/P, 1939, Wood, inner right dial/knobs/pushbuttons, left phono, lift top, front grill, BC, SW, AC $120.00

U127E, Chairside-R/P, 1939, Wood, top dial & pushbuttons, horiz grill bars, lift top, inner phono, storage, AC $135.00

V-100, Table-R/P, 1939, Two-tone wood, right front dial, left horiz bars, 3 knobs, lift top, inner phono, AC $35.00

V-101, Table-R/P, 1940, Wood, right front slide rule dial, left grill, 3 knobs, lift top, inner phono, AC $35.00

V-135, Table-R/P, 1941, Wood, right front dial, left grill w/horiz bars, 3 knobs, lift top, inner phono, AC $35.00

V-225, Console-R/P, 1941, Wood, inner dial & knobs, pull-out phono, lower grill w/cut-outs, BC, 2SW, AC $150.00

X-551, Table, 1947, Plastic, semi-circular grill bars around center dial, right side knob, top strip, BC $40.00

X552, Table, 1951, Plastic, semi-circular grill bars around center dial, side knob, top strip, BC, AC/DC $40.00

X711, Table, 1951, Plastic, upper dotted slide rule dial over large grill area, 3 knobs, AM, FM, AC/DC $25.00

REALTONE
Realtone Electronics,
184 Fifth Ave., New York, New York

TR-555, Portable, 1960, Transistor, upper window dial, lower tear-drop-shaped perforated grill, AM, bat $45.00

TR-801, Portable, 1960, Transistor, upper window dial, lower tear-drop-shaped perforated grill, AM, bat $45.00

RECORDIO
Wilcox-Gay Corp.,
604 Seminary Street, Charlotte, Michigan

1J10, Table-R/Rec, 1951, Suitcase-style, inner left dial, disc recorder, lift top, handle, BC, AC $25.00

6B10, Table-R/P/Rec/PA, 1946, Inner dial/controls/oval grill, fold-down front, lift top, inner phono, BC, AC $35.00

6B30M, Table-R/P/Rec/PA, 1946, Front dial/controls/oval grill, lift top, inner phono, BC, AC $40.00

7D44, Console-R/P/Rec, 1948, Wood, tilt-out dial, 4 knobs, pushbuttons, right pull-out phono drawer, BC, FM, AC $80.00

7E40, Console-R/P/Rec, 1948, Wood, inner right slide rule dial, 4 knobs, left pull-out phono drawer, BC, FM, AC $70.00

8J10, Table-R/P/Rec, 1949, Leatherette, inner left dial, grill, 3 knobs, phono, lift top, handle, BC, AC $25.00

9G10, Table-R/P/Rec, 1950, Leatherette, inner left dial, grill, 5 knobs, phono, lift top, handle, BC, FM, AC $25.00

9G40M, Console-R/P/Rec, 1950, Wood, inner right slide rule dial, 5 knobs, left phono, fold-down front, BC, FM, AC $55.00

9H40B, Console-R/P, 1950, Wood, inner right slide rule dial, 5 knobs, inner left phono, drop-front, BC, FM, AC $60.00

REGAL
Regal Electronics Corp.,
20 West 20th Street, New York, New York

205, Table, 1947, Plastic, upper slanted slide rule dial, lower horiz louvers, 2 knobs, BC, AC/DC $45.00

271, Table, 1953, Plastic, large front half-moon dial over horizontal louvers, 2 knobs, BC, AC/DC $30.00

575, Table, 1953, Plastic, upper slide rule dial, horiz louvers wrap-around right side, BC, AC/DC $45.00

747, Portable, 1947, Metal & plastic, inner slide rule dial, 2 knobs, flip-up front, BC, AC/DC/battery **$45.00**

777, Portable, 1949, Metal & plastic, inner slide rule dial, 2 knobs, flip-up front, BC, AC/DC/bat **$45.00**

1049, Table, 1947, Plastic, slide rule dial, horizontal wrap-around louvers, 3 knobs, BC, SW, AC/DC **$40.00**

1107, Table, 1948, Plastic, upper slanted slide rule dial, 3 rows of grill circles, 2 knobs, BC, AC/DC **$50.00**

1500, Table, 1948, Plastic, upper slanted slide rule dial, lower horiz louvers, 3 knobs, BC, SW, bat **$30.00**

1749, Table, 1947, Plastic, upper slide rule dial, horiz wrap-around louvers, 3 knobs, BC, SW, AC/DC **$45.00**

1877, Portable, 1952, Plastic, lower front dial, upper checkered grill, handle, 2 knobs, BC, AC/DC/bat **$40.00**

7152, Table, 1949, Plastic, upper slanted slide rule dial, horizontal louvers, 3 knobs, BC, SW, AC/DC **$35.00**

7162, Table, 1949, Plastic, upper slanted slide rule dial, raised lattice grill, 3 knobs, BC, SW, AC/DC **$45.00**

7163, Table, 1949, Plastic, slanted slide rule dial, raised lattice grill, 3 knobs, BC, 2SW, AC/DC **$45.00**

7251, Table, 1948, Plastic, lower slide rule dial, lattice grill, slant-down top, 2 knobs, BC, AC/DC **$50.00**

BP48, Portable, 1948, Inner slide rule dial, horizontal grill bars, 2 knobs, flip-open top, handle, BC, bat **$35.00**

C-527, Table-C, 1952, Plastic, lower front dial, upper alarm clock face, side & top bars, 5 knobs, BC, AC **$35.00**

FM78, Table, 1949, Plastic, upper slanted slide rule dial, raised lattice grill, 3 knobs, BC, FM, AC/DC **$45.00**

L-76, Table, 1946, Wood, upper slanted slide rule dial, cloth grill w/ horiz bar, 2 knobs, BC, AC/DC **$35.00**

P-175, Portable, 1952, "Alligator", inner right dial, left grill, fold-down lid, handle, BC, SW, AC/DC/bat **$30.00**

W800, Table, 1947, Plastic, upper slide rule dial, lower horizontal louvers, 2 knobs, BC, AC/DC **$35.00**

W801, Table, 1947, Plastic, upper slide rule dial, lower horizontal louvers, 2 knobs, BC, AC/DC **$35.00**

W901, Table, 1947, Plastic, upper slide rule dial, lower horizontal louvers, 3 knobs, BC, SW, AC/DC **$35.00**

REGENCY
Regency Div., I. D. E. A., Inc.,
7900 Pendleton Pike,
Indianapolis, Indiana

TR-1, Portable, 1955, Plastic, 1st transistor, round dial, dotted grill, thumbwheel knob, AM, battery ....$250.00+

TR-1G, Portable, 1958, Transistor, right dial knob, lower perforated grill, thumbwheel knob, AM, battery **$85.00**

TR-4, Portable, 1958, Transistor, round dial knob, lower perforated grill, thumbwheel knob, AM, bat **$65.00**

TR-5, Portable, 1958, Leather case, transistor, right dial, left lattice grill, handle, AM, battery **$40.00**

TR-5C, Portable, 1958, Leather case w/grill, transistor, right front dial knob, strap, AM, battery **$40.00**

TR-6, Portable, 1957, Leather case w/perforated front grill, transistor, side knobs, handle, AM, battery **$45.00**

TR-11, Portable, 1959, Plastic, transistor, upper front round dial knob, horizontal grill bars, AM, bat **$30.00**

TR-22, Portable, 1959, Leather case, transistor, right side dial, front grill, handle, stand, AM, battery **$30.00**

TR-99 "World Wide", Portable, 1960, Transistor, upper left round dial, lower perforated grill, swing handle, AM, bat **$30.00**

XR-2A, Portable, 1958, Plastic, transistor, round dial, vertical bars, top switch, earphone, AM, battery **$35.00**

REMLER
Remler Co., Ltd.,
2101 Bryant Street,
San Francisco, California

10, Cathedral, 1932, Wood, lower dial, upper cloth grill w/cut-outs, fluted columns, 6 tubes, 3 knobs **$250.00**

15, Cathedral, 1932, Wood, lower dial, upper cloth grill w/cut-outs, fluted columns, 7 tubes, 3 knobs, AC **$275.00**

15-C, Console, 1932, Wood, lowboy, upper dial, lower cloth grill w/ cut-outs, 7 tubes, 3 knobs, AC **$140.00**

19, Console, 1932, Wood, lowboy, upper dial, lower cloth grill w/cut-outs, 10 tubes, BC, SW, AC **$155.00**

21-3, Cathedral, 1932, Wood, flat top, lower front curved dial, upper cloth grill w/cut-outs, 3 knobs, AC $195.00

40 "Scottie", Table, 1936, Bakelite, center front grill w/Deco cut-outs & Scottie logo, 2 knobs, BC, AC **$200.00**

41 "Scottie", Table, 1936, Bakelite, center front grill w/Deco cut-outs & Scottie logo, 2 knobs, BC, SW, AC **$200.00**

43 "Esquire", Table, 1936, Wood, center black dial surrounded by grill cut-outs, 7 tubes, 4 knobs, BC, SW, AC **$70.00**

51 "Skipper", Table, 1936, Bakelite, center front dial surrounded by grill cut-outs, 4 tubes, 2 knobs, BC, AC **$60.00**

54, Table, 1938, Plastic, right front dial, "Venetian blind" louvered grill, 5 tubes, 2 knobs, AC **$55.00**

61, Table, 1938, Wood, upper slide rule dial, pushbuttons, wrap-around horiz louvers, 3 knobs, AC **$55.00**

62 "Grenadier", Table, 1936, Wood, center black dial surrounded by grill cut-outs, 5 tubes, 4 knobs, BC, SW, AC **$70.00**

88, Tombstone, 1936, Wood, shouldered, lower dial, upper grill w/ cut-outs, 10 tubes, 2 speakers, BC, SW, AC **$160.00**

471, Table-R/P, 1940, Wood, left front slide rule dial, right grill w/horiz bars, lift top, inner phono, AC **$35.00**

5300BI "Scottie", Table-R/P, 1947, Two-tone plastic, slide rule dial, louvers, 2 knobs, lift top, Scottie logo, BC, AC **$135.00**

5310BL, Table-R/P, 1948, Blonde, inner right dial, 4 knobs, phono, lift top, front grill doors, BC, AC **$25.00**

5310M, Table-R/P, 1948, Mahogany, inner right dial, 4 knobs, phono, lift top, front grill doors, BC, AC **$25.00**

5400 "Scottie", Portable, 1948, Leatherette/walnut plastic, slide rule dial, handle, 2 knobs, Scottie logo, BC, AC/DC/bat **$85.00**

5410 "Scottie", Portable, 1948, Leatherette/white plastic, slide rule dial, handle, 2 knobs, Scottie logo, BC, AC/DC/bat **$85.00**

5500 "Scottie Pup", Table, 1948, Walnut plastic, slide rule dial, horiz grill bars w/Scottie logo, handle, BC, AC/DC **$150.00**

5505 "Scottie Pup", Table, 1948, Ebony & white plastic, slide rule dial, horiz grill bars w/Scottie logo, handle, BC, AC/ DC **$165.00**

5510 "Scottie Pup", Table, 1948, White plastic, slide rule dial, horiz grill bars w/Scottie logo, handle, BC, AC/DC **$150.00**

5515 "Scottie Pup", Table, 1948, Red & white plastic, slide rule dial, horiz grill bars w/Scottie logo, handle, BC, AC/DC **$175.00**

5520 "Scottie Junior", Table, 1947, Walnut plastic, "upholstered-style", slide rule dial, grill bars w/Scottie logo, BC, AC/ DC **$125.00**

5530 "Scottie Junior", Table, 1947, White plastic, "upholstered-style", slide rule dial, grill bars w/Scottie logo, BC, AC/ DC **$125.00**

5535 "Scottie Pup", Table, 1948, Red & white plastic, slide rule dial, horiz grill bars w/Scottie logo, handle, BC, AC/DC **$175.00**

6000 "Scottie", Table, 1949, Plastic, "waterfall" dial, horizontal grill bars, 2 knobs, Scottie logo, BC, AC/DC **$100.00**

MP5-5-3 "Scottie", Table, 1946, Plastic, "waterfall" dial, horizontal louvers, 2 knobs, handle, Scottie logo, BC, AC **$135.00**

RESAS
Resas, Inc.,
112 Chambers St., New York, New York

5T "Tone-A-Dyne", Table, 1925, Wood, low rectangular case, 3 dial front panel, 5 tubes, battery **$135.00**

REVERE
Revere Camera Company,
320 East 21st Street, Chicago, Illinois

400, Portable, 1955, Leather, purse-style, inner aluminum panel w/dial & grill, lift-up lid, strap, BC, AC/DC/bat $125.00

Minuette, Cathedral, 1932, Wood, small case w/flat top, lower dial, upper grill w/cut-outs, 4 tubes, 3 knobs **$135.00**

Patrician, Table, 1931, Wood, window dial w/escutcheon, upper built-in speaker w/arched top, 5 tubes, BC, AC **$165.00**

ROBIN
Capital Appliance Distributor,
1201 W. Washington St.,
Indianapolis, Indiana

TR-605, Portable, 1960, Transistor, upper front dial, lower grill area, thumbwheel knobs, AM, battery **$30.00**

ROLAND
Roland Radio Corp.,
12-30 Anderson Avenue,
Mt Vernon, New York

4C2, Table-C, 1957, Plastic, right side dial knob, left front alarm clock, right lattice grill, BC, AC **$20.00**

4T1, Table, 1953, Plastic, right round dial, horizontal wrap-around grill bars, 2 knobs, BC, AC/DC **$35.00**

4TR, Portable, 1959, Transistor, top thumbwheel dial & knob, front horizontal bars, handle, AM, bat **$20.00**

5C1, Table-C, 1953, Plastic, right side dial knob, center front clock over horizontal louvers, BC, AC **$35.00**

5C2, Table-C, 1953, Plastic, right front dial, left alarm clock, center vertical bars, 6 knobs, BC, AC **$25.00**

5C5, Table-C, 1956, Plastic, right side dial knob, left front alarm clock, right lattice grill, BC, AC **$20.00**

5P2, Portable, 1954, Plastic, right front round dial knob, center lattice grill, handle, BC, AC/DC/bat **$25.00**

5P5, Portable, 1956, Plastic, right round dial, vertical grill bars, handle, side knob, BC, AC/DC/bat **$20.00**

5T1E, Table, 1953, Plastic, right front round dial over large recessed grill, 2 knobs, feet, BC, AC/DC **$30.00**

5T1V, Table, 1953, Plastic, right round dial over large recessed grill, 2 knobs, feet, BC, AC/DC **$30.00**

5T2M, Table, 1953, Plastic, center front round dial over large recessed grill, 2 knobs, BC, AC/DC **$30.00**

5T3, Table, 1954, Plastic, center front round dial knob over horizontal louvers, BC, AC/DC **$20.00**

5T5, Table, 1954, Plastic, right diagonal half-moon dial, recessed lattice grill, 2 knobs, BC, AC/DC **$25.00**

5T6, Table, 1956, Plastic, right front diagonal half-moon dial, left lattice grill, feet, BC, AC/DC **$30.00**

5X6U, Table-R/P, 1956, Wood, right front vert slide rule dial, 3 side knobs, lift top, inner phono, BC, AC **$25.00**

6P2, Portable, 1954, Plastic, upper front slide rule dial, lattice grill, handle, 2 knobs, BC, AC/DC/bat **$25.00**

6T1M, Table, 1953, Plastic, lower front slide rule dial, upper woven grill, 2 knobs, AM, AC/DC **$25.00**

6TR, Portable, 1957, Leather case w/front grill, transistor, controls on case top, strap, AM, battery **$35.00**

8FT1M, Table, 1953, Plastic, lower front slide rule dial, upper woven grill, 2 knobs, AM, FM, AC/DC **$25.00**

8XF1, Table-R/P, 1953, Upper front slide rule dial, lower grill, 3 knobs, lift top, inner phono, BC, FM, AC **$25.00**

10TF1, Table, 1954, Wood, lower front slide rule dial, upper grill, 4 knobs, AM, FM, AC **$25.00**

10XF1, Console-R/P, 1955, Wood, right front slide rule dial, 5 knobs, left pull-out phono drawer, BC, FM, AC **$50.00**

51-481, Portable, 1960, Plastic, transistor, step-back top w/thumbwheel dial, horiz grill bars, swing handle, AM, bat **$25.00**

54B, Table, Plastic, center front round dial over large horizontal louvers, 1 lower knob, BC $20.00

61-482, Portable, 1960, Transistor, upper left thumbwheel dial, front horiz grill bars, handle, AM, battery $25.00

71-483, Portable, 1959, Transistor, thumbwheel dial, right & left round grills, twin speakers, AM, battery $35.00

71-486, Portable, 1959, Plastic, transistor, slide rule dial, perforated grill, handle, wire stand, AM, bat $40.00

TC-10, Portable-C, 1960, Transistor, lower right dial, lattice grill area, left clock, handle, AM, battery $30.00

TR-8, Portable, 1960, Transistor, top dial & controls, large front perforated grill area, strap, AM, bat $30.00

ROYAL
Royal Radio Co.,
10 West 33rd Street,
New York, New York

AN150, Portable, 1952, Leatherette, inner right dial, 3 knobs, fold-in front, handle, BC, 2SW, AC/DC/bat $30.00

SENTINEL
Sentinel Radio Corp.,
2020 Ridge Avenue, Evanston, Illinois

1U312PW, Portable, 1950, Plastic, slanted slide rule dial, 2 thumbwheel knobs, flex handle, BC, AC/DC/bat $35.00

1U339-K, Table-R/P, 1950, Wood, center front dial, right & left grills, 4 knobs, lift top, inner phono, BC, AC $30.00

1U340-C, Console-R/P, 1951, Wood, upper right dial, 4 knobs, left pull-out phono drawer, lower grill, BC, AC $60.00

1U-345, Portable, 1952, Plastic, right front round dial, diagonal grill w/vertical bars, handle, BC, battery $40.00

1U-355P, Portable, 1956, Plastic, right side round dial knob, front horizontal bars, handle, BC, AC/DC/bat $25.00

1U-360, Table, 1956, Plastic, right side dial knob, right & left side horiz bars, front grill, BC, AC/DC $25.00

1U-363, Table, 1956, Plastic, right side dial, center front recessed horizontal grill bars, BC, AC/DC $25.00

8, Console, 1930, Two-tone walnut, highboy, window dial, lower grill w/oyster shell cut-out, 8 tubes $145.00

11, Console, 1930, Walnut, lowboy, Deco, window dial, octagonal grill w/cut-outs, 7 tubes, 2 knobs $165.00

76AC, Console, 1937, Wood, upper front round dial, lower cloth grill w/center vertical bars, 5 knobs $145.00

108A, Console, 1931, Wood, lowboy, upper window dial, scrolled grill cut-outs, stretcher base, 7 tubes $150.00

109, Console, 1931, Wood, upper window dial, lower grill with cut-outs, bowed front legs, 8 tubes $150.00

111, Cathedral, 1931, Wood, lower front window dial, upper cloth grill with cut-outs, 4 tubes $250.00

125AC-CB, Console-R/P, 1939, Wood, slide rule dial, pushbuttons, grill w/vertical bars, lift top, inner phono, AC $165.00

137UT, Table, 1939, Plastic, upper right streamlined thumbwheel dial, horizontal grill bars, AC $75.00

168BC, Console, 1939, Wood, upper slanted slide rule dial, lower vertical grill bars, 5 tubes, 4 knobs, bat $90.00

168BT, Table, 1939, Walnut, right front slide rule dial, left grill w/ vertical bars, 4 knobs, battery $40.00

170BL, Portable, 1939, Brown leatherette, right front dial, left grill, handle, 2 knobs, AC/DC/battery $25.00

175BCT, Console, 1939, Walnut, upper slanted dial, lower grill w/vert bars, front fluting, 4 tubes, bat $85.00

175BT, Table, 1939, Walnut bakelite, right front dial, left wrap-around louvers, 2 knobs, battery $30.00

175BTW, Table, 1940, Walnut, right front square dial, left horizontal louvers, 2 knobs, feet, battery $30.00

176BT, Table, 1939, Walnut, right front dial, left horizontal grill bars, fluted top & base, battery $40.00

177UT, Table, 1939, Bakelite, upper right streamlined thumbwheel dial, horizontal grill bars, AC/DC $75.00

178BL, Portable, 1939, Striped cloth covered, right front dial, left grill, handle, 2 knobs, battery $25.00

180XL, Portable, 1939, Cloth covered, inner right dial, left grill, fold-down front, handle, AC/DC/battery $30.00

181BL-CB, Portable-R/P, 1940, Cloth covered, inner right dial, fold-down front, lift top, inner crank phono, bat $35.00

194UTI, Table, 1940, Ivory bakelite, right front dial, left horiz wrap-around louvers, 2 knobs, AC/DC $45.00

194UTR, Table, 1940, Red bakelite, right front dial, left horiz wrap-around louvers, 2 knobs, AC/DC $55.00

194UTW, Table, 1940, Walnut bakelite, right front dial, left horiz wrap-around louvers, 2 knobs, AC/DC $45.00

195ULT, Table, 1939, Catalin, right dial, left wrap-around louvers, step-down top w/pushbuttons, AC $1,000.00

195ULTO, Table, 1940, Bakelite, right dial, left wrap-around louvers, step-down top w/pushbuttons, AC $70.00

195ULTWD, Table, 1940, Walnut, right front dial, left wrap-around grill, top pushbuttons, handle, AC/DC $65.00

198ALCE, Console, 1940, Walnut w/inlay, upper dial, pushbuttons, lower grill w/vertical bars, 8 tubes, AC $145.00

198ALT, Table, 1940, Wood, right front dial, pushbuttons, left grill w/ vertical center bar, 4 knobs, AC $60.00

199ACE, Console, 1939, Wood, slanted slide rule dial, pushbuttons, lower grill w/3 vert bars, 11 tubes, AC $160.00

203ULT, Table, 1940, Wood, right dial, left grill w/2 horiz bars, step-down top w/pushbuttons, AC/DC $65.00

204AT, Table, 1940, Two-tone walnut, right front dial, left wrap-around grill, raised right top, AC $50.00

205BL, Portable, 1940, Cloth covered, lower front slide rule dial, upper grill, handle, 2 knobs, battery $25.00

220T460, Table, Wood, right slide rule dial w/plastic escutcheon, left grill w/four bars, 4 knobs $50.00

228P, Portable, 1942, Luggage-style, dial & knobs on top of case, front grill, handle, AC/DC/battery **$30.00**

238-V, Table-N, 1941, Looks like stack of books, maroon leatherette, inner dial/grill/knobs, BC, AC **$165.00**

243T, Table, 1940, Wood, right slide rule dial, left cloth grill w/center vert bar, 5 knobs, fluted base **$45.00**

249-I, Table, 1941, Ivory plastic, right front dial, left horizontal wrap-around louvers, 2 knobs, AC **$50.00**

284GA, Table-R/P, 1947, Wood, outer slide rule dial, grill, 2 knobs, lift-top, inner phono, BC, AC **$35.00**

284I, Table, 1946, Plastic, upper slide rule dial, horizontal wrap around louvers, 2 knobs, BC, AC/DC **$50.00**

284-NB, Table, 1946, Catalin, upper slanted slide rule dial, large lower cloth grill, 2 knobs, AC/DC .... **$900.00+**

284-NI, Table, 1946, Catalin, upper slanted slide rule dial, lower inset curved horiz grill bars, AC/DC .... **$900.00+**

284-NR, Table, 1946, Catalin, upper slanted slide rule dial, lower inset curved horiz louvers, 2 knobs .... **$900.00+**

285P, Portable, 1946, Leatherette, luggage-style, upper slide rule dial, 3 knobs, handle, BC, AC/DC/bat **$30.00**

286P, Portable, 1947, Inner right thumbwheel dial, center metal grill, flip-up lid, BC, AC/DC/battery **$45.00**

286PR, Portable, 1947, Plastic, inner thumbwheel dial, center metal grill, flip-up lid, BC, AC/DC/battery **$45.00**

289-T, Table, 1946, Wood, upper front slide rule dial, large lower cloth grill, 2 knobs, BC, battery **$30.00**

292K, Table-R/P, 1947, Wood, outer slide rule dial, 3 knobs, lift top, inner phono, BC, AC **$30.00**

293-CT, Console-R/P, 1947, Wood, front slide rule dial, 3 knobs, woven grill, lift top, inner phono, BC, AC **$50.00**

293W, Table, 1946, Plastic, upper slanted slide rule dial, lower horiz louvers, 3 knobs, BC, AC/DC **$45.00**

294-I, Table, 1946, Plastic, upper front slanted slide rule dial, vertical grill bars, 4 knobs, AC/DC **$50.00**

294T, Table, 1946, Wood, upper slide rule dial with plastic trim, lower grill, 4 knobs, BC, SW, AC/DC **$40.00**

295-T, Table, 1947, Wood, large slide rule dial, contrasting veneer, 4 knobs, BC, 4 SW, AC/battery **$70.00**

296-M, Console-R/P, 1948, Wood, right slide rule dial, 2 knobs, pushbuttons, left pull-out phono, BC, FM, AC **$70.00**

302-I, Table, 1948, Plastic, upper slanted slide rule dial, vertical grill bars, 4 knobs, BC, FM, AC/DC **$50.00**

305-I, Table, 1948, Plastic, upper slanted slide rule dial, vertical grill bars, 4 knobs, BC, 3SW, AC/DC **$50.00**

309-I, Table, 1948, Ivory plastic, right front dial, horizontal wrap-around grill design, 2 knobs, AC **$45.00**

309-W, Table, 1947, Plastic, right dial, horizontal wrap-around grill design, 2 knobs, BC, AC/DC **$45.00**

313W, Table, 1948, Plastic, upper slanted slide rule dial, lower horiz louvers, 3 knobs, BC, AC/DC **$45.00**

314W, Table, 1948, Plastic, upper slide rule dial, lower wrap-around louvers, 2 knobs, BC, AC/DC **$45.00**

315-W, Table, 1948, Plastic, upper slanted slide rule dial, horizontal louvers, 3 knobs, AM, FM, AC/DC **$45.00**

316PM, Portable, 1948, Plastic, lower slide rule dial, vertical grill bars, handle, 2 knobs, BC, AC/DC/bat **$35.00**

331, Table, Plastic, right front half-round dial, large cloth grill area, 2 knobs, BC **$50.00**

332, Table, 1949, Plastic, upper slide rule dial w/large numerals, center cloth grill, BC **$35.00**

335PM, Portable, 1950, Plastic, lower front slide rule dial, flex handle, 2 knobs, BC, AC/DC/battery **$30.00**

338-W, Table, 1951, Plastic, recessed right dial/left checkered grill, 2 knobs, feet, BC, AC/DC **$30.00**

344, Table, 1953, Plastic, large center front half-moon dial w/inner louvers, 3 knobs, BC **$45.00**

5721, Tombstone, 1934, Two-tone wood, lower front dial, upper cloth grill w/Deco cut-outs, BC, SW, AC **$110.00**

Duotrola, Chairside-R/P-2 Piece, 1930, Wood, chairside case contains tuning controls, speaker case contains phono, AC **$225.00**

Portrola, Chairside, 1930, Wood, top dial & controls, cloth grill with cut-outs, case finished on all sides, 8 tubes **$200.00**

_____

SETCHELL-CARLSON
Setchell Carlson Inc.,
2233 University Avenue, St Paul, Minnesota

23, Tombstone, 1939, Wood, center dial, upper grill w/3 horizontal bars, 4 tubes, 4 knobs, BC, SW, bat..............................$75.00

63, Tombstone, 1939, Wood, center dial, upper grill w/3 horizontal bars, 5 tubes, 4 knobs, BC, SW, bat..............................$75.00

408, Table, 1948, Plastic, raised top slide rule dial, lower horizontal louvers, 2 thumbwheel knobs $80.00

413, Table, 1941, Wood, right front slide rule dial, left horizontal louvers, 5 knobs, 4 band, AC $55.00

416, Table, 1946, Plastic, raised slide rule dial, horizontal louvers, 2 thumbwheel knobs, BC, AC/DC $80.00

427, Table, 1947, Plastic, raised slide rule dial, horizontal louvers, 2 thumbwheel knobs, BC, AC/DC $80.00

437, Table, 1948, Plastic, raised top slide rule dial, horizontal louvers, 2 thumbwheel knobs, BC, DC $80.00

447, Portable, 1948, Leatherette, slide rule dial, plastic grill bars, handle, 2 knobs, BC, AC/DC/battery $45.00

469, Table, 1950, Plastic, raised slide rule dial, horizontal louvers, 2 thumbwheel knobs, FM, AC/DC $65.00

570, Table, 1950, Cylindrical shape, slide rule dial, handle, 2 knobs, removable speaker, BC, AC/DC $80.00

588, Table, 1939, Two-tone wood, right front dial, left 2-section wraparound grill, 3 knobs, AC $45.00

4130, Table, 1940, Wood, lower front slide rule dial, large upper cloth grill, 3 knobs $40.00

WT16 "Whisper-Tone", Radio/Bed Lamp, Painted wood, cylindrical shape, slide rule dial, horiz grill bars, light, 2 knobs $80.00

SHARP
Continental Merchandise Co., Inc.,
236 Fifth Ave., New York, New York

TR-182, Portable, 1959, Transistor, right front window dial over large perforated grill, side knobs, AM, bat $30.00

SIGNAL
Signal Electronics, Inc.,
114 East 16th Street,
New York, New York

141, Portable, 1948, Inner metal grill w/2 thumbwheel knobs, flip-open door, handle, BC, battery $45.00

241 "Concerto", Table, 1948, Wood w/metal front, lower slide rule dial, "sunburst" grill, 2 knobs, BC, AC/DC $65.00

341-A, Portable, 1948, "Snakeskin", slide rule dial, large perforated grill, handle, 2 knobs, BC, AC/DC/bat..........................$35.00

341T, Portable, 1947, "Snakeskin", slide rule dial, perforated grill, handle, 2 knobs, BC, AC/DC/battery $30.00

AF252, Table, 1948, Plastic, upper slide rule dial, raised vert metal grill bars, 3 knobs, BC, FM, AC/DC $70.00

SILVER MARSHALL
Silver-Marshall, Inc.,
848 West Jackson Boulevard, Chicago, Illinois

Q, Console, 1931, Wood, lowboy, inner dial & knobs, double doors, 6 legs, stretcher base, 13 tubes, BC, SW $210.00

Z-10, Console, 1933, Wood, lowboy, upper half-round dial, lower grill w/cut-outs, 7 tubes, 5 knobs, 6 legs, AC $200.00

SILVERTONE
Sears, Roebuck & Co.,
925 South Homan Street, Chicago, Illinois

1, Table, 1950, Metal, right round dial, horizontal grill bars, 2 contrasting knobs, BC, AC/DC $75.00

2, Table, 1950, Metal, right round dial, horizontal grill bars, 2 contrasting knobs, BC, AC/DC $75.00

4-233, Table-N, 1951, Plastic, left cowboy & horse, right round dial w/either gun or hat pointer, AC $300.00

10, Table-C, 1951, Plastic, right round dial, left alarm clock, crisscross grill, 6 knobs, BC, AC $30.00

13, Table, Mahogany plastic, large center front metal dial ring over perforated grill, AC/DC $30.00

14, Table, Ivory plastic, large center front metal dial ring over perforated grill, AC/DC $30.00

15, Table, 1951, Plastic, right dial, left diagonal grill bars w/stylized "S", 3 knobs, BC, AC/DC **$30.00**

18, Table, 1951, Plastic, right front dial, left diagonal grill bars, "S" logo, 3 knobs, BC, FM, AC **$30.00**

24, Table, 1929, Metal, low rectangular case, off-center window dial, 2 knobs, switch, battery **$60.00**

25, Console, 1929, Wood, lowboy, inner front window dial & knobs, fold-down front, lift top, battery **$110.00**

26, Console, 1929, Wood, inner dial & knobs, speaker grill, double doors, lower battery storage, bat **$120.00**

33, Table-R/P, 1950, Leatherette, inner right dial, 3 knobs, phono, lift top, handle, BC, AC **$15.00**

58, Console-R/P, 1930, Walnut, lowboy, inner front window dial, double doors, lift top, inner phono, AC **$135.00**

60, Console-R/P, 1930, Wood, lowboy, inner front window dial, double doors, lift top, inner phono, AC **$150.00**

64, Console-R/P, 1950, Wood, right tilt-out slide rule dial, inner left phono, lower storage, BC, FM, AC **$70.00**

72, Table-R/Rec, 1951, Suitcase-style, inner left dial, disc recorder, lift top, handle, BC, AC **$20.00**

100.195, Table, 1939, Wood, right dial, 5 pushbuttons, tuning eye, left cloth grill w/2 vert bars, 4 knobs **$65.00**

101.571, Table, 1939, Wood, right front slide rule dial, 5 pushbuttons, left horizontal bars, 4 knobs **$55.00**

101.585, Table, Wood, step-down top w/dial & 4 pushbuttons, rounded base w/"X" grill cut-outs **$130.00**

109.356, Table, Plastic, right front gold dial, left horizontal wrap-around louvers, 2 knobs, AC **$65.00**

117, Console, Wood, upper window dial w/escutcheon, lower grill w/cut-outs, stretcher base **$150.00**

132.857, Table, 1949, Plastic, right front dial knob w/ stylized "S" logo, graduated vertical grill bars **$45.00**

132.881, Table, 1951, Plastic, large center front metal dial w/inner perforated grill, 2 knobs, BC **$40.00**

206, Portable, 1960, Plastic, transistor, upper quarter-moon window dial, lower "V" louvers, AM, bat **$20.00**

208 "500", Portable, 1960, Plastic, transistor, right dial knob, horizontal grill bars, handle, AM, battery **$20.00**

210, Portable, 1950, Plastic, right side dial knob, front vertical grill bars, handle, BC, battery **$35.00**

211, Portable, 1960, Black plastic, transistor, wedge-shaped dial, perforated grill, swing handle, AM, bat **$25.00**

212, Portable, 1960, Coral plastic, transistor, wedge-shaped dial, perforated grill, swing handle, AM, bat **$25.00**

213, Portable, 1960, Blue plastic, transistor, wedge-shaped dial, perforated grill, swing handle, AM, bat **$25.00**

214, Portable, 1960, Transistor, right thumbwheel window dial, left horizontal grill bars, AM, battery **$20.00**

215, Portable, 1950, Plastic, right dial, horizontal grill bars, flex handle, 2 knobs, BC, AC/DC/battery **$35.00**

217 "600", Portable, 1960, Leather case, transistor, right dial knob, front lattice cut-outs, handle, AM, bat **$25.00**

220, Portable, 1950, Plastic, upper dial, lower checkered grill, flex handle, 2 knobs, BC, AC/DC/bat **$35.00**

220 "700", Portable, 1960, Leather case, transistor, right side dial knob, front lattice grill, handle, AM, bat **$20.00**

222 "800", Portable, 1960, Leather case, transistor, slide rule dial, lattice grill, telescoping antenna, BC, SW, bat **$25.00**

225, Portable, 1950, "Alligator", upper slide rule dial, lower grill, handle, 2 knobs, BC, AC/DC/battery **$30.00**

640, Table, Plastic, small upper front dial, large lower horiz louvers, 2 thumbwheel knobs **$60.00**

1032, Table-R/P, 1952, Plastic, right front round dial, left vertical grill bars, open top phono, BC, AC **$40.00**

1053, Console-R/P, 1952, Wood, upper front dial, center pull-out phono drawer, lower grill, 3 knobs, BC, AC **$40.00**

1055, Console-R/P, 1952, Wood, upper dial, center pull-out phono drawer, lower grill, 3 knobs, BC, FM, AC **$40.00**

1260, Console-R/P, 1931, Walnut, lowboy, front window dial, grill w/cut-outs, lift top, inner phono, AC **$125.00**

1261, Console-R/P, 1931, Walnut, lowboy, front window dial, grill w/cut-outs, lift top, inner phono, AC **$125.00**

1290, Cathedral, 1931, Wood, window dial, upper grill w/cut-outs, fluted columns, 6 tubes, 4 knobs, bat **$150.00**

1291, Cathedral, 1931, Wood, window dial, upper grill w/cut-outs, fluted columns, 6 tubes, 4 knobs, bat **$150.00**

1292, Console, 1931, Walnut, lowboy, window dial, lower grill w/cut-outs, stretcher base, 6 tubes, bat **$85.00**

1293, Console, 1931, Walnut, lowboy, window dial, lower grill w/cut-outs, stretcher base, 6 tubes, bat **$85.00**

1330, Console-R/P, 1931, Walnut, lowboy, window dial, grill w/cut-outs, lift top, inner phono, 7 tubes, AC **$120.00**

1331, Console-R/P, 1931, Walnut, lowboy, window dial, grill w/cut-outs, lift top, inner phono, 7 tubes, AC **$120.00**

1403, Cathedral, 1931, Wood, lower window dial, upper cloth grill w/cut-outs, fluted columns, 3 knobs **$235.00**

1580, Console, Wood, lowboy, upper quarter-round dial, lower grill w/cut-outs, 7 tubes, 5 knobs **$145.00**

1582, Cathedral, Wood, lower quarter-round dial, upper cloth grill w/cut-outs, 7 tubes, 5 knobs **$260.00**

1589, Cathedral, 1932, Burl walnut, lower quarter-round dial, upper grill w/cut-outs, 7 tubes, 5 knobs **$260.00**

1631, Console, 1931, Wood, lowboy, quarter round dial, grill cut-outs, 9 tubes, 6 legs, 5 knobs, AC **$150.00**

1641, Console, 1932, Walnut, quarter round dial, large grill w/cut-outs, 12 tubes, 5 knobs, BC, SW, AC **$175.00**

1650, Console, 1935, Wood, lowboy, upper quarter-round dial, lower cloth grill w/cut-outs, 6 legs $150.00

1660, Cathedral, Wood, lower window dial w/escutcheon, upper grill w/cut-outs, 5 tubes, 3 knobs $185.00
1743, Tombstone, Two-tone wood, lower round dial, upper cloth grill w/cut-outs, 5 tubes, 3 knobs $100.00

1745, Tombstone, Wood, double step-down top, center round dial, grill cut-outs, 6 tubes, 5 knobs $150.00
1822, Console, 1935, Wood, Deco, upper dial, lower cloth grill w/ curled cut-outs, 14 tubes, 5 knobs, AC $185.00
1835, Console, 1935, Wood, upper round dial, lower grill w/cut-outs, 9 tubes, 5 knobs, BC, SW, AC $135.00
1850, Tombstone, 1935, Wood, Deco, round dial, upper grill w/cut-outs, step-down top, 6 tubes, 5 knobs, bat $100.00
1851, Console, 1935, Wood, upper round dial, lower grill w/cut-outs, 6 tubes, 4 knobs, BC, SW, bat...$90.00
1852, Upright Table, 1935, Walnut, Deco, center front dial, upper grill w/cut-outs, 6 tubes, 3 knobs, battery $75.00
1905, Tombstone, Wood, lower round dial & escutcheon, upper grill w/cut-outs, 9 tubes, 5 knobs $125.00
1906, Tombstone, Wood, lower round dial, upper cloth grill with cut-outs, 6 tubes, 4 knobs, AC $140.00
1911, Console, 1936, Wood, upper round dial, lower cloth grill w/ vertical bars, 6 tubes, BC, SW, AC $120.00
1921, Tombstone, Wood, shouldered, center round dial, upper grill w/bars, 6 tubes, 4 knobs, battery $125.00
1938, Tombstone, 1936, Wood, lower round dial, upper grill with cut-outs, 6 tubes, 4 knobs, BC, SW, AC $110.00

1945, Console, 1936, Walnut, upper round dial, tuning eye, lower grill w/cut-outs, 9 tubes, BC, SW, AC $140.00
1954, Tombstone, Wood, lower round dial, upper cloth grill w/cut-outs, 6 tubes, 4 knobs, BC, SW $85.00

1955, Tombstone, 1936, Wood, step-down top, lower round dial, upper grill w/3 vert bars, 5 knobs, BC, SW $120.00
1972, Console, Wood, upper front round dial, lower cloth grill w/3 splayed bars, 8 tubes, BC, SW $130.00
2001, Table, 1950, Brown metal, right round dial knob over criss-cross grill, lower left knob, BC, AC $50.00

2002, Table, 1950, Ivory metal, right round dial knob over criss-cross grill, lower left knob, BC, AC $50.00
2003, Table, 1953, Brown plastic, upper slide rule dial, checkered grill, feet, 2 knobs, BC, AC/DC $25.00
2004, Table, 1953, Ivory plastic, upper slide rule dial, checkered grill, feet, 2 knobs, BC, AC/DC $25.00

2005, Table, 1953, Red plastic, upper slide rule dial, checkered grill, feet, 2 knobs, BC, AC/DC $30.00

2006, Table, 1953, Green plastic, upper slide rule dial, checkered grill, feet, 2 knobs, BC, AC/DC $30.00

2007, Table-C, 1953, Plastic, center front round dial, right checkered grill, left alarm clock, BC, AC $25.00

2022, Table, 1953, Plastic, right dial, left diagonal bars, stylized "S" logo, 3 knobs, BC, SW, AC/DC $30.00

2028, Table, 1953, Wood, front half-moon metal dial over large grill, 2 knobs, BC, battery $25.00

2041, Table-R/P, 1953, Leatherette, right front round dial, 3/4 lift top, inner phono, handle, BC, AC $25.00

2056, Console-R/P, 1953, Wood, upper front slide rule dial, 2 knobs, center pull-out phono drawer, BC, AC $60.00

2061, Console-R/P, 1953, Wood, inner right slide rule dial, 2 knobs, pull-out phono, double doors, BC, AC $60.00

2225, Portable, 1953, "Alligator", right front half-moon dial, lattice grill, handle, BC, AC/DC/battery $30.00

2411, Table, Wood, right front rectangular dial, left cloth grill w/fluted bars, 4 knobs, DC $45.00

2761, Table, Wood, slide rule dial, 5 pushbuttons, cloth grill w/3 vert bars, 6 tubes, 5 knobs $60.00

3001, Table, 1954, Brown plastic, upper thumbwheel dial, "V" shaped grill w/cloth & checked panels $35.00

3002, Table, 1954, Ivory plastic, upper thumbwheel dial, "V" shpaed grill w/cloth & checked panels $35.00

3004, Table, 1954, Plastic, right front dial, left horizontal grill bars, 2 knobs, feet, BC, AC/DC $20.00

3032, Table-R/P, 1953, Plastic, right front round dial, vertical grill bars, 3 knobs, open top, BC, AC $40.00

3040, Table-R/P, 1953, Plastic, front half-moon dial/louvers, 2 knobs, 3/4 lift top, inner phono, BC, AC $40.00

3040A, Table-R/P, 1955, Plastic, front half-moon dial/louvers, 2 knobs, 3/4 lift top, inner phono, BC, AC......................................$40.00

3045, Table-R/P, 1953, Wood, front metal half-moon dial over cloth grill, lift top, inner phono, BC, AC $30.00

3068, Console-R/P, 1955, Wood, right tilt-out slide rule dial, 3 knobs, pull-down phono door, BC, FM, AC $55.00

3210, Portable, 1953, Plastic, center round dial, horizontal bars, handle, top thumbwheel knob, BC, bat $25.00

3217, Portable, 1954, Plastic, center round dial, upper horizontal grill bars, handle, BC, AC/DC/battery $25.00

3351, Table, Plastic, Deco, left "candy cane" dial, wrap-over vertical grill bars, 4 pushbuttons $85.00

3461, Table, Plastic, Deco, left "candy cane" dial, wrap-over vertical grill bars, 4 pushbuttons $85.00

4032, Table-R/P, 1955, Plastic, center front half-moon dial, 3 knobs, open top phono, BC, AC $40.00

4068A, Console-R/P, 1955, Wood, inner right slide rule dial, 3 knobs, left fold-down phono door, BC, FM, AC $65.00

4200, Table, 1955, Plastic, right trapezoid dial, left horizontal grill bars, 3 knobs, feet, BC, FM, AC $20.00

4204, Table, Maroon plastic, recessed front with right dial and left cloth grill, 3 knobs, AM/FM $35.00

4206, Table, Ivory plastic, recessed front with right dial and left cloth grill, 3 knobs, AM/FM $35.00

4210, Portable, 1955, Plastic, round dial, horizontal grill bars, top thumbwheel knob, handle, BC, bat $25.00

4212, Portable, 1955, Right front round dial knob, left lattice grill w/ Silvertone logo, BC, battery $25.00

4225, Portable, 1955, Cloth covered, lower front dial, upper grill, handle, 1 knob, BC, AC/DC/battery $30.00

4414, Table, 1936, Plastic, right round dial knob, left cloth grill w/ wrap-over vertical bars, BC, AC $70.00

4418, Table, 1937, Wood, center front round dial, top speaker grill, 4 tubes, 3 knobs, BC, SW, battery $40.00

4464, Tombstone, 1936, Two-tone wood, lower round dial, upper grill w/cut-outs, 6 tubes, 4 knobs, AC $120.00

4485, Console, 1937, Wood, upper round dial, tuning eye, lower grill w/vertical bars, 8 tubes, BC, SW, AC $130.00

4486, Console, 1937, Wood, upper round dial, tuning eye, lower grill w/vert bars, 10 tubes, BC, SW, AC $130.00

4487, Console, 1937, Wood, upper round self-tuning dial, lower grill w/cut-outs, 11 tubes, BC, SW, AC **$160.00**

4500, Table, 1936, Plastic, right front round dial knob, left vertical wrap-over grill bars, AC **$70.00**

4500A, Table, 1936, Plastic, right front round dial knob, left cloth grill w/vertical wrap-over bars **$70.00**

4522, Tombstone, Wood, lower round dial w/globe, upper cloth grill with cut-outs, 6 tubes, 3 knobs **$100.00**

4563, Table, Wood, rounded left, right round dial, left grill w/Deco cut-outs, 6 tubes, 4 knobs **$60.00**

4565, Table, Wood, off-center round dial, tuning eye, right/left grill cut-outs, 5 knobs, BC, SW **$65.00**

4612, Table, 1939, Wood, right front dial, left cloth grill with 2 vertical bars, 4 knobs, battery **$30.00**

4619, Tombstone, 1939, Walnut w/maple inlays, center front dial, upper grill w/bars, 6 tubes, battery **$80.00**

4652, Console, 1939, Wood, upper front dial, lower cloth grill w/3 vertical bars, fluted columns, bat **$95.00**

4660, Table, 1937, Two-tone wood, left front dial, tuning eye, right grill w/cut-outs, 3 knobs, AC **$60.00**

4663, Upright Table, 1938, Wood, center dial, tuning eye, curved top w/vert grill bars, 4 knobs, BC, SW, AC **$90.00**

4664, Table, 1937, Wood, right dial, tuning eye, step-down left w/ Deco grill cut-outs, 4 knobs, AC **$55.00**

4665, Tombstone, 1938, Wood, slanted slide rule dial, tuning eye, 3 vert grill bars, 4 knobs, BC, SW, AC **$90.00**

4666, Table, 1938, Wood, rolling slide rule dial, center automatic tuning, side grill bars, 10 tubes, BC, SW, AC **$100.00**

4668, Table-R/P, 1938, Wood, right dial, left vertical grill bars, fluted columns, lift top, inner phono, AC **$40.00**

4685, Console, 1938, Wood, upper slide rule dial, tuning eye, vertical grill bars, 8 tubes, BC, SW, AC **$120.00**

4686, Console, 1938, Wood, slide rule dial, automatic tuning, tuning eye, vert grill bars, 10 tubes, BC, SW, AC **$150.00**

4688, Console, 1938, Wood, slide rule dial, pushbuttons, tuning eye, vert grill bars, 14 tubes, BC, SW, AC **$185.00**

4763, Table, Wood, off-center raised dial panel, tuning eye, left grill cut-outs, 6 tubes, 4 knobs, BC, SW **$65.00**

5016, Table, 1956, Mahogany, right front plastic half-moon dial, left cloth grill, 3 knobs, BC **$20.00**

5017, Table, 1956, Blonde oak, right front plastic half-moon dial, left cloth grill, 3 knobs, BC **$20.00**

5227, Portable, 1955, Leatherette, flip-up front w/map, telescope antenna, BC, 2SW, LW, AC/DC/bat **$65.00**

6002, Table, 1946, Metal, midget, small upper right dial, 2 knobs, horizontal louvers, BC, AC/DC **$75.00**

6012, Table, 1947, Plastic, lower slide rule dial, vertical grill bars, 2 knobs, handle, BC, AC/DC **$45.00**

6016, Table, 1947, Plastic, right square dial, vertical grill bars front/ side/top, handle, BC, AC/DC **$60.00**

6018, Table-C, 1956, Plastic, right round dial, left alarm clock, side louvers, step-down top, BC, AC **$30.00**

6020, Table-C, 1956, Brown plastic, trapezoid shape, slide rule dial, large upper clock, 5 knobs, AC **$25.00**

6021, Table-C, 1956, Ivory plastic, trapezoid shape, slide rule dial, large upper clock, 5 knobs, AC **$25.00**

6042, Table, 1939, Two-tone wood, right front dial, left oval grill w/ filigree bar, 2 knobs, battery **$35.00**

6043, Table, 1939, Two-tone wood, right front dial, left oval grill w/ filigree bar, 2 knobs, battery **$35.00**

6050, Table, 1947, Wood, right round dial, left cloth grill w/3 horizontal bars, 3 knobs, BC, AC/DC **$45.00**

6051, Table, 1947, Two-tone wood, right front dial, left & upper grill w/cut-outs, 2 knobs, battery **$35.00**

6052, Tombstone, 1939, Walnut w/inlays, lower dial, pushbuttons, grill cut-outs, 6 tubes, 2 knobs, bat **$65.00**

6053, Tombstone, 1939, Walnut w/inlays, lower dial, pushbuttons, grill cut-outs, 6 tubes, 2 knobs, bat **$65.00**

6057A, Console-R/P, 1956, Wood, upper front slide rule dial, center pull-out phono drawer, storage, 2 knobs **$65.00**

6064, Console, 1939, Wood, slanted dial, lower grill with 3 vertical bars, 4 knobs, 6 tubes, battery **$85.00**

6065, Console, 1939, Wood, slanted dial, lower grill with 3 vertical bars, 4 knobs, 6 tubes, battery **$85.00**

6068, Console, 1939, Walnut, slanted slide rule dial, lower grill w/2 vertical bars, 7 tubes, BC, SW, bat **$85.00**

6069, Console, 1939, Walnut, slanted slide rule dial, lower grill w/2 vertical bars, 7 tubes, BC, SW, bat **$85.00**

6071, Table-R/P, 1947, Wood, right front round dial, grill w/3 horizontal bars, 3 knobs, lift top, BC, AC **$30.00**

6072, Table-R/P, 1947, Wood, slide rule dial, cloth grill w/3 horizontal bars, 2 knobs, lift top, BC, AC **$30.00**

6093, Console, 1946, Wood, upper slide rule dial, 4 knobs, pushbuttons, lower cloth grill, BC, SW, AC **$110.00**

6100, Console-R/P, 1946, Wood, upper slanted dial, 4 knobs, pull-out drawer w/phono, cloth grill, BC, AC **$100.00**

6105, Console-R/P, 1946, Wood, slanted slide rule dial, 4 knobs, pushbuttons, pull-out phono, BC, SW, AC **$100.00**

6111, Console-R/P, 1946, Wood, inner slide rule dial, 4 knobs, pushbuttons, fold-down door, BC, SW, AC **$85.00**

6111A, Console-R/P, 1947, Wood, inner slide rule dial, 4 knobs, 8 pushbuttons, pull-out phono, BC, SW, AC **$85.00**

6122, Table, Wood, slide rule dial, grill cut-outs, tuning eye, 6 pushbuttons, 7 tubes, 4 knobs, BC, SW **$75.00**

6177-A, Table, Plastic, Deco, right round dial knob, rounded left w/ horizontal louvers, BC, AC/DC **$70.00**

6200A, Table, 1946, Plastic, lower slide rule dial, large upper cloth grill, 2 knobs, BC, battery **$30.00**

6220A, Table, 1946, Wood, lower slide rule dial, upper cloth grill w/ cut-outs, 2 knobs, BC, battery **$35.00**

6230A, Table, 1947, Wood, lower slide rule dial, cloth grill w/scrolled sides, 2 knobs, BC, 3SW, bat **$45.00**

6327, Table, Wood, slanted slide rule dial, criss-cross grill, 5 pushbuttons, 6 tubes, 2 knobs **$65.00**

6372, Table, 1940, Wood, right front vertical slide rule dial, left cloth grill, handle, 2 knobs, battery **$35.00**

6402, Table, Plastic, midget, small right dial over horizontal wrap-around louvers, 2 knobs...................................**$75.00**

6425, Table, Wood, lower slide rule dial, upper horiz louvers, pushbuttons, tuning eye, BC, SW **$65.00**

6437, Console, 1940, Wood, slanted slide rule dial, pushbuttons, criss-cross grill, 12 tubes, BC, SW, AC **$160.00**

6491-A, End Table, Round end table on 4 legged base, inner dial & 4 knobs, double front doors **$185.00**

6821, Portable, Cloth covered, right dial, left metal perforated grill, handle, 2 knobs, BC, AC/DC/bat **$35.00**

7021, Table, 1947, Plastic, lower slide rule dial, large grill, 4 knobs, 4 pushbuttons, BC, AC/DC **$45.00**

7025, Table, 1947, Plastic, Deco, left "candy cane" dial, wrap-over vertical grill bars, 4 pushbuttons **$85.00**

7031A, Table, Two-tone wood, upper slanted slide rule dial, lower grill w/2 horiz bars, 4 knobs **$50.00**

7038, Table, 1941, Wood, lower slide rule dial, upper horiz grill bars, 6 pushbuttons, 4 knobs, BC, SW **$60.00**

7039, Table, 1941, Wood, Deco, lower slide rule dial, 8 pushbuttons, upper horiz bars, 2 knobs, BC, SW **$85.00**

7054, Table, 1947, Wood, lower slide rule dial, 4 knobs, 4 pushbuttons, rounded corners, BC, AC/DC **$35.00**

7070, Table-R/P, 1947, Plastic, front dial, vertical louvers, 2 knobs, curved side, open top, BC, AC **$40.00**

7085, Table-R/P/Rec, 1947, Wood, inner slide rule dial, 4 knobs, phono, lift top, horizontal grill bars, BC, AC **$25.00**

7086, Table-R/P/Rec, 1947, Wood, inner slide rule dial, 4 knobs, phono, lift top, horizontal grill bars, BC, AC **$25.00**

7090, Console, 1947, Wood, slanted dial, 2 vertical grill bars, 4 knobs, 4 pushbuttons, BC, SW, AC/DC **$95.00**

7100, Console-R/P, 1947, Wood, inner right slide rule dial, 4 knobs, phono, lift top, cloth grill, BC. AC **$95.00**

7102, Console-R/P/Rec, 1947, Wood, inner slide rule dial, 4 knobs, phono, lift top, criss-cross grill, BC, AC **$60.00**

7103, Console-R/P/Rec, 1947, Wood, inner dial/knobs/phono, lift top, lower criss-cross grill, feet, BC, AC **$60.00**

7104, Table, Wood, upper slanted slide rule dial, lower cloth grill w/ 2 vert bars, 3 knobs, bat **$30.00**

7111, Console-R/P, 1947, Wood, inner right slide rule dial, 4 knobs, door, phono inside lift top, BC, 2SW, AC **$70.00**

7115, Console-R/P, 1947, Wood, slanted slide rule dial, 4 knobs, pull-out phono drawer, BC, FM, AC **$90.00**

7116, Console-R/P, 1947, Wood, right tilt-out slide rule dial & knobs, inner phono, lower grill, BC, FM, AC **$75.00**

7166, Portable, 1946, Inner right dial, horizontal grill bars, flip-up lid, handle, BC, AC/DC/battery **$35.00**

7204, Table, 1957, Brown plastic, 2 right slide rule dials, left cloth grill, Hi-Fi, 3 knobs, AM/FM **$25.00**

7206, Table, 1957, Ivory plastic, 2 right slide rule dials, left cloth grill, Hi-Fi, 3 knobs, AM/FM **$25.00**

7210, Table-R/P, 1948, Wood, front slide rule dial, 2 knobs, lift top, inner phono, side crank, BC, battery **$35.00**

7222, Portable, 1957, Upper slide rule dial, large front grill w/1 knob & logo, handle, BC, AC/DC/bat **$20.00**

7226, Table, 1948, Wood, lower slanted slide rule dial, upper cloth grill, 2 knobs, BC, AC/DC/battery **$45.00**

7228, Portable, 1958, Transistor, lower front slide rule dial, upper grill, handle, 1 front knob, AM, bat **$30.00**

7244, Table-R/P, Wood, right front dial, left cloth grill w/2 horiz bars, lift top, inner phono, AC **$35.00**

8000, Table, 1948, Upper right dial, perforated wrap-under grill, 2 knobs, flared base, BC, AC/DC **$60.00**

8003, Table, 1949, Metal, midget, small upper right dial, horizontal louvers, 2 knobs, BC, AC/DC **$75.00**

8005, Table, 1948, Plastic, upper slide rule dial, aluminum grill, slanted sides, 2 knobs, BC, AC/DC **$35.00**

8010, Table-C, 1948, Plastic, slide rule dial lights through grill, upper clock face, 5 knobs, BC, AC **$60.00**

8020, Table, 1948, Plastic, upper slide rule dial, metal grill, flared sides, 4 knobs, BC, FM, AC/DC **$35.00**

8021, Table, 1949, Plastic, upper slide rule dial, lower grill, flared base, 4 knobs, AM, FM, AC/DC **$35.00**

8024, Table, 1949, Plastic, lower slide rule dial, upper vertical grill bars, 2 knobs, AM, FM, AC **$30.00**

8041, Table-R/P, 1958, Top raised slide rule dial, "waterfall" grill, 3 knobs, lift top, handle, BC, AC **$35.00**

8050, Table, 1948, Two-tone wood, slide rule dial, cloth grill w/cut-outs, 2 knobs, BC, AC/DC $30.00

8051, Table, 1948, Two-tone wood, lower slide rule dial, cloth grill w/cut-out, 2 knobs, BC, AC/DC $30.00

8052, Table, 1949, Wood, lower slide rule dial, 4 knobs, raised pushbuttons, cloth grill, BC, AC/DC $40.00

8055B, Console-R/P, 1959, Wood, left front dial & 5 knobs, right lift top, inner phono, legs, HiFi, BC, FM, AC $40.00

8057, Console-R/P, 1958, Wood, inner dial & 4 speed phono, lift top, large front grill, legs, BC, AC $40.00

8073, Table-R/P, 1950, Plastic, right front round dial knob, vertical grill bars, open top phono, BC, AC $35.00

8080, Table-R/P, 1948, Plastic, upper front slide rule dial, 4 knobs, lift top, inner phono, BC, AC $40.00

8086A, Table-R/P/Rec, 1949, Inner right vertical slide rule dial, 4 knobs, lift top, horizontal louvers, BC, AC $25.00

8090, Console, 1948, Wood, upper slanted slide rule dial, 4 knobs, cloth grill w/2 vert bars, BC, AC/DC $60.00

8100, Console-R/P, 1948, Wood, inner right slide rule dial, 4 knobs, phono, lift top, cloth grill, BC, AC/DC $50.00

8101C, Console-R/P, 1949, Wood, inner right vertical slide rule dial, phono, lift top, criss-cross grill, BC, AC $50.00

8103, Console-R/Rec, 1949, Wood, inner right slide rule dial, 4 knobs, lift top, criss-cross grill, BC, AC $50.00

8105A, Console-R/P, 1948, Wood, inner dial, 4 knobs, pushbuttons, phono, lift top, cloth grill, BC, SW, AC $65.00

8108A, Console-R/P, 1949, Wood, right tilt-out slide rule dial, 4 knobs, pull-out phono drawer, BC, FM, AC $65.00

8115B, Console-R/P, 1949, Wood, tilt-out slide rule dial, 4 knobs, pushbuttons, pull-out phono, BC, FM, AC $60.00

8127, Console-R/P/Rec, 1948, Wood, inner dial, 4 knobs, pushbuttons, fold-down phono door, BC, 2SW, FM, AC $65.00

8200, Table, 1949, Plastic, upper slide rule dial, lower grill, flared base, 2 knobs, BC, battery $25.00

8210, Table-R/P, 1949, Wood, lift top, inner left dial, 2 knobs, front grill cut-outs, crank phono, BC, bat $35.00

8220, Portable, 1958, Transistor, right side dial, vertical front grill bars, top antenna, handle, AM, bat $40.00

8228, Portable, 1958, Plastic, transistor, lower right dial, large grill w/center divider, handle, AM, bat $25.00

8230, Table, 1949, Wood, lower slide rule dial, criss-cross grill, 2 knobs, BC, 3SW, AC/DC/battery $40.00

8270A, Portable, 1949, Plastic & metal, slide rule dial, 2 thumbwheel knobs, flip-up top, BC, AC/DC/bat $30.00

9000, Table, 1949, Plastic, right dial knob, diagonally divided front w/graduated vert bars, BC, AC/DC $45.00

9002, Table, 1959, Brown plastic, modern, lower right dial knob, recessed horiz bars, feet, BC, AC/DC $30.00

9003, Table, 1959, Ivory plastic, modern, lower right dial knob, recessed horiz bars, feet, BC, AC/DC $30.00

9005, Table, 1949, Plastic, upper V-shaped slide rule dial, lower vert grill bars, 2 knobs, BC, AC/DC $30.00

9006, Table, 1959, Plastic, off-center round dial, horizontal front bars, twin speakers, BC, AC/DC $15.00

9012, Table, 1959, Plastic, right front vertical slide rule dial, large center grill, 4 knobs, BC, AC/DC $20.00

9014, Table, 1959, Ivory & brown plastic, transistor, raised dial, swirled perforated grill, twin speakers, BC, bat $25.00

9015, Table, 1959, Ivory plastic, transistor, raised dial, swirled perforated grill, twin speakers, BC, bat $25.00

9016, Table, 1959, Ivory & blue plastic, transistor, raised dial, swirled perforated grill, twin speakers, BC, bat $25.00

9022, Table, 1949, Plastic, right square dial, wrap-around vertical grill bars, 3 knobs, BC, FM, AC $40.00

9028, Table-C, 1959, Plastic, lower front dial, upper alarm clock, right & left grills, feet, BC, AC $15.00

9049, Console-R/P, 1959, Wood, inner dial & 4 speed phono, lift top, large front grill, legs, BC, FM, AC $40.00

9054, Table, 1949, Wood, right vertical slide rule dial, criss-cross grill, 4 knobs, multi-band, AC $35.00

9061, Console-R/P, 1959, Wood, controls under top "piano-lid" cover, pull-out front phono, legs, BC, FM, AC $40.00

9073B, Table-R/P, 1950, Plastic, right front round dial knob, vertical grill bars, open top phono, BC, AC $35.00

9082, Table-R/P, 1950, Wood, center front clear dial, large grill, 4 knobs, lift top, inner phono, BC, AC $25.00

9105, Console-R/P, 1950, Wood, upper front dial, 3 knobs, center pull-out phono drawer, BC, FM, AC $50.00

9201, Table, 1959, Plastic, right FM dial, left AM dial, cloth grill, 2 speakers, 4 knobs, AM, FM, AC $25.00

9202, Portable, 1959, Plastic, transistor, upper front half-moon dial, lower lattice grill, AM, battery $25.00

9204, Portable, 1959, Plastic, transistor, upper triangle dial, lower perforated grill, stand, AM, bat $25.00

9205, Portable, 1959, Plastic, transistor, upper triangle dial, lower perforated grill, stand, AM, bat $25.00

9222, Portable, 1959, Leather case w/ perforated grill, transistor, side knobs, handle, AM, battery $25.00

9226, Portable, 1960, Transistor, lift top w/map, inner 9 band slide rule dial, telescoping antenna, bat $65.00

9260, Portable, 1948, Plastic, lower dial, vertical metal grill bars, handle, 2 knobs, BC, AC/DC/bat $35.00

9270, Portable, 1950, Leatherette, slide rule dial, criss-cross grill, handle, 2 knobs, BC, AC/DC/bat $20.00

9280, Portable, 1950, "Alligator", luggage-style, front slide rule dial, handle, 2 knobs, BC, AC/DC/bat $20.00

Neutrodyne, Table, 1924, Wood, low rectangular case, 3 dial black front panel, lift top, 5 tubes, battery $120.00

SIMPLEX
Simplex Radio Co., Sandusky, Ohio

H, Cathedral, 1930, Wood, center front window dial, upper round cloth grill w/cut-outs, 2 knobs **$160.00**

N, Cathedral, 1932, Two-tone wood, center window dial, upper grill w/cut-outs, 5 tubes, 3 knobs **$160.00**

NT, Console, 1936, Wood, oval black dial, tuning eye, 3 vert grill bars, 10 tubes, 4 knobs, BC, SW, AC **$150.00**

NT, Table, 1936, Wood, center front dial, tuning eye, horizontal grill bars, 10 tubes, BC, SW, AC **$65.00**

P, Tombstone, 1932, Two-tone wood, window dial, lyre grill cut-out, 5 tubes, 3 knobs, BC, SW, AC **$150.00**

Q, Cathedral, 1932, Two-tone wood, center window dial, upper grill w/cut-outs, 5 tubes, 3 knobs **$200.00**

RF, Table, 1925, Wood, low rectangular case, 2 dial front panel, 4 tubes, battery **$135.00**

RS5, Table, 1925, Mahogany, low rectangular case, 3 dial front panel, 5 tubes, battery **$110.00**

RX, Table, 1925, Wood, square case, 2 dial front panel, 4 tubes, battery **$110.00**

SR8, Table, 1925, Mahogany, low rectangular case, slanted 3 dial front panel, 5 tubes, battery **$120.00**

SIMPLON
Industrial Electronic Corp.,
505 Court Street, Brooklyn, New York

CA-5, Table-R/P, 1947, Front slanted dial, lower grill, 2 knobs, handle, 3/4 lift-top, inner phono, BC, AC **$20.00**

WVV2, Table, 1947, Wood, slanted slide rule dial, cloth grill w/4 horizontal bars, 2 knobs, BC, AC/DC **$40.00**

SKY KNIGHT
Butler Bros.,
Randolph & Canal Streets, Chicago, Illinois

CB-500-P, Table, 1947, Wood, right square dial, left cloth grill w/2 horizontal cut-outs, 2 knobs, BC, bat **$30.00**

SKYROVER
Butler Bros.,
Randolph & Canal Streets, Chicago, Illinois

C-10, Cathedral, Wood, lower window dial w/escutcheon, upper cloth grill w/cut-outs, 3 knobs **$200.00**

N5-RD-250, Table, 1946, Plastic, slide rule dial, horizontal louvers w/ 3 vertical bars, 3 knobs, BC, AC/DC **$40.00**

N5-RD295, Table, 1947, Plastic, right front square dial, left vertical grill bars, 2 knobs, BC, AC/DC **$50.00**

SMOKERETTE
Porto Products, Inc.

SR-600W "Smokerette", Table-N, 1947, Plastic, combination radio/pipe rack/humidors/ashtray, slide rule dial, AC .. **$275.00**

SONIC
Sonic Industries, Inc.,
19 Wilbur Street, Lynbrook, New York

415, Table-R/P, 1958, Plastic, inner right front dial, 3 speed record player, lift top, handle, BC, AC **$15.00**

465, Table-R/P, 1958, Center front dial, horizontal grill bars, 3/4 lift top, inner phono, handle, BC, AC **$20.00**

TR-500, Portable, 1958, Leather case w/"brick" front cut-outs, transistor, side knobs, handle, AM, battery **$40.00**

SONORA
Sonora Radio & Television Corp.,
325 North Hoyne Avenue, Chicago, Illinois

100, Table, 1948, Plastic, upper curved slide rule dial, lower horiz louvers, 2 knobs, BC, AC/DC **$35.00**

101, Portable, 1948, Plastic, lower slide rule dial, curved horizontal louvers, handle, 2 knobs, BC, bat **$30.00**

102, Portable, 1949, Plastic, slide rule dial, curved horizontal louvers, handle, 2 knobs, BC, AC/DC/bat **$30.00**

171, Table, 1950, Plastic, lower slide rule dial, upper horizontal grill bars, 2 knobs, BC, AC/DC **$35.00**

306, Table-R/P, 1950, Wood, center front dial, criss-cross grill, 4 knobs, lift top, inner phono, BC, AC **$30.00**

379, Table, 1954, Plastic, large front dial with decorative lines, 2 knobs, BC, AC/DC **$20.00**

401, Console-R/P, 1948, Wood, small cabinet, upper square dial, 4 knobs, lift top, inner phono, BC, AC **$55.00**

538, Portable, 1957, Plastic, 2 upper thumbwheel knobs, perforated grill w/"V", handle, BC, AC/DC/bat **$30.00**

568, Table, 1957, Wood, lower front slide rule dial, large upper cloth grill, 2 knobs, BC, AC/DC **$25.00**

610, Portable, 1958, Plastic, transistor, right dial, left lattice grill, thumbwheel knob, AM, battery **$30.00**

C-22, Table, 1938, Plastic, right slide rule dial over vertical bars, left lattice grill, 2 knobs, BC, AC **$50.00**

D-12, Table, 1938, Wood, right slide rule dial, rounded left w/horiz louvers, 4 knobs, BC, SW, AC/DC **$60.00**

D-800, Table, 1926, Wood, low rectangular case, two window dials w/thumbwheel tuning, lift top, bat **$150.00**

KF, Table, 1941, Two-tone plastic, small case, top right dial knob, front vertical grill bars **$60.00**

KG-132 "Brownie", Portable, 1941, Brown plastic, personal portable, left side knobs, handle, BC, battery **$40.00**

KM "Coronet", Table, 1941, Catalin, right front square dial, left horizontal grill bars, handle, 2 knobs, AC/DC ....**$800.00+**

KNF-148, Table-R/P, 1941, Walnut, right square dial, left lattice grill, 3 knobs, lift top, inner phono, AC **$30.00**

KT "Cameo", Table, 1941, Plastic, right front airplane dial, left horizontal louvers, 2 knobs, AC/DC **$40.00**

KZ-111, Table, 1940, Wood, right front dial, left cloth grill with vertical bars, 2 knobs, battery **$30.00**

LD-93, Table, 1941, Walnut, right front square dial, left horizontal louvers, 3 knobs, BC, SW, AC/DC **$40.00**

LR-147 "Triple Play", Portable, 1941, Leatherette, front slide rule dial, walnut horizontal grill bars, handle, AC/DC/bat **$35.00**

171

PL-29 "Playboy", Portable, 1939, Striped cloth covered square case, front dial & grill, handle, battery **$35.00**

RBU-175, Table, 1946, Plastic, upper slanted slide rule dial, lower horiz louvers, 2 knobs, BC, AC/DC **$40.00**

RBU-176, Table, 1946, Plastic, upper slanted slide rule dial, lower horiz louvers, 2 knobs, BC, AC/DC **$40.00**

RBU-177, Table, 1946, Plastic, upper slanted slide rule dial, lower horiz louvers, 2 knobs, BC, AC/DC **$40.00**

RCU-208, Table, 1946, Wood, right square dial, curved left side w/ vertical louvers, 3 knobs, BC, AC/DC **$60.00**

RDU-209, Table, 1946, Wood, upper slide rule dial, waterfall front, horiz louvers, 3 knobs, BC, AC/DC **$45.00**

RET-210, Table, 1947, Wood, right front square dial, left cloth grill, 4 knobs, BC, SW, AC **$40.00**

RGMF-230, Table-R/P, 1947, Wood, right front dial, left horiz grill bars, 2 knobs, lift top, inner phono, BC, AC **$25.00**

RKRU-215, Table-R/P, 1946, Wood, upper front dial, lower grill, lift top, inner phono, BC, AC/DC **$25.00**

RMR-219, Console-R/P, 1947, Wood, inner right dial, 4 knobs, pushbuttons, left phono, lift top, BC, SW, AC **$80.00**

RQU-222, Table, 1946, Plastic, upper slide rule dial, lower horizontal louvers, 3 knobs, feet, BC, AC/DC **$50.00**

RX-223, Table, 1947, Wood, right square dial, plastic escutcheon, cloth grill w/ bars, 2 knobs, BC, bat **$35.00**

TH-46, Table, 1940, Wood, right front vertical slide rule dial, left cloth grill, battery **$25.00**

TK-44, Table, 1939, Walnut w/gold band overlay, right dial, left horizontal grill bars, 2 knobs, AC **$45.00**

TN-45, Table, 1939, Wood, right dial panel over large wrap-around front grill, 2 knobs, BC, AC/DC **$50.00**

TNE-60, Table-R/P, 1939, Wood, square dial, right & left cloth grills, 2 knobs, lift top, inner phono, AC **$30.00**

TSA-105 "Cosmo", Table, 1939, Plastic, right half-moon dial, left horiz louvers, "Air Magnet" antenna, AC/DC **$45.00**

TT-52, Table, 1939, Walnut, front slide rule dial, pushbuttons, wrap-over top cloth grill, 2 knobs, AC **$60.00**

TV-48, Table, 1939, Plastic, small case, center round dial knob, right & left vert grill bars, AC/DC **$60.00**

TW-49 "Pee-Wee", Table, 1939, Plastic, top right dial knob, right pushbuttons, horiz wrap-around bars, AC/DC **$125.00**

TZ-56, Console, 1939, Wood, slanted slide rule dial, pushbuttons, vert grill bars, 4 knobs, BC, SW, AC **$125.00**

WAU-243, Table, 1947, Plastic, large round dial on top of case, Deco design, 4 pushbuttons, BC, AC/DC **$125.00**

WBRU-239, Table-R/P, 1948, Wood, right front dial, 4 knobs, left grill, lift top, inner phono, BC, AC **$30.00**

WCU-246, Radio/Lamp, 1948, Plastic, controls on sides of case, top grill bars, front lamp switch, BC, AC/DC **$70.00**

WDU-233, Portable, 1947, Two-tone, upper slide rule dial, perforated grill, 2 knobs, handle, BC, AC/DC/bat **$40.00**

WDU-249, Portable, 1948, Mottled case w/plastic accents, slide rule dial, handle, 2 knobs, BC, AC/DC/bat **$40.00**

WEU-262, Table, 1948, Plastic, right dial, left cloth grill, decorative case lines, 4 knobs, BC, FM, AC/DC **$50.00**

WGFU-241, Table-R/P, 1947, Plastic, right round dial, left vertical grill bars, 2 knobs, open top, BC, AC **$45.00**

WGFU-242, Table-R/P, 1947, Plastic, right round dial, left vertical grill bars, 2 knobs, open top, BC, AC **$45.00**

WJU-252, Table, 1948, Plastic, midget, controls on top of case, grill bars - front, back & top, BC, AC/DC **$90.00**

WKRU-254A, Console-R/P, 1948, Wood, inner right dial, 4 knobs, left phono, lift top, BC, FM, AC **$60.00**

WLRU-219A, Console-R/P, 1948, Wood, inner half-moon dial, 4 knobs, phono, lift top, front burl panel, BC, FM, AC **$75.00**

YB-299, Table, 1950, Plastic, right front half-moon dial, left horizontal louvers, 2 knobs, BC, AC/DC **$45.00**

SONY
Delmonico International Sales Corp.,
42-24 Orchard Street,
Long Island City, New York

TFM-151, Portable, 1960, Transistor, top dial & knobs, telescoping antenna, swing handle, AM, FM, battery **$50.00**

TR-63, Portable, 1958, Plastic, transistor, left front round dial, lower perforated grill, AM, battery **$150.00**

TR-86, Portable, 1959, Plastic, transistor, upper right dial, lower perforated grill, swing handle, AM, bat **$45.00**

TR-610, Portable, 1959, Plastic, transistor, front window dial, side thumbwheel knob, handle, battery **$35.00**

SOUND, INC.
Sound, Inc.,
221 East Cullerton Street, Chicago, Illinois

5R2, Table, 1947, Two-tone, right front dial, horiz grill bars, 2 bullet pointer knobs, BC, AC/DC **$40.00**

SPARTON
Sparks-Withington Company,
Jackson, Michigan

The Sparks-Withington Company began to manufacture radios in 1926. In the 1930's, their introduction of Deco-styled mirrored radios, along with other Deco cabinet designs by Walter Dorwin Teague, helped make the Sparton line one of the most sought after by collectors of Deco radios. The company produced one of the most expensive to own radios in today's collecting world - the 1936 Nocturne, a Deco, 46" round mirrored console which currently sells for over $25,000.

4-6, Tombstone, Wood, lower front round airplane dial, upper cloth grill w/cut-outs, 5 knobs **$135.00**

4AW17, Table, 1948, Plastic, slanted slide rule dial, vertical grill bars, 2 knobs, raised top, BC, bat **$30.00**

4AW17-A, Table, 1948, Plastic, slanted slide rule dial, vertical grill bars, raised top, 2 knobs, BC, bat **$30.00**

5, Cathedral, 1931, Wood, lower window dial, upper round grill w/cut-outs, 5 tubes, 2 knobs, AC **$225.00**

5AI16, Table, 1947, Plastic, upper slanted slide rule dial, lower vert grill bars, 2 knobs, BC, AC/DC **$45.00**

5AM26-PS, Table-R/P, 1946, Wood, upper slide rule dial, lower grill, 2 knobs, lift top, inner phono, BC, AC **$30.00**

5AW06, Table, 1946, Plastic, upper slide rule dial, horizontal wrap around louvers, 2 knobs, BC, AC/DC **$45.00**

6-66A, Table, 1948, Leatherette, right rectangular dial, left cloth grill, 3 knobs, BC, SW, AC/DC **$25.00**

6AM06, Portable, 1948, Leatherette, inner dial, 2 knobs, fold-down front, handle, BC, AC/DC/battery **$30.00**

6AW26PA, Table-R/P, 1947, Wood, inner slide rule dial, 4 knobs, phono, lift top, BC, SW, AC **$25.00**

7-46, Console-R/P, 1946, Wood, upper slide rule dial, lower grill w/ 2 vert bars, tilt-out phono, 4 knobs **$100.00**

7AM46, Console, 1946, Wood, upper black slide rule dial, lower grill, 4 knobs, BC, Police, SW, AC **$100.00**

9, Console, 1931, Wood, lowboy, front window dial, lower grill w/cut-outs, 5 tubes, 2 knobs, AC **$125.00**

10, Tombstone, 1931, Wood, shouldered, lower half-moon dial, grill cut-outs, 7 tubes, 3 knobs, AC **$125.00**

10AM76-PA, Console-R/P, 1947, Wood, inner right dial, door, left pull-out phono, lower horiz louvers, BC/SW/FM **$90.00**

10BW76PA, Console-R/P, 1947, Wood, inner slide rule dial, 4 knobs, pull-out phono drawer, BC, SW, FM, AC **$85.00**

15, Console, 1931, Wood, lowboy, half-round dial, lower cloth grill w/ cut-outs, front carvings, AC **$150.00**

25, Console, 1931, Wood, upper half-round dial, lower grill w/cut-outs, 10 tubes, 3 knobs, AC **$140.00**

28 "**Triolian**", Console, 1932, Wood, inner dial & knobs, double front doors, front & side cut-outs, 13 tubes, AC **$275.00**

53, Table, 1934, Wood, round dial, center grill w/cut-outs, 5 tubes, 2 knobs, BC, SW, AC/DC **$75.00**

62, Table, 1926, Wood, low rectangular case, center front window dial w/escutcheon, 3 knobs, AC **$125.00**

65, Tombstone, 1934, Wood, lower round dial, upper grill w/cut-outs, 6 tubes, 5 knobs, BC, SW, AC **$125.00**

67, Tombstone, 1934, Wood, center round dial, upper cloth grill with cut-outs, 6 tubes, BC, SW, AC **$125.00**

75A, Tombstone, 1933, Wood, half-moon dial, upper cloth grill w/cut-outs, 8 tubes, 5 knobs, 5 bands, AC **$130.00**

83, Console, 1934, Wood, upper round dial, lower cloth grill with cut-outs, feet, 8 tubes, BC, SW, AC **$130.00**

89-A, Console, 1928, Wood, inner window dial, lower grill w/cut-outs, 3 knobs, doors, 9 tubes, AC **$185.00**

100, Table, 1948, Plastic, upper slide rule dial, horizontal wrap-around louvers, 2 knobs, BC, AC/DC **$45.00**

104, Console, 1934, Wood, upper round dial, lower grill with cut-outs, 10 tubes, 4 knobs, BC, SW, AC **$150.00**

132, Table, 1950, Plastic, modern, oval case, half-moon dial, metal perforated grill, 2 knobs, BC, AC/DC **$60.00**

141A, Table, 1950, Wood, lower front slanted slide rule dial, 4 knobs, large upper grill, AM, FM, AC **$30.00**

141XX, Table, 1951, Wood, lower front slide rule dial, large upper grill, 4 knobs, AM, FM, AC $40.00

152, Portable, 1950, Plastic, inner dial, 2 knobs, checkered grill, flip-up front, handle, BC, AC/DC/bat **$35.00**

232, Table, 1953, Plastic, oval case, right half-moon dial, center woven grill, 2 knobs, BC, AC/DC **$60.00**

235, Console-R/P, 1930, Carved walnut, lowboy, inner dial & knobs, lower drawer, double front doors, bat **$175.00**

301 "Equasonne", Console, 1929, Carved wood, highboy, inner dial & knobs, double front doors, lower back panel, AC **$575.00**

309, Portable, 1953, Plastic covered, left front dial, center lattice grill, handle, BC, AC/AC/battery **$35.00**

345, Table, 1953, Left front round dial, center circular louvers over perforated grill, BC, AC/DC **$30.00**

409-GL, Table, 1938, Blue mirror glass, 7 sided Deco case, right dial, left grill, 2 black feet, AC/DC .... **$2,000.00+**

410 "Junior", Upright Table, 1930, Wood, center window dial, upper grill w/cut-outs, arched top, finials, feet **$225.00**

500 "Cloisonne", Table, 1939, Catalin & chrome, Deco, right dial, left round grill w/horiz bars, 2 knobs, AC/DC .... **$2,000.00+**

500-C "Cloisonne", Table, 1939, Catalin & chrome, Deco, right dial, left round grill w/horiz bars, 2 knobs, AC/DC .... **$2,000.00+**

500-DG, Table, 1939, Wood w/inlay, right dial, left vert grill bars, step-down sides, 2 knobs, AC/DC **$45.00**

506 "Bluebird", Table, 1936, Round blue or peach mirror glass w/ chrome, round dial, feet, 3 knobs, BC, SW, AC .... $3,500.00+

537, Upright table, 1936, Wood, large center front round dial, top grill, 4 knobs, BC, SW, AC **$85.00**

557, Table, 1936, Blue or peach mirror glass, Deco, chrome fins, black or brown base, 3 knobs, AC/DC .... **$2,500.00+**

558, Table, 1937, Blue or peach mirror glass, Deco, chrome fins, black or brown base, 4 knobs, AC/DC .... **$2,500.00+**

567, Console, 1936, Wood, upper front round dial, lower cloth grill with cut-outs, 5 tubes, BC, SW, AC **$125.00**

577, Console, 1936, Wood, upper round dial, lower grill w/center vertical bars, 5 tubes, BC, SW, AC **$120.00**

616, Tombstone, 1935, Wood, Deco, lower square dial, upper cloth grill w/cut-outs, 4 knobs, BC, SW, AC **$150.00**

617, Tombstone, 1936, Wood, center front round dial, upper grill w/ horizontal bars, 4 knobs, BC, SW, AC **$150.00**

637, Tombstone, 1937, Wood, lower front round dial, upper cloth grill w/4 horizontal bars, 4 knobs **$150.00**

667, Console, 1936, Wood, large upper round dial, lower 3-section grill, 6 tubes, 4 knobs, BC, SW, AC **$160.00**

678, Console, 1937, Wood, upper dial, pushbuttons, lower grill w/ vertical bars & circles, BC, SW, AC **$200.00**

768, Console, 1937, Wood, upper dial, tuning eye, lower grill w/lattice bars, 7 tubes, 5 knobs, BC, SW, AC **$150.00**

867, Console, 1936, Wood, upper round dial, lower grill w/vertical center bars, 8 tubes, BC, SW, AC **$175.00**

930 "Equasonne", Console, 1929, Wood, inner dial, grill w/cut-outs, double front doors, stretcher base, 9 tubes, AC **$135.00**

931, Console, 1929, Wood, inner front dial, lower grill w/cut-outs, double doors, stretcher base, AC **$135.00**

987, Console, 1936, Wood, large round dial, tuning eye, splayed grill bars, 9 tubes, 6 knobs, BC, SW, AC **$195.00**

997, Tombstone, 1937, Wood, center front round dial, upper cloth grill w/4 horizontal bars, 5 knobs **$150.00**

1003, Console-R/P, 1949, Wood, "chest of drawers", inner right dial, left phono, 13 tubes, BC, SW, FM, AC **$175.00**

1006, Console-R/P, 1947, Wood, inner slide rule dial, 2 knobs, door, pull-out phono drawer, BC, FM, AC **$70.00**

1010, Console-R/P, 1948, Wood, inner dial, 4 knobs, phono, lift top, criss-cross grill, 7 tubes, BC, AC **$65.00**

1030, Console-R/P, 1948, Wood, inner slide rule dial, 4 knobs, phono, fold-down front, BC, AC **$60.00**

1040, Console-R/P, 1949, Wood, right tilt-out slide rule dial, 2 knobs, pull-out phono drawer, BC, FM, AC **$70.00**

1051, Console-R/P, 1949, Wood, left slide rule dial, 4 knobs, lower left tilt-out phono, storage, BC, SW, AC **$50.00**

1059, Console-R/P, 1949, Wood, left slide rule dial, 4 knobs, tilt-out phono, lower right storage, BC, FM, AC **$50.00**

1068, Console, 1937, Wood, upper dial, pushbuttons, 2 tuning eyes, 4 vert grill bars, 5 knobs, BC, SW, AC **$180.00**

1160, Console, 1939, Wood, slide rule dial, pushbuttons, horiz grill bars, 11 tubes, 5 knobs, BC, SW, AC **$175.00**

1167, Console, 1936, Wood, upper round dial, lower cloth grill with vertical bars, 11 tubes, BC, SW, AC **$175.00**

1186 "Nocturne", Console, 1936, Blue mirror glass & chrome, round Deco design, tuning eye, 5 knobs, BC, SW, AC ... **$25,000.00+**

1288-P, Console-R/P, 1937, Wood, center dial, pushbuttons, tuning eyes, side storage, inner phono, 12 tubes, AC **$225.00**

1567, Console, 1936, Wood, upper slanted round dial, lower grill w/ vertical bars, 15 tubes, BC, SW, AC **$250.00**

1867 "Triolian", Console, 1936, Leather & walnut, Deco, round dial, tuning eye, 18 tubes, 3 speakers, 5 knobs, BC, SW, AC **$450.00**

SPLITDORF
Splitdorf Radio Corporation,
Newark, New Jersey

Abbey, Table, 1927, Wood, front window dial w/escutcheon, single dial tuning, lift top, 6 tubes, bat **$160.00**

R-100, Table, 1925, Wood, low rectangular case, 3 dial front panel, 5 tubes, battery **$145.00**

R-500, Table, 1924, Wood, low rectangular case, metal 3 dial front panel, lift top, 5 tubes, battery **$145.00**

R-V-695, Table, 1926, Two-tone wood, rectangular case, metal 2 dial slant front panel, 6 tubes, battery **$130.00**

ST. REGIS

676X, Table, Wood, black dial, upper grill w/horiz bars and rectangular cut-out, 2 knobs, BC **$55.00**

STANDARD
**Standard Radio Corp.,
Worcester, Massachusetts
Standard Polyrad Corp.,
521 Broadway, Cincinnati, Ohio**

B5 "Standardyne", Table, 1925, Wood, low rectangular case, 3 dial black front panel, lift top, 5 tubes, battery **$115.00**
BH "Standardyne", Console, 1925, Wood, inner 3 dial black panel, upper speaker grill w/cut-outs, lower storage, bat **$165.00**
SR-F22, Portable, 1959, Transistor, left thumbwheel dial knob, lower divided perforated grill, AM, battery **$30.00**

STARK
**Stark Sound Engineering,
2131 Fairfield Avenue,
Fort Wayne, Indiana**

1010, Table, 1950, Plastic, right front dial, left cloth grill w/horizontal bars, 2 knobs, BC, AC/DC **$40.00**

1020, Table, 1950, Plastic, raised upper slide rule dial, horizontal wrap-around louvers, BC, AC/DC **$50.00**

STARR
**Starr Equipment Corp.,
366 Hamilton Ave., Brooklyn, New York**

D "Starr-Harmonic", Table, 1925, Wood, low rectangular case, 3 dial front panel, 5 tubes, battery **$135.00**

STEELMAN
**Steelman Radio & Phonograph Co., Inc.,
12-30 Anderson Avenue,
Mt Vernon, New York**

3A16U, Table-R/P, 1956, Right front slide rule dial, 3 right side knobs, lift top, inner phono, handle, BC, AC **$20.00**
3AR1, Table-R/P, 1953, Right front round dial, left square grill, lift top, inner phono, handle, BC, AC **$25.00**
3AR4, Table-R/P, 1956, Right front round dial, left grill, lift top, inner phono, handle, BC, AC **$20.00**
3AR5U, Table-R/P, 1956, Right side dial and knobs, front grill, lift top, inner phono, handle, BC, AC **$15.00**
3RP1, Table-R/P, 1953, Center front round dial knob over grill, lift top, inner phono, BC, AC **$20.00**
3RP4, Table-R/P, 1956, Inner right dial, 3 knobs, 3 speed record player, lift top, handle, BC, AC **$20.00**
3RP8, Table-R/P, 1959, Inner front dial, phono, lattice grill, 2 knobs, lift top, handle, BC, AC **$15.00**
4AR12, Console-R/P, 1959, Wood, inner dial & 4 speed phono, lift top, large front grill, legs, BC, AC **$40.00**
4RP7, Table-R/P, 1958, Inner front half-round dial, 4 speed record player, lift top, handle, BC, AC **$15.00**
450, Table-C, 1952, Plastic, right round dial, left round alarm clock, center vertical grill bars, BC, AC **$25.00**
517, Table-R/P, 1952, Suitcase-style, inner right dial, 4 knobs, phono, lift top, handle, BC, AC **$20.00**
595, Table-R/P, 1952, Leatherette, outer right side dial, 4 knobs, "Steelman" grill, lift top, BC, AC **$20.00**
597, Table-R/P, 1952, Leatherette, right half-round dial, left grill, 4 knobs, lift top, inner phono, BC, AC **$20.00**
601, Portable, 1952, Center front round dial knob over perforated grill, handle, side knob, BC, battery **$20.00**
602, Portable, 1952, Leatherette, front dial, lower perforated grill, handle, 1 side knob, BC, AC/DC/bat **$25.00**
5101, Table, 1952, Wood, off-center dial over large front grill, 3 knobs, BC, AC/DC **$20.00**
6000, Portable, 1952, Inner center front round dial, 2 knobs, fold-in front, handle, BC, AC/DC/bat **$25.00**
AF1100, Table, 1952, Lower front slide rule dial, large upper grill area, 4 knobs, BC, FM, AC **$20.00**

STEINITE
Steinite Labs., Atchinson, Kansas

102, Console-R/P, 1929, Walnut, lowboy, ornate, inner window dial, lower grill, double front doors, AC **$190.00**
605, Table-R/P, 1931, Wood, upright table, center front window dial, lower pull-out phono drawer, AC **$175.00**
700, Cathedral, 1931, Wood, center front window dial, upper grill w/ cut-outs, 5 tubes, 3 knobs, AC **$250.00**
712, Console, 1931, Wood, lowboy, upper window dial, lower grill w/ cut-outs, 8 tubes, 3 knobs **$160.00**

991, Table, 1927, Wood, two window dials w/large escutcheons, voltmeter, 4 knobs, 6 tubes, AC **$165.00**

STELLAR

Roger Maris/Mickey Mantle, Table-N, 1959, Wood, left baseball batter, right round dial w/"diamond" pointer, "autographed" **$575.00**

STERLING
The Sterling Manufacturing Co., Cleveland, Ohio

5 "Concertone", Tombstone, 1931, Wood, small case, shouldered, right front dial, 3 section cloth grill, 5 tubes **$135.00**
7 "Concertone", Tombstone, 1931, Wood, shouldered, center front quarter-round dial, upper 3 section grill, 3 knobs **$165.00**
8 "Concertone", Console, 1931, Wood, lowboy, upper quarter-round dial, lower grill w/cut-outs, 8 tubes, 3 knobs **$165.00**
A-2-60 "Troubador", Console, 1929, Walnut, highboy, inner window dial, lower round grill, double doors, 8 tubes, AC **$165.00**
A-3-60 "Serenader", Console, 1929, Walnut, lowboy, upper front window dial, lower cloth grill w/cut-outs, AC **$145.00**
B-2-60 "Imperial", Console, 1929, Walnut, upper window dial w/ escutcheon, lower square grill w/cut-outs, AC **$150.00**

LS-4, Portable, Plastic, right thumbwheel dial, diagonal perforated chrome grill, handle, battery **$40.00**

STEWART-WARNER
Stewart-Warner Corp.,
1826 Diversey Parkway, Chicago, Illinois

01-5D9, Table-R/P, 1939, Wood, right front dial, left grill w/horiz bars, 2 knobs, lift top, inner phono, AC **$35.00**
01-5H7, Console, 1939, Walnut, upper slide rule dial, pushbuttons, 2 lower vertical grill bars, BC, SW, AC **$135.00**
01-6A7, Console, 1939, Wood, slanted slide rule dial, pushbuttons, 3 vert grill bars, 4 knobs, BC, SW, AC **$150.00**
01-6B9, Console-R/P, 1939, Walnut, upper slide rule dial, pushbuttons, lower vertical grill bars, 8 tubes, AC **$150.00**
01-6C9, Console-R/P, 1939, Wood, upper front pushbuttons, lower grill w/vertical bars, inner "magic" dial, AC **$160.00**
01-8A7, Console, 1939, Wood, upper slide rule dial, pushbuttons, lower grill w/vertical bars, BC, SW, AC **$150.00**
01-8B7, Console, 1940, Wood, upper slide rule dial, 8 pushbuttons, lower grill with vertical bars, 8 tubes **$150.00**
01-611, Table, 1939, Wood, right front dial, pushbuttons, raised left horizontal louvers, 4 knobs, AC **$60.00**
01-817, Console, 1939, Wood, upper slide rule dial, pushbuttons, 2 vert grill bars, 11 tubes, 2 band, AC **$160.00**
02-411, Portable, 1939, Striped cloth covered, right front dial, left grill, handle, 2 knobs, battery **$25.00**
03-5C1, Table, 1939, Two-tone wood, right front dial, left horizontal louvers, 2 knobs, AC/DC **$40.00**
03-5C1-WT, Table, 1939, Wood, right front dial, left horizontal louvers, 2 knobs, feet **$40.00**
03-5E1, Upright Table, 1939, Plastic, center front dial, pushbuttons, upper horizontal louvers, 2 knobs, AC/DC **$110.00**
03-5K3 "The Magician", Table, 1939, Plastic, streamlined, right dial knob, rounded left w/wrap-around louvers, AC **$85.00**
07-5B3Q "Dionne Quints", Table-N, 1939, Plastic, decals of quints, right front dial, rounded left w/"wavy" grill, AC/DC **$1,000.00**
07-5R3, Table, 1940, Ivory plastic, right slide rule dial, rounded left w/wrap-around louvers, 2 knobs **$75.00**

07-51H, Table, 1940, Plastic, streamline, right front dial, rounded left w/horiz wrap-around louvers **$85.00**
07-55BK, Bed, 1939, Waterfall headboard w/built-in radio, large dial knob, grill cut-outs, BC, AC/DC **$250.00**
07-512 "Campus", Table, 1939, Plastic, right dial, rounded left w/ wrap-around louvers, school letters available **$85.00**
07-513 "Gulliver's Travels", Table-N, 1939, Plastic, decals of Gulliver's Travels, right front dial, rounded left w/louvers, AC **$625.00**
07-513Q "Dionne Quints", Table-N, 1938, Plastic, decal of quints, right dial, rounded left w/wrap-around louvers, AC/DC **$1,000.00**
07-514, Table, 1939, Two-tone wood, Deco, right front dial, squared left wrap-around louvers, AC/DC **$65.00**
07-516 "Fireside", Chairside, 1939, Wood, slanted front dial, horizontal louvers, lower magazine shelf, 4 legs, AC/DC **$100.00**

4B4, Table, 1940, Two-tone wood, center front dial, right & left grills w/horiz bars, 2 knobs, bat $45.00

4D1, Table, 1940, Two-tone wood, right dial, left cloth grill w/cut-outs & vertical bars, 2 knobs $45.00

5V9, Table-R/P, 1941, Wood, right front dial, left grill w/horizontal bars, lift top, inner phono, AC $30.00

6T8, Table-R/P, 1940, Inner right dial, left grill, fold-down front, lift top, inner phono, handle, BC, AC $30.00

50, Console, 1932, Wood, lowboy, upper dial, lower cloth grill w/cut-outs, 11 tubes, 6 legs, BC, SW $160.00

51, Console, 1932, Wood, lowboy, inner dial/knobs, doors, grill w/cut-outs, 11 tubes, 5 legs, BC, SW $165.00

51T56, Table, 1948, Wood, upper slanted slide rule dial, large lower grill, 2 knobs, BC, AC/DC $30.00

51T136, Table, 1947, Wood, upper slanted slide rule dial, large lower cloth grill, 2 knobs, BC, AC/DC $30.00

61T16, Table, 1946, Wood, upper slide rule dial, lower cloth grill, 3 knobs, BC, AC/DC $35.00

61T26, Table, 1946, Plastic, upper slide rule dial, lower horizontal louvers, 3 knobs, BC $35.00

62T36, Table, 1946, Catalin, upper slanted slide rule dial, lower inset horiz louvers, 3 knobs, BC, SW $700.00

62TC18, Table, 1946, Wood, right slide rule dial, pushbuttons, left cloth wrap-around grill, 4 knobs $60.00

72CR26, Console-R/P, 1947, Wood, inner slide rule dial, 4 knobs, 6 pushbuttons, pull-out phono, BC, SW, AC $75.00

91-511, Table, 1938, Walnut, right front slide rule dial, pushbuttons, left horizontal louvers, 2 knobs $50.00

91-513, Table, 1938, Wood, triangular case, slide rule dial, 4 pushbuttons, grills on all four side, AC $185.00

91-514, Table, 1938, Black & ivory, Deco, right slide rule dial, pushbuttons, left vertical bars, AC $75.00

91-531, Upright Table, 1938, Walnut, center front dial, pushbuttons, upper grill w/vertical bars, BC, SW, AC $75.00

91-621, Table, 1938, Wood, right front dial, pushbuttons, tuning eye, raised left louvers, BC, SW, AC $75.00

91-1117, Console, 1938, Wood, slide rule dial, pushbuttons, tuning eye, horiz grill bars, 11 tubes, BC, SW, AC $155.00

97-521, Table, 1938, Walnut, right front slide rule dial, pushbuttons, left horizontal louvers, 2 knobs $50.00

97-562, Table, 1938, Plastic, right "magic keyboard" dial, pushbuttons, raised left w/"wavy" grill, AC $100.00

300, Table, 1925, Wood, low rectangular case, front panel w/3 half-moon dials, 5 tubes, battery $115.00

303, Table, Wood, rectangular case, slanted black 3 dial front panel, lift top, 5 tubes $130.00

305 **"Aeromaster"**, Table, 1925, Wood, rectangular case, slanted 3 dial front panel, 5 tubes, battery $130.00

325, Table, 1925, Wood, low rectangular case, metal front panel w/3 pointer dials, lift top, bat $115.00

385, Table, 1927, Wood, low rectangular case, center front dial, 6 tubes, battery $95.00

520, Console, 1927, Wood, lowboy, upper window dial, lower metal grill w/dancing girls, 3 knobs $300.00

900, Console, 1929, Walnut, lowboy, front window dial w/escutcheon, round grill w/cut-outs, AC $135.00

900AC, Table, 1929, Metal, low rectangular case, center front window dial, lift-off top, 3 knobs, AC $85.00

1264, Console, 1934, Wood, lowboy, upper front dial, lower cloth grill, 6 legs, BC, SW, AC $140.00

1883 AC/DC, Chairside, 1937, Walnut, half-round case, top dial under glass, front grill w 3/horiz bars, AC/DC $185.00

3041, Table, 1937, Rectangular case lays down or stands up, slide rule dial, wrap-around grill & bars $75.00

9000-B, Table, 1947, Wood, slanted slide rule dial, lower horizontal louvers, 3 knobs, BC, SW, AC/DC $45.00

9001-E, End Table, 1946, Wood, drop leaf style, slide rule dial, 4 knobs, pushbuttons, drop-front, BC, SW, AC $125.00

9001-F, End Table, 1946, Wood, inner slide rule dial, pushbuttons, fold-down front, lower 4-legged base $135.00

9002-A, Table, 1948, Plastic, upper curved slide rule dial, large center grill, 3 knobs, BC, AC/DC $40.00

9002-B, Table, 1948, Plastic, upper slanted slide rule dial, lower horizontal louvers, 3 knobs, BC $40.00

9007-F, Portable, 1946, Luggage-style, inner right dial, 3 knobs, fold down front, handle, BC, AC/DC/bat $25.00

9150-D, Console-R/P, 1950, Wood, right tilt-out dial & knobs, left pull-down phono door, storage, AC/DC/bat $110.00

9151-A, Table, 1950, Plastic, right raised see-through dial, checkered grill, 2 knobs, AM, FM, AC/DC $35.00

9152-A, Table, 1950, Plastic, right round dial over vertical front grill bars, 2 knobs, feet, BC, AC/DC $30.00

9152-B, Table, 1950, Plastic, right round dial over vertical front grill bars, 2 knobs, feet, BC, AC/DC $30.00

9153-A, Table, 1950, Plastic, lower right dial knob, large perforated metal grill, BC, AC/DC/battery $35.00

9154-C, Console-R/P, 1951, Wood, upper front dial, 3 knobs, center pull-out phono drawer, lower grill, BC, AC $50.00

9160-AU, Table, 1952, Plastic, right front round dial, left lattice grill, 2 knobs, BC, AC/DC $30.00

9162-A, Table-C, 1952, Plastic, left half-round dial over alarm clock, right horiz bars and circles, BC, AC $30.00

9165-A, Table, 1953, Plastic, right half-round dial over dotted grill, left speaker, 2 knobs, feet, BC $25.00

9170-B "Gadabout", Portable, 1954, Plastic, top dial & volume, 2-section front lattice grill, handle, BC, AC/DC/bat $30.00

9170-D, Portable, 1954, Plastic, top dial & volume, 2-section front lattice grill, handle, BC, AC/DC/bat $30.00

9178-C, Table-R/P, 1955, Wood, right side dial knob, large front grill, lift top, inner phono, BC, AC $25.00

9180-H, Table, 1954, Plastic, right front raised round dial, left recessed lattice grill, 2 knobs $30.00

A6-1Q "Dionne Quints", Table-N, 1938, Plastic, decals of Dionne quints, top thumbwheel knobs, wrap-over vertical bars $1,000.00

A51T3 "Air Pal", Table, 1947, Plastic, upper slide rule dial, lower vert grill bars, 2 large top knobs, BC, AC/DC $100.00

A61CR3, Console, 1948, Wood, recessed slanted dial, 3 knobs, front pull-out phono drawer, BC, AC $50.00

A61P1, Portable, 1948, Leatherette, inner right dial, 3 knobs, fold-down front, handle, BC, AC/DC/bat $40.00

A72T3, Table, 1948, Wood, upper slanted slide rule dial, criss-cross grill, 3 knobs, BC, FM, AC/DC $35.00

A92CR6, Console-R/P, 1947, Wood, inner slide rule dial, 4 knobs, 6 pushbuttons, pull-out phono, BC, FM, AC $75.00

B51T2 "Air Pal", Table, 1949, Plastic, upper slanted slide rule dial, 2 thumbwheel knobs on case top, BC, AC/DC $75.00

B61T1, Table, 1949, Plastic, slide rule dial, left wrap-around horizontal louvers, 3 knobs, BC, AC/DC $40.00

B72CR1, Console-R/P, 1948, Wood, curved top slide rule dial, 4 knobs, front pull-out phono drawer, BC, FM, AC $60.00

B92CR1, Console-R/P, 1949, Wood, inner right recessed dial, 3 knobs, left lift top, inner phono, BC, FM, AC $55.00

C51T1, Table, 1948, Plastic, right half-moon dial, left horiz wrap-around louvers, 2 knobs, BC, AC/DC $55.00

C51T2, Table, 1948, Plastic, right half-moon dial, left horizontal louvers, base, 2 knobs, BC, AC/DC $55.00

R-109-A, Cathedral, 1932, Wood, window dial, upper grill w/cut-outs, fluted columns, 6 tubes, 3 knobs, AC $215.00

R-110-AT, Table, 1933, Wood, center window dial w/escutcheon, right & left grills w/cut-outs, 10 tubes, 2 speakers, 4 knobs $100.00

R-116-AH, Table, 1933, Wood, right dial & volume windows, left grill w/horizontal bars, 2 knobs, AC $70.00

R-172-A, Table, Wood, right front round black dial, left front & side grills w/vert bars, 4 knobs $55.00

R-192 "Good Companion", Table, 1936, Deco round case design with base, half round dial, center circular grill, 4 tubes $400.00

R-1235A, Tombstone, 1935, Wood, lower dial & knobs, upper cloth grill w/aluminum cut-outs, top fluting $185.00

R-1301-A, Tombstone, Wood, rounded/step-down top, lower round dial, upper grill w/cut-outs, 4 knobs $160.00

R-1725-A, Console, Wood, upper front round black dial, lower cloth grill w/3 bars, 4 knobs, BC, SW $135.00

R-3043-A, Table, Wood, rectangular case lays down or stands up, slide rule dial, grill bars, 2 knobs $75.00

STRATOVOX
Grossman Music Co.,
210 Prospect Street, Cleveland, Ohio

579-1-58A, Table, 1946, Wood, slanted slide rule dial, cloth grill w/ 2 horizontal bars, 2 knobs, BC, AC/DC $40.00

STROMBERG-CARLSON
Stromberg-Carlson Company,
100 Carlson Road, Rochester, New York

Stromberg-Carlson was formed in 1894 for the production of telephone equipment. The company began to make radio parts and by 1923 was producing complete radios. Stromberg-Carlson was well-known for its commitment to quality and their products are often called the "Rolls Royce" of radios.

1-A, Table, 1924, Wood, low rectangular case, 3 dial black bakelite front panel, 5 tubes, battery $175.00

2, Console, 1924, Wood, inner three dial panel, fold-down front, lower storage, 5 tubes, battery $250.00

19, Console, 1931, Wood, lowboy, upper front window dial, lower cloth grill w/cut-outs, 9 tubes, AC $165.00

20, Console, 1931, Wood, highboy, window dial, cloth grill w/cut-outs, stretcher base, 9 tubes, AC $170.00

22, Console, 1931, Wood, upper window dial, lower cloth grill w/cut-outs, 10 tubes, carved legs, AC $190.00

24, Console-R/P, 1932, Wood, inner window dial, lower grill cut-outs, doors, phono, 10 tubes, 6 legs, AC $190.00

25, Console, 1931, Wood, lowboy, window dial, lower grill w/cut-outs, stretcher base, 8 tubes, AC $150.00

38, Console, 1932, Wood, lowboy, upper window dial, lower grill w/ cut-outs, 9 tubes, bowed legs, AC $170.00

39, Console, 1932, Wood, lowboy, inner window dial, grill w/cut-outs, doors, 10 tubes, stretcher base, AC $180.00

41, Console-R/P, 1932, Wood, lowboy, inner window dial, grill w/cut-outs, doors, 9 tubes, 6 legs, AC $175.00

49, Console, 1933, Wood, lowboy, inner window dial, circular grill cut-outs, doors, 11 tubes, AC $200.00

51, Console-R/P, 1933, Wood, lowboy, inner dial/grill, doors, lift top, inner phono, 11 tubes, 6 legs, AC $200.00

52, Console, 1933, Wood, inner window dial, grill w/cut-outs, doors, carved legs, 12 tubes, 4 knobs, AC $200.00

54, Console-R/P, 1933, Wood, lowboy, inner window dial, grill w/cut-outs, doors, lift top, 12 tubes, 6 legs, AC $200.00

55 "Te-lek-tor-et", Table/2 Piece, 1933, Wood, chest-type table set w/dial & knobs, separate speaker case, 8 tubes, AC . $275.00

56, Console, 1933, Wood, Deco, hinged front door hids controls, Deco grill cut-outs, 8 tubes, AC $225.00

58-L, Console, 1935, Wood, upper octagonal dial, lower grill w/cut-outs, 6 tubes, 3 knobs, BC, SW, AC $135.00

58-T, Tombstone, 1935, Wood, lower octagonal dial, upper grill w/ vertical cut-outs, 3 knobs, BC, SW, AC $145.00

58-W, Console, 1935, Wood, upper octagonal dial, lower grill w/cut-outs, 6 tubes, 3 knobs, BC, SW, AC $135.00

60-M, Console, 1935, Wood, lowboy, upper front dial, lower cloth grill w/cut-outs, 6 legs, BC, SW, AC $145.00

61-H, Table, 1935, Wood, Deco design, off-center octagonal dial, right & left grills, 3 knobs, AC $75.00

61-L, Console, 1936, Wood, upper octagonal dial, lower grill w/cut-outs, 6 tubes, 3 knobs, BC, SW, AC $140.00

61-T, Tombstone, 1935, Wood, lower octagonal dial, upper grill w/ vertical cut-outs, 3 knobs, BC, SW, AC $135.00

61-U, Tombstone, 1935, Wood, step-down top, lower octagonal dial, upper 5-section grill, 3 knobs $140.00

61-W, Console, 1935, Wood, upper octagonal dial, lower grill w/cut-outs, 6 tubes, 3 knobs, BC, SW, AC $140.00

62, Console, 1935, Wood, upper octagonal dial, lower grill w/cut-outs, 8 tubes, 5 knobs, BC, SW, AC $160.00

63, Console, 1935, Wood, upper octagonal dial, lower grill w/cut-outs, 8 tubes, 5 knobs, BC, SW, AC $160.00

64, Console, 1934, Wood, Deco design, upper dial, lower grill w/cut-outs, 8 tubes, 5 knobs, feet, AC $160.00

65 "Te-lek-tor-et", Table/2 Piece, 1935, Wood, chest-type table set w/dial & knobs, separate speaker case, 9 tubes, AC $275.00

68, Console, 1934, Wood, Deco, octagonal dial, tuning eye, grill cut-outs, 10 tubes, 4 knobs, BC, SW, AC $190.00

68R, Console, 1935, Wood, lowboy, inner dial & knobs, double front doors, 6 legs, stretcher base, AC $150.00

70, Console, 1935, Wood, lowboy, inner dial, doors, lower grill w/cut-outs, 13 tubes, 2 speakers, 6 legs, BC, SW, AC $225.00

72, Console-R/P, 1935, Wood, inner dial & phono, doors, lower vert grill bars, 13 tubes, 2 speakers, 6 feet, BC, SW, AC ..$225.00

74, Console-R/P, 1935, Wood, inner dial & phono, doors, lower vert grill bars, 16 tubes, 2 speakers, 6 feet, BC, SW, AC ..$300.00

82, Console, 1935, Walnut, octagonal dial, tuning eye, cloth grill w/cut-outs, 10 tubes, 4 knobs, BC, SW, AC $185.00

83, Console, 1935, Walnut, octagonal dial, tuning eye, cloth grill w/cut-outs, 11 tubes, 4 knobs, BC, SW, AC $185.00

84, Console, 1935, Wood, octagonal dial, tuning eye, cloth grill w/cut-outs, 12 tubes, 6 feet, BC, SW, AC $200.00

115, Console, 1936, Wood, upper octagonal dial, lower cloth grill with cut-outs, 7 tubes, BC, SW, bat $110.00

125-H, Table, 1936, Two-tone wood, right front octagonal dial, left grill w/vertical bars, BC, SW, AC $60.00

130-H, Table, 1936, Wood, off-center octagonal dial, left cloth grill w/crossed bars, BC, SW, AC $60.00

130-J, Table, 1937, Wood, Deco case, right octagonal dial, left vertical grill bars, 4 knobs, BC, SW $110.00

130-L, Console, 1936, Wood, upper octagonal dial, lower grill w/2 vertical bars, 7 tubes, BC, SW, AC $135.00

130-M, Console, 1936, Wood, upper octagonal dial, lower cloth grill w/vertical bars, 8 tubes, BC, SW, AC $140.00

130-R, Table, 1936, Two-tone wood, Deco, off-center octagonal dial, left grill, 4 knobs, BC, SW, AC $85.00

130-U, Tombstone, 1936, Wood, lower octagonal dial, upper cloth grill w/horiz bars, 4 knobs, BC, SW, AC $135.00

140-H, Table, 1936, Wood, rectangular case, octagonal dial, left horiz louvers, 4 knobs, BC, SW, AC $75.00

140-L, Console, 1936, Wood, upper octagonal dial, vertical grill bars, 9 tubes, 4 knobs, BC, SW, AC $160.00

140-M, Console, 1937, Wood, upper octagonal dial, tuning eye, center horizontal grill bars, 4 knobs, feet $185.00

140-P, Console, 1936, Wood, octagonal dial, tuning eye, lower cloth grill w/vertical bars, 9 tubes, feet $175.00

145-L, Console, 1936, Wood, upper front dial, lower cloth grill w/cut-outs, 10 tubes, feet, BC, SW, AC $180.00

145-SP, Console-R/P, 1937, Wood, front dial w/escutcheon, tuning eye, cloth grill w/center vert bars, 5 knobs, inner phono $170.00

150-L, Console, 1936, Wood, upper front dial, lower cloth grill, 12 tubes, 5 knobs, feet, BC, SW, AC $200.00

160-L, Console, 1936, Wood, upper front dial, lower "U" shaped cloth grill, 14 tubes, feet, BC, SW, AC $225.00

225-H, Table, 1937, Wood, right front octagonal dial, left cloth grill w/floral cut-outs, 4 knobs, BC, SW, AC $85.00

228-H, Table, 1937, Wood, right front octagonal dial, left cloth grill w/vert bars, 4 knobs, BC, SW, AC $75.00

228-L, Console, 1937, Wood, upper octagonal dial, lower grill w/vert bars, 6 tubes, 4 knobs, BC, SW, AC $135.00

229-P, Console-R/P, 1937, Wood, front octagonal dial, lower grill with horizontal bars, 8 tubes, BC, SW, AC $150.00

230-H, Table, 1937, Wood, rectangular case, right front dial, left horizontal grill bars, BC, SW, AC $75.00

230-L, Console, 1937, Wood, upper dial, lower cloth grill with center vertical bars, 7 tubes, BC, SW, AC $130.00

231-F, Chairside, 1937, Wood, Deco, mirrored top with dial & knobs, horizontal side louvers, BC, SW, AC $250.00

231-R, Chairside, 1937, Wood, Deco, half-round, mirrored top with dial & knobs, front louvers, BC, SW, AC $275.00

235-H, Table, 1938, Wood, right dial, 6 pushbuttons, tuning eye, left grill w/cut-outs, BC, SW, AC $85.00

240-H, Table, 1937, Wood, front dial, rounded left side with horiz wrap-around louvers, BC, SW, AC $85.00

240-L, Console, 1937, Wood, upper dial, center horiz bars & semi-circular cut-out, 11 tubes, BC, SW, AC $180.00

240-M, Console, 1937, Wood, upper front dial, tuning eye, lower cloth grill w/bars, 11 tubes, BC, SW, AC $175.00

240-R, Console, 1937, Wood, half-round case, inner dial, pushbuttons, doors, horiz louvers, BC, 4SW, AC $375.00

240-W, Console/Desk, 1937, Wood, looks like Governor Winthrop desk, inner dial & controls, 11 tubes, AC $400.00

250-L, Console, 1937, Wood, upper dial, lower grill w/curved horizontal bars, 13 tubes, feet, BC, SW, AC $200.00

255-L, Console, 1937, Wood, upper dial, tuning eye, lower grill w/curved horiz bars, 13 tubes, 6 knobs, feet, BC, SW ...$200.00

260-L, Console, 1937, Wood, upper dial, tuning eye, center horizontal grill bars, 16 tubes, BC, SW, AC $250.00

320-H, Table, 1938, Two-tone wood, right front dial, left grill with Deco cut-outs, 3 knobs, BC, SW $50.00

320-T, Drop Leaf Table, 1938, Wood, drop leaf Duncan Phyfe design, inner dial & controls, 5 tubes, BC, SW, AC $130.00

325-J, Table, 1938, Walnut, center dial, right pushbuttons, left "horseshoe" grill, 3 knobs, BC, SW, AC $65.00

325-N, Side Table, 1938, Mahogany, Chippendale design side table, inner dial & controls, 5 tubes, BC, SW, AC $135.00

325-S, Drop Leaf Table, 1938, Maple, Early American design, inner dial & controls, 5 tubes, BC, SW, AC $135.00

335-H, Table, 1939, Wood, rectangular case, right dial, pushbuttons, left cloth grill w/center cut-out $75.00

335-L, Console, 1939, Walnut, upper dial, lower grill w/cut-outs, pushbuttons, tuning eye, 7 tubes, BC, SW, AC $140.00

336-P, Chairside, 1939, Walnut, top dial/5 knobs/6 pushbuttons/ tuning eye, front vert bars, storage, BC, SW, AC **$130.00**

337-H, Table, 1939, Walnut, right dial, 3 vertical grill bars, 6 pushbuttons, tuning eye, 4 knobs, BC, SW, AC **$80.00**

337-L, Console, 1939, Walnut, upper dial, lower grill bars, 6 pushbuttons, tuning eye, 7 tubes, BC, SW, AC **$160.00**

340-F, Console, 1939, Maple, square dial, vertical grill bars, pushbuttons, tuning eye, 9 tubes, 4 knobs, BC, SW, AC **$160.00**

340-H, Table, 1939, Rosewood, right dial, left horiz grill bars, pushbuttons, tuning eye, 4 knobs, BC, SW, AC **$85.00**

340-M, Console, 1939, Walnut, square dial, vert grill bars, pushbuttons, tuning eye, 9 tubes, BC, SW, AC **$165.00**

340-P, Console-R/P, 1939, Walnut, square dial, horiz grill bars, pushbuttons, tuning eye, inner phono, 9 tubes, BC, SW, AC **$165.00**

340-V, Console/Corner, 1939, Maple, corner-style, square dial, vert grill bars, pushbuttons, tuning eye, 9 tubes, 4 knobs, BC, SW, AC **$160.00**

340-W, Console/Corner, 1938, Walnut, corner-style, square dial, vert grill bars, pushbuttons, tuning eye, BC, SW, AC **$185.00**

341-R, Chairside, 1938, Walnut, half-round, top dial/8 pushbuttons/ tuning eye, horiz grill bars, 9 tubes, BC, SW, AC **$185.00**

345-F, Console, 1939, Mahogany, square dial, center grill cut-outs, pushbuttons, tuning eye, 10 tubes, BC, SW, AC **$180.00**

345-M, Console, 1939, Walnut, square dial, grill cut-outs, pushbuttons, tuning eye, 10 tubes, BC, SW, AC **$185.00**

345-X, Console, 1939, Wood, upper dial, pushbuttons, lower grill w/ 3 vert column cut-outs, 10 tubes **$200.00**

350-M, Console, 1938, Walnut, upper dial, center horiz grill bars, 8 pushbuttons, tuning eye, 11 tubes, BC, SW, AC **$200.00**

350-V, Console/Corner, 1939, Walnut, corner-style, upper dial, vert grill bars, pushbuttons, tuning eye, 11 tubes, BC, SW, AC **$200.00**

360-M, Console, 1939, Walnut, upper dial, grill cut-outs, pushbuttons, tuning eye, 12 tubes, BC, SW, AC **$200.00**

370-M, Console, 1939, Walnut, inner dial & controls, double doors, vert grill bars, 14 tubes, BC, SW, AC **$235.00**

400-N, Side Table, 1939, Mahogany, Chippendale design side table, inner dial & controls, 5 tubes, BC, AC **$135.00**

411-PT, Table-R/P, 1939, Walnut, front slide rule dial, Deco grill, 4 knobs, lift top, inner phono, BC, SW, AC **$50.00**

480-M, Console, 1940, Wood, inner dial & pushbuttons, doors, grill w/cut-outs, 20 tubes, BC, FM, SW, AC **$350.00**

505-H, Table, 1940, Wood, step-down top, right front dial, left vertical grill bars, 9 tubes, 3 knobs **$75.00**

515-M, Console, 1940, Wood, upper dial, lower grill bars, pushbuttons, 17 tubes, 2 chassis, BC, FM, AC **$300.00**

535-M, Console, 1940, Wood, inner dial & pushbuttons, double doors, lower grill w/vertical bars, BC, SW **$160.00**

600-H, Table, 1941, Plastic, right front dial, rounded left side w/ horizontal bars, 3 knobs, AC/DC **$50.00**

635 "Treasure Chest", Table, 1928, Two-tone walnut, low rectangular case, window dial with escutcheon, 7 tubes, AC **$135.00**

636-A, Console, 1928, Wood, inner window dial, fold-down front, stretcher base, 7 tubes, 3 knobs, AC **$175.00**

638, Console, 1929, Wood, highboy, center window dial, upper speaker grill w/vert bars, 8 tubes, AC **$180.00**

641, Table, 1929, Walnut finish, low rectangular case, front window dial, 5 tubes, AC **$100.00**

642, Console, 1929, Wood, highboy, carving, window dial, cloth grill w/carved cut-outs, 3 knobs, AC **$170.00**

900-H, Table, 1941, Two-tone plastic, right front dial, left vertical grill bars, 3 knobs, BC, AC/DC **$45.00**

920-L, Console, 1941, Wood, upper front slide rule dial, pushbuttons, lower criss-cross grill, BC, SW **$150.00**

955-PF "Georgian", Console-R/P, 1941, Wood, inner slide rule dial, pushbuttons, phono, scalloped base, BC, FM, SW, AC **$200.00**

1000-H, Table, Plastic, right square dial, left vertical grill bars, arched top, 3 knobs, BC, AC/DC **$45.00**

1100-H, Table, 1947, Wood, right square dial, left cloth grill, 3 knobs, rounded left, BC, AC/DC **$30.00**

1101, Table, Plastic, raised top, slide rule dial, wrap-around louvers, 2 knobs, rear hand-hold **$50.00**

1101HPW, Table-R/P, 1948, Wood, right front curved dial, Deco grill, 3 knobs, lift top, inner phono, BC, AC **$60.00**

1101-HW, Table, 1946, Wood, small upper slide rule dial, lower horizontal louvers, 2 knobs, BC, AC/DC **$40.00**

1105, Portable, 1947, Cloth covered, window dial, Deco grill design, 2 knobs, handle, BC, AC/DC/bat $45.00

1110-HW, Table, 1947, Wood, center vertical dial, horizontal louvers, 4 knobs, 6 pushbuttons, BC, SW, AC **$60.00**

1121LW, Console, 1946, Wood, slide rule dial, 4 knobs, pushbuttons, criss-cross grill, 11 tubes, BC, 3SW, AC **$120.00**

1135PLW, Console-R/P, 1947, Wood, inner slide rule dial, 2 knobs, pushbuttons, phono drawer, 16 tubes, BC, 2SW, FM, AC **$150.00**

1202, Table-R/P, 1949, Wood, right front curved dial, Deco grill, 3 knobs, lift top, inner phono, BC, AC **$60.00**

1204, Table, 1948, Plastic, lower slide rule dial, upper slanted cloth grill, 4 knobs, BC, FM, AC/DC **$50.00**

1210-H "Courier", Table, 1948, Two-tone wood, 3 right front dials, left cloth criss-cross grill, 4 knobs, BC, 2FM **$45.00**

1210PLM, Console-R/P, 1948, Wood, inner right dials, 4 knobs, left pull-out phono drawer, 11 tubes, BC, 2FM, AC **$110.00**

1220-PL, Console-R/P, 1948, Wood, inner right slide rule dial, phono, lift top, criss-cross grill, BC, SW, AC **$60.00**

1235-PLM, Console-R/P, 1948, Wood, inner slide rule dial, pushbuttons, pull-out phono, 16 tubes, BC, 2SW, 2FM, AC **$125.00**

1400, Table, 1949, Plastic, raised top, slide rule dial, horizontal wrap-around louvers, BC, AC/DC **$50.00**

1407PFM, Console-R/P, 1949, Wood, inner right slide rule dial, left pull-out phono, double doors, BC, FM, AC $95.00

1409PGM, Console-R/P, 1949, Wood, inner right slide rule dial, left pull-out phono drawer, 14 tubes, BC, FM, AC $110.00

1500, Table, 1951, Plastic, raised top, slide rule dial, horiz wrap-around louvers, 2 knobs, BC, AC/DC $50.00

1500-HR "Dynatomic", Table, Maroon plastic, curved slide rule dial, horiz wrap-around louvers, 2 knobs, AC $50.00

1507, Console-R/P, 1951, Wood, inner right slide rule dial, 4 knobs, left pull-out phono drawer, BC, FM, AC $80.00

1608, Console-R/P, 1951, Wood, inner right slide rule dial, 5 knobs, left phono, double doors, BC, AC $60.00

AWP-8, Portable, 1956, Inner slide rule dial, fold-up front w/world map, handle, 8 band, AC/DC/battery $60.00

C-1, **Table-C, 1951, Plastic, right front dial, left alarm clock, center checkered panel, BC, AC** $40.00

C-3, Table-C, 1955, Plastic, right side dial, front clock, circular louvers, center vertical bars, BC, AC $35.00

C-5 Deluxe, Table-C, 1955, Plastic, right side dial knob, left front alarm clock, right perforated grill, BC, AC $25.00

EP-2, Portable, 1955, Top dial and controls, front decorative "V", rounded sides, handle, BC, AC/DC/bat $40.00

FR-711M, Console-R/P, 1958, Wood, inner right dial, left phono, lift top, large front grill, legs, BC, FM, AC $40.00

SR-407, Table, 1957, Center front slide rule dial, 4 knobs, 4 pushbuttons, high fidelity, AM, FM, AC $25.00

T-4, Table, 1955, Plastic, right side dial & volume knobs, front brick-like grill panel, BC, AC/DC $35.00

SUPEREX
Superex Electronics Corp.,
4-6 Radford Place, Yonkers, New York

TR-66, Portable, 1960, Plastic, transistor, thumbwheel dial, perforated grill, swing handle, AM, battery $25.00

SUPREME (LIPAN)
Aim Industries,
41 Union Square, New York, New York

750, Table-R/P, 1949, Leatherette, inner right dial, 4 knobs, phono, lift top, BC, AC $15.00

SYLVANIA
Sylvania Electric Products (Colonial Radio & TV),
254 Rand Street, Buffalo, New York

5P11R, Portable, 1960, Plastic, transistor, left window dial, front horizontal grill bars, AM, battery $35.00

7P12T, Portable, 1960, Transistor, right front window dial, large perforated grill, swing handle, AM, bat $25.00

7P13, Portable, 1960, Leather case, transistor, right dial knob over large front grill, handle, AM, bat $20.00

430L, Portable, 1952, Center front panel w/ round dial, woven grill, handle, 2 knobs, BC, AC/DC/battery $25.00

510B, Table, 1950, Plastic, right round dial, horizontal louvers, raised top, 2 knobs, BC, AC/DC $35.00

511B, Table, 1952, Plastic, right front round dial, bowed front panel with checkered grill, BC, AC/DC $20.00

540MA "Tune-Riser", Table-C, 1951, Plastic, right slide rule dial, horizontal grill bars, center clock, 4 knobs, BC, AC $35.00

542GR, Table-C, 1952, Plastic, lower front slide rule dial, large upper alarm clock face, BC, AC $30.00

1102, Table, 1957, Plastic, right side dial knob, front lattice grill, left side switch, BC, AC/DC $25.00

2108, Table-C, 1959, Plastic, right side round dial knob, left front clock, right horizontal bars, BC, AC $15.00

2109TU, Table-C, 1959, Plastic, lower right round dial knob, left clock, right horizontal louvers, BC, AC $20.00

2302H, Table-C, 1957, Plastic, lower center front dial, upper alarm clock w/day-date, side knob, BC, AC $25.00

3204TU, Portable, 1958, Plastic, transistor, right front round dial, lattice grill, handle, AM, battery $30.00

3204YE, Portable, 1958, Plastic, transistor, right front round dial, lattice grill, handle, AM, battery $30.00

3303TA, Portable, 1957, Leather case, front round compass over grill, side knobs, strap, BC, AC/DC/bat $45.00

3305BL, Portable, 1958, Plastic, transistor, right front round dial, lattice grill, handle, AM, battery $30.00

3406 Series, Portable-C, 1960, Leather case, transistor, off-center dial knob over grill, right clock, AM, battery $30.00

4501, Table-R/P, 1958, Two-tone, top center dial, wrap-over front grill, 3 knobs, lift top, handle, BC, AC $20.00

SYMPHONY

200, Table-Lamp-Planter, 1948, Lamp radio w/trapezoid-shaped case, right dial, left trapezoid grill, plants pots, BC, AC/DC $125.00

200L-R, Table-Lamp-Planter, 1948, Lamp radio w/trapezoid-shaped case, right dial, left trapezoid grill, plants pots, BC, AC/DC $125.00

250, Portable, 1949, Upper right dial, left volume, lower round grill w/ vert cut-outs, handle, BC, bat $40.00

260, Table-N, 1948, Striped grasscloth w/painted palm trees, left front dial panel, handle, battery $150.00
348, Portable, 1948, Leatherette, inner right dial, round grill w/horiz bars, flip-up front, AC/DC/bat..$40.00
401, Cathedral, Two-tone wood, lower front round dial, upper cloth grill w/cut-outs, 3 knobs $225.00

L. TATRO
L. Tatro Products Corporation, Decorah, Iowa

L. Tatro began business in 1930 as a producer of battery radios for the rural consumer. Because of the Depression and the increase of rural electrification, sales declined and in 1939 the company was sold to the Eckstein Radio & Television Company.

H465, Tombstone, Wood, lower round dial, upper grill w/cut-outs, 4 tubes, 3 knobs, BC, battery $115.00
M4616, Tombstone, Wood, lower round dial, upper cloth grill w/cut-outs, 4 tubes, 3 knobs, BC, bat $115.00
O4626, Tombstone, Wood, lower dial w/escutcheon, upper grill w/cut-outs, 4 tubes, 4 knobs, BC, SW, bat $120.00
S6636, Console, Wood, round dial, lower cloth grill w/cut-outs, 6 tubes, 4 knobs, BC, SW, battery $120.00

TELE-TONE
Tele-tone Radio Corp.,
609 West 51st Street, New York, New York

109, Table, 1946, Plastic, lower slide rule dial, upper grill w/circle cut-outs, 2 knobs, BC, AC/DC $45.00
111, Table, 1948, Wood, slanted lower slide rule dial, upper criss-cross grill, 2 knobs, BC, AC/DC $35.00
117, Table, 1946, Wood, slant front, lower slide rule dial, upper perforated grill, 2 knobs, BC $35.00
117A, Table, 1946, Wood, slant front, lower slide rule dial, upper perforated grill, 2 knobs, BC, AC/DC $35.00
133, Table-R/P, 1947, Wood, lower front dial, upper grill, 4 knobs, lift top, inner phono, BC, AC $25.00
134, Table-R/P, 1947, Inner front slide rule dial, 4 knobs, fold-up front, inner phono w/lid, BC, AC $30.00
135, Table, 1947, Plastic, upper front slide rule dial, lower lattice grill, 2 knobs, BC, AC/DC $40.00
138, Table, 1947, Two-tone plastic, lower slide rule dial, upper lattice grill, BC, SW, AC/DC $40.00
145, Portable, 1947, Leatherette, slide rule dial, horizontal grill bars, 2 knobs, handle, BC, AC/DC/bat $30.00

148, Table, 1947, Plastic, upper slide rule dial, lower lattice grill, 2 knobs, BC, AC/DC $35.00
150, Table, 1948, Plastic, upper slide rule dial, lower lattice grill, 2 knobs, BC, AC/DC $30.00
152, Portable, 1948, Plastic, inner slide rule dial, horiz grill bars, 2 knobs, fold-down front, handle $35.00
156, Portable, 1948, Plastic, slide rule dial, recessed horizontal louvers, handle, 2 knobs, BC, battery $35.00
157, Table, 1948, Plastic, upper slide rule dial, lower horizontal louvers, 2 knobs, BC, AC/DC $40.00
159, Table, 1948, Plastic, lower slide rule dial, upper vertical grill bars, 2 knobs, BC, AC/DC $35.00
160, Table, 1948, Two-tone plastic, lower dial & 2 knobs, recessed checkered grill, BC, AC/DC $40.00
165, Table, 1948, Plastic, right front round dial, horizontal grill bars, 2 knobs, BC, AC/DC $25.00
166, Table, 1948, Plastic, upper slide rule dial, lower horiz louvers, handle, 2 knobs, BC, AC/DC $35.00
185, Portable, 1948, Plastic, inner dial, 2 knobs, lattice grill, fold-down front, handle, BC, AC/DC/bat $30.00
190, Portable, 1949, Plastic, lower slide rule dial, upper lattice grill, handle, 2 knobs, BC, AC/DC/bat $25.00
195, Table, 1949, Plastic, right front round dial, horizontal grill bars, 2 knobs, BC, AC/DC $25.00
198, Table, 1949, Wood, lower slanted slide rule dial, upper slanted grill area, 2 knobs, AM, FM, AC $30.00
201, Table, 1949, Plastic, lower raised slide rule dial, recessed checkered grill, 2 knobs, BC, AC/DC $40.00
205, Table, 1949, Plastic, right front round dial, horizontal grill bars, 2 knobs, BC, AC/DC $25.00
206, Table, 1951, Plastic, lower front slide rule dial, upper vertical grill bars, 2 knobs, AM, FM, AC $30.00
228, Portable, 1951, Plastic, lower front half-moon dial/louvers, flex handle, 2 knobs, BC, AC/DC/bat $40.00
235, Console-R/P, 1951, Wood, upper front slide rule dial, 2 knobs, pull-out phono drawer, BC, FM, AC $45.00

TELECHRON
Telechron, Inc.,
Ashland, Massachusetts

8H59 "Musalarm", Table-C, 1950, Plastic, trapezoid shape, lower right dial, circular grill cut-outs, left clock, AC $45.00
8H67 "Musalarm", Table-C, 1948, Plastic, top thumbwheel dial, left clock, right horiz wrap-around louvers, BC, AC $35.00

TELE KING
Tele King Corp.,
601 West 26th Street, New York, New York

RK41, Table, 1953, Plastic, right front round dial, center checkered grill panel, 2 knobs, BC, AC/DC $35.00

RK51A, Table, 1953, Plastic, right front round dial, center checkered grill panel, 2 knobs, BC, AC/DC **$35.00**

RKP-53-A, Portable, 1954, Plastic, right side dial knob, large front checkered grill, handle, BC, AC/DC/bat **$30.00**

TELEVOX
Televox, Inc.,
451 South 5th Avenue, Mt Vernon, New York

RP, Table-R/P, 1947, Wood, inner right vertical slide rule dial, 4 knobs, phono, lift top, BC, AC **$25.00**

TEMPLE
Templetone Radio Mfg. Corp.,
New London, Connecticut

E-511, Table-R/P, 1947, Wood, front slanted slide rule dial, horizontal louvers, 2 knobs, lift top, BC, AC **$35.00**

E-514, Table, 1946, Wood, upper slanted slide rule dial, lower cloth grill, 2 knobs, BC, AC/DC **$30.00**

F-611, Portable, 1946, Luggage-style, upper slanted slide rule dial, 2 knobs, handle, BC, AC/DC/battery **$30.00**

F-616, Table, 1946, Wood, slanted slide rule dial, large cloth warparound grill, 2 knobs, BC, AC/DC **$35.00**

F-617, Table-R/P, 1947, Wood, front slanted slide rule dial, 4 knobs, horizontal louvers, lift top, BC, AC $35.00

G-410, Portable, 1947, Leatherette, upper slide rule dial, lower grill, 2 knobs, handle, BC, AC/DC/battery **$30.00**

G-415, Portable, 1948, Inner dial, vertical grill bars, flip-open door, handle, BC, AC/DC/battery **$40.00**

G-418, Table, 1947, Right dial, left horizontal graduated louvers, curved top, 2 knobs, BC, AC/DC **$65.00**

G-419, Table, 1947, Metal, right dial, left horiz graduated louvers, curved top, 2 knobs, BC, AC/DC **$65.00**

G-513, Table, 1947, Plastic, upper front slide rule dial, lower horizontal louvers, 2 knobs, BC, AC/DC **$35.00**

G-515, Table, 1947, Wood, upper slanted slide rule dial, large lower cloth grill, 2 knobs, BC, AC/DC **$35.00**

G-516, Table-R/P, 1947, Wood, front slanted slide rule dial, 4 knobs, criss-cross grill, lift top, BC, AC **$30.00**

G-518, Table-R/P, 1947, Wood, front slanted slide rule dial, 4 knobs, lift top, inner phono, BC, AC **$30.00**

G-521, Portable, 1947, Upper slide rule dial, lattice grill, telescoping antenna, strap, BC, SW, AC/DC/bat **$35.00**

G-522, Table, 1947, Wood, upper slanted slide rule dial, large lower grill, 4 knobs, BC, SW, AC/DC **$30.00**

G-619, Table, 1947, Wood, upper slanted slide rule dial, large lower cloth grill, 2 knobs, BC, AC/DC **$30.00**

G-722, Console-R/P, 1947, Wood, slide rule dial, 4 knobs, 6 pushbuttons, pull-out phono drawer, BC, SW, AC **$85.00**

G-724, Table, 1948, Wood, upper slide rule dial, lower criss-cross grill, 4 knobs, BC, FM, AC/DC **$30.00**

G-725, Console-R/P, 1948, Wood, inner right slide rule dial, 4 knobs, left pull-out phono drawer, BC, FM, AC **$70.00**

G-1430, Console-R/P, 1948, Wood, inner slide rule dial, 4 knobs, pull-out phono drawer, 14 tubes, BC, SW, FM, AC **$100.00**

H-411 "Playmate", Portable, 1948, Oblong panel w/right round dial & horizontal louvers, handle, 2 knobs, BC, battery **$40.00**

H-622, Console-R/P, 1948, Wood, inner slide rule dial, 4 knobs, criss-cross grill, pull-out phono, BC, SW, AC **$65.00**

TEMPOTONE
Barker Bros. Co.,
7th & Figueroa Street, Los Angeles, California

500E, Table, 1946, Plastic, left front round dial, right horizontal louvers, 2 knobs, BC, AC/DC **$30.00**

THERMIODYNE
Thermiodyne Radio Corp.,
Plattsburgh, New York

TF6, Table, 1925, Wood, rectangular case, front panel w/center dial, lift top, 6 tubes, battery **$140.00**

THOMPSON
R. E. Thompson Mfg. Co.,
30 Church St., New York, New York

S-60 "Parlor Grand", Table, 1925, Wood, rectangular case, slanted 3 dial front panel, 5 tubes, battery **$125.00**

S-70 "Concert Grand", Table, 1925, Wood, rectangular case, slanted 3 dial front panel, 6 tubes, battery **$135.00**

V-50 "Grandette", Table, 1925, Wood, rectangular case, 3 dial front panel, 5 tubes, battery **$120.00**

THOROLA

50 "Islodyne", Table, 1925, Leatherette, low rectangular case, 3 dial black front panel, 5 tubes, battery **$150.00**

TOSHIBA
Excel Corp. of America,
9 Rockefeller Plaza, New York, New York

6TP-304, Portable, 1960, Transistor, left vertical slide rule dial, lower right perforated grill, AM, battery $35.00

6TP-309Y, Portable, 1960, Transistor, window dial in "V" wedge, perforated grill, thumbwheel knobs, AM, bat $40.00

6TP-314, Portable, 1960, Transistor, left window dial, lower horiz grill bars, thumbwheel knobs, AM, bat $30.00

6TP-354, Portable, 1960, Transistor, right thumbwheel dial, large perforated grill area, AM, battery $35.00

6TR-92, Portable-N, 1959, Transistor, round case w/floral design, top dial, bottom grill, handle, AM, bat $125.00

6TR-186, Portable, 1959, Transistor, right thumbwheel dial, left patterned grill, AM, battery $45.00

7TP-352S, Portable, 1960, Transistor, upper slide rule dial, lower grill, telescoping antenna, BC, SW, bat $45.00

8TM-294, Portable, 1960, Transistor, top wrap-over dial, large front lattice grill, side knobs, AM, battery $50.00

TRANCEL
Excel Corp. of America,
9 Rockefeller Plaza, New York, New York

T-7, Portable, 1959, Transistor, upper slide rule dial, lower diagonal grill bars, AM, battery $30.00

TRAV-LER
Trav-ler Karenola Radio & Television Corp.,
571 West Jackson Boulevard,
Chicago, Illinois

41, Table, 1936, Wood, small case, center front dial, finished on both front & back, AC/DC $65.00

53, Tombstone, 1935, Wood, lower front dial, upper cloth grill w/cut-outs, BC, SW, AC $125.00

55-37, Table, 1955, Plastic, right front dial, left checkered grill, 2 knobs, BC, AC/DC $30.00

55-38, Table, 1955, Plastic, center front half-moon dial over checkered grill, 2 knobs, BC, AC/DC $30.00

55C42, Table-C, 1955, Plastic, right side dial knob, center alarm clock, left checkered grill, BC, AC $25.00

125, Console, 1937, Wood, round airplane dial, cloth grill w/3 splayed bars, 5 tubes, BC, SW, battery $100.00

131, Console, 1937, Walnut, oval dial, cloth grill w/center vert bars, 11 tubes, 5 knobs, BC, SW, AC $150.00

135, Console, 1937, Wood, upper round dial, lower grill w/cut-outs, 7 tubes, 4 knobs, BC, SW, AC $125.00

135-M, Console, 1937, Wood, round dial, lower grill w/cut-outs, tuning eye, 7 tubes, 4 knobs, BC, SW, AC $135.00

149, Console, 1937, Wood, upper round dial, lower grill w/3 splayed bars, 6 tubes, 3 knobs, BC, SW, AC $125.00

149-M, Console, 1937, Wood, round dial, grill w/3 splayed bars, tuning eye, 6 tubes, 3 knobs, BC, SW, AC $135.00

173, Console, 1937, Wood, upper round dial, lower grill w/cut-outs, 7 tubes, 4 knobs, BC, SW, battery $100.00

315, Table, 1939, Two-tone wood, right wrap-around dial, round grill w/horiz bars, 2 knobs, AC/DC $85.00

325, Table-C, 1939, Walnut & leatherette, center slide rule dial, right clock, left grill, 2 knobs, AC $50.00

431, Table, 1937, Two-tone wood, right front dial, left 3-section grill, 4 tubes, 2 knobs, BC, AC/DC $45.00

431-SW, Table, 1937, Two-tone wood, right dial, left 3-section grill, 4 tubes, 2 knobs, BC, SW, AC/DC $50.00

442, Table, 1936, Walnut, small oval case, center front dial, feet, AC $85.00

502, Table, 1937, Wood, rectangular front panel w/center dial, side grills, 2 knobs, BC, AC/DC $60.00

522, Table, 1937, Wood, right front oval dial, left oval grill w/"H" cut-out, 3 knobs, BC, SW, AC/DC $60.00

525, Tombstone, 1937, Wood, lower round dial, upper grill w/curved cut-outs, 5 tubes, 3 knobs, BC, bat $85.00

539-M, Table, 1939, Wood, slide rule dial, pushbuttons, tuning eye, upper wrap-over grill, BC, SW, AC $75.00

549, Table, 1937, Wood, right front round dial, left grill w/Deco cut-outs, 3 knobs, BC, SW, AC $65.00

550, Table, 1937, Wood, right front round dial, left grill w/Deco cut-outs, 3 knobs, BC, SW, AC/DC $65.00

635, Table, 1937, Wood, right round dial, left grill w/horizontal bars, 7 tubes, 4 knobs, BC, SW, AC $65.00

635-M, Table, 1937, Wood, right round dial, left grill w/horiz bars, tuning eye, 7 tubes, 4 knobs, BC, SW, AC $70.00

701, Table, 1937, Wood, center front panel w/half-moon dial, side grills, tuning eye, BC, SW, AC/DC $75.00

733, Tombstone, 1937, Wood, lower round dial, upper grill w/cut-outs, 7 tubes, 4 knobs, BC, SW, battery $85.00

802, Table, 1940, Wood, right front slide rule dial, pushbuttons, left horiz grill bars, 3 knobs, AC $60.00

5000I, Table, 1947, Plastic, step-down top, right dial, left cloth grill w/ horizontal bars, BC, AC/DC $40.00

5002, Table, Plastic, right front dial w/red pointer, left vertical wrap-over grill bars, 2 knobs $45.00

5008, Table, 1946, Two-tone wood, slanted slide rule dial, lower cloth grill, 3 knobs, BC, AC/DC $40.00

5009, Table, 1946, Wood, upper front slide rule dial, lower cloth grill, 3 knobs, BC, AC/DC $40.00

5010, Table, 1946, Wood, upper front dial, lower cloth grill w/"X", 4 knobs, BC, SW, AC/DC $40.00

5015, Table, 1948, Plastic, raised top, slanted slide rule dial, horizontal louvers, 2 knobs, BC, AC/DC $45.00

5019, Portable, 1947, Leatherette, right front dial, left cloth grill, 2 knobs, handle, BC, battery $20.00

5020, Portable, 1947, Luggage-style, right square dial, left cloth grill, 3 knobs, handle, BC, AC/DC/bat $25.00

5021, Table, 1948, Plastic, right square dial, left vertical wrap-over louvers, 2 knobs, BC, battery $35.00

5022, Portable, 1950, "Alligator" & plastic, right dial, horiz. grill bars, handle, 3 knobs, BC, AC/DC/bat $45.00

5027, Portable, 1948, Leatherette, upper dial, lower horizontal louvers, 3 knobs, handle, BC, AC/DC/bat $30.00

5028, Portable, 1948, Leatherette, right dial, left horizontal grill bars, handle, 3 knobs, BC, AC/DC/bat $30.00

5028-A, Portable, 1947, "Snakeskin", right dial, left plastic grill bars, handle, 3 knobs, BC, AC/DC/bat $45.00

5029, Portable, 1948, Leatherette, right front dial, left horizontal grill bars, handle, 2 knobs, BC, bat $30.00

5036, Table-R/P, 1949, Right front slide rule dial, left cloth grill, 3 knobs, lift top, inner phono, BC, AC $25.00

5049, Portable, 1948, "Alligator", right front dial, left horizontal louvers, handle, 2 knobs, BC, battery $30.00

5051, Table, 1948, Plastic, right square dial, left vertical wrap-over grill bars, 2 knobs, BC, AC/DC $45.00

5054, Table, 1948, Midget case, right front gold dial, left horizontal grill bars, 2 knobs, BC, AC/DC $80.00

5056-A, Table, 1950, Plastic, right front round dial knob, left checkered grill, BC, AC/DC $30.00

5060, Table, 1950, Plastic, slanted slide rule dial, horizontal grill bars, 2 top knobs, BC, AC/DC $45.00

5061, Table, 1950, Plastic, slanted slide rule dial, wrap-around horiz grill bars, 2 top knobs, BC, AC/DC $45.00

5066, Table, 1948, Plastic, upper slanted slide rule dial, lower lattice grill, 2 knobs, BC, AC/DC $40.00

5170, Table-C, 1952, Wood, right front square dial, left square alarm clock, 5 knobs, BC, AC $30.00

5300, Portable, 1953, Plastic, left thumbwheel dial, right thumbwheel knob, lower grill, handle, BC, bat $25.00

5301, Portable, 1953, "Alligator" & plastic, right dial over horiz bars, handle, 3 knobs, BC, AC/DC/bat $30.00

5310, Table-R/P, 1953, Wood, front slanted slide rule dial/grill/4 knobs, lift top, inner phono, BC, AC $30.00

5372, Portable-R/P, 1954, Leatherette, inner right square dial, 3 knobs, phono, lift top, handle, BC, AC $25.00

5510, Table-R/P, 1956, Wood, front half-moon dial over grill, 4 knobs, lift top, inner phono, BC, AC $30.00

6040, Table-R/P, 1948, Wood, upper slide rule dial, lower grill, 4 knobs, lift top, inner phono, BC, AC $25.00

6050, Table-R/P, 1949, Wood, upper front slanted slide rule dial, 4 knobs, lift top, inner phono, BC, AC $25.00

6300, Portable, 1957, Plastic, right and left side thumbwheel knobs, lower grill, handle, BC, battery $20.00

6528, Console-R/P, 1959, Wood, inner dial & controls, 4 speed phono, lift top, large front grill, legs, BC, AC $40.00

M, Cathedral, Two-tone wood, center front dial, upper grill w/ cut-outs, scalloped top, 3 knobs $175.00

T-200, Table, 1959, Plastic, off-center "steering wheel" dial over recessed horiz bars, BC, AC/DC $40.00

T-201, Table, 1959, Plastic, off-center "steering wheel" dial over recessed horiz bars, BC, AC/DC $40.00

TR-280, Portable, 1959, Transistor, upper round dial, lower perforated grill, top knob, stand, AM, battery $30.00

TROPHY

Baseball, Table-N, 1941, Molded cardboard, shaped like a baseball, right & left side knobs, top grill, base $600.00

TROY
Troy Radio & Television Co.,
1142-1144 South Olive Street,
Los Angeles, California

The Troy Radio Manufacturing Company began in 1933; the company being named for the USC Trojans. Troy made radios for many other companies besides their own. As a result of strict quality control standards, Troy was one of the few West coast radio manufacturers to receive the Underwriters approval. With the beginning of WW II, Troy ceased radio production.

45-M, Table, 1938, Peach, silver, blue or green etched mirror glass, right dial, left grill, 2 knobs, AC .... $2.000.00+

75PC, Table-R/P, 1936, Wood, center front dial, right & left horiz grill bars, lift top, inner phono, AC $40.00

100, Table, 1937, Wood, rounded right front w/telephone dial, left grill w/horiz bars, 3 knobs, AC $65.00

113-AW, Table, 1938, Two-tone wood, right front dial, left horiz wrap-around louvers, 4 knobs, BC, SW $60.00

825, Table, 1938, Wood, off-center dial, 6 pushbuttons, wrap-around grill w/cut-outs, 3 knobs, AC $65.00

C-170-PC, Console-R/P, 1938, Wood, inner right dial & knobs, left phono, lift top, front vertical grill bars, AC $125.00

TRUETONE
Western Auto Supply Co.,
2107 Grand Avenue, Kansas City, Missouri

636, Table, 1938, Plastic, Deco, right front dial, 5 pushbuttons, left wrap-around bars, side knob $110.00

D-910-B, Table, 1939, Wood, slide rule dial, upper vertical grill bars, pushbuttons, 4 knobs, BC, SW, AC $65.00

D941, Table, 1937, Plastic, Deco, right wrap-over dial, pushbuttons, left wrap-around louvers, side knob $90.00

D-1014 , Table, Wood, right front dial, pushbuttons, left wrap-around horizontal louvers, 3 knobs $55.00

D1046A, Console-R/P, 1950, Wood, upper slide rule dial, 4 knobs, lower pull-out phono drawer, BC, FM, AC $90.00

D1117, Table, Wood, slide rule dial, grill w/cut-outs, decorative molding, side handles, 3 knobs $60.00

D1240A, Console-R/P, 1952, Wood, inner right slide rule dial, 4 knobs, pull-out phono, double doors, BC, FM, AC $65.00

D-1612, End Table, 1947, Wood, tilt-out radio, slide rule dial, 3 knobs, pie-crust top edge, BC, AC $125.00

D1644, Console-R/P, 1947, Wood, upper slide rule dial, 4 knobs, lower open storage, lift top, BC, AC $40.00

D-1645, Console-R/P, 1946, Wood, upper slanted slide rule dial, 4 knobs, lower grill w/cut-out, BC, SW, AC $125.00

D1752, Console-R/P, 1948, Wood, slide rule dial, horizontal grill bars, 4 knobs, pull-out phono, BC, FM, AC $110.00

D1835, Console, 1948, Wood, slanted slide rule dial, lower cloth grill, 6 tubes, 4 knobs, BC, SW, AC $110.00

D1836A, Console-R/P, 1948, Wood, upper slide rule dial, 4 knobs, lower cloth grill, tilt-out phono, BC, FM, AC $110.00

D1840, Console-R/P, 1948, Wood, right front dial, 5 knobs, left lift top, inner phono, lower storage, BC, AC $60.00

D-1845, Console-R/P, 1948, Wood, upper slanted slide rule dial, 4 knobs, pull-out phono drawer, BC, SW, AC $115.00

D-1846A, Console-R/P, 1948, Wood, upper slanted slide rule dial, 4 knobs, pull-out phono drawer, BC, FM, AC $120.00

D1850, Console-R/P, 1948, Wood, inner right slide rule dial, 4 knobs, left pull-out phono drawer, BC, FM, AC $95.00

D1949, Console-R/P, 1949, Wood, inner right "sun burst" dial, 3 knobs, left pull-out phono drawer, BC, FM, AC $85.00

D2018, Table, 1950, Plastic, Modern, right black half-moon dial, left cloth grill, 2 knobs, BC, AC/DC $125.00

D-2020, Table, 1950, Plastic, slide rule dial, grill bars w/vertical center strip, 2 knobs, AM, FM, AC/DC $40.00

D2025A, Table, 1950, Wood, right square dial, left grill, mitered front corners, 4 knobs, AM, FM, AC $25.00

D-2026, Table, Plastic, lower black dial over checkered panel, upper cloth grill, 2 knobs, BC, FM $40.00

D2027A, Table, 1950, Wood, right square dial, left grill, mitered front corners, 4 knobs, AM, FM, AC $30.00

D2145, Table-R/P, 1953, Wood, right square dial, left grill, 2 knobs, 2/3 lift top, inner phono, BC, AC $30.00

D-2205, Table-C, 1953, Plastic, right half-moon dial, left alarm clock, center perforated grill, BC, AC $25.00

D2210, Table, 1941, Metal w/wood finish, right dial, left wrap-around grill, chrome trim, 2 knobs $100.00

D2214A, Table, 1953, Plastic, large front dial w/ inner horizontal grill bars, 2 knobs, BC, AC/DC $20.00

D2226, Table, 1953, Plastic, center front half-moon dial over perforated grill, 2 knobs, AM, FM, AC $25.00

D2237A, Table, 1952, Plastic, center dial, large cloth grill, slanted sides, 2 knobs, feet, BC, AC/DC $25.00

D2255, Table-R/P, 1953, Wood, center front dial, right & left grills, lift top, inner phono, BC, AC $30.00

D2263, Table, 1953, Wood, lower center front slide rule dial, large upper grill, 2 knobs, BC, battery $25.00

D2270, Table-R/P, 1953, Leatherette, inner left dial, phono, lift top, outer front grill, handle, BC, AC $20.00

D2325-A, Table, 1953, Plastic, lower front half-moon dial w/crest over large grill, 2 knobs, BC, AC $25.00

D2383, Table, 1953, Plastic, upper slide rule dial, lower horizontal louvers, 2 knobs, BC, SW, AC/DC $35.00

D2389, Table-C, 1954, Plastic, right side round dial knob, front alarm clock over checkered grill, BC, AC $25.00

D2483, Table, 1954, Plastic, lower front slide rule dial, 1 side knob, 2 front knobs, BC, SW, AC/DC $35.00

D2556A, Table-R/P, 1955, Wood, right side dial knob, large front grill, lift top, inner phono, BC, AC $25.00

D2560A, Table-R/P, 1955, Two-tone leatherette, round dial, 2 knobs, lift top, inner phono, handle, BC, AC $25.00

D2582A, Table, 1955, Plastic, right front half-moon dial, horizontal grill bars, 2 knobs, feet, BC, AC/DC $20.00

D2612, Table, 1946, Plastic, upper slide rule dial, lower criss-cross metal grill, 3 knobs, BC, AC/DC $35.00

D2613, Table, 1947, Plastic, slanted slide rule dial, criss-cross grill w/bars, 3 knobs, BC, SW, AC/DC $35.00

D2615 "Stratoscope", Table, 1946, Plastic, streamline, right slide rule dial, 6 pushbuttons, left vert bars, BC, AC/DC $110.00

D2616, Table, 1946, Plastic, slide rule dial, curved louvers, 2 knobs, 6 pushbuttons, BC, AC/DC $50.00

D-2616B, Table, 1948, Plastic, upper slide rule dial, curved louvers, 2 knobs, 6 pushbuttons, BC, AC/DC $50.00

D-2619, Table, 1947, Wood, lower slanted slide rule dial, large upper grill, 2 knobs, BC, AC/DC $35.00

D2620, Table, 1946, Wood, upper slide rule dial, lower cloth grill with 2 bars, 2 knobs, BC, AC/DC $50.00

D-2621, Table, 1946, Two-tone wood, upper slide rule dial, lower grill w/cutouts, 2 knobs, BC, battery $50.00

D-2622, Table, 1947, Wood, upper slide rule dial, large cloth grill, 2 knobs, rounded corners, BC, AC/DC $45.00

D2623, Table, 1947, Wood, upper front slide rule dial over large cloth grill, 4 knobs, BC, AC $40.00

D2624, Table, 1946, Wood, upper slide rule dial, metal criss-cross grill, 3 knobs, BC, SW, AC/DC $40.00

D2626, Table, 1948, Wood, upper slanted slide rule dial, criss-cross grill, 4 knobs, BC, SW, AC $40.00

D2630, Table, 1946, Plastic, upper slanted slide rule dial, lower cloth grill, 3 knobs, BC, SW, AC/DC $35.00

D-2634, Table, 1947, Wood, lower front clear slide rule dial over large cloth grill, 3 knobs, BC, AC $35.00

D2637A, Table, 1956, Plastic, center recessed round dial, horizontal front bars, 2 knobs, BC, AC/DC $25.00

D2640, Table-R/P, 1948, Wood, front slanted slide rule dial, criss-cross grill, 2 knobs, lift top, BC, AC $25.00

D2642, Table-R/P, 1947, Wood, front slide rule dial, 3 vertical grill bars, 2 knobs, lift top, BC, AC $30.00

D2644, Table, 1947, Wood, upper slanted slide rule dial, lower criss-cross grill, 2 knobs, BC, battery $30.00

D2645, Table-R/P, 1946, Wood, front slide rule dial, 4 knobs, cloth grill, lift top, inner phono, BC, SW, AC $30.00

D2661, Table, 1946, Plastic, large center square dial, horizontal side louvers, 2 knobs, BC, battery $75.00

D2663, Table, 1947, Wood, upper slanted slide rule dial, cloth grill w/ center bar, 2 knobs, BC, battery $30.00

D2665, Table, 1947, Wood, center front dial, right & left cloth grills, 2 knobs, BC, battery $30.00

D-2692, Table, 1948, Wood, upper slanted slide rule dial, lower horiz grill bars, 2 knobs, BC, AC/DC $30.00

D-2709, Table, 1947, Plastic, right front dial, left lattice grill, small base, 2 knobs, BC, AC/DC $45.00

D2710, Table, 1947, Plastic, upper slanted slide rule dial, large lower grill, 2 knobs, BC, AC/DC $40.00

D2718, Table, 1947, Plastic, upper slide rule dial, lower wrap-around louvers, 3 knobs, BC, SW, AC/DC $45.00

D2718B, Table, 1947, Plastic, upper slide rule dial, lower wrap-around louvers, 3 knobs, BC, SW, AC/DC $45.00

D-2743, Table-R/P, 1947, Wood, right front dial, left horizontal louvers, 3 knobs, open top, BC, AC $20.00

D2748, Table-R/P, 1947, Wood, front slide rule dial, horiz louvers, 4 knobs, lift top, inner phono, BC, AC $25.00

D2806, Table, 1948, Plastic, right front round dial, horizontal louvers, 2 knobs, BC, AC/DC $35.00

D2807, Table, 1948, Plastic, right front round dial, horizontal louvers, 2 knobs, BC, AC/DC $35.00

D2810, Table, 1948, Plastic, upper slanted slide rule dial, lower cloth grill, 2 knobs, BC, AC/DC $40.00

D2815, Table, 1948, Plastic, slanted slide rule dial, curved louvers, 2 knobs, pushbuttons, BC, AC/DC $50.00

D2819, Table, 1948, Plastic, upper slide rule dial, horizontal wrap-around louvers, 4 knobs, BC, FM, AC $45.00

D2819E, Table, 1948, Plastic, upper slide rule dial, horizontal wrap-around louvers, 4 knobs, BC, FM, AC $45.00

D2836A, Table, 1958, Plastic, right front AM dial, left front FM dial, center horizontal bar, AM, FM, AC $20.00

D-2851, Table-R/P, 1948, Wood, front slanted slide rule dial, 4 knobs, lift top, inner phono, BC, AC $30.00

D2857A, Table, 1958, Plastic, right side dial, horizontal front louvers w/divider & crest, BC, AC/DC $20.00

D2907, Table, 1949, Plastic, right front round dial, raised center lattice grill, 2 knobs, BC, AC/DC $30.00

D2910, Table, 1949, Plastic, upper slide rule dial, large lower recessed grill, 2 knobs, BC, AC/DC $40.00

D2919, Table, 1949, Plastic, slide rule dial, lattice grill, lower horiz bands, 4 knobs, BC, FM, AC/DC $45.00

D2923, Table, Wood, upper slide rule dial over large grill area, 6 tubes, 4 knobs, BC, SW, AC $40.00

D2963, Table, 1949, Plastic, upper slanted slide rule dial, large lower grill area, 2 knobs, BC, battery $30.00

D3120A, Portable, 1953, Plastic, upper slide rule dial, horiz front bars, side knobs, handle, BC, AC/DC/bat $30.00

D3210A, Portable, 1953, Plastic, right side dial knob, front & rear horizontal bars, handle, BC, AC/DC/bat $35.00

D3265A, Portable, 1952, Leatherette & plastic, center front round dial, handle, 2 knobs, BC, AC/DC/bat $30.00

D3300, Portable, 1953, Plastic, center front round dial over horizontal bars, handle, side knob, BC, bat $30.00

D3490, Portable, 1955, Plastic, center front round dial over horizontal grill bars, handle, BC, battery $30.00

D3600A, Portable, 1957, Plastic, 2 side thumbwheel knobs/left is dial, lower front grill, handle, BC, bat $35.00

D3614A "Deluxe", Portable, 1957, Plastic, transistor, right dial, center lattice design, thumbwheel knob, AM, bat $35.00

D3615, Portable, 1947, Leatherette, dial/2 knobs/lattice grill form a "T", handle, BC, AC/DC/battery $35.00

D3619, Portable, 1946, Luggage-style, slide rule dial, 2 knobs on top of case, handle, BC, AC/DC/battery $30.00

D3630, Portable, 1947, Luggage-style, inner dial, louvers, 2 knobs, drop front, handle, BC, AC/DC/battery $35.00

D3715A "Deluxe", Portable, 1958, Turquoise plastic, transistor, right front dial, center checkered grill, AM, battery $35.00

D3716A, Portable, 1957, Leather case w/front lattice grill, transistor, 2 front knobs, handle, AM, battery $30.00

D3716B, Portable, 1958, Leather case w/lattice grill, transistor, left front dial knob, handle, AM, battery $30.00

D-3720, Portable, 1947, Leatherette, small right round dial, woven grill, handle, BC, AC/DC/battery $25.00

D-3721, Portable, 1948, Leatherette, front slide rule dial, plastic grill, handle, 2 knobs, BC, AC/DC/bat $25.00

D3722, Portable, 1948, Leatherette, plastic escutcheon & louvers, handle, 2 knobs, BC, AC/DC/battery $30.00

D3780A, Portable, 1957, Leather case w/grill, round dial knob, right thumbwheel knob, handle, BC, battery $25.00

D3784A, Portable, 1957, Leather case w/grill, right front round dial, side knob, handle, BC, AC/DC/battery $25.00

D3789A, Portable, 1957, Leather case w/grill, right side dial knob, left side knob, handle, BC, AC/DC/bat $25.00

D3809, Portable, 1948, Plastic, small upper slide rule dial, recessed louvers, handle, 2 knobs, BC, bat $30.00

D-3810, Portable, 1948, Plastic, slide rule dial, recessed horiz louvers, handle, 2 knobs, BC, AC/DC/bat $30.00

D3811, Portable, 1948, Inner right dial, lattice grill, radio plays when flip-up lid opens, BC, AC/DC/bat $35.00

D3840, Portable, 1948, Leatherette, inner right dial, 3 knobs, fold-down front, handle, BC, AC/DC/bat $35.00

D3910, Portable, 1949, Plastic, lower right dial, horizontal grill bars, handle, 2 knobs, BC, AC/DC/bat $25.00

DC2980A, Table, 1959, Plastic, right center front round dial, left lattice grill, BC, AC/DC $15.00

DC2989A, Table-C, 1959, Plastic, right front dial, left alarm clock, center horizontal bars, feet, BC, AC $15.00

DC3050, Portable, 1959, Plastic, transistor, upper dial, perforated grill, telescope antenna, AM, SW, bat $35.00

DC3052, Portable, 1960, Transistor, right wedge-shaped window dial over large perforated grill, AM, bat $25.00

DC3084A, Portable, 1960, Leather case w/lattice grill, transistor, right front dial knob, handle, AM, battery $25.00

DC3088A, Portable, 1960, Transistor, right front half-moon dial, large perforated grill area, handle, AM, bat $25.00

DC3090, Portable, 1960, Transistor, small window dial, round perforated grill, swing handle, AM, battery $45.00

DC3800, Portable, 1959, Plastic, front perforated grill, side knobs, handle, BC, AC/DC/battery $20.00

DC3884, Portable, 1959, Leather case, transistor, right dial, circular grill cut-outs, handle, AM, battery $30.00

DC3886A, Portable, 1958, Leather case, transistor, right round dial, circular grill cut-outs, handle, AM, bat $30.00

DC5987A, Table-R/P, 1959, Right front dial, upper patterned perforated grill, lift top, handle, BC, AC $25.00

TUSKA
The C. D. Tuska Co.,
Hartford, Connecticut

222, Table, 1922, Wood, low rectangular case, 2 dial black front panel, battery **$350.00**
224, Table, 1922, Wood, rectangular case, 2 dial black front panel, 1 tube, battery **$350.00**

225 - Single Panel, Table, 1923, Mahogany, low rectangular case, 2 dial black bakelite single panel, 3 tubes, bat $300.00
225 - Double Panel, Table, 1923, Mahogany, low rectangular case, 2 dial black bakelite double panel, 3 tubes, bat **$600.00**
228 "Superdyne", Table, 1924, Wood, low rectangular case, 2 dial black front panel, 4 tubes, battery **$225.00**
301 "Junior", Table, 1925, Wood, low rectangular case, 3 dial front panel, 3 tubes, battery **$235.00**
305 "Superdyne", Table, 1925, Wood, low rectangular case, black front panel w/2 window dials, 4 tubes, battery **$195.00**

401, Table, Wood, low rectangular case, black front panel w/3 gold half-moon pointer dials **$250.00**

20TH CENTURY
Electronic Devices Corp.,
601 West 26th Street, New York, New York

100X, Table, 1947, Wood, right square black dial, cloth grill w/vertical bars, 3 knobs, BC, SW, AC/DC **$35.00**
101, Table, 1947, Wood, right square black dial, left criss-cross grill, 3 knobs, BC, SW, AC/DC **$40.00**

U. S. TELEVISION
U. S. Television Mfg. Co.,
3 West 61st Street, New York, New York

5C66, Table, 1947, Wood, right front square dial, left grill w/horizontal bars, 2 knobs, BC, AC/DC **$40.00**

5D66MPA, Table-R/P, 1947, Wood, right front square dial over criss-cross grill, lift top, inner phono, BC, AC **$30.00**
8-16M, Console-R/P, 1947, Wood, inner slide rule dial, 4 knobs, phono, doors, criss-cross grill, BC, 2SW, AC **$75.00**

ULTRADYNE
Regal Electronics Corp.,
20 West 20th Street, New York, New York

L-46, Table, 1946, Wood, right front half-moon dial, left curved grill bars, 2 knobs, BC, AC/DC **$40.00**

ULTRATONE
Audio Industries, Michigan City, Indiana

355, Table-R/P, 1956, Right side dial knob, large front grill, lift top, inner phono, handle, BC, AC **$20.00**

UNITONE
Union Electronics Corp.,
38-01 Queens Boulevard,
Long Island City, New York

88, Table, 1946, Wood, upper slide rule dial, lower grill w/2 horiz bars, 2 knobs, BC, SW, AC/DC **$40.00**

UNIVERSAL
Universal Battery Co., Chicago, Illinois

72A6, Tombstone, 1935, Wood, lower front round dial, upper cloth grill with cut-outs, battery **$80.00**
73A6, Console, 1935, Wood, upper front round dial, lower cloth grill with cut-outs, battery **$110.00**
7222, Console, 1935, Wood, upper front round dial, lower cloth grill with cut-outs, battery **$110.00**
7232, Tombstone, 1935, Wood, lower front round dial, upper cloth grill with cut-outs, battery **$80.00**

VAN CAMP
Van Camp Hardware & Iron Co.,
401 West Maryland Street,
Indianapolis, Indiana

576-1-6A, Table, 1946, Wood, upper slide rule dial, lower cloth grill w/vertical bars, 3 knobs, BC, AC/DC **$55.00**

VIEWTONE
Viewtone Co.,
203 East 18th Street, New York, New York

RRC-201, Table-R/P, 1947, Inner slanted dial, 4 knobs, phono, lift top, handle, BC, AC **$20.00**

VIKING
Viking Radio Laboratories,
433 N. LaSalle St, Chicago, Illinois

5-A, Table, Wood, low rectangular case, 3 dial front panel, 5 tubes, battery **$115.00**
5-A, Console, Wood, inner three dial panel, fold-down front, speaker grill, storage, 5 tubes, bat **$185.00**

599, Table, 1926, Two-tone cardboard & leatherette, large center front dial, lift back, 5 tubes **$125.00**

VIZ
Molded Insulation Co.,
335 East Price Street, Philadelphia, Pennsylvania

RS-1, Table, 1947, Plastic, right front round dial, left horizontal wrap-around louvers, BC, AC/DC **$60.00**

VOGUE
Sheridan Electronics Corp.,
2850 South Michigan Avenue, Chicago, Illinois

2554R, Table, 1946, Plastic, right half-moon dial, left wrap-around louvers, 2 knobs, handle, BC, AC/DC **$45.00**

WARE
Ware Radio Corporation,
529-549 West 42nd Street, New York, New York

B1 "Bantam", Cathedral, 1931, Wood, right dial, upper grill w/ scalloped cut-outs, fluted columns, finials, 7 tubes **$275.00**
L, Table, 1925, Wood, low rectangular case, 3 dial black front panel, lift top **$110.00**
T, Table, 1924, Mahogany, high rectangular case, slanted 2 dial panel with 3 exposed tubes, bat **$275.00**
TU, Console, 1926, Wood, inner top panel w/3 exposed tubes, lift top, inner front speaker grill, bat **$325.00**
W, Table, 1924, Walnut, low rectangular case, 3 dial front panel, 5 tubes, battery **$125.00**
WU, Console, 1925, Wood, inner slanted 3 dial panel, double doors, upper speaker grill, battery **$175.00**
X, Table, 1925, Walnut, low rectangular case, 3 dial front panel, meter, 5 tubes, battery **$120.00**

WATTERSON
Watterson Radio Mfg. Co.,
2700 Swiss Avenue, Dallas, Texas

601, Table, 1958, Wood, transistor, recessed right front dial, left grill, 3 knobs, feet, AM, battery **$25.00**
4581, Table, 1946, Wood, right front square dial, horizontal louvers, 2 knobs, BC, AC/DC **$35.00**
4582, Table, 1946, Wood, right front square dial, left cloth grill w/ horizontal bars, 2 knobs, BC, bat **$25.00**
4782, Table, 1947, Two-tone wood, right black dial, left cloth grill w/ cut-outs, 2 knobs, BC, bat **$35.00**
4790, Table, 1947, Wood, right black dial, left cloth grill w/ horizontal bars, 3 knobs, BC, AC/DC **$35.00**
4800, Table, 1948, Wood, center oblong slide rule dial, large grill, 4 knobs, BC, FM, AC/DC **$30.00**
ARC-4591A, Table-R/P, 1947, Wood, front slide rule dial, 4 knobs, horizontal louvers, lift top, BC, AC **$25.00**
RC-4581, Table-R/P, 1947, Wood, right front square dial, horizontal grill bars, 3 knobs, lift top, BC, AC **$25.00**

WECCO
William E. Cheever Co.,
Providence, Rhode Island

Junior, Table, Wood, crystal set, 1 dial black front panel . **$150.00**

WELLS
Wells Mfg. Co., Fond du Lac, Wisconsin

24, Table, 1925, Wood, low rectangular case, black front panel w/ center dial, 4 tubes, battery **$130.00**

WESTERN COIL & ELECTRICAL
Western Coil & Electrical Co.,
313 Fifth St., Racine, Wisconsin

WC-5B "Radiodyne", Table, 1925, Wood, low rectangular case, 3 dial front panel, 4 tubes, battery **$125.00**
WC-11 "Radiodyne", Console, 1925, Wood, inner three dial panel, fold-up front, speaker grill, double doors, 4 tubes, bat **$175.00**
WC-11B "Radiodyne", Table, 1925, Wood, high rectangular case, 3 dial front panel, 4 tubes, battery **$140.00**
WC-12 "Radiodyne", Console, 1925, Wood, inner three dial panel, fold-up front, speaker grill, double doors, 6 tubes, bat **$200.00**
WC-12B "Radiodyne", Table, 1925, Wood, high rectangular case, 3 dial front panel, 6 tubes, battery **$160.00**
WC-15-JR "Radiodyne", Table, 1926, Wood, slanted 2 dial front panel, open top w/5 exposed tubes, side knob, battery **$200.00**

WESTERN ELECTRIC

Western Electric began business in 1872 making the Bell Telephone equipment. In the early 1920's, the company began radio production - at first only for commercial interests.

4B, Table, 1923, Wood, low rectangular case, 2 dial black front panel, battery **$450.00**

4C, Table, C1924, Wood, low rectangular case, 2 dial front panel, battery **$450.00**
4D, Table, C1924 ..Wood, low rectangular case, 2 dial front panel, battery **$400.00**
7A, Table/Amp, Wood, amplifier, black bakelite panel with 3 exposed tubes **$250.00**

WESTINGHOUSE
Westinghouse Electric Corp., Home Radio Division., Sunbury, Pennsylvania

The Westinghouse company sold its line of radios through RCA until 1930. The company is well-known for its slogan - "You Can Be Sure If It's Westinghouse".

Aeriola Jr, Table, 1922, Wood, square case, crystal set, lift top, inner panel, storage **$295.00**
Aeriola Sr, Table, 1922, Wood, square case, inner 1 dial panel w/1 exposed tube, lift top, battery **$175.00**
H-103, Table, 1946, Wood, curved top, recessed front, slide rule dial, horiz louvers, 4 knobs, BC, 2SW **$55.00**
H-104, Table, 1946, Wood, curved top, recessed front, slide rule dial, horiz louvers, 6 pushbuttons, 4 knobs, BC, 2SW, AC ..**$65.00**
H-117, Console-R/P, 1947, Wood, slide rule dial, pull-out phono, 4 knobs, 6 pushbuttons, 14 tubes, BC, SW, FM, AC **$110.00**
H-119, Console-R/P, 1947, Wood, inner slide rule dial, 4 knobs, 6 pushbuttons, pull-out phono, 14 tubes, BC, SW, FM, AC **$110.00**
H-122, Table-R/P, 1946, Front radio unit detaches from phono, dial on radio case top, 4 knobs, BC, AC/DC **$75.00**
H-124 "Little Jewel", Portable, 1945, "Refrigerator", plastic w/metal center strip, vert bars, fold-down handle, AC **$95.00**
H-125 "Little Jewel", Portable, 1946, "Refrigerator", plastic w/metal center strip, vert bars, fold-down handle, AC/DC **$95.00**
H-126 "Little Jewel", Portable, 1948, "Refrigerator", plastic w/metal center strip, vert bars, fold-down handle, AC **$95.00**
H-127 "Little Jewel", Portable, 1948, "Refrigerator", plastic w/metal center strip, vert bars, fold-down handle, AC/DC **$95.00**
H-130, Table, 1946, Wood, top recessed slide rule dial, center front cloth grill, 4 knobs, AC/DC **$45.00**

H-133, Table, 1947, Two-tone wood, upper slanted slide rule dial, lower cloth grill, 3 knobs, BC, bat **$35.00**

H-138, Console-R/P, 1946, Wood, slide rule dial, horizontal louvers, 4 knobs, 6 pushbuttons, BC, SW, AC $100.00
H-142, Table, 1948, Wood, slanted multi-band dial, large grill area, top fluting, 5 knobs, BC, 4SW, AC **$65.00**
H-147, Table, 1948, Plastic, right geometric dial, recessed cloth grill, 2 knobs, BC, AC/DC **$40.00**
H-148, Portable, 1947, Small upper right dial, criss-cross grill, 2 knobs, handle, BC, AC/DC/battery **$25.00**
H-157, Table, 1948, Wood, lower slide rule dial, large upper recessed cloth grill, 2 knobs, BC, AC/DC **$30.00**

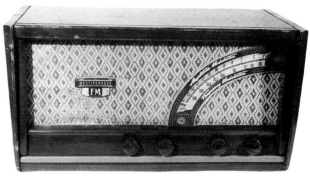

H-161, Table, 1948, Wood, right front "rainbow" dial over large cloth grill, 4 knobs, AM, FM, AC $45.00
H-165, Portable, 1948, Inner dial, horizontal grill bars, 4 knobs, fold-down front, handle, BC, AC/DC/bat **$25.00**
H-166, Console-R/P, 1948, Wood, inner slide rule dial, 4 knobs, pull-out phono drawer, 12 tubes, BC, FM, AC **$95.00**
H-168, Console-R/P, 1948, Wood, inner "rainbow" dial, 4 knobs, phono, lift top, grill cut-outs, BC, FM, AC **$65.00**
H-168A, Console-R/P, 1948, Wood, inner "rainbow" dial, 4 knobs, phono, lift top, grill cut-outs, BC, FM, AC **$65.00**
H-169, Console-R/P, 1948, Wood, inner slide rule dial, 4 knobs, pull-out phono, 14 tubes, BC, 2SW, FM, AC **$125.00**

H-171, Console-R/P, 1948, Wood cabinet, plastic radio unit can be removed, slide rule dial, 4 knobs, BC, AC **$145.00**

H-178, Table, 1948, Two-tone wood, upper slanted slide rule dial, lower cloth grill, 3 knobs, BC, bat **$30.00**

H-182, Table, 1949, Plastic, upper slanted slide rule dial, lower metal bands, 3 knobs, AM, FM, AC/DC **$35.00**

H-183A, Console-R/P, 1948, Wood, inner right dial, 4 knobs, left pull-out phono drawer, storage, BC, AC **$75.00**

H-185, Portable, 1949, Plastic, upper slide rule dial, horiz grill bars, handle, 2 knobs, BC, AC/DC/battery **$30.00**

H-186, Console-R/P, 1949, Wood, inner slide rule dial, pushbuttons, pull-out phono drawer, 12 tubes, BC, FM, AC **$85.00**

H-188, Table, 1948, Plastic, oriental design, right dial, left grill w/ cut-outs, 2 knobs, BC, AC/DC **$80.00**

H-190, Console-R/P, 1949, Wood, upper slide rule dial, 4 knobs, front pull-out phono drawer, BC, FM, AC **$65.00**

H-191, Console-R/P, 1949, Wood, inner right slide rule dial, 4 knobs, left pull-out phono, doors, BC, FM, AC **$70.00**

H-191A, Console-R/P, 1949, Wood, inner right slide rule dial, 4 knobs, left pull-out phono, doors, BC, FM, AC **$75.00**

H-195, Portable, 1949, Leatherette, upper slide rule dial, cloth grill, handle, 2 knobs, BC, AC/DC/battery **$30.00**

H-198, Table, 1949, Wood, lower slide rule dial, large upper cloth grill, 2 knobs, AM, FM, AC **$25.00**

H-199, Console-R/P, 1949, Wood, slide rule dial, 2 knobs, cloth grill, right & left lower storage, BC, FM, AC **$70.00**

H-202, Table, 1948, Plastic, upper slanted slide rule dial, large lower grill area, 3 knobs, BC, FM **$40.00**

H-203, Console-R/P, 1949, Wood, inner right slide rule dial, left pull-out phono drawer, storage, BC, FM, AC **$70.00**

H-204, Table, 1948, Plastic, upper slide rule dial, lower grill w/ "oriental" cut-outs, 3 knobs, BC, SW **$60.00**

H-204A, Table, 1948, Plastic, upper slide rule dial, lower grill w/ "oriental" cut-outs, 3 knobs, BC, FM, SW **$60.00**

H-210, Portable, 1949, Plastic, vertical slide rule dial, horizontal louvers, handle, 2 knobs, BC, AC/DC **$45.00**

H-211, Portable, 1949, Plastic, vertical slide rule dial, horizontal louvers, handle, 2 knobs, BC, AC/DC **$45.00**

H-214A, Console-R/P, 1949, Wood, upper slide rule dial, 2 knobs, front pull-out phono drawer, storage, BC, AC **$50.00**

H301T5, Table, 1950, Plastic, lower front slide rule dial, large upper grill, 2 knobs, BC, AC/DC **$25.00**

H-302P5, Portable, 1950, Leatherette, luggage-style, front slide rule dial, handle, 2 knobs, BC, AC/DC/bat **$30.00**

H303P4, Portable, 1950, Plastic, front slide rule dial, horizontal louvers, handle, 2 knobs, BC, AC/DC/bat **$35.00**

H-307T7, Table, 1950, Plastic, lower front slide rule dial, horizontal louvers, 2 knobs, AM, FM, AC/DC **$35.00**

H-309P5, Portable, 1950, Plastic, front slide rule dial, horizontal louvers, handle, 2 knobs, BC, AC/DC/bat **$30.00**

H-311T5, Table, 1950, Plastic, lower front slide rule dial, upper vertical grill bars, 2 knobs, BC, AC/DC **$40.00**

H-312P4, Portable, 1950, Plastic, front slide rule dial, contrasting grill, handle, 2 knobs, BC, AC/DC/bat **$35.00**

H-316C7, Console-R/P, 1950, Wood, inner left slide rule dial, 2 knobs, right pull-out phono, doors, BC, FM, AC **$55.00**

H318T5, Table, 1950, Plastic, modern slanted case, round dial over perforated grill, 2 knobs, BC, AC/DC **$40.00**

H-324T7U, Table, 1950, Plastic, lower front slide rule dial, horizontal louvers, 2 knobs, AM, FM, AC/DC **$35.00**

H-333P4U, Portable, 1952, Plastic, slide rule dial, lattice grill, 2 thumbwheel knobs, handle, BC, AC/DC/bat **$35.00**

H-334T7UR, Table, 1951, Plastic, center front clear curved dial w/ large pointer & "W", 2 knobs, BC, AC/DC **$40.00**

H-336T5U, Table, 1951, Plastic, modern slanted case, round dial over perforated grill, 2 knobs, BC, AC/DC **$40.00**

H-341T5U, Table, 1951, Plastic, right front round dial over woven grill w/ "W" logo, 2 knobs, BC, AC/DC **$35.00**

H-343P5U, Portable, 1951, Plastic, front round dial, 2 thumbwheel knobs, fold-back handle, BC, AC/DC/bat **$55.00**

H-350T7, Table, 1951, Plastic, center front curved dial over large grill, 2 knobs, AM, FM, AC/DC **$40.00**

H-354C7, Console-R/P, 1952, Wood, upper front half-moon dial, lower pull-out phono, storage, BC, FM, AC **$75.00**

H-355T5, Table-C, 1952, Maroon plastic, right thumbwheel dial, left clock, center grill w/horiz bar, BC, AC **$50.00**

H-356T5, Table-C, 1952, Ivory plastic, right thumbwheel dial, left clock, center grill w/horiz bar, BC, AC **$50.00**

H-357C10, Console-R/P, 1952, Wood, inner left curved dial, right pull-out phono, double front doors, BC, FM, AC **$80.00**

H-357T5, Table-C, 1952, Brown plastic, right thumbwheel dial, left clock, center grill w/horiz bar, BC, AC **$50.00**

H-359T5, Table, 1953, Plastic, center front half-moon dial w/inner checkered grill, 2 knobs, BC, AC/DC **$30.00**

H-361T6, Table, 1952, Plastic, right front round dial over large cloth grill, "W" logo, 2 knobs, BC, AC/DC **$30.00**

H-365T5, Table, 1952, Brown plastic, right front round dial, left recessed "tic/tac/toe" grill, 2 knobs **$30.00**

H-366T5, Table, 1952, Ivory plastic, right front round dial, left recessed "tic/tac/toe" grill, 2 knobs **$30.00**

H-374T5, Table-C, 1952, Plastic, right front thumbwheel dial knob, lattice grill, left alarm clock, BC, AC **$30.00**

H-379T5, Table, 1953, Plastic, right front dial knob, lower horizontal grill bars, feet, BC, AC/DC **$25.00**

H-381T5, Table, 1953, Plastic, right clear round dial, lower horizontal grill bars, feet, BC, AC/DC **$25.00**

H-388T5, Table-C, 1953, Plastic, right front dial, left alarm clock, center "woven" panel, 4 knobs, BC, AC **$25.00**

H-393T6, Table, 1953, Plastic, center front half-moon dial w/inner checkered grill, 2 knobs, BC, AC/DC **$35.00**

H-397T5, Table-C, Maroon plastic, small tombstone style, slide rule dial, upper alarm clock, lower grill, BC, AC **$45.00**

H-398T5, Table-C, Ivory plastic, small tombstone-style, slide rule dial, upper alarm clock, lower grill, BC, AC **$45.00**

H-409P4, Portable, 1954, Plastic, upper thumbwheel dial, large front lattice grill, handle, BC, battery **$30.00**

H-417T5, Table, 1954, Maroon plastic, center round dial over large front lattice grill, wire stand, BC, AC **$35.00**

H-418T5, Table, 1954, Ivory plastic, center round dial over large front lattice grill, wire stand, BC, AC $35.00

H-434T5, Table, 1955, Black plastic, raised slide rule dial, horizontal bars w/center divider, BC, AC/DC $35.00

H-435T5, Table, 1955, Ivory plastic, raised slide rule dial, horizontal bars w/center divider, BC, AC/DC $35.00

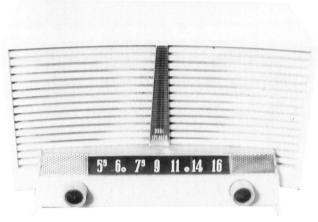

H-435T5A, Table, 1955, Ivory plastic, raised slide rule dial, horizontal bars w/center divider, BC, AC/DC $35.00

H-436T5, Table, 1955, Maroon plastic, raised slide rule dial, horiz bars w/center divider, BC, AC/DC $35.00

H-437T5, Table, 1955, Tan plastic, raised slide rule dial, horizontal bars w/center divider, BC, AC/DC $35.00

H-437T5A, Table, 1955, Tan plastic, raised slide rule dial, horizontal bars w/center divider, BC, AC/DC $35.00

H-438T5, Table, 1955, Green plastic, raised slide rule dial, horizontal bars w/center divider, BC, AC/DC $35.00

H-438T5A, Table, 1955, Green plastic, raised slide rule dial, horizontal bars w/center divider, BC, AC/DC $35.00

H-440T5, Table, 1955, Grey plastic, raised slide rule dial, horizontal bars w/center divider, BC, AC/DC $35.00

H-447T4, Table, 1955, Brown plastic, right front dial, lower horizontal grill bars, lower left knob, feet $25.00

H-448T4, Table, 1955, Grey plastic, right front dial, lower horizontal grill bars, lower left knob, feet $25.00

H-449T4, Table, 1955, Aqua plastic, right front dial, lower horizontal grill bars, lower left knob, feet $25.00

H-471T5, Table-C, 1955, Grey plastic, right dial knob, raised left w/ alarm clock, checkered grill, BC, AC $30.00

H-472T5, Table-C, 1955, Ivory plastic, right dial knob, raised left w/ alarm clock, checkered grill, BC, AC $30.00

H-473T5, Table-C, 1955, Rose plastic, right dial knob, raised left w/ alarm clock, checkered grill, BC, AC $30.00

H-474T5, Table-C, 1955, Green plastic, right dial knob, raised left w/ alarm clock, checkered grill, BC, AC $30.00

H-482PR5, Table-R/P, 1955, Front slide rule dial, large grill w/ "W" logo, lift top, side knob, handle, BC, AC $25.00

H-486T5, Table-C, 1955, Ivory plastic, lower slide rule dial, large clock face, metal bezel & knobs, BC, AC $30.00

H-487T5, Table-C, 1955, Maroon plastic, lower slide rule dial, large clock face, metal bezel & knobs, BC, AC $30.00

H-488T5, Table-C, 1955, Black plastic, lower slide rule dial, large clock face, metal bezel & knobs, BC, AC $30.00

H-489T5, Table-C, 1955, Grey plastic, lower slide rule dial, large clock face, metal bezel & knobs, BC, AC $30.00

H-499T5A, Table, 1955, Black plastic, right front dial knob over plaid metal perforated grill, BC, AC/DC $30.00

H-500T5A, Table, 1955, Red plastic, right front dial knob over plaid metal perforated grill, BC, AC/DC $30.00

H-501T5A, Table, 1955, Brown plastic, right front dial knob over plaid metal perforated grill, BC, AC/DC $30.00

H-502T5A, Table, 1955, Green plastic, right front dial knob over plaid metal perforated grill, BC, AC/DC $30.00

H-503T5A, Table, 1955, Grey plastic, right front dial knob over plaid metal perforated grill, BC, AC/DC $30.00

H-536T6, Table, 1956, Plastic, lower front slide rule dial, horizontal grill bars, 2 knobs, BC, AC/DC $30.00

H-537P4, Portable, 1957, Grey plastic, top dial knob, front metal grill, handle, feet, BC, AC/DC/battery $35.00

H-557P4, Portable, 1957, Two-tone green plastic, top dial, front metal grill, feet, handle, BC, AC/DC/bat $35.00

H-558P4, Portable, 1957, White & sand plastic, top dial, front metal grill, handle, feet, BC, AC/DC/battery $35.00

H-559P4, Portable, 1957, Grey & black plastic, top dial, front metal grill, handle, feet, BC, AC/DC/battery $35.00

H-562P4, Portable, 1957, Plastic, right dial, thumbwheel knobs, lower perforated grill, side strap, BC, bat $25.00

H-574T4, Table, 1956, Black plastic, raised right front w/round dial knob, left criss-cross grill, feet $20.00

H-575T4, Table, 1956, Ivory plastic, raised right front w/round dial knob, left criss-cross grill, feet $20.00

H-576T4, Table, 1956, Pink plastic, raised right front w/round dial knob, left criss-cross grill, feet $25.00

H-577T4, Table, 1956, Red plastic, raised right front w/round dial knob, left criss-cross grill, feet $25.00

H-602P7, Portable, 1958, Leather case, transistor, right front dial, perforated metal grill, handle, bat $30.00

H-610P5, Portable, 1957, Charcoal grey plastic, transistor, right dial knob, left lattice grill, AM, battery $25.00

H-611P5, Portable, 1957, Blue plastic, transistor, right dial knob, left lattice grill, AM, battery $25.00

H-612P5, Portable, 1957, Yellow plastic, transistor, right dial knob, left lattice grill, AM, battery $25.00

H-621P6, Portable, 1958, Charcoal plastic, transistor, "laydown" style, thumbwheel dial, horizontal grill bars, handle, AM, bat $50.00

H-622P6, Portable, 1958, Yellow & white plastic, transistor, "laydown" style, thumbwheel dial, horizontal grill bars, handle, AM, bat $50.00

H-627T6U, Table, 1951, Plastic, lower slide rule dial, upper geometric grill design, 2 knobs, BC, AC/DC $40.00

H-636T6, Table, 1958, Ivory/white plastic, slide rule dial, large grill area, 2 knobs, feet, BC, AC/DC $20.00

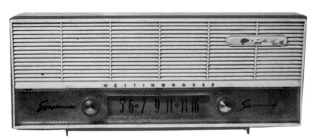

H-637T6, Table, 1958, Coral/white plastic, slide rule dial, large grill area, 2 knobs, feet, BC, AC/DC $20.00

H-652P6, Portable, 1958, Plastic, transistor, right dial, upper checkered/lower perforated grill, AM, bat $25.00

H-653P6, Portable, 1958, Plastic, transistor, right dial, upper checkered/lower perforated grill, AM, bat $25.00

H-655P5, Portable, 1959, Transistor, round dial, checkered grill, right side knob, swing handle, AM, battery $30.00

H-656P5, Portable, 1959, Transistor, round dial, checkered grill, right side knob, swing handle, AM, battery $30.00

H-666P5, Portable, 1959, Right round dial, lower lattice grill, telescope antenna, handle, BC, AC/DC/bat $20.00

H-678T4, Table-C, 1959, Plastic, right side round dial knob, large front alarm clock, feet, BC, AC $15.00

H-681T5, Table, 1959, Plastic, lower right round dial knob, horizontal front bars, side knob, BC, AC/DC $15.00

H-685P8, Table-C, 1959, Brown plastic, transistor, right dial over perforated grill, left alarm clock, AM, bat $20.00

H-686P8, Table-C, 1959, Pink plastic, transistor, right dial over perforated grill, left alarm clock, AM, bat $20.00

H-690P5, Portable, 1960, Plastic, transistor, right dial over perforated front, handle, side knob, AM, bat $25.00

H-694P8, Portable, 1959, Transistor, right front dial, upper checkered/lower perforated grill, AM, battery $30.00

H-697P7, Portable, 1960, Plastic, transistor, thumbwheel dial over checkered grill, swing handle, AM, bat $30.00

H-698P7, Portable, 1960, Plastic, transistor, thumbwheel dial over checkered grill, swing handle, AM, bat $30.00

H-713P9, Portable, 1960, Transistor, 2 slide rule dials, telescoping antenna, side knobs, handle, BC, SW, bat $35.00

H-726P6, Portable, 1960, Plastic, transistor, right dial over perforated front, handle, side knob, AM, bat $25.00

H-730P7, Portable, 1960, Leather case, transistor, right front dial, left grill, side knob, handle, AM, battery $20.00

H-742TA, Table, 1956, Aqua plastic, raised right front w/round dial knob, left criss-cross grill, feet $25.00

H-743TA, Table, 1956, Shadow white plastic, raised right front w/ round dial knob, left criss-cross grill $25.00

H-744TA, Table, 1956, Carnation pink plastic, raised right front w/ round dial, left criss-cross grill $25.00

H769P7A, Portable, 1960, Tan leather, transistor, right round dial, left checkered grill, handle, BC, bat $25.00

H770P7A, Portable, 1960, Gray leather, transistor, right round dial, left checkered grill, handle, BC, bat $25.00

H-1251 "Little Jewel", Portable, 1946, "Refrigerator", plastic w/ metal center strip, vert bars, fold-down handle, AC/DC $95.00

HR102BN, Console-R/P, 1958, Wood, inner right round dial, 6 knobs, left phono, lift top, large grill, BC, FM, AC $45.00

HR-109DP, Console-R/P, 1958, Wood, inner left round dial, 6 knobs, sliding right top, inner phono, BC, FM, AC $45.00

RC (RA/DA), Table, 1922, Wood, 2 boxes - receiver & amp, black front panels, 1 dial, 3 tubes, lift top $235.00

WR-8 "Columaire", Grandfather Clock, 1931, Wood, Deco case, right side dial/knobs/switch, front clock face, 9 tubes, AC $250.00

WR-8-R "Columnaire", Grandfather Clock, 1931, Wood, Deco case, right side dial/knobs/switch, front clock face, remote control, 9 tubes, AC $275.00

WR-12X7, Table, 1941, Wood, right front slide rule dial, pushbuttons, left grill, 4 knobs, BC, SW, AC/DC $60.00

WR-14, Cathedral, 1931, Two-tone wood, lower window dial, upper 3-section cloth grill, 4 tubes, 3 knobs $170.00

WR-15, Grandfather Clock, 1931, Wood, Deco case, center window dial, 3 knobs, upper gold clock face, 9 tubes, AC $250.00

WR-20, Table, 1934, Wood, center cloth grill with cut-outs, 2 knobs, right & left vertical fluting, BC $60.00

WR-21, Table, 1934, Wood, center cloth grill with horizontal bars, 5 tubes, 2 knobs, BC, SW, AC $60.00

WR-22, Tombstone, 1934, Wood, center window dial, upper cloth grill w/Deco cut-outs, 5 tubes, BC, SW, AC **$100.00**

WR-23, Tombstone, 1934, Wood, quarter-round dial, upper vertical cut-outs, 7 tubes, 4 knobs, BC, SW, AC **$130.00**

WR-24, Console, 1934, Wood, upper quarter-round dial, 3 vertical grill bars, 7 tubes, 4 knobs, BC, SW, AC **$130.00**

WR-27, Table, 1934, Wood, center front cloth grill with cut-outs, 4 tubes, 2 knobs, BC, AC **$65.00**

WR-28, Tombstone, 1934, Wood, center round dial, upper grill w/cut-outs, 6 tubes, 4 knobs, BC, SW, AC **$140.00**

WR-29, Console, 1934, Wood, upper round dial, lower grill w/cut-outs, 6 tubes, 4 knobs, BC, SW, AC **$150.00**

WR-30, Console, 1934, Wood, quarter-round dial, lower grill w/cut-outs, medallions, 10 tubes, 4 knobs, BC, SW, AC **$175.00**

WR-62K1 "Carryette", Portable, 1941, Striped cloth covered, right front dial, left grill, 3 knobs, handle, AC/DC/battery **$35.00**

WR-62K2 "Carryette", Portable, 1941, Leatherette, inner right dial, left grill, fold-down front, handle, AC/DC/battery **$35.00**

WR-100, Tombstone, 1935, Wood, small case, lower front round dial, upper grill w/cut-outs, 5 tubes, BC, AC **$100.00**

WR-101, Table, 1935, Wood, small case, lower dial, upper cloth grill w/cut-outs, 4 knobs, BC, SW, AC **$60.00**

WR-120, Table, 1937, Plastic, right dial, left grill, slightly rounded top, 6 tubes, 3 knobs, BC, SW, AC **$45.00**

WR-152, Table, 1939, Wood, right front dial, left grill w/horizontal bars, 6 tubes, 2 knobs, BC, AC/DC **$45.00**

WR-154, Table, 1939, Wood, right front dial, left wrap-around grill w/ horizontal bars, BC, AC/DC **$50.00**

WR-162, Table, 1939, Wood, right slide rule dial, left wrap-around grill, tuning eye, BC, SW, AC/DC **$60.00**

WR-165M, Table, 1939, Wood, right front round ivory dial, left wrap-around louvers, brown knobs, BC, AC **$50.00**

WR-166, Table, 1938, Plastic, right front round dial, left horizontal wrap-around louvers, 2 knobs, AC **$50.00**

WR-166A, Table, 1939, Plastic, right front round dial, left horiz wrap-around louvers, 2 knobs, BC, AC **$50.00**

WR-168, Table, 1939, Wood, right slide rule dial, left cloth grill w/ horizontal bars, 2 knobs, BC, AC/DC **$50.00**

WR-169, Table, 1939, Wood, right slide rule dial, vertical grill bars, 5 pushbuttons, 2 knobs, BC, AC/DC **$60.00**

WR-170, Table, 1939, Wood, slide rule dial, wrap-around grill w/horiz bars, pushbuttons, 2 knobs, BC, SW, AC/DC **$65.00**

WR-172, Table, 1939, Wood, slide rule dial, wrap-around grill w/horiz bars, pushbuttons, 4 knobs, BC, SW, AC/DC **$65.00**

WR-182, Table, 1940, Two-tone walnut, right front dial, left grill, 3 knobs, BC, SW, AC/DC **$50.00**

WR-201, Tombstone, 1935, Wood, small case, lower round dial, upper grill w/cut-outs, 5 tubes, BC, SW, AC **$95.00**

WR-203, Tombstone, 1935, Wood, center round dial, upper grill w/ cut-outs, 6 tubes, 4 knobs, BC, SW, AC **$125.00**

WR-204, Tombstone, 1935, Wood, center round dial, upper grill w/ cut-outs, 7 tubes, 4 knobs, BC, SW, AC **$135.00**

WR-205, Tombstone, 1935, Wood, center round dial, upper grill w/ cut-outs, 8 tubes, 4 knobs, BC, SW, AC **$135.00**

WR-207 "Trumpter", Table, 1935, Wood, front off-center round black & white dial, left grill, 4 knobs, BC, SW, AC **$40.00**

WR-208 "Jubileer", Table, 1935, Wood, front off-center round black & white dial, left grill, 4 knobs, BC, SW, AC **$40.00**

WR-209, Table, 1936, Wood, right front oval dial, left cloth grill w/ horiz bars, 3 knobs, BC, SW, AC **$50.00**

WR-212, Table, 1936, Wood, right round dial, tuning eye, left grill w/ vertical bars, 4 knobs, BC, SW, AC **$80.00**

WR-214, Tombstone, 1936, Wood, large lower oval dial, upper cloth grill w/cut-outs, 10 tubes, 4 band, AC **$150.00**

WR-217, Table, 1937, Wood, right oval dial, left grill w/horseshoe cut-outs, 5 tubes, BC, SW, AC **$55.00**

WR-222, Table, 1937, Wood, large right front dial, left grill w/vertical cut-outs, 4 knobs, BC, SW, AC **$50.00**

WR-224, Table, 1937, Wood, large right front dial, left cloth grill w/ horiz bars, 4 knobs, BC, SW, AC **$50.00**

WR-228, Table, 1937, Walnut, right dial, left wrap-around grill w/ bars, tuning eye, 4 knobs, BC, SW, AC **$55.00**

WR-256, Table, 1939, Wood, right front dial, left wrap-around grill w/ horizontal bars, 5 tubes, AC **$50.00**

WR-258, Table, 1938, Wood, lower slide rule dial, pushbuttons, upper oblong grill w/horizontal bars, AC **$60.00**

WR-262, Table, 1939, Wood, slide rule dial, wrap-around grill w/horiz bars, pushbuttons, tuning eye, BC, SW, AC **$60.00**

WR-264, Table, 1939, Wood, lower slide rule dial, upper grill w/vert bars, tuning eye, pushbuttons, BC, SW, AC **$65.00**

WR-270, Table, 1939, Wood, right slide rule dial, left wrap-around grill w/horiz bars, pushbuttons, 2 knobs, AC **$60.00**

WR-272, Table, 1939, Wood, right slide rule dial, left wrap-around grill w/horiz bars, pushbuttons, 4 knobs, BC, SW, AC .. **$60.00**

WR-274, Table, 1939, Wood, right slide rule dial, left wrap-around grill, pushbuttons, tuning eye, 4 knobs, BC, SW, AC **$70.00**

WR-303, Console, 1935, Wood, upper round dial, lower grill w/cut-outs, 6 tubes, 4 knobs, BC, SW, AC **$135.00**

WR-304, Console, 1935, Wood, upper round dial, lower grill w/cut-outs, 7 tubes, 4 knobs, BC, SW, AC **$135.00**

WR-305, Console, 1935, Wood, upper round dial, lower grill w/cut-outs, 8 tubes, 4 knobs, BC, SW, AC **$140.00**

WR-306, Console, 1935, Wood, upper round dial, lower grill w/cut-outs, 10 tubes, 4 knobs, BC, SW, AC **$150.00**

WR-326, Console, 1937, Walnut, upper dial, lower grill w/vertical bar, tuning eye, 7 tubes, BC, SW, AC **$145.00**

WR-328, Console, 1937, Wood, upper square dial, lower grill w/vert bars, tuning eye, 8 tubes, BC, SW, AC **$145.00**

WR-336, Console, 1937, Wood, upper automatic tuning dial, lower grill w/vert bar, 12 tubes, BC, SW, AC **$180.00**

WR-338, Chairside, 1939, Wood, slanted front dial, lower grill, tuning eye, 7 tubes, 4 knobs, BC, SW, AC **$130.00**

WR-342, Console, 1939, Wood, automatic tuning dial, vertical grill bars, tuning eye, 8 tubes, BC, SW, AC **$150.00**

WR-366, Console, 1939, Wood, slide rule dial, vertical grill bars, tuning eye, pushbuttons, 8 tubes, 4 knobs, BC, SW, AC **$140.00**

WR-368, Console, 1938, Wood, slide rule dial, vert grill bars, pushbuttons, tuning eye, 10 tubes, 4 knobs, BC, SW, AC $165.00

WR-370, Console, 1939, Wood, slide rule dial, splayed grill bars, pushbuttons, tuning eye, 12 tubes, 4 knobs, BC, SW, AC **$165.00**

WR-372, Console, 1939, Wood, slide rule dial, vert grill bars, 6 pushbuttons, 6 tubes, 4 knobs, BC, SW, AC **$130.00**

WR-373, Console, 1939, Wood, slide rule dial, vert grill bars, pushbuttons, tuning eye, 7 tubes, 4 knobs, BC, SW, AC $130.00

WR-373-Y, Console, 1939, Wood, slide rule dial, vert grill bars, pushbuttons, tuning eye, 6 tubes, 4 knobs, BC, SW, AC/DC $130.00

WR-374, Console, 1939, Wood, slide rule dial, vert grill bars, pushbuttons, 8 tubes, 4 knobs, BC, SW, AC $140.00

WR-388, Console, 1940, Wood, upper slide rule dial, lower grill w/bars, 6 pushbuttons, 9 tubes, AC $140.00

WR-468, Table-R/P, 1939, Wood, right front dial, left grill w/horizontal bars, lift top, inner phono, AC $30.00

WR-470, Table-R/P, 1939, Wood, slide rule dial, pushbuttons, wrap-around grill, lift top, inner phono, AC $40.00

WR-472, Table-R/P, 1939, Wood, left front dial, right grill w/horizontal bars, lift top, inner phono, BC, AC $35.00

WR-473, Console-R/P, 1939, Wood, inner dial/knobs/phono, lift top, front grill w/vertical bars, BC, SW, AC $100.00

WR-474, Console-R/P, 1939, Wood, inner dial/knobs/phono, lift top, front grill w/vertical bars, BC, SW, AC $100.00

WR-478, Table-R/P, 1948, Two-tone wood, right front dial, left grill, 3 knobs, lift top, inner phono, BC, AC $30.00

WR-602, Tombstone, 1935, Wood, center front round dial, upper cloth grill w/cut-outs, BC, SW, battery $80.00

WR-675 "Carryette", Portable, 1939, Cloth covered, right front slide rule dial, left grill, 2 knobs, handle, battery $30.00

WR-675A "Carryette", Portable, 1939, Cloth covered, right front slide rule dial, left grill, 2 knobs, handle, battery $30.00

WR-678 "Carryette", Portable, 1940, Brown & white cloth covered, upper front dial, handle, 2 knobs, AC/DC/battery $30.00

WR-679 "Carryette", Portable, 1940, Two-tone leatherette, upper front dial, lower grill, handle, 2 knobs, AC/DC/bat $30.00

WILCOX-GAY
Wilcox-Gay Corp., Charlotte, Michigan

5A6-75, Tombstone, 1935, Wood, center front round airplane dial, upper cloth grill w/cut-outs, 4 knobs, bat $95.00

A-17, Table, 1936, Circular case w/ebony finish, Deco, center front dial, decorative vertical bars, 4 knobs, BC, SW, AC/DC $400.00

A-32, Table, 1937, Two-tone wood, slide rule dial, center automatic tuning, right & left grills, AC $60.00

A-37, Console, 1937, Wood, upper slide rule dial & telephone dial, lower grill, 6 tubes, 5 knobs, AC $165.00

A-51, Table, 1938, Plastic, right dial, left round grill w/wrap-around horiz bars, 2 knobs, AC/DC $50.00

A-52, Table, 1938, Walnut, right slide rule dial, left cloth grill with Deco cut-outs, 5 tubes, BC, AC $55.00

A-53 "Thin Man", Table, 1939, Plastic, thin case, right front dial, left cloth grill with horizontal bars, AC $50.00

A-54, Table, 1939, Walnut, right slide rule dial, pushbuttons, left wrap-around louvers, BC, SW, AC $60.00

A-55, Console, 1939, Walnut, upper slide rule dial, pushbuttons, lower horiz bars, 7 tubes, BC, SW, AC $130.00

A-58, Table, 1940, Wood, right front slide rule dial, left cloth grill with cut-outs, 4 tubes $45.00

A-69, Console-R/P, 1939, Walnut, inner dial/knobs/phono, lift top, front grill, lower record storage, AC $150.00

A-111, Console-R/P, 1941, Wood, inner left slide rule dial, lower storage, right grill w/bars, 9 tubes, BC, FM, AC $135.00

A-114, Console-R/P, 1941, Wood, inner left slide rule dial, lower storage, door, right front grill, BC, FM, AC $135.00

Chameleon, Wall Radio, 1937, Two-tone metal case 3" thick, Deco with stripes, lower front dial, center grill $125.00

TXF-67, Console-R/P, 1940, Wood, inner right dial & knobs, left phono, lift top, lower front grill , BC, SW, AC $130.00

WILMAK
Wilmak Corp.,
RR 3, Benton Harbor, Michigan

W-446 "Denchum", Table, 1947, Wood, top plastic handle w/built-in thumbwheel dials, side louvers, BC, AC/DC $70.00

WOOLAROC
Phillips Petroleum Co.,
Bartlesville, Oklahoma

3-1A, Table, 1946, Plastic, streamline, upper slide rule dial, horizontal louvers, 3 knobs, BC, AC/DC $65.00

3-2A, Table, 1946, Plastic, streamline, upper slide rule dial, lower horizontal louvers, BC, AC/DC $65.00

3-3A, Table, 1946, Wood, upper slanted slide rule dial, lower cloth grill, 4 knobs, BC, 2SW, AC $50.00

3-5A, Table, 1947, Two-tone plastic, square dial, horizontal grill bars, handle, 2 knobs, BC, AC/DC $65.00

3-6A/5, Table, 1947, Wood, upper slanted slide rule dial, large lower cloth grill, 2 knobs, BC, AC/DC $35.00

3-9A, Table, 1946, Ivory plastic, upper slide rule dial, criss-cross grill, 2 knobs, feet, BC, AC/DC $45.00

3-10A, Table, 1946, Plastic, upper slide rule dial, lower criss-cross grill, 2 knobs, feet, BC, AC/DC $45.00

3-11A, Table, 1946, Wood, upper slanted slide rule dial, lower criss-cross grill, 4 knobs, BC, SW, AC $35.00

3-12A/3, Portable, 1947, Leatherette, upper slide rule dial, metal grill, 2 knobs, handle, BC, AC/DC/battery $25.00

3-15A, Table, 1948, Metal, right dial, left horizontal louvers, scalloped sides, 2 knobs, BC, AC/DC $60.00

3-17A, Table, 1948, Plastic, upper slide rule dial, lower horizontal louvers, 2 knobs, BC, AC/DC $45.00

3-20A, Table-R/P, 1947, Wood, front slanted slide rule dial, horizontal louvers, 4 knobs, lift top, BC, AC $35.00

3-29A, Table-R/P, 1946, Wood, top vert slide rule dial, 3 knobs, front grill, lift top, inner phono, BC, AC $25.00

3-70A, Console-R/P, 1948, Wood, inner right slide rule dial, 4 knobs, left pull out phono drawer, BC, SW, AC $85.00

3-71A, Console-R/P, 1948, Wood, inner right slide rule dial, 4 knobs, left pull-out phono drawer, BC, FM, AC $85.00

WORKRITE
The Workrite Manufacturing Co.,
1806 East 30th Street,
Cleveland, Ohio

Workrite began business manufacturing automobile equipment. Following WW I, they began to produce radio parts and by 1924 were producing complete radio sets. Due to increasing competition, by 1929 the company was out of business.

17, Table, 1927, Wood, low rectangular case, center front window dial, 6 tubes, 2 knobs, battery $90.00

18, Table, 1928, Wood, rectangular case, slanted front panel w/center window dial, 2 knobs, AC $110.00

38, Table, 1928, Wood, rectangular case, center window dial w/escutcheon, 9 tubes, 2 knobs, AC $130.00

Air Master, Table, 1924, Wood, high rectangular case, slanted 3 dial front panel, 5 tubes, battery $120.00

Aristocrat, Console, 1924, Mahogany, inner slanted 3 dial panel, fold-down front, inner left speaker, battery $235.00

Chum, Table, 1924, Wood, high rectangular case, slanted 2 dial front panel, 3 exposed tubes, battery $225.00

Radio King, Table, 1924, Mahogany, slanted 3 dial front panel, upper enclosed speaker w/grill, 5 tubes, bat $210.00

WURLITZER LYRIC
Rudolph Wurlitzer Mfg. Co., North Tonawanda, New York

C-4-LI, Table, 1934, Wood, right front dial, center grill with horizontal louvers, 4 tubes, 2 band $55.00

M-4-L, Table, 1934, Wood, right front dial, left horizontal wrap-around louvers, 4 tubes $55.00

M-4-LI, Table, 1934, Two-tone wood, right front dial, left horizontal wrap-around louvers, 4 tubes $60.00

SA-5-L, Tombstone, 1934, Wood, shouldered, center window dial, upper grill w/gothic cut-outs, 5 tubes, 3 knobs $120.00

SA-46, Console, 1934, Wood, lowboy, upper front window dial, lower cloth grill with cut-outs, 4 tubes $120.00

SA-99, Console, 1934, Wood, lowboy, upper window dial, lower cloth grill with cut-outs, 9 tubes, 6 legs $145.00

SA-120, Console, 1934, Wood, lowboy, upper half-moon dial, lower grill w/cut-outs, 12 tubes, 6 legs $180.00

SA-133, Console, 1934, Wood, upper front quarter-round dial, lower cloth grill with cut-outs, 13 tubes $200.00

SU-5 "Duncan Phyfe", Side Table, 1934, Wood, Duncan Phyfe style, inner dial & knobs, fold-down front, 5 tubes, AC/DC $150.00

SU-5 "Queen Ann", Side Table, 1934, Wood, Queen Ann style, inner dial & knobs, fold-down front, 5 tubes, AC/DC .. $150.00

SW-88, Tombstone, 1934, Two-tone wood, center window dial, upper horiz grill bars, 8 tubes, 4 knobs, BC, SW $100.00

SW-89, Console, 1934, Wood, upper window dial, lower cloth grill with cut-outs, 8 tubes, 6 legs, 4 knobs $135.00

ZANEY-GILL
Zaney-Gill Corp., Los Angeles, California

2445 "Music Box", Cathedral, 1930, Wood, center window dial, upper grill w/cut-outs, fluted columns, 6 tubes, 3 knobs $225.00

Clarion, Cathedral, 1930, Mahogany, lower front dial w/bronze-finish escutcheon, upper round grill, AC $250.00

Legionair, Cathedral, 1930, Mahogany, modernistic, center front dial, upper grill w/"sun-burst" cut-outs, peaked top, AC $265.00

ZENITH
Zenith Radio Corporation, Chicago, Illinois

The Zenith company began as Chicago Radio Labs in 1918 and the name "Zenith" came from the station call letters of its founders - 9ZN. Commander Eugene McDonald built Zenith into one of the most successful and prolific of radio manufacturers. Some of the most sought after Zenith sets today are the black dial sets of the 1930's.

3-R, Table, 1923, Wood, low rectangular case, 1 dial front panel, 4 tubes, battery $375.00

4-B-131, Tombstone, 1937, Wood, lower round dial, upper grill w/cut-outs, fluting, 4 tubes, BC, battery $130.00

4-B-132, Table, 1937, Wood, cube-shaped, front round dial, top grill,

step-down sides, 4 tubes, BC, bat $100.00

4-B-231, Table, 1937, Wood, lower black dial, upper cloth grill w/ vertical cut-outs, 4 tubes, BC, bat $55.00

4-B-313, Table, 1939, Plastic, right dial, rounded left w/wrap-around horizontal bars, 2 knobs, BC, bat $75.00

4-B-314, Table, 1939, Plastic, right dial, rounded left w/wrap-around horiz bars, 5 pushbuttons, 2 knobs, BC, bat $85.00

4-B-317, Table, 1939, Wood, right dial, rounded left w/vert glass bars, pushbuttons, 2 knobs, BC, bat $125.00

4-B-355, Console, 1939, Wood, upper slanted dial, lower cloth grill w/vertical bars, 4 tubes, BC, battery $110.00

4-B-422, Table, 1940, Plastic, right dial, rounded left w/wrap-around horizontal bars, 2 knobs, BC, bat $75.00

4-B-437, Table, 1940, Walnut, right dial, left wrap-around grill w/horiz bars, 4 tubes, 2 knobs, BC, bat $40.00

4-B-466, Console, 1940, Wood, upper dial, lower cloth grill w/vertical bars, 4 tubes, BC, battery $110.00

4-B-515, Table, 1941, Plastic, right dial, rounded/raised left w/horiz louvers, 4 tubes, 2 knobs, BC, bat $110.00

4-B-535, Table, 1941, Walnut, right dial, left wrap-around grill w/horiz bars, 4 tubes, 2 knobs, BC, bat $40.00

4-B-536, Table, 1941, Wood, right dial, left cloth grill w/cut-outs, 4 tubes, 2 knobs, BC, battery $45.00

4-B-639, Table, 1942, Wood, right black dial, left cloth grill w/bars, 4 tubes, 2 knobs, BC, battery $45.00

4-F-227, Table, 1937, Wood, lower front black dial, upper cloth grill w/vertical cut-outs, BC, battery $55.00

4-G-800, Portable, 1948, Plastic, inner dial, lattice grill, flip-up front, handle, 2 knobs, BC, AC/DC/bat $35.00

4-G-800Z, Portable, 1948, Plastic, inner left dial, metal lattice grill, flip-up front, handle, BC, AC/DC/bat $35.00

4-K-016, Table, 1946, Mottled plastic, inverted V-shaped dial & louvers, 2 knobs, handle, BC, battery $40.00

4-K-035, Table, 1946, Walnut veneer, right black dial, left grill, rounded top corners, 2 knobs, BC, bat $35.00

4-K-035G, Table, 1946, Limed walnut, right black dial, left cloth grill w/horiz bars, 2 knobs, battery $35.00

4-K-331, Table, 1939, Wood, right front dial, left wrap-around grill w/ horizontal bars, 2 "Z" knobs, bat $50.00

4-K-402D, Portable, 1940, Cloth-covered, inner right dial, left grill, fold-down front, handle, 4 tubes, bat $35.00

4-K-402L, Portable, 1940, Leather, inner right dial, left grill, fold-down front, handle, 4 tubes, battery $35.00

4-K-402M, Portable, 1940, Cloth-covered, inner right dial, left grill, fold-down front, handle, 4 tubes, bat $35.00

4-K-402Y, Portable, 1940, Cloth-covered, inner right dial, left grill, fold-down front, handle, 4 tubes, bat $35.00

4-K-422, Table, 1940, Plastic, right front dial, rounded left w/horizontal bars, 4 tubes, 2 knobs, bat $75.00

4-K-435, Table, 1940, Wood, right dial, left wrap-around grill w/horiz bars, 4 tubes, 2 knobs, BC, bat $35.00

4-K-465, Console, 1940, Wood, upper front dial, lower grill w/2 vertical bars, 2 knobs, BC, battery $110.00

4-K-515, Table, 1941, Plastic, right dial, rounded/raised left w/horiz louvers, 4 tubes, 2 knobs, BC, bat $110.00

4-K-570, Console, 1941, Wood, large upper dial, lower cloth grill w/ vertical bars, 4 tubes, BC, battery $115.00

4-K-600 "Pocketradio", Portable, 1942, Black plastic, inner dial, curved louvers & "Z" logo, plays when cover opens, bat $65.00

4-K-600G "Pocketradio", Portable, 1942, Two-tone green plastic, inner dial, curved louvers, plays when cover opens, bat $75.00

4-K-600P "Pocketradio", Portable, 1942, Red/white/blue plastic, inner dial, curved louvers, plays when cover opens, bat $75.00

4-K-600R "Pocketradio", Portable, 1942, Red plastic, inner dial, curved louvers & "Z" logo, plays when cover opens, battery $75.00

4-K-600W "Pocketradio", Portable, 1942, Ivory plastic, inner dial, curved louvers & "Z" logo, plays when cover opens, bat $65.00

4-K-616, Table, 1942, Plastic, half-round black dial, concentric circular louvers, 2 knobs, BC, battery $45.00

4-K-635, Table, 1942, Wood, right "boomerang" black dial, left grill w/ 2 vertical bars, 2 knobs, BC, bat $40.00

4-K-640, Table, 1942, Wood, "boomerang" black dial, diagonal grill bars, pushbuttons, 3 knobs, BC, SW, bat $45.00

4-K-658, Console, 1942, Wood, "boomerang" black dial, vertical grill bars, 4 tubes, 2 knobs, BC, battery $120.00

4-R, Table, 1923, Wood, low rectangular case, 1 dial brown front panel, 4 tubes, battery $375.00

4-T-26, Tombstone, 1935, Wood, lower black dial, upper cloth grill w/ Deco cut-outs, 4 tubes, BC, SW, AC $125.00

4-T-51, Console, 1935, Wood, upper black dial, lower cloth grill w/ cut-outs, 4 tubes, BC, SW, AC $120.00

4-V-31, Tombstone, 1935, Wood, lower round black dial, upper grill w/cut-outs, 4 tubes, 3 knobs, BC, bat $120.00

4-V-59, Console, 1935, Wood, upper black dial, lower cloth grill w/ cut-outs, 4 tubes, 3 knobs, BC, bat $110.00

5-A-10, Table, Wood, center front black dial, right & left horiz louvers, curved sides, 2 knobs $100.00

5-C-01 "Consoltone", Table, 1946, Plastic, right front black dial, left cloth grill, off/on bull's eye, 5 tubes $45.00

5-D-011 "Consoltone", Table, 1946, Swirl plastic, right black dial, left cloth grill, off/on bull's eye, BC, AC/DC $45.00

5-D-011W "Consoltone", Table, 1946, White plastic, right front black dial, left cloth grill, off/on bull's eye, AC/DC $45.00

5-D-011Y "Consoltone", Table, 1946, Black plastic, right front black dial, left cloth grill, off/on bull's eye, AC/DC $45.00

5-D-027 "Consoltone", Table, 1946, Two-tone wood, right front black dial, left cloth grill, fluted sides, AC/DC $50.00

5-D-610, Table, 1941, Brown plastic, right square black dial over horizontal louvers, 2 knobs, BC, AC/DC $45.00

5-D-610W, Table, 1941, Ivory plastic, right square black dial over horizontal louvers, 2 knobs, BC, AC/DC $45.00

5-D-611, **Table, 1942, Brown plastic, right black dial over large horiz front louvers, 2 knobs, AC/DC** $40.00

5-D-611W, Table, 1942, Ivory plastic, right black dial over large horiz front louvers, 2 knobs, AC/DC $40.00

5-D-627, Table, 1942, Wood, right black dial, left cloth grill w/cut-outs, 5 tubes, 2 knobs, BC, AC/DC $50.00

5-D-810, Table, 1949, Plastic, right front round dial over recessed metal grill, feet, 2 knobs, BC, AC/DC $35.00

5-F-134, Tombstone, 1941, Wood, lower round black dial, upper cloth grill w/horiz cut-outs, BC, 3SW, bat $150.00

5-F-233, Table, 1937, Wood, off-center round black dial, left grill cut-outs, feet, 3 knobs, BC, SW, bat.........................$80.00

5-F-251, Console, 1937, Wood, upper round black dial, lower cloth grill w/vertical bars, BC, SW, battery $115.00

5-G-003, Portable, 1947, Plastic, left dial, semi-circular grill bars, handle, center knob, BC, AC/DC/bat $40.00

5-G-003ZZ, Portable, 1947, Plastic, left dial part of semi-circular louvers, handle, 1 knob, BC, AC/DC/bat $40.00

5-G-03, Table-C, 1950, Plastic, oblong case, right dial, left clock, center perforated grill w/crest, BC, AC $40.00

5-G-036, Table, 1947, Wood, right black dial, left cloth grill w/ horizontal bars, 2 knobs, BC, AC/DC/bat $45.00

5-G-40 "Transoceanic", Portable, 1950, Leatherette, inner black dial, fold-up front w/crest, handle, 6 bands, AC/DC/bat $80.00

5-G-401D, Portable, 1940, Cloth, inner right dial, left grill, 2 knobs, fold-down front, handle, AC/DC/bat $35.00

5-G-401L, Portable, 1940, Leather, inner right dial, left grill, 2 knobs, fold-down front, handle, AC/DC/bat $35.00

5-G-401M, Portable, 1940, Cloth, inner right dial, left grill, 2 knobs, fold-down front, handle, AC/DC/bat $35.00

5-G-401Y, Portable, 1940, Cloth, inner right dial, left grill, 2 knobs, fold-down front, handle, AC/DC/bat $35.00

5-G-441, Table, 1940, Wood, slide rule dial, wrap-around grill w/horiz bars, pushbuttons, AC/DC/bat $60.00

5-G-442, Table, 1940, Wood, slide rule dial, upper grill w/vert bars, pushbuttons, 3 bands, AC/DC/bat $70.00

5-G-484M, Portable-R/P, 1940, Cloth covered, right front dial, left grill, lift top, inner phono, BC, AC/DC/bat $30.00

5-G-500, Portable, 1941, Cloth, inner right black dial, left grill, 2 "Z" knobs, fold-down front, handle $30.00

5-G-500L, Portable, 1941, Leather, inner right dial, left grill, 2 knobs, fold-down front, handle, AC/DC/bat $30.00

5-G-500M, Portable, 1941, Cloth, inner right dial, left grill, 2 knobs, fold-down front, handle, AC/DC/bat $30.00

5-G-572, Console, 1941, Wood, upper black dial, lower horizontal grill bars, 5 tubes, BC, SW, AC/DC/bat $145.00

5-G-603M, Portable, 1942, Striped cloth, right black dial, left grill w/ sailboat, handle, 2 knobs, AC/DC/bat $45.00

5-G-617, Table, 1942, Plastic, "boomerang" dial, concentric "horse-shoe" louvers, pushbuttons, 2 knobs, BC, AC/DC/bat ..$55.00

5-G-636, Table, 1942, Wood, "boomerang" dial, Deco grill cut-outs, pushbuttons, 2 knobs, BC, AC/DC/bat $75.00

5-G-2617, Table, 1942, Wood, black "boomerang" dial, vertical grill cut-outs, pushbuttons, 2 knobs, BC $65.00

5-J-217, Table, 1937, Wood, cube-shaped, center round black dial, top grill cut-outs, 3 knobs, BC, SW, AC/bat $150.00

5-J-247, Chairside, 1937, Wood, step-down top w/round black dial, lower storage, 3 knobs, BC, SW, AC/bat $140.00

5-J-255, Console, 1937, Wood, upper round black dial, lower cloth grill w/vertical bars, BC, SW, AC/bat **$100.00**

5-K-637, Table, 1942, Wood, right "boomerang" dial, left grill w/Deco cut-outs, pushbuttons, 2 knobs, BC, bat **$45.00**

5-R-086, Table-R/P, 1946, Wood, right front black dial, left grill, lift top, inner phono, BC, AC **$30.00**

5-R-135, Tombstone, 1937, Wood, lower front black dial, upper cloth grill with 3 vertical bars, 3 knobs, AC **$100.00**

5-R-216, Table, 1937, Wood, off-center round black dial, left front grill cut-outs, feet, 2 knobs, BC, AC **$120.00**

5-R-216W, Table, 1937, Bone white, off-center round black dial, left grill cut-outs, feet, 2 knobs, BC, AC **$120.00**

5-R-226, Table, 1937, Wood, child's console, round black dial, lower 2 section grill, 2 knobs, BC, AC **$200.00**

5-R-236, Chairside, 1937, Wood, top round dial & knobs, vertical grill bars, lower & side storage, feet, AC **$120.00**

5-R-236W, Chairside, 1937, Bone white, top round dial & knobs, vert grill bars, lower & side storage, feet, AC **$120.00**

5-R-312, Table, 1938, Plastic, right dial, rounded/raised left w/ horiz louvers, pushbuttons, 2 knobs, BC, AC **$150.00**

5-R-316, Table, 1938, Wood, right dial, left grill w/horizontal grill bars, pushbuttons, 2 knobs, BC, AC **$60.00**

5-R-317, Table, 1937, Wood, Deco, right dial, rounded left w/glass rods, 5 pushbuttons, 2 knobs, BC, AC **$175.00**

5-R-337, Chairside, 1938, Wood, top dial/knobs/pushbuttons, front grill, lower open storage, 5 tubes, BC, AC **$110.00**

5-R-680, Table-R/P, 1942, Wood, right front black dial, left grill w/cut-outs, lift top, inner phono, BC, AC **$35.00**

5-R-686, Table-R/P, 1942, Wood, right front black dial, left grill w/vert bars, lift top, inner phono, BC, AC **$35.00**

5-S-29, Tombstone, 1935, Wood, round black dial, cloth grill w/cut-outs, 5 tubes, 4 knobs, BC, 2SW, AC **$165.00**

5-S-56, Console, 1936, Wood, round black dial, vertical grill cut-outs, 5 tubes, 4 knobs, BC, SW, AC **$140.00**

5-S-126, Table, 1936, Wood, cube-shaped, round black dial, top grill w/cut-outs, 4 knobs, BC, SW, AC **$145.00**

5-S-127, Tombstone, 1936, Wood, lower round black dial, upper horiz grill bars, 5 tubes, 4 knobs, BC, SW, AC **$150.00**

5-S-128, Table, 1937, Wood, cube-shaped, round black dial, top grill w/cut-outs, 4 knobs, BC, SW, AC **$140.00**

5-S-150, Console, 1937, Wood, upper round black dial, lower vert grill bars, 5 tubes, 4 knobs, BC, SW, AC **$150.00**

5-S-151, Console, 1936, Wood, upper round black dial, lower grill w/ vertical bar, 5 tubes, 4 knobs, BC, SW, AC **$175.00**

5-S-161, Console, 1937, Wood, upper round black dial, lower grill w/ vertical bars, 5 tubes, BC, SW, AC **$160.00**

5-S-218, Table, 1937, Wood, off-center round black dial, left grill cut-outs, feet, 3 knobs, BC, SW, AC **$140.00**

5-S-220, Table, 1937, Wood, large center front round black dial, feet, 5 tubes, 3 knobs, BC, SW, AC **$150.00**

5-S-220Y, Table, 1937, Ebony finish, large center round black dial, feet, 5 tubes, 3 knobs, BC, SW, AC **$175.00**

5-S-228, Tombstone, 1937, Wood, lower round black dial, grill cut-outs, feet, 5 tubes, 3 knobs, BC, SW, AC **$150.00**

5-S-228W, Tombstone, 1937, Bone white, round black dial, grill cut-outs, feet, 5 tubes, 3 knobs, BC, SW, AC **$150.00**

5-S-237, Chairside, 1938, Wood, Deco, step-down top w/round dial, front vert grill bars, shelf, BC, SW, AC **$150.00**

5-S-237W, Chairside, 1938, Bone white, Deco, step-down top w/round dial, front vert bars, shelf, BC, SW, AC **$150.00**

5-S-237Y, Chairside, 1938, Ebony, Deco, step-down top w/round dial, front vert grill bars, shelf, BC, SW, AC **$160.00**

5-S-250, Console, 1937, Wood, upper round black dial, lower grill w/ vertical bars, 5 tubes, BC, SW, AC **$125.00**

5-S-252, Console, 1937, Wood, round black dial, lower grill w/vertical bars, 5 tubes, 3 knobs, BC, SW, AC **$130.00**

5-S-319, Table, 1937, Wood, oval gold dial, 5 pushbuttons, left wrap-around grill, 4 knobs, BC, SW, AC **$85.00**

5-S-320, Table, 1939, Wood, oval gold dial, 5 pushbuttons, left wrap-around grill, 4 knobs, BC, SW, AC **$95.00**

5-S-327, Tombstone, 1938, Wood, oval gold dial, horiz grill bars, pushbuttons, 5 tubes, 4 knobs, BC, SW, AC **$175.00**

5-S-338, Chairside, 1939, Wood, top dial/knobs/pushbuttons, front grill, lower open storage, BC, SW, AC **$150.00**

5-S-339M, Chairside, 1939, Wood, top dial/knobs/pushbuttons, front grill, lower open storage, BC, SW, AC **$140.00**

5-X-230, Tombstone, 1939, Wood, lower round black dial, upper grill cut-outs, 5 tubes, 4 knobs, BC, SW, bat **$130.00**

5-X-248, Chairside, 1939, Wood, half-round, top round black dial, front grill, 5 tubes, 4 knobs, BC, SW, bat **$110.00**

5-X-274, Console, 1939, Wood, upper round black dial, lower grill w/ vertical bars, 5 tubes, BC, SW, bat **$115.00**

6-A-40 "Transoceanic", Portable, Black leatherette, inner multi-band dial, lattice grill, fold-up front, handle **$80.00**

6-B-129, Tombstone, 1936, Wood, lower front black dial, upper cloth grill w/cut-outs, 6 tubes, BC, SW, bat **$150.00**

6-B-164, Console, 1937, Wood, upper black dial, lower cloth grill w/ vertical bars, 6 tubes, BC, SW, bat **$145.00**

6-B-321, Table, 1939, Wood, round dial, wrap-around grill, pushbuttons, 6 tubes, 4 knobs, BC, SW, bat **$60.00**

6-D-014, Table, 1946, Swirl plastic, inverted "V" dial, inverted "V" louvers, handle, 2 knobs, BC, AC/DC **$50.00**

6-D-014W, Table, 1946, White plastic, inverted "V" dial, inverted "V" louvers, handle, 2 knobs, BC, AC/DC **$55.00**

6-D-014Y, Table, 1946, Black plastic, inverted "V" dial, inverted "V" louvers, handle, 2 knobs, BC, AC/DC **$50.00**

6-D-015 "Consoltone", Table, 1946, Walnut plastic, quarter-moon dial, half-moon louvers, metal center strip, handle, AC/DC **$65.00**

6-D-015W "Consoltone", Table, 1946, Ivory plastic, quarter-moon dial, half-moon louvers, metal center strip, handle, AC/DC **$65.00**

6-D-015Y, Table, 1946, Black plastic, quarter-moon dial, half-moon louvers, metal center strip, handle, AC/DC **$65.00**

6-D-029, Table, 1946, Walnut, front inverted "V" shaped black dial, lower cloth grill, 2 knobs, BC, AC/DC **$50.00**

6-D-029G, Table, 1946, Lime walnut, inverted "V" shaped black dial, lower cloth grill, 2 knobs, BC, AC/DC **$50.00**

6-D-030, Table, 1946, Walnut, right quarter-moon black dial, cloth grill, metal center strip, BC, AC/DC **$55.00**

6-D-030E, Table, 1946, Mahogany, right quarter-moon black dial, cloth grill, metal center strip, AC/DC **$55.00**

6-D-116, Table, 1936, Wood, right black dial w/"Z" pointer, left grill w/ horiz bars, 3 knobs, BC, SW, AC **$55.00**

6-D-117, Table, 1936, Black & walnut, right round black dial, left horiz grill bars, 3 knobs, BC, SW, AC **$75.00**

6-D-118, Table, 1936, Wood, right front round black dial, left grill w/ vertical bars, 6 tubes, AC **$55.00**

6-D-219, Table, 1937, Wood, Deco, rounded right side w/round dial, left grill, 3 knobs, BC, SW, AC/DC **$65.00**

6-D-219W, Table, 1937, Bone white, Deco, rounded right w/round dial, left grill, 3 knobs, BC, SW, AC/DC **$65.00**

6-D-219Y, Table, 1937, Ebony, Deco, rounded right side w/round dial, left grill, 3 knobs, BC, SW, AC/DC **$65.00**

6-D-221, Table, 1937, Wood, right round black dial, left grill w/cut-outs, feet, 3 knobs, BC, SW, AC/DC **$65.00**

6-D-311, Table, 1938, Bakelite, Deco, right half-moon dial, left wrap-around louvers, "Z" knob, BC, AC **$150.00**

6-D-312, Table, 1938, Plastic, right dial, rounded/raised left w/horiz louvers, pushbuttons, 2 knobs, BC, AC/DC **$150.00**

6-D-315, Table, 1938, Bakelite, Deco, right half-moon dial, left wrap-around louvers, handle, "Z" knob **$160.00**

6-D-316, Table, 1938, Wood, right dial, left grill w/horizontal bars, pushbuttons, 2 knobs, BC, AC/DC **$60.00**

6-D-317, Table, 1938, Wood, Deco, right dial, rounded left w/glass rods, 5 pushbuttons, 2 knobs, AC/DC **$175.00**

6-D-326, Table, 1938, Wood, child's console, half-round dial, cloth grill w/vert bar, "Z" knob, BC, AC **$200.00**

6-D-336, Chairside, 1938, Wood, top dial & knobs, slanted grill w/ horizontal cut-outs, 6 tubes, legs, BC, AC **$110.00**

6-D-337, Chairside, 1938, Wood, top dial/knobs/pushbuttons, front grill, lower open storage, 6 tubes, BC, AC/DC **$115.00**

6-D-410, Table, 1939, Brown plastic, right front dial, left wrap-over vertical grill bars, BC, AC/DC **$50.00**

6-D-411, Table, 1939, Plastic, right front dial, left wrap-over vertical grill bars, 6 tubes, BC, AC/DC **$50.00**

6-D-413, Table, 1939, Brown plastic, center dial, left wrap-over vert grill bars, 5 pushbuttons, AC/DC **$55.00**

6-D-414, Table, 1939, Plastic, center dial, left wrap-over vertical grill bars, 5 pushbuttons, AC/DC **$55.00**

6-D-425, Table, 1939, Wood, right front dial, left cloth grill w/vertical cut-outs, 6 tubes, BC, AC/DC **$50.00**

6-D-426, Table, 1939, Wood, right dial, left grill w/vert spiral bars, pushbuttons, 6 tubes, BC, AC/DC **$70.00**

6-D-427, Table, 1939, Wood, right dial, left grill w/horizontal bars, pushbuttons, 6 tubes, BC, AC/DC **$65.00**

6-D-446, Chairside, 1940, Wood, top dial/knobs/pushbuttons, front grill, right & left storage, BC, AC/DC **$125.00**

6-D-455, Bookcase, 1940, Wood, inner dial & controls, fold-down front, 2 lower shelves, BC, AC/DC **$200.00**

6-D-510, Table, 1941, Brown plastic, black dial over horiz wrap-around grill bars, handle, 2 knobs, AC/DC **$45.00**

6-D-510W, Table, 1941, Ivory plastic, black dial over horiz wrap-around grill bars, handle, 2 knobs, AC/DC **$45.00**

6-D-516, Table, 1940, Plastic, right black dial over horizontal wrap-around grill bars, 2 knobs **$45.00**

6-D-520, Table, 1941, Plastic, right black dial over lattice grill, handle, 6 tubes, 2 knobs, BC, AC/DC **$45.00**

6-D-520W, Table, 1941, Plastic, right black dial over lattice grill, handle, 6 tubes, 2 knobs, BC, AC/DC **$45.00**

6-D-525, Table, 1941, Wood, right black dial, left raised horizontal grill bars, 2 "Z" knobs, BC, AC/DC **$70.00**

6-D-526, Table, 1941, Walnut, "waterfall" front, right dial, left vertical grill bars, 6 tubes, BC, AC/DC **$65.00**

6-D-538, Table, 1941, Wood, right black dial, left grill cut-outs, decorative case lines, 2 knobs, BC, AC/DC **$65.00**

6-D-612, Table, 1942, Brown plastic, right recessed black dial, horiz wrap-around bars, handle, 2 knobs, BC, AC/DC **$45.00**

6-D-612W, Table, 1942, White plastic, right recessed black dial, horiz wrap-around bars, handle, 2 knobs, BC, AC/DC . **$45.00**

6-D-614, Table, 1942, Plastic, half-round black dial, concentric circular louvers, handle, 2 knobs, BC, AC/DC **$55.00**

6-D-614W, Table, 1942, Black & white plastic, half-round black dial, concentric circular louvers, 2 knobs, BC, AC/DC **$65.00**

6-D-615, Table, 1942, Plastic, half-round dial, concentric "horseshoe" louvers, pushbuttons, handle, 2 knobs, BC, AC/DC **$65.00**

6-D-615W, Table, 1942, Plastic, half-round dial, concentric "horseshoe" louvers, pushbuttons, handle, 2 knobs, BC, AC/DC **$65.00**

6-D-620, Table, 1942, Plastic, half-round black dial, concentric circular louvers, handle, 2 knobs, BC, SW, AC/DC **$55.00**

6-D-620W, Table, 1942, Black & white plastic, half-round black dial, concentric circular louvers, handle, 2 knobs, BC, SW, AC/DC **$65.00**

6-D-628, Table, 1942, **Wood, right black dial, left grill w/cut-outs, 6 tubes, 2 knobs, BC, AC/DC** **$50.00**
6-D-629, Table, 1942, Wood, black "boomerang" dial, center grill w/cut-outs, 6 tubes, 2 knobs, BC, AC/DC **$45.00**
6-D-630, Table, 1942, Wood, black "boomerang" dial, center round grill, pushbuttons, 2 knobs, BC, AC/DC **$55.00**
6-D-644, Table, 1942, Wood, black "boomerang" dial, center grill w/cut-outs, 6 tubes, 2 knobs, BC, SW, AC/DC **$45.00**

6-D-815, Table, 1949, **Plastic, dial numbers on clear lucite, metal grill, flex handle, 2 knobs, BC, AC/DC** **$35.00**

6-D-2615, Table, 1942, **Wood, arched top, "bomerang" dial, vert grill cut-outs, pushbuttons, 2 knobs, BC** **$60.00**
6-G-001Y "Universal", Portable, 1946, Luggage style, right quarter moon dial, flip up lid, handle, 2 knobs, BC, AC/DC/bat .**$50.00**

6-G-001YX "Universal", Portable, 1946, Black leatherette, inner right quarter-round dial, flip-up lid, handle, 2 knobs **$50.00**
6-G-004Y "Universal", Portable, 1947, Luggage-type, semi-circular louvers, flip-up lid, handle, BC, SW, AC/DC/battery **$50.00**

6-G-005TZ1 "Transoceanic", Portable, Black leatherette, inner black dial, fold-up front w/"Z" logo, handle, AC/DC/bat **$80.00**
6-G-038, Table, 1948, Wood, black dial, pushbuttons, 2 knobs, telescoping antenna, BC, 2SW, AC/DC/bat **$50.00**
6-G-560, Console, 1941, Wood, upper black dial, lower grill w/center vertical bar, 6 tubes, BC, AC/DC/bat **$130.00**
6-G-601D, Portable, 1942, Cloth, inner black dial, grill with sailboat, fold-down front, handle, AC/DC/bat **$55.00**

6-G-601L, Portable, 1942, Leatherette, inner wood panel w/ black dial & sailboat grill, fold-down front, handle, BC, AC/DC/bat **$65.00**
6-G-601M, Portable, 1941, Cloth, inner black dial, grill with sailboat, fold-down front, handle, BC, AC/DC/bat **$55.00**
6-G-601MH, Portable, 1941, Cloth, inner black dial, grill with sailboat, fold-down front, handle, BC, AC/DC/bat **$55.00**

6-G-601ML, Portable, 1942, "Snakeskin", inner black dial, grill with sailboat, fold-down front, handle, BC, AC/DC/bat **$65.00**

6-G-601Y, Portable, 1948, Black leatherette, inner front dial, semi-circular louvers, fold-up front, handle **$50.00**

6-G-638, Table, 1942, Wood, right black dial, left grill cut-outs, pushbuttons, 2 knobs, BC, SW, AC/DC/bat **$60.00**

6-G-660, Console, 1942, Wood, slanted black dial, vertical grill bars, pushbuttons, 2 knobs, BC, SW, AC/DC/bat **$150.00**

6-G-801, Portable, 1949, Plastic, fold-open front doors, handle, pull-up antenna, 2 knobs, BC, AC/DC/bat **$45.00**

6-J-230, Tombstone, 1937, Wood, lower round black dial, upper grill w/cut-outs, 4 knobs, BC, SW, bat/AC **$160.00**

6-J-257, Console, 1937, Wood, upper round black dial, lower grill w/center bars, 4 knobs, BC, SW, bat/AC **$165.00**

6-J-322, Table, 1939, Wood, step-down top, right dial, rounded left w/horiz bars, pushbuttons, 4 knobs, BC, SW, bat/AC ..**$85.00**

6-J-357, Console, 1939, Wood, round black dial, vertical grill bars, pushbuttons, 6 tubes, BC, SW, bat/AC **$185.00**

6-J-436, Tombstone, 1940, Wood, round black dial, upper grill w/wrap-over vert bars, pushbuttons, BC, SW, bat/AC ...**$135.00**

6-J-463, Console, 1940, Wood, round black dial, lower vertical grill bars, pushbuttons, BC, SW, bat/AC **$185.00**

6-L-03, Table-C, 1953, Plastic, right front dial, left alarm clock, center horiz bars w/crest, feet, 5 knobs **$35.00**

6-P-416, Table, 1940, Brown plastic, rounded top, right dial, left wrap-over vertical bars, BC, AC **$65.00**

6-P-417, Table, 1940, Plastic, rounded top, right front dial, left wrap-over vertical bars, BC, AC **$65.00**

6-P-418, Table, 1940, Brown plastic, right dial, left wrap-around louvers, pushbuttons, handle, BC, AC **$60.00**

6-P-419, Table, 1940, Plastic, right dial, left wrap-around horiz louvers, pushbuttons, handle, BC, AC **$60.00**

6-P-429, Table, 1940, Wood, right front dial, left grill w/horizontal bars, pushbuttons, BC, AC **$65.00**

6-P-430, Table, 1940, Step-down top, right dial, rounded left w/horiz louvers, pushbuttons, BC, AC **$100.00**

6-R-060, Console, 1946, Walnut, upper slanted black dial, large lower cloth grill w/3 vertical bars, BC, AC **$130.00**

6-R-084, Table-R/P, 1947, Wood, front black "boomerang" dial, cloth grill, lift top, inner phono, feet, BC, AC **$40.00**

6-R-087, Console-R/P, 1946, Walnut, upper black dial, 2 knobs, cloth grill, pull-down phono door, BC, AC **$135.00**

6-R-480, Table-R/P, 1940, Wood, right front dial, left grill w/cut-outs, feet, open top phono, BC, AC **$40.00**

6-R-481, Table-R/P, 1940, Wood, right front dial, rounded left w/grill cut-outs, lift top, inner phono, BC, AC **$100.00**

6-R-583, Table-R/P, 1941, Walnut, right front black dial, left grill, 3 knobs, lift top, inner phono, BC, AC **$35.00**

6-R-631, Table, 1941, Wood, black "boomerang" dial, center grill w/cut-outs, pushbuttons, 2 knobs, BC, AC **$75.00**

6-R-683, Table-R/P, 1942, Wood, right front black dial, left grill w/vert bars, lift top, inner phono, BC, AC **$40.00**

6-R-684, Table-R/P, 1942, Wood, right front black dial, left grill w/vert bars, lift top, inner phono, BC, AC **$40.00**

6-R-687, Console-R/P, 1942, Walnut, inner black dial & phono, lift top, front grill w/lyre cut-out, BC, AC **$150.00**

6-R-687R, Console-R/P, 1942, Mahogany, inner black dial & phono, lift top, front grill w/lyre cut-out, BC, AC **$150.00**

6-R-688 "Modern", Console-R/P, 1942, Blonde, inner black dial & phono, lift top, front cloth grill w/cut-outs, BC, AC **$130.00**

6-R-886, Table-R/P, 1948, Wood, front round dial/louvers, 2 knobs, 2 switches, fold-back cover, BC, AC **$40.00**

6-S-27, Tombstone, 1935, Wood, round black dial, cloth grill w/fleur-de-lis cut-out, 4 knobs, BC, SW, AC **$225.00**

6-S-52, Console, 1935, Wood, upper round black dial, lower cloth grill w/cut-outs, 4 knobs, BC, SW, AC **$225.00**

6-S-128, Tombstone, 1936, Wood, lower front black dial, upper cloth grill w/horizontal bars, BC, SW, AC **$225.00**

6-S-137 "Zephyr", Tombstone, 1936, Wood, lower black dial, upper wrap-around horizontal louvers, BC, SW, AC **$185.00**

6-S-147 "Zephyr", Chairside, 1936, Wood, Deco, top dial & knobs, wrap-around horizontal louvers, BC, SW, AC **$160.00**

6-S-152, Console, 1936, Wood, upper black dial, lower cloth grill w/3 vertical bars, 6 tubes, BC, SW, AC **$195.00**

6-S-157 "Zephyr", Console, 1936, Wood, upper black dial, wrap-around horizontal louvers, 6 tubes, BC, SW, AC **$200.00**

6-S-203, Chairside-R/P, 1937, Wood, round top dial, lift top, inner phono, lower record storage, BC, SW, AC **$145.00**

6-S-203W, Chairside-R/P, 1937, Bone white, round top dial, lift top, inner phono, lower record storage, BC, SW, AC **$145.00**

6-S-222, Table, 1937, Wood, cube-shaped, center round black dial, top grill w/cut-outs, 4 knobs, BC, SW, AC **$140.00**

6-S-223, Table, 1937, Wood, right round black dial, left grill w/cut-outs, feet, 4 knobs, BC, SW, AC **$130.00**

6-S-229, Tombstone, 1937, Wood, lower round black dial, upper grill w/Deco cut-outs, 4 knobs, BC, SW, AC **$160.00**

6-S-238, Chairside, 1938, Wood, half-round, top dial & knobs, vertical grill bars, 6 tubes, BC, SW, AC **$150.00**

6-S-239, Chairside, 1937, Wood, step-down top w/round black dial, 4 knobs, lower storage, BC, SW, AC **$140.00**

6-S-241, Chairside/Bar, 1937, Wood, top round dial, front grill w/vertical bars, storage & bar area, BC, SW, AC **$160.00**

6-S-241W, Chairside/Bar, 1937, Bone white, top round dial, front grill w/bars, storage & bar area, BC, SW, AC **$160.00**

6-S-249, Chairside, 1938, Wood, half-round, top triangular dial & knobs, horiz wrap-around louvers, BC, SW, AC **$160.00**

6-S-254, Console, 1937, Wood, large black triangular dial, lower cloth grill w/vertical bars, BC, SW, AC **$225.00**

6-S-254H, Console, 1937, Wood, large black triangular dial, lower cloth grill w/vertical bars, BC, SW, AC **$225.00**

6-S-254W, Console, 1937, Bone white, large black triangular dial, lower grill w/vertical bars, BC, SW, AC **$225.00**

6-S-254Y, Console, 1937, Ebony finish, large black triangular dial, lower grill w/vertical bars, BC, SW, AC **$230.00**

6-S-256, Console, 1937, Wood, black triangular dial, recessed grill w/vertical bar, 4 knobs, BC, SW, AC **$235.00**

6-S-275, Console, 1938, Wood, upper round black dial, lower grill w/ vertical bars, 6 tubes, BC, SW, AC **$225.00**

6-S-301, Chairside-R/P, 1939, Wood, top dial, inner phono, rounded front w/wrap-around horiz louvers, BC, SW, AC **$185.00**

6-S-305, Console-R/P, 1938, Wood, inner left dial, pushbuttons, right phono, lift top, front grill w/vertical bar **$160.00**

6-S-321, Table, 1937, Wood, right dial, rounded left w/wrap-around grill bars, pushbuttons, 4 knobs, BC, SW, AC **$75.00**

6-S-322, Table, 1939, Wood, right dial, step-down left w/rounded grill, pushbuttons, 4 knobs, BC, SW, AC **$90.00**

6-S-330, Tombstone, 1938, Wood, round black dial, upper grill w/ Deco cut-outs, pushbuttons, BC, SW, AC **$170.00**

6-S-340, Chairside, 1939, Wood, half-round, top dial/knobs/ pushbuttons, vertical grill bars, BC, SW, AC **$180.00**

6-S-341, Chairside, 1938, Wood, Deco, half-round, top dial & knobs, horiz wrap-around louvers, BC, SW, AC **$175.00**

6-S-361, Console, 1939, Wood, round black dial, vertical grill bars, pushbuttons, 6 tubes, BC, SW, AC **$215.00**

6-S-362, Console, 1939, Wood, round black dial, vertical grill bars, pushbuttons, 6 tubes, BC, SW, AC **$215.00**

6-S-439, Table, 1940, Wood, black slide rule dial, pushbuttons, horiz wrap-around grill bars, BC, SW **$100.00**

6-S-469, Console, 1940, Wood, upper slide rule dial, lower grill w/ vertical bars, pushbuttons, BC, SW, AC **$140.00**

6-S-511, Table, 1941, Brown plastic, right dial, horiz wrap-around grill bars, pushbuttons, handle, BC, SW, AC $55.00

6-S-511W, Table, 1941, Ivory plastic, right dial, horiz wrap-around grill bars, pushbuttons, handle, BC, SW, AC **$55.00**

6-S-527, Table, 1941, Wood, right front dial, left horizontal grill bars, pushbuttons, 3 knobs, BC, SW, AC **$60.00**

6-S-528, Table, 1941, Wood, right black dial, left grill w/vert bars, pushbuttons, 3 knobs, BC, SW, AC **$75.00**

6-S-532, Table, 1941, Wood, large black dial, wrap-around grill, pushbuttons, 2 knobs, BC, SW, AC **$75.00**

6-S-546, Chairside, 1941, Wood, top black dial/knobs/pushbuttons, front grill, casters, BC, SW, AC **$135.00**

6-S-556, Console, 1941, Wood, upper black dial, lower grill w/vertical bars, pushbuttons, BC, SW, AC **$175.00**

6-S-580, Table-R/P, 1941, Wood, front black dial, vert grill bars, pushbuttons, lift top, inner phono, BC, SW, AC **$65.00**

6-S-632, Table, 1942, Wood, black dial, wrap-around grill w/horiz bars, pushbuttons, 2 knobs, BC, SW, AC **$55.00**

6-S-646, Chairside, 1942, Walnut, top black dial/knobs/pushbuttons, front grill w/cut-outs, BC, SW, AC **$145.00**

6-S-646R, Chairside, 1942, Mahogany, top black dial/knobs/ pushbuttons, front grill w/cut-outs, BC, SW, AC **$145.00**

6-S-656, Console, 1942, Wood, black dial, lower grill w/vertical bars, pushbuttons, 2 knobs, BC, SW, AC **$145.00**

6-S-892, Table, Wood, black dial, pushbuttons, left wrap-around grill w/2 bars, 2 knobs, BC, SW **$65.00**

6-V-27, Tombstone, 1935, Wood, round black dial, upper grill w/fleur-de-lis cut-out, 4 knobs, BC, SW, bat **$135.00**

6-V-62, Console, 1936, Wood, upper round black dial, lower grill w/ cut-outs, 4 knobs, BC, SW, bat.................................**$150.00**

7-D-126, Table, 1936, Wood, cube-shaped, round black dial, top grill cut-outs, 4 knobs, BC, SW, AC/DC **$140.00**

7-D-127, Tombstone, 1937, Wood, lower round black dial, upper grill w/horiz bars, 4 knobs, BC, SW, AC/DC **$145.00**

7-D-138 "Zephyr", Tombstone, 1936, Wood, round black dial, upper grill w/wrap-around horiz louvers, BC, SW, AC/DC **$185.00**

7-D-148 "Zephyr", Chairside, 1936, Wood, Deco, top dial & knobs, sliding glass cover, wrap-around louvers, BC, SW, AC **$160.00**

7-D-168 "Zephyr", Console, 1936, Wood, upper round black dial, lower wrap-around horiz louvers, BC, SW, AC/DC **$200.00**

7-D-203, Chairside-R/P, 1937, Wood, top round black dial, lift top, inner phono, record storage, BC, SW, AC/DC **$140.00**

7-D-203W, Chairside-R/P, 1938, Bone white, top round dial, lift top, inner phono, record storage, BC, SW, AC/DC **$140.00**

7-D-222, Table, 1937, Wood, cube-shaped, round black dial, top grill cut-outs, 4 knobs, BC, SW, AC/DC **$135.00**

7-D-223, Table, 1937, Wood, right round black dial, left grill w/cut-outs, feet, 4 knobs, BC, SW, AC/DC **$100.00**

7-D-229, Tombstone, 1937, Wood, lower round black dial, upper grill w/cut-outs, 4 knobs, BC, SW, AC/DC **$160.00**

7-D-239, Chairside, 1937, Wood, step-down top w/round black dial, 4 knobs, lower storage, BC, SW, AC **$140.00**

7-D-241, Chairside/Bar, 1937, Wood, top round dial & knobs, front grill w/3 vert bars, storage & bar area, BC, SW, AC/ DC **$160.00**

7-D-241W, Chairside/Bar, 1937, Bone white finish, top round dial, front grill w/bars, storage & bar area, BC, SW, AC/DC**$160.00**

7-D-243, Chairside, 1937, Wood, Deco, step-down top w/dial, large horiz wrap-around louvers, BC, SW, AC/DC **$160.00**

7-D-253, Console, 1937, Wood, upper round black dial, lower cloth grill w/vertical bars, BC, SW, AC/DC **$195.00**

7-G-605 "Transoceanic", Portable, 1941, Leatherette, inner grill w/airplane or sailboat, fold-down front, handle, 6 bands, AC/DC/bat **$200.00**

7-H-820, Table, 1948, Plastic, center front round concentric louvers and dial, 3 knobs, BC, FM, AC/DC **$45.00**

7-H-822, Table, 1949, Plastic, left diagonal black dial, right perforated grill, 3 knobs, BC, FM, AC/DC **$30.00**

7-H-918, Table, 1949, Plastic, center front round dial w/inner perforated grill, 2 knobs, FM, AC/DC **$30.00**

7-H-920, Table, 1949, Plastic, center front round concentric louvers and dial, 3 knobs, BC, FM, AC/DC **$45.00**

7-H-921, Table, 1949, Plastic, left diagonal black dial, right perforated grill, 3 knobs, AM, FM, AC/DC **$30.00**

7-H-921Z, Table, 1949, Plastic, left diagonal black dial, right perforated grill, 3 knobs, AM, FM, AC/DC **$30.00**

7-H-922, Table, 1950, Plastic, left diagonal black dial, perforated grill, handle, 3 knobs, BC, FM, AC/DC $30.00

7-J-232 "Walton's", Tombstone, 1937, Wood, large case, black "robot" dial, upper 2-section grill, 4 knobs, BC, SW, AC/bat $375.00

7-J-259, Console, 1937, Wood, upper black "robot" dial, lower cloth grill w/vertical bars, BC, SW, bat/AC $250.00

7-J-328, Table, 1939, Wood, black "robot" dial, wrap-around grill w/horiz bars, pushbuttons, BC, SW, bat/AC $95.00

7-J-368, Console, 1939, Wood, black "robot" dial, low horiz wrap-around louvers, pushbuttons, BC, SW, bat/AC $275.00

7-R-070, Table-R/P, 1948, Leatherette, right dial, plastic "automobile" grill, lift top, inner phono, BC, AC $50.00

7-R-887, Console-R/P, 1949, Wood, upper slanted black dial, pushbuttons, front fold-down phono door, BC, AC $135.00

7-S-28, Tombstone, 1936, Wood, lower round black dial, upper cloth grill w/cut-outs, 4 knobs, BC, SW, AC $165.00

7-S-53, Console, 1936, Wood, upper round black dial, lower grill w/vertical bars, 4 knobs, BC, SW, AC $225.00

7-S-232 "Walton's", Tombstone, 1937, Wood, large case, black "robot" dial, upper 2-section grill, 4 knobs, BC, SW, AC $425.00

7-S-240, Chairside, 1938, Wood, Deco, top black "robot" dial, side grill w/vert bars, storage, BC, 2SW, AC $225.00

7-S-240W, Chairside, 1937, Bone white, Deco, top "robot" dial, side grill w/vert bars, storage, BC, 2SW, AC $225.00

7-S-242, Chairside, 1937, Wood, Deco, top "robot" dial, large horizontal wrap-around louvers, BC, SW, AC $250.00

7-S-242W, Chairside, 1937, Ebony finish, Deco, top "robot" dial, large horiz wrap-around louvers, BC, SW, AC $250.00

7-S-258, Console, 1937, Wood, upper black "robot" dial, tuning eye, lower grill w/vert bars, BC, SW, AC $250.00

7-S-258W, Console, 1937, Bone white, upper black "robot" dial, tuning eye, lower grill w/vert bars, BC, SW, AC $250.00

7-S-260, Console, 1937, Wood, upper black "robot" dial, lower cloth grill w/vertical bars, BC, SW, AC $250.00

7-S-260Y, Console, 1937, Ebony finish, upper black "robot" dial, lower cloth grill w/vert bars, BC, SW, AC $250.00

7-S-261, Console/Bookcase, 1937, Wood, black "robot" dial, lower grill w/cut-outs, side book shelves, BC, SW, AC $300.00

7-S-323, Table, 1939, Wood, round black dial, wrap-around grill w/horiz bars, pushbuttons, BC, SW, AC $75.00

7-S-342, Chairside, 1939, Wood, top black dial/knobs/pushbuttons, front grill, side shelves, BC, SW, AC $140.00

7-S-343, Chairside, 1939, Wood, top black dial/knobs/pushbuttons, front scalloped grill, feet, BC, SW, AC $145.00

7-S-363, Console, 1939, Wood, round black dial, vertical grill bars, pushbuttons, tuning eye, BC, SW, AC $200.00

7-S-364, Console, 1939, Wood, upper slanted round black dial, vertical grill bars, pushbuttons, BC, SW, AC $185.00

7-S-366, Console, 1939, Wood, large ornate cabinet w/carvings, round black dial, pushbuttons, BC, SW, AC $250.00

7-S-432, Table, 1939, Wood, right dial, rounded left w/wrap-around horiz bars, pushbuttons, BC, SW, AC $85.00

7-S-433, Table, 1939, Wood, lower slide rule dial, upper grill w/horiz bars, pushbuttons, BC, SW, AC $90.00

7-S-434, Table, 1939, Wood, slide rule dial, upper half-round grill w/cut-outs, pushbuttons, columns, BC, SW, AC $125.00

7-S-449, Chairside, 1939, Wood, top dial/knobs/pushbuttons, front grill w/curved vertical bars, BC, SW, AC $150.00

7-S-453, Table, 1939, Wood, round black dial, left wrap-around grill w/horiz bars, pushbuttons, BC, SW $135.00

7-S-458, Console, 1939, Wood, upper slide rule dial, lower grill w/vertical bars, pushbuttons, BC, SW, AC $160.00

7-S-461, Console, 1939, Wood, upper slide rule dial, lower grill w/vertical bars, pushbuttons, BC, SW, AC $175.00

7-S-462, Console/Bookcase, 1939, Wood, slide rule dial, vert grill bars, pushbuttons, 2 drawers, shelves, BC, SW, AC .$250.00

7-S-529, Table, 1941, Wood, black dial, left wrap-around grill w/horiz bars, pushbuttons, 2 knobs, BC, SW, AC $85.00

7-S-530, Table, 1941, Wood, right black dial, left grill w/vertical bars, pushbuttons, BC, SW, AC $85.00

7-S-547, Chairside, 1941, Wood, top black dial/knobs/pushbuttons, front grill, casters, 7 tubes, BC, SW, AC $160.00

7-S-557, Console, 1941, Wood, upper black dial, lower grill w/vertical bars, pushbuttons, BC, SW, AC $200.00

7-S-558, Console, 1941, Wood, upper black dial, lower grill w/vertical bars, pushbuttons, BC, SW, AC $225.00

7-S-581, Chairside-R/P, 1941, Wood, top dial/knobs/pushbuttons, lift top, inner phono, horiz grill bars, BC, SW, AC $160.00

7-S-582, Console-R/P, 1941, Wood, black dial, pull-out phono, lower vert grill bars, pushbuttons, BC, SW, AC $175.00

7-S-585, Console-R/P, 1941, Wood, step-down top, black dial, center phono, vert grill bars, pushbuttons, BC, SW, AC $190.00

7-S-598, Console-R/P, 1941, Wood, inner right dial/knobs/pushbuttons, inner left phono, doors, BC, SW, AC $135.00

7-S-633, Table, 1942, Wood, black dial, wrap-around grill w/horiz bars, pushbuttons, 2 knobs, BC, SW, AC $85.00

7-S-634, Table, 1942, Wood, right black dial, left grill w/spiral bars, pushbuttons, 2 knobs, BC, SW, AC $85.00

7-S-634R, Table, 1942, Wood, right black dial, left grill w/spiral bars, pushbuttons, 2 knobs, BC, SW, AC $85.00

7-S-635, Table, 1942, Wood, black dial, wrap-around grill w/horiz bars, pushbuttons, 2 knobs, BC, SW, AC $75.00

7-S-657, Console, 1942, Wood, slanted black dial, vertical grill bars, pushbuttons, 2 knobs, BC, SW, AC $145.00

7-S-681 "Westchester", Chairside-R/P, 1941, Wood, top black dial/knobs/pushbuttons, front pull-out phono unit, BC, SW, AC $120.00

7-S-682 "Beverly", Console-R/P, 1942, Wood, black dial, pull-out phono, vertical grill bars, pushbuttons, 2 knobs, BC, SW, AC $140.00

7-S-685 "Carleton", Console-R/P, 1942, Wood, inner right black dial, left lift top, innner phono, vert grill bars, BC, SW, AC $120.00

7-X-075, Console-R/P, 1946, Wood, upper front black dial, center tilt-out phono, lower cloth grill, BC, battery $100.00

8-A-02, Console, 1940, Wood, upper round black dial, lower grill w/horizontal bars, pushbuttons, BC, SW $200.00

8-D-363, Console, 1939, Wood, round black dial, vert grill bars, pushbuttons, tuning eye, BC, SW, AC/DC $210.00

8-D-510, Table, Plastic, right black dial w/white pointer over horiz grill bars, handle, 2 knobs $45.00

8-G-005 "Transoceanic", Portable, 1946, Black leatherette, inner dial, telescoping antenna, handle, 6 bands, AC/DC/bat $80.00

8-G-005Y "Transoceanic", Portable, 1946, Black leatherette, inner dial, telescoping antenna, handle, 6 bands, AC/DC/bat $80.00

8-G-005YT "Transoceanic", Portable, 1946, Black leather-ette, inner dial, telescoping antenna, handle, 6 bands, AC/DC/ bat $80.00

8-G-005YTZ1 "Transoceanic", Portable, 1949, Black leather-ette, inner dial, telescoping antenna, handle, 6 bands, AC/DC/ bat $80.00

8-H-023, Table, 1946, Plastic, black half moon dial, semi-circular louvers, 2 knobs, BC, FM, AC/DC $45.00

8-H-034, Table, 1946, Wood, center curved black dial, lower cloth grill, 4 feet, 2 knobs, BC, 2FM, AC/DC $50.00

8-H-061, Console, 1946, Wood, slanted black dial, lower vert grill bars, pushbuttons, 2 knobs, BC, FM, AC $135.00

8-H-832, Table, 1948, Wood, black dial, brass & plastic escutcheon, 3 knobs, 6 pushbuttons, BC, 2FM, AC $50.00

8-S-129, Tombstone, 1936, Wood, lower black dial, upper grill w/ vertical bars, 8 tubes, BC, SW, AC $170.00

8-S-154, Console, 1936, Wood, step-down top, upper black dial, lower grill w/vertical bars, BC, SW, AC $225.00

8-S-443, Table, 1940, Wood, step-down top, round black dial, vertical wrap-over bars, pushbuttons, BC, SW, AC $125.00

8-S-451, Chairside, 1940, Wood, top round black dial/knobs/ pushbuttons, front grill w/cut-outs, BC, SW, AC $145.00

8-S-463, Console, 1939, Wood, upper round black dial, lower grill w/ vert bars, pushbuttons, BC, SW, AC $225.00

8-S-531, Table, 1941, Wood, step-down top, round black dial, vert grill bars, pushbuttons, BC, SW, AC $125.00

8-S-548, Chairside, 1941, Wood, top round black dial/knobs/ pushbuttons, front grill w/cut-outs, BC, SW, AC $150.00

8-S-563, Console, 1941, Wood, upper round black dial, lower horizontal grill bars, pushbuttons, BC, SW, AC $200.00

8-S-563X, Console, 1942, Wood, upper round black dial, lower vertical grill bars, pushbuttons, BC, SW, AC $200.00

8-S-647, Chairside, 1942, Wood, top black dial/knobs/pushbuttons, front grill w/vertical bars, 3 bands, AC $150.00

8-S-661, Console, 1942, Wood, upper black dial, lower cloth grill w/ vertical bar, pushbuttons, BC, SW, AC $140.00

8-T-01C "Universal", Portable, Black leatherette, inner right dial, semi-circular louvers, fold-up front, handle $50.00

9-H-079, Console-R/P, 1946, Wood, large black dial & knobs on top of case, pull-down phono door, BC, 2FM, AC $125.00

9-H-081, Console-R/P, 1946, Wood, large black dial & knobs on top of case, pull-down phono door, BC, 2FM, AC $130.00

9-H-088R, Console-R/P, 1946, Wood, right tilt-out black dial, left pull-out phono, criss-cross grill, BC, FM, AC $100.00

9-H-881, Console-R/P, 1948, Wood, slanted black dial, criss-cross grill, 3 knobs, pull-out phono, BC, 2FM, AC $85.00

9-H-984LP, Console-R/P, 1949, Wood, dial on case top, front fold-down phono door, criss-cross grill, BC, FM, AC $85.00

9-H-995, Console-R/P, 1949, Wood, right tilt-out black dial, left fold-down phono door, 9 tubes, BC, FM, AC $80.00

9-S-30, Tombstone, 1936, Wood, lower round black dial, upper grill w/cut-outs, 4 knobs, BC, SW, AC $175.00

9-S-54, Console, 1936, Wood, upper round black dial, lower cloth grill w/cut-outs, 4 knobs, BC, SW, AC $225.00

9-S-55, Console, 1936, Wood, shouldered, round black dial, lower grill w/cut-outs, 4 knobs, BC, SW, AC $240.00

9-S-204, Console-R/P, 1937, Wood, upper black "robot" dial, lower grill w/vert bars, inner phono, BC, SW, AC $275.00

9-S-232 "Walton's", Tombstone, 1937, Wood, large case, black "robot" dial, upper 2-section grill, 4 knobs, BC, SW, AC $550.00

9-S-242, Chairside, 1937, Wood, step-down top w/"robot" dial, large horiz wrap-around louvers, BC, SW, AC $250.00

9-S-242Y, Chairside, 1937, Ebony finish, step-down top w/"robot" dial, large horiz wrap-around louvers, BC, SW, AC $250.00

9-S-244, Chairside, 1937, Wood, Deco, top "robot" dial, front cloth grill with vertical bars, BC, SW, AC $200.00

9-S-262, Console, 1937, Wood, black "robot" dial, lower grill w/2 vertical bars, tuning eye, BC, SW, AC $275.00

9-S-263, Console, 1937, Wood, upper black "robot" dial, lower cloth grill w/vertical bars, BC, SW, AC $255.00

9-S-264, Console, 1937, Wood, low cabinet, upper black "robot" dial, lower grill w/vert bars, BC, SW, AC $260.00

9-S-319, Table, 1938, Wood, oblong gold dial, left wrap-around grill w/horiz bars, pushbuttons, BC, SW $100.00

9-S-324, Table, 1939, Wood, black "robot" dial, wrap-around grill w/ horiz bars, pushbuttons, BC, SW, AC $175.00

9-S-344, Chairside, 1939, Wood, top "robot" dial/knobs/pushbuttons, wrap-around horiz louvers, BC, SW, AC $220.00

9-S-365, Console, 1939, Wood, upper black "robot" dial, lower grill w/ vert bars, pushbuttons, BC, SW, AC $260.00

9-S-367, Console, 1938, Wood, upper round black "robot" dial, low horiz louvers, pushbuttons, BC, SW, AC $275.00

9-S-369, Console, 1939, Wood, step-down top, "robot" dial, vertical grill bars, pushbuttons, BC, SW, AC $295.00

10, Console, 1931, Wood, lowboy, upper front dial, lower cloth grill w/cut-outs, 8 tubes $165.00

10-A-3R, Console, Wood, round black dial, lower grill w/center cut-outs, pushbuttons, BS, SW, AC $250.00

10-H-562, Console, 1941, Wood, spinet piano-style, slanted round black dial, "lyre" grill, pushbuttons, spindles, BC, SW $350.00

10-H-571, Console, 1941, Wood, spinet piano-style, slanted round black dial, "lyre" grill, pushbuttons, spindles, BC, SW, FM $350.00

10-H-573, Console, 1941, Wood, slanted round black dial, lower grill w/cut-outs, pushbuttons, BC, FM, SW, AC $230.00

10-S-130, Tombstone, 1936, Wood, lower black dial, upper cloth grill w/cut-outs, 10 tubes, BC, SW, AC $190.00

10-S-147 "Zephyr", Chairside, 1936, Wood, top black dial & knobs, large lower wrap-around louvers, BC, SW, AC $185.00

10-S-153, Console, 1936, Wood, upper black dial, lower cloth grill w/splayed bars, 10 tubes, BC, SW, AC $230.00

10-S-155, Console, 1936, Wood, upper black dial, lower cloth grill w/vertical bars, 10 tubes, BC, SW, AC $225.00

10-S-156, Console, 1936, Wood, upper black dial, lower cloth grill w/vertical bars, 10 tubes, BC, SW, AC $220.00

10-S-157 "Zephyr", Console, 1936, Wood, upper black dial, tuning eye, large horiz wrap-around grill bars, BC, SW, AC ..$250.00

10-S-160, Console, 1936, Wood, upper front black dial, lower cloth grill w/center vertical bar, BC, SW, AC $250.00

10-S-443, Table, 1940, Wood, step-down top, black dial, vert wrap-over grill bars, pushbuttons, BC, SW, AC $125.00

10-S-452, Chairside, 1940, Wood, half-round, top black dial/knobs/pushbuttons, wrap-around horiz grill bars, BC, SW, AC $220.00

10-S-464, Console, 1940, Wood, upper round black dial, lower grill w/horiz bars, pushbuttons, BC, SW, AC $235.00

10-S-470, Console, 1940, Wood, upper round black dial, lower concave grill area, pushbuttons, BC, SW, AC $250.00

10-S-491, Console-R/P, 1940, Wood, black dial, pull-out phono, horiz wrap-around louvers, pushbuttons, BC, SW, AC$220.00

10-S-531, Table, 1941, Wood, step-down top, black dial, vertical grill bars, pushbuttons, BC, SW, AC $125.00

10-S-549, Chairside, 1940, Walnut, top round black dial/knobs/pushbuttons, grill cut-outs, storage, BC, SW, AC $200.00

10-S-566, Console, 1941, Wood, upper round black dial, lower vertical grill bars, pushbuttons, BC, SW, AC $245.00

10-S-567, Console, 1941, Wood, upper round black dial, lower vertical grill bars, pushbuttons, BC, SW, AC $245.00

10-S-568, Console, 1941, Wood, upper round black dial, lower vertical grill bars, pushbuttons, BC, SW, AC $225.00

10-S-589, Console-R/P, 1941, Wood, black dial, pull-out phono, horiz wrap-around louvers, pushbuttons, BC, SW, AC$230.00

10-S-599, Console-R/P, 1941, Wood, inner right black dial, left phono, lower vertical grill bars, BC, SW, AC $175.00

10-S-669, Console, 1939, Wood, upper black dial, lower grill w/vertical bars, pushbuttons, BC, SW, AC $175.00

10-S-690 "Wilshire", Console-R/P, 1942, Wood, inner right black dial, left pull-out phono, vertical grill bars, BC, SW, AC $145.00

11, Console, 1931, Wood, lowboy, upper window dial, lower grill w/cut-outs, stretcher base, 8 tubes $150.00

11, Table, 1927, Walnut, rectangular case, slanted front panel w/center thumbwheel dial, battery $120.00

11-E, Table, 1927, Wood, rectangular case, slanted front panel w/thumbwheel tuning, 2 knobs, AC $120.00

11-S-474, Console, 1940, Wood, upper round black dial, lower grill w/vert bars, pushbuttons, BC, SW, AC $250.00

12, Console, 1931, Wood, highboy, inner front dial, double doors, arched stretcher base, 8 tubes $150.00

12, Table, 1927, Mahogany, rectangular case, slanted front panel w/off-center dial, 6 tubes, bat $120.00

12-A-57, Console, 1936, Wood, upper round black dial, lower cloth grill w/cut-outs, 4 knobs, BC, SW, AC $300.00

12-A-58, Console, 1936, Wood, round black dial, lower grill w/"torch" cut-out, 2 speakers, 4 knobs, BC, SW, AC $375.00

12-H-090, Console-R/P, 1946, Wood, right tilt-out drawer w/radio, left fold-down phono door, BC, SW, FM, AC $120.00

12-H-092R, Console-R/P, 1946, Mahogany, right tilt-out black dial, left fold-down phono door, BC, SW, 2FM, AC $120.00

12-H-093R, Chairside-R/P, 1946, Wood, top dial/pushbuttons, front pull-out phono, grill cut-outs, BC, FM, SW, AC $130.00

12-H-094E, Console-R/P, 1946, Bleached mahogany, tilt-out dial, pushbuttons, pull-out phono, BC, SW, FM, AC $120.00

12-H-650 "Lenox", Chairside, 1942, Wood, center top black dial/pushbuttons, front vertical grill bars, BC, SW, AC $140.00

12-H-670 "Newport", Console, 1941, Wood, slanted black dial, lower grill w/vert bars, pushbuttons, BC, SW, FM, AC $190.00

12-H-679, Console-R/P, 1941, Wood, inner dial, knobs & phono, double front doors, lower grill, BC, SW, FM, AC $150.00

12-H-689 "Kenwood", Console-R/P, 1942, Wood, black dial, pull-out phono, lower grill, pushbuttons, storage, BC, SW, AC $150.00

12-H-695 "Williamsburg", Console-R/P, 1942, Wood, inner right black dial, left lift top, inner phono, lower grill, storage, BC, SW, AC $150.00

12-H-696 "Georgetown", Console-R/P, 1942, Wood, inner right black dial, left lift top, inner phono, lower grill, storage, BC, SW, AC $150.00

12-S-205, Console-R/P, 1937, Wood, wide cabinet, inner "robot" dial & phono, lift top, grill cut-outs, storage, BC, SW, AC $300.00

12-S-232 "Walton's", Tombstone, 1937, Wood, large case, black "robot" dial, upper 2-section grill, 4 knobs, BC, SW, AC $750.00

12-S-245, Chairside, 1937, Wood, recessed center w/black "robot" dial, side grill w/horiz bars, BC, SW, AC $250.00

12-S-265, Console, 1937, Wood, large upper black "robot" dial, lower cloth grill w/vert bar, BC, SW, AC $275.00

12-S-266, Console, 1937, Wood, large upper black "robot" dial, lower arched grill w/cut-outs, BC, SW, AC $275.00

12-S-267, Console, 1937, Wood, large upper black "robot" dial, lower cloth grill w/vert bars, BC, SW, AC $285.00

12-S-268, Console, 1937, Wood, wide cabinet, upper black "robot" dial, lower grill w/cut-outs, BC, SW, AC $300.00

12-S-345, Chairside, 1939, Wood, top black "robot" dial, pushbuttons, circular grill w/horiz bars, BC, SW, AC $250.00

12-S-370, Console, 1939, Wood, upper black "robot" dial, lower grill w/horiz bars, pushbuttons, BC, SW, AC $320.00

12-S-371, Console, 1939, Wood, half-round, black "robot" dial, vertical grill bars, pushbuttons, BC, SW, AC $350.00

12-S-445, Table, 1940, Wood, step-down top, "robot" dial, wrap-over grill bars, pushbuttons, BC, SW, AC $195.00

12-S-453, Chairside, 1945, Wood, top "robot" dial, bowed front grill, pushbuttons, storage, feet, BC, SW, AC $175.00

12-S-471, Console, 1940, Wood, black "robot" dial, center vert grill panel w/cut-outs, pushbuttons, BC, SW, AC $300.00

12-S-568, Console, 1940, Wood, step-down top, "robot" dial, vertical grill bars, pushbuttons, BC, SW, AC $325.00

12-S-569, Console, 1940, Wood, upper slanted "robot" dial, vertical grill bars, pushbuttons, BC, SW, AC $325.00

12-U-158, Console, 1936, Wood, upper round black dial, lower cloth grill w/vertical bars, BC, SW, AC $250.00

12-U-159, Console, 1936, Wood, upper round black dial, center vertical grill panel w/cut-outs, BC, SW, AC $250.00

14-H-789, Console-R/P, 1948, Wood, tilt-out dial, 3 knobs, pushbuttons, fold-down phono door, BC, SW, 2FM, AC $120.00

15-S-372, Console, 1939, Wood, black "robot" dial, lower grill w/vertical bars, pushbuttons, BC, SW, AC $550.00

15-S-373, Console, 1939, Wood, black "robot" dial, lower wrap-around horiz louvers, pushbuttons, BC, SW, AC **$575.00**

15-U-269, Console, 1937, Wood, upper black "robot" dial, lower cloth grill w/bars, 15 tubes, BC, SW, AC **$550.00**

16-A-61 "Stratosphere", Console, 1936, Wood, round black dial, lower concave grill area, 2 speakers, 16 tubes, BC, SW, AC **$4,000.00**

16-A-63 "Stratosphere", Console, 1936, Wood, ornate cabinet, inner black dial, double doors, lower concave grill area, 2 speakers, 16 tubes, BC, SW, AC **$5,500.00**

17E, Console, 1927, Wood, looks like spinet piano, slanted front w/ window dial & escutcheon, AC **$175.00**

27, Table, 1927, Wood, rectangular case, slanted panel w/2 pointer dials, right/left eliminator storage, AC **$225.00**

33-X, Table, 1928, Wood, rectangular case, slanted thumbwheel dial w/escutcheon, rosettes, AC **$145.00**

35-PX, Console, 1928, Wood, highboy, inner front thumbwheel dial, grill w/cut-outs, double doors, AC **$175.00**

40-A, Console-R/P, 1929, Wood, massive ornate carved cabinet, upper front dial, pushbuttons, AC **$600.00**

52, Console, 1929, Wood, lowboy, upper dial w/escutcheon, lower grill w/cut-outs, pushbuttons, AC **$275.00**

54, Console, 1929, Wood, upper dial w/escutcheon, lower grill cut-outs, pushbuttons, 9 tubes, AC **$325.00**

55, Console, 1929, Wood, lowboy, ornate case, upper dial, lower grill cut-outs, pushbuttons, remote control, 9 tubes, AC **$325.00**

61, Console, 1930, Wood, lowboy, upper front dial, lower cloth grill w/cut-outs, 9 tubes, AC **$275.00**

64, Console, 1930, Wood, lowboy, upper dial, lower grill w/cut-outs, inner pushbuttons, 9 tubes, AC **$350.00**

67, Console, 1930, Wood, massive ornate lowboy, upper dial, inner pushbuttons, cloth grill w/cut-outs, 9 tubes, AC **$400.00**

210 "Zenette", Cathedral, 1932, Wood, lower window dial, upper scalloped grill w/cut-outs, 7 tubes, 3 knobs **$275.00**

210-5 "Zenette", Cathedral, 1932, Wood, lower window dial, upper scalloped grill w/cut-outs, 7 tubes, 3 knobs, LW **$275.00**

215 "Zenette", Cathedral, 1932, Wood, center window dial, upper scalloped grill w/cut-outs, fluted columns, 4 knobs **$275.00**

220, Console, 1932, Wood, lowboy, upper window dial, lower grill w/ cut-outs, 7 tubes, 3 knobs, AC **$150.00**

230 "Zenette", Cathedral, 1930, Wood, peaked top, center window dial, upper grill cut-outs, 8 tubes, 4 knobs, AC **$220.00**

240, Console, 1932, Wood, lowboy, upper window dial, lower grill w/ cut-outs, 8 tubes, 6 legs, AC **$160.00**

245, Console, 1932, Wood, lowboy, window dial, inner pushbuttons, grill cut-outs, 8 tubes, 4 knobs, 6 legs, AC **$235.00**

250, Cathedral, 1932, Wood, window dial, upper scalloped grill w/ cut-outs, 9 tubes, 5 knobs, BC, SW, AC **$295.00**

260, Console, 1932, Wood, upper window dial, lower grill w/cut-outs, 9 tubes, 6 legs, BC, SW, AC **$175.00**

270, Console-R/P, 1932, Wood, lowboy, window dial, grill cut-outs, lift top, inner phono, 7 tubes, AC **$160.00**

288, Tombstone, 1934, Two-tone wood, Deco, center window dial, upper grill, 8 tubes, 6 knobs, BC, SW, AC $175.00

293, Console, 1933, Wood, lowboy, upper window dial, lower grill cut-outs, 7 tubes, 6 legs, BC, SW, AC **$160.00**

672, Console, 1930, Wood, massive ornate lowboy, upper dial, inner pushbuttons, cloth grill w/cut-outs, 9 tubes, AC **$400.00**

701, Table, 1933, Inlaid wood, right dial, center grill cut-outs, side fluting, 5 tubes, 2 knobs, AC/DC **$100.00**

705, Table, 1933, Wood, flared base, window dial, center grill w/ cut-outs, side fluting, 2 knobs, AC $115.00

706, Table, 1933, Wood, right window dial, center cloth grill w/horiz bars, 6 tubes, 2 knobs, AC **$75.00**

707, Table, 1933, Inlaid wood, right window dial, center grill w/cut-outs, 6 tubes, 2 knobs, AC **$115.00**

711, Table, 1933, Wood, flared base, right window dial, center cloth grill w/cut-outs, 2 knobs, AC **$115.00**

715, Tombstone, 1933, Wood, shouldered, center front window dial, upper grill w/cut-outs, 4 knobs, AC **$240.00**

755, Console, 1933, Wood, lowboy, upper window dial, lower grill w/ cut-outs, 8 tubes, 6 legs, AC **$175.00**

760, Console, 1933, Wood, upper dial w/escutcheon, lower grill w/ scrolled cut-outs, 9 tubes, 6 feet, AC **$210.00**

765, Console, 1933, Wood, inner dial, grill w/scrolled cut-outs, double sliding doors, 9 tubes, 6 legs **$250.00**

801, Table, 1934, Wood, step-down top, right dial, center grill cut-outs, 5 tubes, 2 knobs, AC **$100.00**
805, Cathedral, 1934, Two-tone wood, window dial, upper grill w/cut-outs, 5 tubes, 4 knobs, BC, SW, AC **$200.00**
807, Tombstone, 1933, Wood, lower round dial, upper grill with cut-outs, 5 tubes, 4 knobs, BC, SW, AC **$165.00**
808, Tombstone, 1934, Wood, lower round dial, upper grill w/vertical bars, 6 tubes, 4 knobs, BC, SW, AC **$165.00**

809, Tombstone, 1934, Wood, round dial, upper grill w/Deco chrome cut-outs, 6 tubes, 4 knobs, BC, SW, AC ...$250.00

827, Table, 1934, Wood, right window dial, center grill w/vert chrome bars, 7 tubes, 4 knobs, BC, SW, AC $145.00
829, Tombstone, 1934, Wood, lower round dial, upper grill w/chrome cut-outs, 7 tubes, BC, SW, AC **$250.00**
845, Console, 1934, Wood, upper window dial, lower grill w/cut-outs, 5 tubes, 4 knobs, BC, SW, AC **$140.00**
847, Console, 1934, Wood, upper window dial, lower grill w/cut-outs, 5 tubes, 4 knobs, BC, SW, AC **$140.00**
850, Console, 1934, Wood, upper round dial, lower cloth grill w/cut-outs, 5 tubes, BC, SW, AC **$150.00**
860, Console, 1934, Wood, round airplane dial, lower grill w/cut-outs, 6 tubes, BC, SW, AC **$150.00**
861, Console, 1934, Wood, upper round dial, lower cloth grill w/cut-outs, 6 tubes, BC, SW, AC **$160.00**
970, Console, 1934, Wood, upper round dial, lower grill with scrolled cut-outs, 9 tubes, 6 legs, AC **$180.00**
975, Console, 1934, Wood, upper round dial, lower grill w/cut-outs, sliding doors, 9 tubes, 6 legs, AC **$180.00**
1000-Z "Stratosphere", Console, 1934, Wood, round black dial, lower concave grill, 25 tubes, 3 speakers, BC, SW, AC ....**$7,000.00+**
1117, Tombstone, 1935, Wood, lower front round dial, upper cloth grill with cut-outs, 6 tubes, BC, SW, AC **$150.00**

5808, Console, 1940, Wood, round black dial, lower curved grill w/ vert bars, pushbuttons, 8 tubes, BC, SW **$235.00**
A-519Y, Table-C, 1958, Plastic, center front dial, right clock, left checkered grill w/crest, feet, BC, AC **$20.00**

A-600 "Transoceanic", Portable, 1957, Black leatherette, inner multi-band dial, fold-up front, telescope antenna, handle $80.00
A-600L "Transoceanic", Portable, 1957, Leatherette, multi-band slide rule dial, fold-up front, telescope antenna, handle **$160.00**
B-513V, Table, 1959, Plastic, lower right dial, large center oval grill w/vert bars & crest, BC, AC/DC **$15.00**
B-514F, Table-C, 1959, Plastic, right front oval dial, left oval clock, checkered grill w/crest, BC, AC **$20.00**
B-600 "Transoceanic", Portable, Leatherette, multi-band slide rule dial, fold-up front, telescope antenna, handle **$80.00**
B-615F, Table, 1959, Plastic, lower right front dial knob, oval grill bars, side knob, feet, BC, AC/DC **$15.00**
B-728C, Table-C, 1959, Plastic, lower front slide rule dial, upper center clock, 4 knobs, AM, FM, AC/DC **$20.00**
C-730, Table, 1955, Wood, lower front slide rule dial, large upper plastic woven grill, 2 knobs, AM, FM **$25.00**
G-500 "Transoceanic", Portable, 1950, Leatherette, inner dial, fold-up front, telescope antenna, BC, 5SW, AC/DC/bat **$80.00**

G-503, Portable, 1950, Leatherette, flip-over dial w/crest, 2 knobs, woven grill, handle, BC, AC/DC/bat $45.00
G-511, Table, 1950, Plastic, right round dial over large metal criss-cross grill, flex handle, 2 knobs **$35.00**

G-516, Table-C, 1950, Plastic, right round metal dial, left alarm clock, dotted grill, 5 knobs, BC, AC $40.00

G-516W, Table-C, 1950, Plastic, oblong case, right dial, left alarm clock, perforated grill w/crest, BC, AC $40.00

G-615, Table, 1950, Plastic, round metal center dial over recessed grill, handle, 2 knobs, BC, AC/DC $35.00

G-660, Table-R/P, 1950, Plastic, front square dial/perforated grill, 4 knobs, lift top, inner phono, BC, AC $45.00

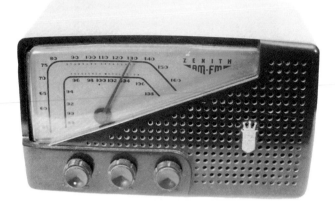

G-723, Table, 1950, Plastic, left diagonal dial, right perforated grill w/crest, 3 knobs, AM, FM, AC/DC $30.00

G-724, Table, 1950, Plastic, left diagonal dial, perforated grill, flex handle, 4 knobs, AM, FM, AC/DC $30.00

G-725, Table, 1950, Plastic, large front round dial w/crest, flex handle, 3 knobs, BC, FM, AC/DC $40.00

G-882, Console-R/P, 1950, Wood, inner right dial, 3 knobs, storage, left fold-down phono door, BC, FM, AC $70.00

H-401, Portable, 1952, Plastic, center round dial w/inner perforations, handle, 2 knobs, BC, AC/DC/bat $35.00

H-500 "Transoceanic", Portable, 1951, Black leatherette, inner dial, fold-back front, handle, BC, SW, AC/DC/battery ... $80.00

H-503, Portable, 1951, Black leatherette, inner round dial, 2 knobs, fold-back front, handle, AC/DC/bat $45.00

H-511, Table, 1951, Plastic, oblong case, right half-moon dial, center metal strip, 2 knobs, BC, AC/DC $50.00

H-511W, Table, 1951, Plastic, oblong case, right half-moon dial, center metal strip, 2 knobs, BC, AC/DC $50.00

H-615, Table, 1951, Plastic, front square dial/inner criss-cross grill, flex handle, 2 knobs, BC, AC/DC $40.00

H-615W, Table, 1951, Plastic, front square dial/inner criss-cross grill, flex handle, 2 knobs, BC, AC/DC $40.00

H-615Z, Table, 1952, Plastic, center front square metal dial, 2 knobs, feet, BC, AC/DC $40.00

H-615Z1 , Table, 1952, Plastic, center front square metal dial, flex handle, 2 knobs, feet, BC, AC/DC $40.00

H-661R, Table-R/P, 1951, Leatherette, front round dial w/crest, 4 knobs, lift top, inner phono, BC, AC $30.00

H-664, Table-R/P, 1951, Plastic, front raised round metal dial, 2 knobs, lift top, inner phono, BC, AC $75.00

H-723, Table, 1951, Plastic, round off-center dial, criss-cross grill w/crest, 3 knobs, BC, FM, AC/DC $30.00

H-723Z, Table, 1951, Plastic, large round dial, criss-cross grill w/crest, 3 knobs, feet, BC, FM, AC/DC $30.00

H-724, Table, 1951, Plastic, round metal dial, criss-cross grill w/crest, flex handle, BC, FM, AC/DC $30.00

H-724Z1, Table, 1952, Plastic, off-center round dial, cloth grill w/crest, flex handle, BC, FM, AC/DC $30.00

H-724Z2, Table, 1952, Plastic, off-center round dial, flex handle, right crest, side knob, BC, FM, AC/DC $30.00

H-725, Table, 1951, Plastic, round dial/inner perforated grill, flex handle, 2 knobs, BC, FM, AC/DC $40.00

H-880R, Console-R/P, 1951, Wood, inner right dial w/crest, 3 knobs, left pull-down phono door, BC, FM, AC $75.00

H-880RZ, Console-R/P, 1950, Wood, inner right dial, lower storage, left fold-down phono door, BC, FM, AC $75.00

HF-660F, Table-R/P, 1959, Front left dial, large grill, lift top, inner phono, high fidelity, handle, BC, AC $15.00

HF-772R, Console-R/P, 1957, Wood, inner right slide rule dial, 4 knobs, left phono, lift top, high fidelity, BC, AC $40.00

HF-1286RD, Console-R/P, 1958, Wood, inner right slide rule dial, 5 knobs, pull-out phono, high fidelity, BC, FM, AC $50.00

J-402, Portable, 1952, Plastic, dial on case top, front perforated grill, handle, 2 knobs, BC, AC/DC/bat $35.00

J-420T, Table, 1952, Plastic, oblong case, right half-moon dial, crest, 2 knobs, feet, BC, SW, battery $40.00

J-514, Table-C, 1952, Plastic, oval case, right dial, left alarm clock, center horiz bars, 5 knobs, BC, AM $40.00

J-615, Table, 1952, Plastic, half-round dial w/circular grill louvers & crest, flex handle, 2 knobs, BC $45.00

J-616, Table-C, 1952, Plastic, oblong case, right round dial, left alarm clock, crest, 5 knobs, BC, AC $40.00

J-733, Table-C, 1952, Plastic, modern, left square dial, right clock, "dotted" grill w/crest, 6 knobs, AC $30.00

K-412R, Portable, 1953, Plastic, "owl-eye" knobs, center concentric circular louvers w/crest, handle, BC $50.00

K-412Y, Portable, 1953, Plastic, "owl-eye" knobs, center concentric circular louvers w/crest, handle, BC $50.00

K-510W, Table, 1952, Plastic, center round dial w/inner perforations, rounded corners, 2 knobs, BC, AC/DC $40.00

K-518 "Deluxe" , Table-C, 1952, Plastic, oblong case, right dial, left alarm clock, center horiz bars w/crest, AC $40.00

K-526, Table, 1953, Plastic, center round dial w/inner lattice grill & crest, flex handle, 3 knobs, BC $35.00

K-622, Table-C, 1953, Plastic, modern, right dial, left alarm clock, center horiz bars w/crest, 2 knobs $35.00

K-666R, Table-R/P, 1953, Plastic, step-down front w/half-moon metal dial, 3 knobs, lift top, BC, AC $40.00

K-725, Table, 1953, Plastic, large round metal dial w/crest, flex handle, 2 knobs, BC, FM, AC/DC $40.00

K-777R, Console-R/P, 1953, Wood, inner right slide rule dial, 3 knobs, pull-out phono, fold-down front, BC, AC **$50.00**

L, Cathedral, 1931, Two-tone wood, lower window dial, upper grill w/ cut-outs, 5 tubes, 3 knobs **$200.00**

L-406R, Portable, 1953, Plastic, top dial, front round perforated grill w/crest, handle, BC, SW, AC/DC/bat **$35.00**

L-507, Portable, 1954, Leatherette, inner front dial, flip up front w/ map, handle, BC, 2SW, AC/DC/bat **$50.00**

L-518, Table-C, 1953, Plastic, oblong case, right dial, center horiz louvers, left clock, 2 knobs, BC, AC **$40.00**

L-566, Table-R/P, 1954, Plastic, front half-moon dial, horizontal grill bars, lift top, inner phono, BC, AC **$40.00**

L-622, Table-C, 1953, Plastic, modern, right dial, left alarm clock, center horiz bars w/crest, BC, AC **$35.00**

L-721, Table, 1954, Plastic, left diagonal dial, right dotted grill w/ crest, 3 knobs, AM, FM, AC/DC **$30.00**

L-846, Chairside-R/P, 1954, Wood, top dial & knobs, front louvered pull-out phono drawer, legs, BC, FM, AC **$50.00**

L-880, Console-R/P, 1954, Wood, inner right dial, 3 knobs, left fold-down phono door, BC, FM, AC **$70.00**

LH "Zenette", Cathedral, 1931, Two-tone wood, window dial, upper scalloped grill w/cut-outs, 7 tubes, 3 knobs **$275.00**

LP "Zenette", Cathedral, 1931, Two-tone wood, window dial, upper cloth grill w/cut-outs, 5 tubes, 3 knobs **$200.00**

M-505, Portable, 1953, Plastic, top round dial knob, front gold metal grill w/logo, handle, side knobs **$40.00**

M510-R, Table, 1955, Plastic, oblong case, right round dial, checkered grill, handle, 2 knobs, BC, AC/DC $40.00

R510-G, Table, 1955, Plastic, oblong case, right round dial, checkered grill, handle, 2 knobs, BC, AC/DC **$40.00**

R-510Y, Table, 1955, Plastic, oblong case, right round dial, checkered grill, handle, 2 knobs, BC, AC/DC **$40.00**

R-510Z1, Table, 1955, Plastic, oblong case, right round dial, checkered grill, handle, 2 knobs, BC, AC/DC **$40.00**

R511F, Table, 1955, Plastic, right half-moon dial, center checkered grill w/crest, handle, BC, AC/DC $40.00

R-511V, Table, 1955, Plastic, right half-moon dial, center checkered grill w/crest, handle, BC, AC/DC **$40.00**

Royal 200, Portable, 1959, Plastic, transistor, front dial, checkered grill w/crest, swing handle, BC, battery **$40.00**

Royal 275F, Portable, 1960, Plastic, transistor, upper right dial, large lattice grill, swing handle, AM, bat **$25.00**

Royal 300F, Portable, 1958, Plastic, transistor, round window dial, lower lattice grill, swing handle, AM, bat **$35.00**

Royal 300Y, Portable, 1959, Plastic, transistor, round window dial, lower lattice grill, swing handle, AM, bat **$35.00**

Royal 500, Portable, 1955, Plastic, transistor, upper "owl-eye" knobs, lower round grill, swing handle, AM, bat **$85.00**

Royal 500D, Portable, 1959, Plastic, transistor, upper "owl-eye" knobs, lower round grill, swing handle, AM, bat $60.00

Royal 500RD, Portable, 1959, Transistor, upper "owl eye" knobs, lower round perforated grill, swing handle, AM, bat **$60.00**

Royal 675L, Portable, 1960, Leather case, transistor, dial over large lattice front grill, handle, AM, battery **$40.00**

Royal 700L, Portable, 1959, Leather case, transistor, right front dial, chrome lattice grill, handle, AM, bat **$40.00**

Royal 710L, Portable, 1960, Leather case, transistor, right dial knob over large lattice grill, handle, AM, bat **$25.00**

Royal 750L, Portable, 1959, Leather case, transistor, right dial, checkered grill w/crest, handle, AM, battery $40.00

Royal 755L, Portable, 1960, Leather case, transistor, slide rule dial, lower grill w/crest, handle, AM, bat **$50.00**

Royal 760 "Navigator", Portable, 1959, Leather case, transistor, top "compass", chrome lattice grill, handle, AM, bat **$50.00**

Royal 800, Portable, 1957, Plastic, transistor, right side dial, round front perforated grill, handle, BC, bat **$65.00**

Royal 950 "Golden Triangle", Portable-C-N, 1960, Transistor, triangular case on stand, round dial & clock face, top handle, AM, bat **$100.00**

Royal 1000, Portable, 1958, Transistor, fold-down front w/map, telescope antenna in handle, 8 bands, battery $125.00
Royal 1000-1, Portable, 1958, Leatherette & metal, transistor, fold-down front, telescope antenna in handle, bat $125.00
Royal 1000-D, Portable, 1959, Leatherette & metal, transistor, fold-down front, telescope antenna in handle, bat $125.00

Royal 3000, Portable, Transistor, slide rule dial, fold-down front, telescope antenna in handle, battery $145.00
Royal 3000-1, Portable, Transistor, slide rule dial, fold-down front, telescope antenna in handle, battery $145.00

S-829, Tombstone, 1934, Wood, raised top, lower round dial, upper Deco chrome grill w/cut-outs, 4 knobs $250.00
SF-177E, Console-R/P, 1959, Wood, modern, inner left dial & knobs, pull-out phono, stretcher base, BC, AC $50.00
Super-Zenith VII, Table, 1924, Mahogany, low rectangular case, front panel w/2 pointer dials, 6 tubes, battery $200.00
Super-Zenith VIII, Console, 1924, Model "Super-Zenith VII" on legs, battery storage, 6 tubes, battery $250.00
Super-Zenith IX, Console, 1924, Model "Super-Zenith VII" on legs w/built-in speaker & battery storage, battery $275.00
Super-Zenith X, Console, 1924, Wood, 2 pointer dials, upper built-in speaker w/grill, lower battery storage, bat $275.00

T-545, Table-R/P, 1955, Plastic, square dial w/inner perforations, 4 knobs, lift top, inner phono, BC, AC $40.00

T-600 "Transoceanic", Portable, 1953, Black leatherette, inner multi-band dial, fold-up front, telescope antenna, handle $80.00

T825, Table, 1955, Plastic, center round metal dial w/inner perforations, handle, 2 knobs, BC, FM $40.00
U-723, Table, Plastic, off-center round dial over checkered grill w/ crest, feet, 3 knobs, BC, FM $35.00
WH, Console, 1931, Wood, upper window dial, lower grill w/cut-outs, stretcher base, 3 knobs, AC $135.00
Y-506L, Portable, 1957, Leather case w/front lattice grill & crest, side knobs, handle, BC, AC/DC/battery $30.00

Y-723, Table, 1956, Plastic, off-center round dial over checkered grill, handle, crest, side knob, AM, FM $35.00
Y-724, Table, 1956, Plastic, off-center round dial over checkered grill, handle, crest, side knob, AM, FM $35.00

Z-550G, Table-R/P, 1957, Left side dial, large front grill w/crest, 2 knobs, lift top, handle, BC, AC $20.00

Z-733, Table-C, 1956, Plastic, modern, left dial/right alarm clock over large grill w/crest, AM, FM, AC $35.00

ZEPHYR
Zephyr Radio Co.,
13139 Hamilton Avenue, Detroit, Michigan

35Y12, Console, 1937, Walnut, large round dial, tuning eye, lower grill w/vert bars, 10 tubes, 4 knobs $195.00

41X6, Table, 1937. Walnut w/inlay, right front dial, left grill, 5 tubes, 3 knobs, base $55.00

Radio Clubs

The following is a list of antique radio clubs throughout the country. They are always happy to supply potential members with information about their activities and publications. The two clubs listed first are national organizations; the rest are regional.

Antique Wireless Association
P.O. Box E
Breesport, NY 14816

Antique Radio Club of America
Jim and Barbara Rankin
3445 Adaline Drive
Stow, OH 44224

Alabama Historical Radio Society
4721 Overwood Circle
Birmingham, AL 35222

Antique Radio Club of Illinois
Carl Knipfel
Rt. 3 Vererans Road
Morton, IL 61550

Antique Radio Club of Schenectady
Jack Nelson
915 Sherman Street
Schenectady, NY 12303

Antique Radio Collectors Club of Ft. Smith, Arkansas
Wanda Conatser
7917 Hermitage Drive
Ft. Smith, AR 72903

Antique Radio Collectors of Ohio
Karl Koogle
2929 Hazelwood Avenue
Dayton, OH 45419

Arkansas Antique Radio Club
Tom Burgess
P.O. Box 9769
Little Rock, AR 72219

Arizona Antique Radio Club
Treasurer
8311 E. Via de Sereno
Scottsdale, AZ 85258

Belleville Area Antique Radio Club
Karl Stegman
4 Cresthaven Drive
Belleville, IL 62221

Buckeye Radio and Phonograph Club
Steve Dando
4572 Mark Trail
Copley, OH 44321

California Historical Radio Society
Jim McDowell
2265 Panoramic Drive
Concord, CA 94520

Carolina Antique Radio Society
Carl Shirley
824 Fairwood Road
Columbia, SC 29209

Connecticut Area Antique Radio Collectors
Chris Cuff
(203) 873–8762

Connecticut Vintage Radio Collectors Club
John Ellsworth
665 Arch Street
New Britain, CT 06051

North Valley Chapter, California Historical Radio Society
Norm Braithwaite
P.O. Box 2443
Redding, CA 96099

Cincinnati Antique Radio Collectors
Tom Ducro
6805 Palmetto
Cincinnati, OH 45227

Colorado Radio Collectors
Bruce Young
4030 Quitman Street
Denver, CO 80212

Florida Antique Wireless Group
Paul Currie
Box 738
Chuluota, FA 32766

Greater Boston Antique Radio Collectors
Richard Foster
12 Shawmut Avenue
Cochituate, MA 01778

Greater New York Vintage Wireless Association
Bob Scheps
12 Garrity Avenue
Ronkonkoma, NY 11779

Hawaii Antique Radio Club
Bob or Tina Weipert
98-1438 Koahehe Street, Cpt. C
Parl City, HI 96782

Houston Vintage Radio Association
P.O. Box 31276
Houston, TX 77231-1276

Hudson Valley Antique Radio and Phonograph Society
John or Linda Gramm
(914) 427–2602
Geoff or Judy Shepperd
(914) 561–0132

Indiana Historical Radio Society
245 N. Oakland Avenue
Indianapolis, IN 46201-3360

Louisiana & Mississippi Gulf Coast Area
F.V. Bernauer
1503 Admiral Nelson Dr.
Slidell, LA 70461

Michigan Antique Radio Club
Larry Anderson
3453 Balsam NE
Grand Rapids, MI 49505

Mid–America Antique Radio Club
Carleton Gamet
2307 W. 131 Street
Olathe, KS 66061

Mid–Atlantic Antique Radio Club
Joe Koester
249 Spring Gap South
Laurel, MD 20707

Mid–Hudson New York Radio & Phonograph Club
Campbell Hall Antiques
P.O. Box 207
Campbell Hall, NY 10916

Mississippi Historical Radio & Broadcasting Society
Randy Guttery
2412 C Street
Meridian, MS 39301

Nebraska Radio Collectors Antique Radio Club
Steve Morton
905 West First
North Platte, NE 69101

New England Antique Radio Club
Marty Bunis
RR 1, Box 36
Bradford, NH 03221

Niagara Frontier Wireless Asociation
Art Albion
440 69th Street
Niagara Falls, NY 14304

Northland Antique Radio Club
P.O. Box 18362
Minneapolis, MN 55418

Old Time Radio Club (Minnesota)
Bob Ruble
(507)288–3082

Pittsburgh Antique Radio Society
Richard Harris, Jr.
407 Woodside Road
Pittsburgh, PA 15221

Puget Sound Antique Radio Association
P.O. Box 125
Snohomish, WA 98290-0125

Sacramento Historical Raodio Society
P.O. Box 162612
Sacramento, CA 95816-9998

Southern California Antique Radio Society
Edward Sheldon
656 Gravilla Place
La Jolla, CA 92037

Southern Vintage Wireless Association
Bill Moore
1005 Fieldstone Court
Huntsville, AL 35803

Vintage Radio & Phonograph Society
Larry Lamia
P. O. Box 165345
Irving, TX 75016

Western Wisconsin Antique Radio Collectors Club
Dave Wiggert
1025 Oak Avenue So., #B-21
Onalaska, WI 54650

Other Radio Books and Periodicals

Books

Radio Manufacturers of the 1920's, Vol. I
by Alan Douglas.
The Vestal Press, Ltd.,
Vestal, New York, 1988.

Radio Manufacturers of the 1920's, Vol. II
by Alan Douglas.
The Vestal Press, Ltd.,
Vestal, New York, 1989.

Radio Manufacturers of the 1920's, Vol. III
by Alan Douglas.
The Vestal Press, Ltd.,
Vestal, New York, 1991.

Radios, The Golden Age
by Philip Collins.
Chronicle Books
San Francisco, California, 1987.

Classic Plastic Radios Of The 1930's and 1940's
by Jon Sideli.
E.P. Dutton,
New Yrok, 1990.

Antique Radios: Restoration and Price Guide
by David and Betty Johnson.
Wallace-Homestead Book Company,
Des Moines, Iowa, 1982.

Guide To Old Radios; Pointers, Pictures and Prices
by David and Betty Johnson.
Wallace-Homestead Book Company,
Des Moines, Iowa, 1982.

A Flick Of The Switch
by Morgan E. McMahon.
Vintage Radio,
Palos Verdes Peninsula, California, 1983.

Radio: A Blast From The Past
by H.G. Wolff and I. Jacobson.
Sound of Music,
Stillwater, New Jersey, 1987.

Periodicals

Antique Radio Classified, (published monthly)
by John V. Terrey,
P.O. Box 2-V32,
Carlisle, Massachusetts

Radio Age, (published monthly)
by Donald O. Patterson
636 Cambridge Road
Augusta, Georgia

Schroeder's Antiques Price Guide

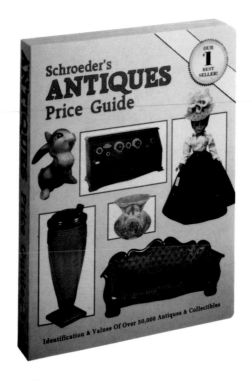

Schroeder's Antiques Price Guide has become THE household name in the antiques & collectibles field. Our team of editors works year-round with more than 200 contributors to bring you our #1 best-selling book on antiques & collectibles.

With more than 50,000 items identified & priced, Schroeder's is a must for the collector & dealer alike. If it merits the interest of today's collector, you'll find it in Schroeder's. Each subject is represented with histories and background information. In addition, hundreds of sharp original photos are used each year to illustrate not only the rare and unusual, but the everyday "fun-type" collectibles as well — not postage stamp pictures, but large close-up shots that show important details clearly.

Our editors compile a new book each year. Never do we merely change prices. Accuracy is our primary aim. Prices are gathered over the entire year previous to publication, from ads and personal contacts. Then each category is thoroughly checked to spot inconsistencies, listings that may not be entirely reflective of actual market dealings, and lines too vague to be of merit. Only the best of the lot remains for publication. You'll find Schroeder's Antiques Price Guide the one to buy for factual information and quality.

No dealer, collector or investor can afford not to own this book. It is available from your favorite bookseller or antiques dealer at the low price of $12.95. If you are unable to find this price guide in your area, it's available from Collector Books, P.O. Box 3009, Paducah, KY 42002-3009 at $12.95 plus $2.00 for postage and handling.

8½ x 11", 608 Pages

$12.95

COLLECTOR BOOKS
A Division of Schroeder Publishing Co., Inc.